Biomineralization '90

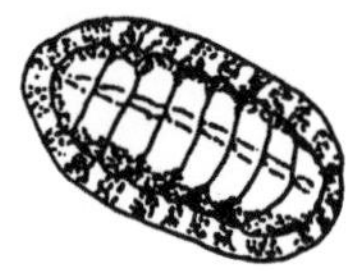

S. Suga H. Nakahara (Eds.)

Mechanisms and Phylogeny of Mineralization in Biological Systems

With 371 Illustrations

Springer-Verlag
Tokyo Berlin Heidelberg
New York London Paris
Hong Kong Barcelona

Shoichi Suga
Professor, Department of Pathology,
The Nippon Dental University School of Dentistry at Tokyo,
Chiyoda-ku, Tokyo, 102 Japan

Hiroshi Nakahara
Associate Professor, Department of Oral Anatomy,
Meikai University School of Dentistry,
Sakado, Saitama, 350-02 Japan

ISBN-13: 978-4-431-68134-2 e-ISBN-13: 978-4-431-68132-8
DOI: 10.1007/978-4-431-68132-8

Softcover reprint of the hardcover 1st edition 1991

Preface

Various kinds of mineralization have been found in many biological systems. Investigations made at a microscopical level using various sophisticated analytical methods and using principles developed in different fields have clarified their mechanisms very much.
Sometimes, very similar phenomena have been found in the mineralized tissues of completely different biological systems. Compilation and comparative investigations of such findings obtained from the many specimens systematically collected contribute a great deal to an understanding of the crucial mechanisms and significance of biomineralization which originated in very primitive organisms and remain in advanced ones.
Previously, the functional significance of mineralized tissues was considered mainly from an anatomical point of view based upon their morphological and structural features. However, the recent advance of investigations has made it possible to interpret the functional significance of biomineralization not only from local and mechanical points of view, but also from a systemic and phylogenetic point of view.
It is also well-known that biomineralization has contributed in various ways to geological and oceanographical conditions of the environment in which the organisms were living. During this process, the mechanisms of biomineralization may have evolved to maintain harmony between organisms and their environments.
The Sixth International Symposium on Biomineralization (Biomineralization '90) was held at Asia Center, Odawara, Kanagawa Prefecture, Japan, on October 8-12, 1990. The aim of this symposium was to provide the opportunity for scientists working on biomineralization in animals and plants in various fields of the basic and applied sciences, to meet and discuss their views on the mechanisms and significance of biomineralization and their relationship to the evolution of various organisms.
At the First Symposium which was held in Mainz, Germany, in 1970, about 60 participants (36 papers) mainly consisted of scientists working in geology and paleontology of invertebrates and plants. At this most recent symposium, 120 participants (80 papers) were composed of specialists from dental science, geology, chemistry, biology, medicine, zoology, paleontology, environmental science, biotechnology, botany, and physics. The mineralized tissue subjects presented were teeth (enamel, enameloid, dentin, and pleromin), molluscan shell and radular tooth, bone, coral, otolith, bacteria, algae, leaf, egg shell, coccolith, and synthetic minerals.
At this symposium, it was apparent that biomineralization research is a typical multidisciplinary science. For example, the scientists who presented papers concerning dental tissues and other related subjects and who belong to dental institutions, were composed not only of dentists, but also of graduates from schools of chemistry (biochemistry), geology, zoology, medicine, fisheries and physics. A similar situation was also true for scientists of other mineralized tissue areas.

During this symposium, we employed several ideas for a better understanding of the information presented by researchers belonging to different research fields and having a different educational background. For example, it was requested that all the papers be accompanied by posters which were displayed during the meeting and night discussions were scheduled in the poster room. In this way, we hoped that we had provided for stimulating discussions, an exchange of knowledge and techniques, and the opportunity for participants to arrange meetings with each other in the future.

We are very grateful to the Science Council of Japan, Kanagawa Prefecture, and Kanagawa Academy of Science and Technology for their sponsorship of the symposium. We are also very much indebted to Professor N. Watabe (University of South Carolina) for his valuable suggestions and advice given during the organization of the symposium.

Shoichi Suga
Hiroshi Nakahara

Contents

3 Trace Elements

Committees

Organizing Committee

Chairman:	S. Suga, Nippon Dental University (Tokyo)
Co-Chairman:	H. Nakahara, Meikai University (Saitama)
Secretary:	I. Kobayashi, Niigata University (Niigata)
Treasurer:	M. Goto, Tsurumi University (Yokohama)
Public relations:	M. Omori, Azabu University (Tokyo)
	M. Okazaki, Tokyo Gakugei University (Tokyo)

Advisory Committee

M. Akiyama (Matsumoto)
R.W. Fearnhead (Yokohama)
Y. Isa (Okinawa)
Y. Kitano (Nagoya)
Y. Moriwaki (Gifu)
M. Shimizu (Yokohama)
S. Takuma (Tokyo)
K. Wada (Mie)
F. Watanabe (Tokyo)
J. Yamada (Hakodate)
K. Yamazato (Okinawa)

International Advisory Committee

M. Chetail (Paris)
M.A. Crenshaw (Chapel Hill)
H.A. Lowenstam (Pasadena)
H. Mutvei (Stockholm)
K. Simkiss (Reading)
N. Watabe (Columbia)

The Organizing Committee wishes to express their deep and sincere thanks to the following organizations and people for their support of the Symposium (in alphabetical order).

Asia Air Survey Co., Ltd.
Bishimetal Exploration Co., Ltd.
Ehime Shinju Yoshoku Gyogyo Kyodo Kumiai
Fuji Bank
Gakken-Shoin Ltd.
G-C Dental Industrial Corp.
General Science Corporation
Hot Springs Research Institute of Kanagawa Prefecture
Ishiyaku Publishers, Inc.
Japan Pearl Promotion Society
Japan Petroleum Exploration Co., Ltd.
JEOL Co.
The Kajima Foundation
Kanagawa Academy of Science and Technology
Lion Corporation
Maruto Instrument Co., Ltd.
Miharu-Shokai

Mitsubishi Bank
Mutsumi Printing Co.
Nagasaki Shinju Yoshoku Gyogyo Kyodo Kumiai
Niigata Science Service Co.
Nikon Corporation
Nikon Corporation, Niigata Branch
Ohken Shoji Co.
Oyo Corporation
Quintessence Publishing Co., Inc., Tokyo
Sankei Co., Ltd.
Shimadzu Seisakusho Co., Ltd.
Shimaken Co.
Shinano Seisakusho Co., Ltd.
Shiseido Research Center
Shofu Co., Ltd.
Softex Co., Ltd.
The late Dr. Shizuo Sone, and his family
Suntory Ltd.
Teikoku Oil Co., Ltd.
Mr. Teisaku Taniguchi
Tokai University Press
The Wednesday Club of Department of Pathology, The Nippon Dental University, Tokyo
Yakuruto Honsha Co., Ltd.
Professor Juro Yamada
Yamamoto Yakuhin Shokai Co.
Professor Kiyoshi Yamazato
Yoshida Foundation for Science and Technology
Zenkoku Shinju Yoshoku Gyogyo Kyodo Kumiai Rengokai

The Organizing Committee also express their appreciation to the following helpers from the Department of Pathology, The Nippon Dental University School of Dentistry at Tokyo, whose dedicated team work made the meeting in Odawara possible (in alphabetical order).

Miss K. Amano
Dr. H. Aoki
Dr. Y. Hirano
Dr. K. Koide
Miss M. Kurimoto
Dr. M. Ogawa
Dr. S. Otake
Dr. A. Sato
Dr. H. Setoguchi
Dr. K. Tanaka
Dr. Y. Taya
Dr. K. Utsugi
Dr. H. Yagishita

1 Organic Matrix

CHAPTER 1.1

S-Layer Similarity in Matrix and Mineral Structure of a *Leptotrix* Species (Bacteria)

H.A. LOWENSTAM

California Institute of Technology, Pasadena, CA 91125

Key words: Organic matrix, Template, Mineralization, Bacteria.

INTRODUCTION

The uniquely shaped mineral coat reported in the literature for a *Leptotrix* species [1] is shown to have an organic matrix with an identical morphology. The same structural design is found among the crystalline cell surface layers of many bacteria. Hence presentation of the *Leptotrix* data is prefaced by pertinent information on similarly shaped crystalline surface layers.

S-LAYER

Crystalline cell surface layers are formed by many Archaebacteria and Eubacteria. These surface layers are widely referred to as S-layers. They constitute regular, periodic macromolecular monolayers composed of a single molecular species of acidic proteins or glycoproteins [2]. The three dimensional structure of the periodic arrays consist of hexagonal, tetragonal, oblique or goblet shaped subunits when seen in high resolution of planar projections of negatively stained S-layer preparations and appear in cross section as a periodic structure [2,3]. Three dimensional image reconstructions show considerably greater complexity in the multidomain structures of S-layers [2,4]. The morphologic units have center to center spacings of 5 to 20 nm and the lattice pores range in diameter from 2 to 5 nm. [2].

Some S-layer bearing Archaebacteria and Eubacteria have been found in laboratory cultures to bind metal ions to their cell surface [5,6]. Information largely obtained from biochemical studies of *Bacillus subtilis* has shown that S-layers can take up many positively charged metal ions, that these seem to occupy select sites, and that carboxylate and phosphodiester groups of the proteins may serve as the binding sites for the metal ions [7,8,9]. TEM micrographs of bacterial cross-sections designed to show the surface distribution of the metal ions have been as a rule too low in magnification to resolve the S-layer structure and to show the distribution pattern of the uptake sites for the metal ions on the S-layer.

One exception is a TEM micrograph (Fig. 1) which shows the distribution in cross section of iron oxide depositional sites on the cell surface of a *Pedomicrobrium*-like budding bacterium from a laboratory culture [5]. The mineral coat shows significant differences in thickness from one side of the cell surface to the other. Its most conspicuous feature is the periodic structure of the mineral deposits which emanates from a continuum of discrete depositional sites with clearly defined lateral borders. They are either in contact with each other or are separated by mineral-free spaces. In the area of least mineral deposition only minute pinpoint mineralization sites can be seen that occur at distances as close as about 80 nm from each other. By contrast, where the iron oxide deposits are thickest, the borders of the discrete depositional areas show a spacing between about 170–830 nm. 18.5 nm center to center spacings have been reported for the subunits of the S-layer in this *Pedomicrobium* strain [5]. There is no information on the planar distribution pattern of the mineral deposits on the S-layer of this bacterial strain.

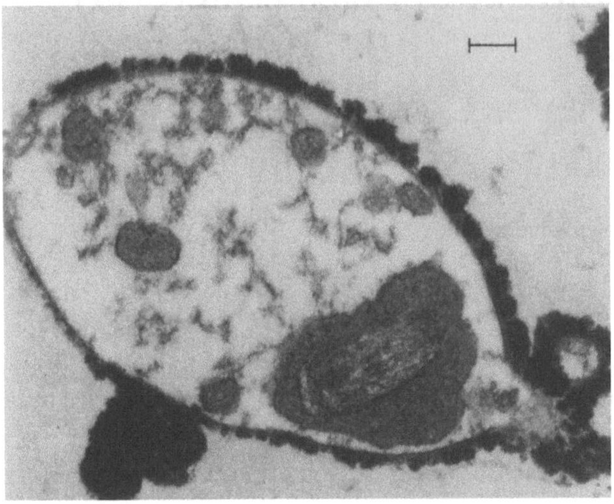

Fig. 1. TEM micrograph of a *Pedomicrobium*-like budding bacterium seen in cross-section showing the spacing relations of the iron oxide deposits on the cell surface. Scale bar: 1 μm. Reproduced from Ghiorse and Hirsch (1979).

The periodicity of the mineralization sites as seen in cross section would seem to be a reflection of the distribution pattern of the nucleation centers of the mineral deposits on the S-layer surface. Seen in this context, the large discrete iron oxide bodies would then represent various stages of progressive lateral expansion in mineralization, leading to the obliteration of the initial nucleation sites as well as of the adjacent original pore spaces in the S-layer structure. However, this does not seem to explain why, in the area of minimal mineralization, center to center spacings of the pin point mineralization sites appear to be at least four times larger than those reported for the S-layer of this bacterial strain. Higher resolution TEM micrographs are required to clarify the extent to which the S-layer of this bacterial strain controls the fabric of the mineral coat.

The feasibility to use bacterial S-layers as templates for nanometer molecular lithography has been explored [10]. S-layer fragments from *Sulfolobus acidocaldarius* which have been adsorbed on an amorphous carbon film were coated with a metal film. Replication of the surface configuration of the S-layer was achieved following ion milling to expose the lattice pores. However, the holes produced in the metal deposit were larger and their outlines were round instead of the hexagonally shaped lattice pores of the S-layer.

LEPTOTRIX SP. MINERALIZATION

Caldwell and Caldwell [1] found in an iron rich pool in Carlsbad Cavern of New Mexico (U.S.A.) a filamentous *Leptotrix* species with a mineral coat composed of an unidentified iron mineral. Iron was determined by using the Prussian blue assay. TEM micrographs of ultra thin sections shows that the mineral coat is tubular in shape and consists of an open grill work-like structure composed of pore bearing hexagons of 0.1 μm in diameter. Although no organic framework was observed, the unique mineral fabric nevertheless suggested that a biologically controlled process is involved in its formation [11].

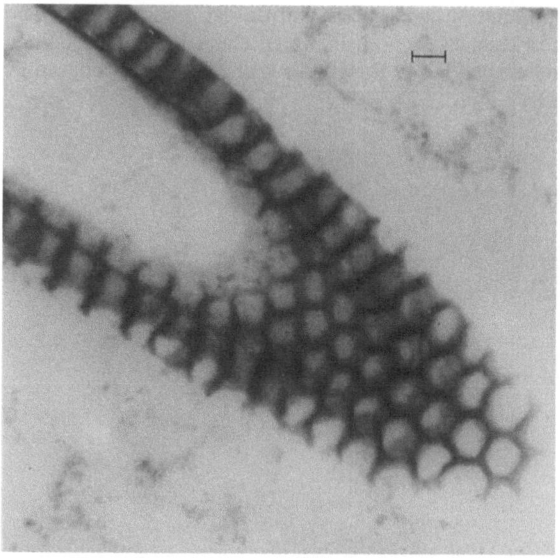

Fig. 2. TEM micrograph of *Leptotrix* sp. indet. showing the grillwork-like structure of its mineral coat and some of the underlying organic framework. Scale bar: 100 nm.

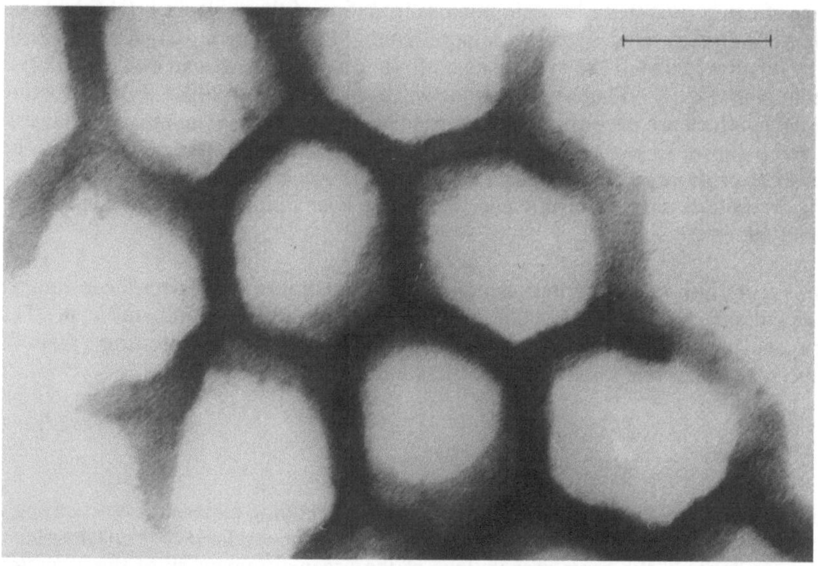

Fig. 3. TEM micrograph of *Leptotrix* sp. indet. showing a higher magnification view of the mineralized structure and underlying organic framework. Scale bar: 100 nm.

An attempt was made to collect additional specimens to determine whether an organic framework is actually associated with the mineral deposits and further to identify the mineralization product.

However, it turned out that the *Leptotrix* population in the Carlsbad Cavern pool, where it had been originally discovered, had disappeared due to changes in water chemistry. This limited further investigation to those specimens still embedded in the blocks that had been prepared by the Caldwells [1] for the original TEM study.

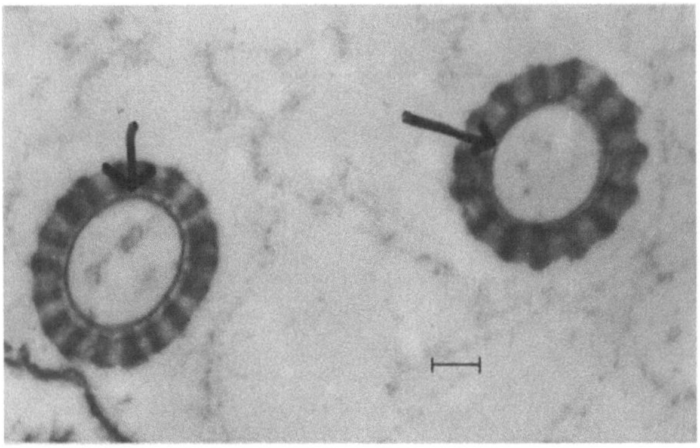

Fig. 4. TEM micrograph of *Leptotrix* sp. indet. showing in cross-section the periodic structure of the organic framework and of the overlying mineral coat. Arrows point to the columnar repeat structures of the organic matrix. Scale bar: 100 nm.

TEM micrographs of newly prepared ultra thin sections from one of these blocks established that an organic framework is associated with the mineral coat. The configuration of the organic component and its relation to that of the mineral coat are shown in planar view in Fig. 2 and Fig. 3 and in cross section in Fig. 4. The organic framework has a grillwork-like sheet structure composed of hexagons which are perforated by hexagonally shaped pores. The mineral coat that occupies its surface is shown to replicate its shape, but as seen in Fig. 3, its strands tend to be somewhat narrower than those of the organic substrate. The smooth margins of the fractured mineral coat (Fig. 3) indicate that the iron oxide precipitate is an amorphous compound rather than polycrystalline material.

In cross-section (Fig. 4), the organic component has the appearance of a periodic columnar structure with the columnar spacings coinciding with those of the overlying mineral coat. The center to center spacings of the hexagonally shaped pores in the organic substrate range from 70 to 140 nm and the pore diameters range from 60 to 100 nm.

DISCUSSION

The morphologic replication of the organic framework by the overlying mineral deposit strongly suggests that the organic component functions as a template for the mineralized coat on its surface. There is a striking similarity in the gross morphology of the organic framework of this *Leptotrix* species with that reported for bacterial S-layers with hexagonal repeat structure. However, the center to center spacings of the pores and also the pore diameters are significantly larger than those reported for bacterial S-layers. Hence, to decide whether the organic component of the mineral coat in this *Leptotrix* species constitutes a significantly larger S-layer type will have to await the rediscovery of this species and thereafter the determination of its chemical structure. The same applies to the identification of the bioinorganic component of the mineral coat.

The numerous mineral coats seen in thin sections consist invariably of pore-bearing grillwork-like sheet structures as shown here in Figs. 2, 3, and 4. There is no evidence that the mineralization products ever built up vertically to any extent nor, significantly, do they ever extend laterally to block the pore surfaces of the mineral coat. This establishes firmly that the mineralization process in this species is rigidly controlled by the organic component of the mineral coat. It is not yet known whether the pores are needed for the cell surface to maintain open communication with the environment.

Turning to the consideration of possible functions, the pore bearing structure of the rigid mineral coat may serve as a strengthening agent of the cell wall and also may be a protective device against bacterial predators. If it should further function as a molecular sieve, the rather large pores of the structure would suggest that it is designed for passage of large molecular structures.

Most monarans form bioinorganic materials by biologically induced mineralization processes [12]. Therefore, the documentation that mineralization in a *Leptotrix* species is biologically controlled is rather significant, since this mineralization process was previously known to occur only in magnetotactic bacteria [13,14] and the cyanobacteria, namely, *Scytonema julianum* [12] and *Geitleria calcaria* [15]. As is the case for the *Leptotrix* species, the two cyanobacteria species belong to the Enbacteria, are sheath bearing and form their mineralization products extra-cellularly, whereas in the magnetotactic bacteria mineral formation occurs intra-cellularly. An increasing number of cyanobacteria have been reported in recent years to form S-layers [3]. As far as bacteria are concerned, *Scytonema julianum*, *Geitleria calcaria* and *Geitleria floridiana* have unique crystal fabrics [16,17,18]. High resolution TEM micrographs should determine whether their mineral coat is the product of S-layer controlled mineralization.

Acknowledgements: Drs. D. and S. Caldwell made available the blocks of the embedded *Leptotrix* specimens used in their original study. Dr. S. Caldwell also attempted to collect for me additional specimens of this species from the Carlsbad Cavern pool. Dr. Z. Mason obtained the TEM micrographs of the *Leptotrix* material presented in this study. Contribution No. 4933 of the Division of Geological and Planetary Sciences, California Institute of Technology.

REFERENCES

[1] Caldwell DE, Caldwell SJ (1980) Geomicrobiol J 2:39–53

[2] Sleytr UB, Messner P (1983) Ann Rev Microbiol 37:3111–339

[3] Sleytr UB, Messner P (1988) In: Sleytr UB, Messner P, Pum D, Sara M (eds) Crystalline Bacterial Cell Surface Layers. Springer-Verlag, Berlin Heidelberg, pp. 160–174.

[4] Taylor KA, Deatherage JF, Amos LA (1982) Nature 299:840–842

[5] Ghiorse WC, Hirsch P (1979) Arch Microbiol 123:213–226

[6] Beveridge TJ (1984) In: Klug MJ, Reddy CA (eds) Current Perspectives in Microbial Ecology. Amer. Soc. Microbiol., Washington DC pp. 601–607

[7] Beveridge TJ, Murray RGE (1976) J Bacteriol 127:1502–1518

[8] Beveridge TJ, Murray RGE (1980) J Bacteriol 141:876–887

[9] Doyle RJ, Matthew TA, Streips UN (1980) J Bacteriol 143:471–480

[10] Douglas K, Clark NA, Rothschild KJ (1986) Appl Phys Lett 48 (10):676–678

[11] Lowenstam HA, Weiner S (1983) In: Westbroek P, deJong EW (eds) Biomineralization and Biological Metal Accumulation. Reidel, Dordrecht, pp. 191–203

[12] Lowenstam HA (1986) In: Leadbeater BSC, Riding R (eds) Biomineralization in Lower Plants and Animals. Systematics Association Spec. Vol. No. 30. Clarendon Press, Oxford, pp. 1–17.

[13] Blakemore RP (1975) Science 190:377–379

[14] Vali H, Kirschvink JL In: Frankel RB, Blakemore RP (eds) Iron Biomineralization. Plenum Press, Newyork, in press.

[15] Lowenstam HA unpublished

[16] Schönleber K (1936) Arch Protistenk 88:36–68

[17] Friedmann EI (1955) Bot Notiser 108:439–445

[18] Friedmann EI (1979) Pl Syst Evol 131:169–178

CHAPTER 1.2

The Organic Matrix of Spicules of the Gorgonian *Leptogorgia virgulata*

N. WATABE[1], M. OISHI[1], and R. J. KINGSLEY[2]

[1]Electron Microscopy Center, The University of South Carolina, Columbia, SC 29208 and [2]Department of Biology, University of Richmond, Richmond, VA23173, U.S.A.

Key Words: Collagen, Insoluble Matrix, Soluble Matrix, Spicules, Gorgonian, Leptogorgia.

Organic matrices of biominerals generally consist of proteins, glycoproteins, carbohydrates, and lipids, and their roles in mineralization have drawn considerable attention in recent years [1]. As in the vertebrates, the matrices of many invertebrate hard tissues such as foraminiferan tests [2], echinoderm hard parts [3, 4], molluscan shells [3, 5, 6, 7] and gorgonian spicules [8] have been shown to consist of water-soluble and insoluble components. However, unlike vertebrate osseous and dental tissues which are made of calcium phosphates, invertebrate hard tissues of calcium carbonate rarely contain collagenous protein components in the insoluble matrices [9].

The gorgonian Leptogorgia virgulata is a cnidarian characterized by the presence of numerous microscopic calcium carbonate (calcite) spicules in the mesoglea. The mature spicules assume elongated spindle-shaped structures which are approximately 100 µm long and surrounded at intervals by bands of wart-like projections (Fig.1). The spicule is initially formed within a Golgi vacuole of the scleroblast in association with an organic matrix [10]. At a later stage of the development, the limiting membrane of the Golgi vacuole fuses with the plasma membrane of the scleroblast. This is followed by formation of a breach in the fused membrane, and the spicule becomes an extracellular entity [10] (Fig.2).

During the past ten years, many aspects of the mechanisms of calcification of the Leptogorgia spicules have been investigated [11], and biochemical and physiological characteristics of the spicule organic matrix have been elucidated in some detail. In this paper, we will discuss our recent findings that these calcium carbonate spicules have a collagenous component in the insoluble matrix and that the collagen content exhibits seasonal variations [12][13].

The insoluble matrix of the Leptogorgia spicules can be observed under the transmission electron microscope after staining thin sections of spicules in uranyl acetate. The spicules are decalcified by this treatment and stained networks of the matrix are revealed (Fig.3). The framework is the insoluble matrix and delineates perforations in which calcite crystallites were present prior to decalcification. Fibers approximately 30 nm in diameter displaying no periodic banding are frequently observed to constitute the framework. The autoradiographic investigation using H^3-aspartic acid as the tracer revealed that the protein moiety of the organic matrix is synthesized in the RER during

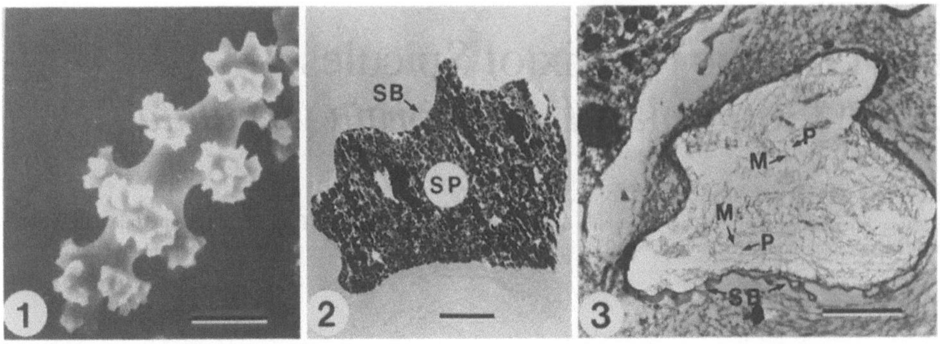

Fig. 1. Isolated mature spicule of ***Leptogorgia virgulata***. SEM. Scale Bar = 10 µm; Fig. 2. Thin section of a spicule *(SP)*. The spicule occupies almost the entire area of the scleroblast. *SB*: Scleroblast plasma membrane (see Fig. 3 for clear view). TEM. Scale Bar = 2 µm; Fig. 3. Spicule section stained/decalcified with uranyl acetate. Insoluble matrix *(M)* framework delineates perforations *(P)* where crystallites were present prior to decalcification. *SB*: Scleroblast plasma membrane. TEM. Scale Bar = 2 µm. (Watabe, submitted for publication)

the 5 min period following the 10 min incubation of the animal in the tracer. In 5 min, the label moves from the RER to the Golgi complexes where the carbohydrate moiety is presumed to be synthesized. From 60 to 120 min, the label is seen in the spicules in the Golgi vacuole [14]. The entire process of the matrix synthesis and transport into the spicule takes approximately 2 h at 21°C in the summer.

Biochemical analysis of the organic matrix was carried out by first isolating spicules by digesting the surrounding tissues with 1 % papain [12]. (The animals were collected from different localities in different seasons.) The spicules were washed with 0.25 M NaCl in 0.2 M NH_4CO_3 buffer, pH 8.0, and decalcified by dialysis with 0.5 M potassium EDTA in 0.05 M NH_4CO_3, pH 8.0 in a 3,500 dalton cut-off tubing. After centrifugation, the supernatant was dialyzed against distilled water and lyophilized. This is the soluble matrix. The precipitate was washed with NH_4CO_3 and distilled water and lyophilized. This is the insoluble matrix. Proteins of the insoluble matrix of summer (July) spicules from North Carolina and South Carolina indicate similar amino acid compositions containing significant amounts of hydroxyproline (Hyp), hydroxylysine (Hyl) and 30-33 % glycine (Gly), typical of a collagen (Table 1). The composition is more similar to the mesogleal collagen of sea anemone (cnidaria) (Table 1, I) than the bovine bone collagen (Table 1, J). However, the Leptogorgia spicule collagen contains less Hyp (74 vs.103/1000) and more acidic amino acids (202 vs. 170/1000) than the sea anemone collagen. The collagenous nature of the spicule matrix has also been confirmed by the presence of the intermolecular reducible cross-link, dihydroxylysinonorleucine [12]. The collagen content decreases in the December samples, as indicated by the values of Hyp-31/1000, Hyl-15/1000, and Gly-225/1000 (Table 1,D). The values further decrease in the March sample, showing very little of collagenous characteristics

Table 1. Amino Acids Compositions of Matrix Protein Fractions From Isolated Calcite Spicules of *Leptogorgia virgulata*

Residues per 1000 Total Residues

	A TIM S. C.	B TIM 3/87	C TIM 7/87	D TIM 12/87	E Pepsin Sol	F Pepsin Res	G SM 12/87	H SM 7/87	I Sea Anemone Collagen[d]	J Bovine Bone[e]
Hyp	74	8	82	31	80	11	1	-	103	98
Asp	93	132	76	130	84	153	516	577	73	45
Thr[a]	36	57	34	48	32	63	18	16	37	17
Ser[a]	43	57	42	53	43	78	16	13	41	34
Glu	109	87	94	92	112	76	34	32	97	74
Pro	66	67	65	59	64	50	17	19	67	123
Gly	295	158	332	225	335	161	210	185	339	337
Ala	83	78	82	76	77	67	130	108	65	109
Val	27	58	26	37	22	55	17	16	25	20
Cys[b]/2	2	16	1	10	-	17	-	-	-	-
Met[c]	4	11	7	10	5	10	-	-	-	5
Ile	18	38	15	26	13	46	6	5	22	11
Leu	27	53	26	42	23	58	7	6	30	25
Tyr	8	24	8	20	6	32	-	2	3	4
Phe	9	31	9	22	7	35	4	4	8	13
His	3	17	4	11	3	11	2	1	1	4
Hyl	33	7	34	15	34	6	-	-	25	6
Lys	15	44	12	33	10	33	12	9	16	26
Arg	56	55	50	53	50	39	8	5	68	50

TIM - Total Insoluble Matrix; S.C. - Summer collection in South Carolina; 3/87, 7/87, 12/87 - Dates of collection in North Carolina; Pepsin Sol. - 7/87 collection, pepsin soluble material; Pepsin Res - 7/87 collection, pepsin insoluble material; SM - Soluble Matrix; a. Uncorrected for hydrolysis; b. half Cys, sum of cysteic acid and cystine; c. sum of methionine sulphoxide and methionine; d. ref. Nowack and Nordwig 1974 [27]; e. ref. Herring 1972 [28]. From Kingsley et al., 1990 [12].

(Hyp-8/1000, Hyl-7/1000, and Gly-158/1000) (Table 1, B). The summer insoluble matrix reveals, after pepsin digestion, a non-collagenous component, the amino acid composition of which is similar to that of the March sample (Table 1, E).

After treatment with chitinase to remove carbohydrate moieties, the insoluble matrix was partially solubilized in n-butanol, followed by homogenization with 8 M guanidine-HCl containing 10 % β-mercaptoethanol and 0.1 M Tris-HCl, pH.7.5. This completely solubilized the insoluble matrix. The homogenate was dialyzed against 5 M urea-1 M thiourea mixture containing 0.17 % β-mercaptoethanol and was fractionated by 5 to 12 % SDS -PAGE following the method of Laemmli and Favre [15]. Treatment with other enzymes such as N-glycanase, chondroitinase, hyaluronidase, or other sorbent such as chloroform, acetone, chloroform-methanol, dichloromethane, or dichloroethane did not resolve as many protein fractions as the combination of chitinase and n-butanol. Figure 4a shows a SDS-PAGE of the summer (September) and spring (March) insoluble matrix. The summer matrix shows bands at 500 kD, 140 kD, 120 kD, and 40 kD. Both the 140 and 120 kD bands actually consist of several closely spaced sub-fractions which are difficult to separate. The spring sample does not show the 500, 140, and 120 kD fractions. However, occasionally, faint bands of 140 and 120 kD were recognizable. The 140 and 120 kD

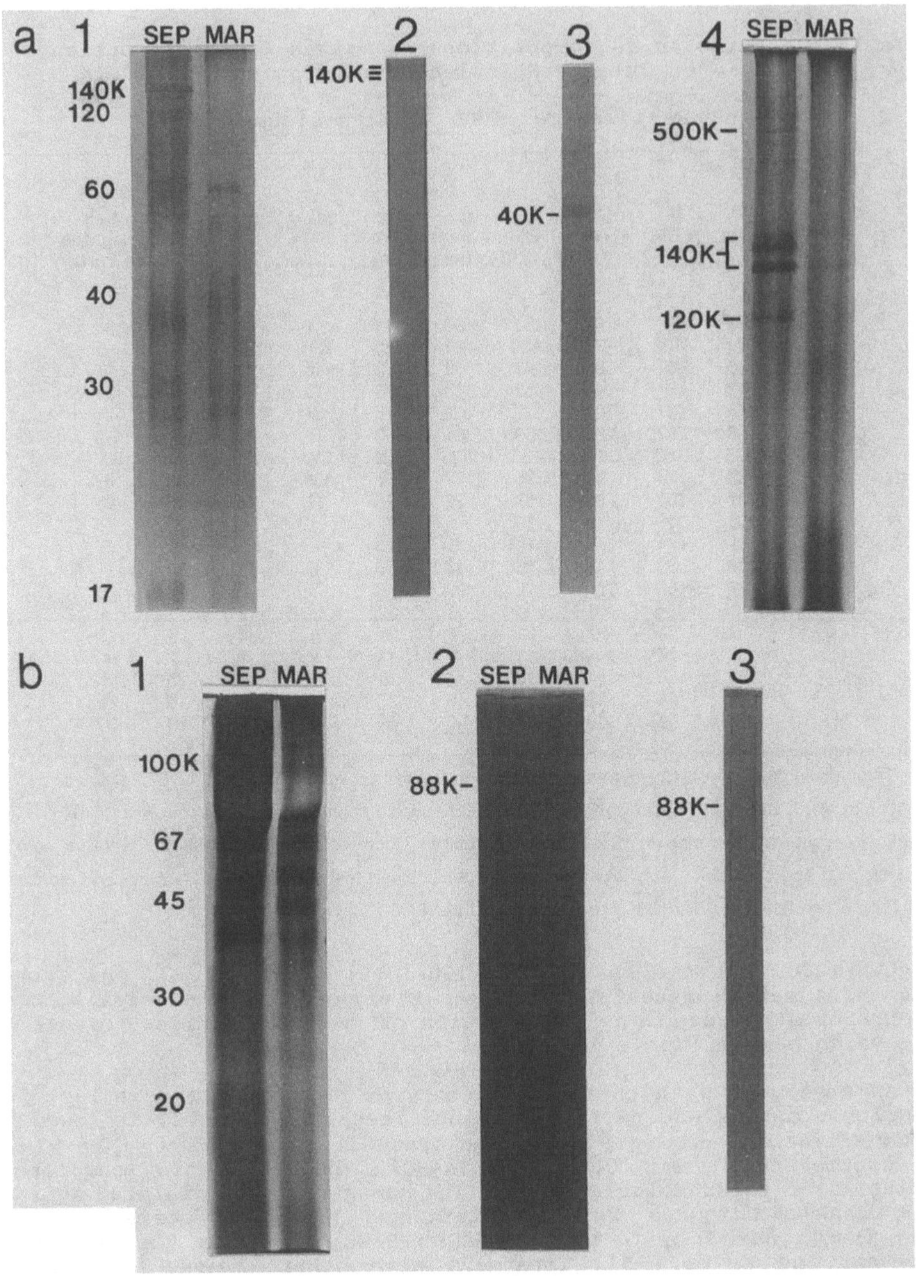

Fig. 4a. SDS-PAGE of the insoluble matrix proteins from gorgonian spicules. 1- 12% gel. SEP: September sample; MAR: March sample. 2- Antibody specificity test against 140 kD proteins (collagen) from the insoluble matrix; 3- Antibody specificity test against 40 kD protein (non-collagenous) from the insoluble matrix. 4- 5% gel of the high molecular proteins (over 100 kD).
Fig. 4b. SDS-PAGE of the soluble matrix proteins from the gorgonian spicules. 1- 12% gel, silver stain; 2- 12% gel, PAS stain; 3- Specificity of the anti-88 kD protein (soluble) antiserum.

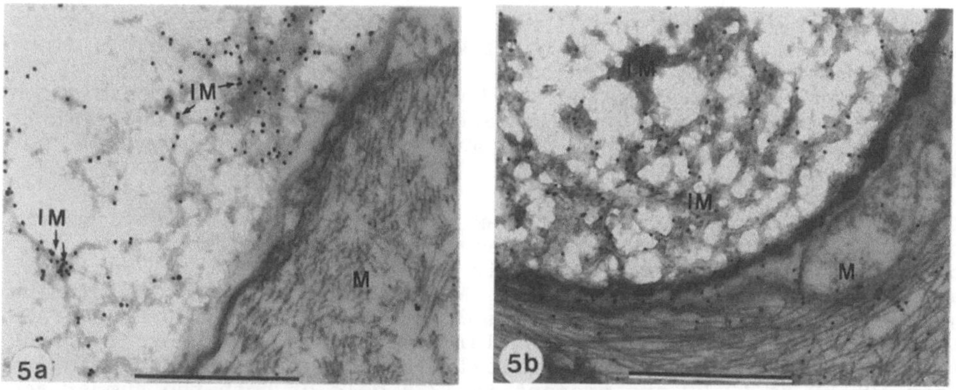

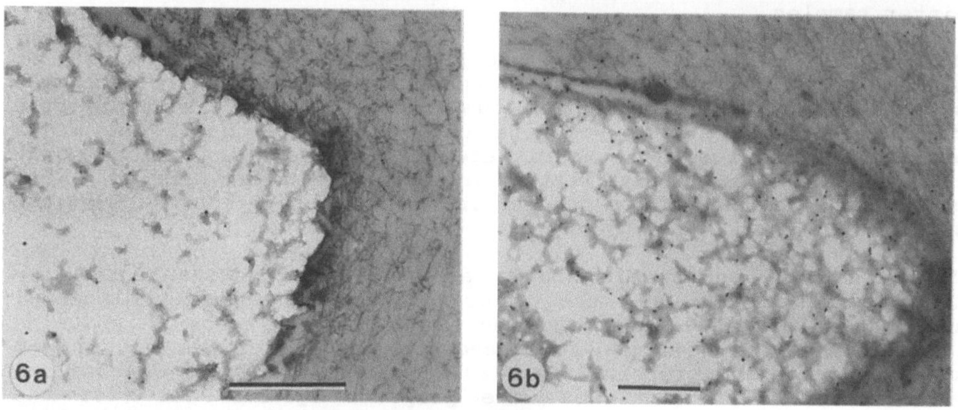

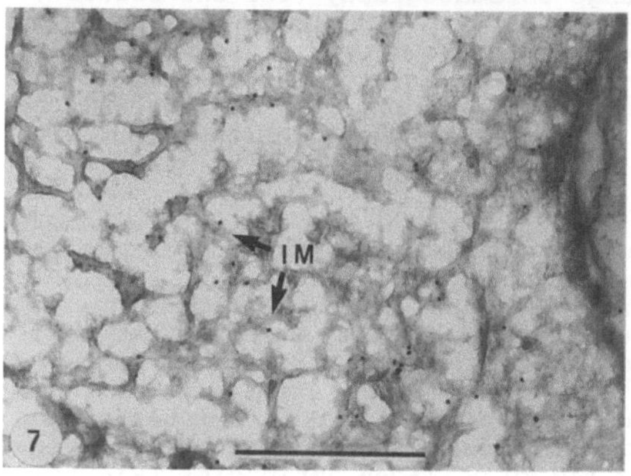

Fig. 5a. Insoluble matrix framework (IM) of summer spicules labeled with antibodies against the collagenous 140 kD protein. Mesoglea fibers (M) are labeled very little. scale bar = 1 µm. Fig. 5b. Summer spicule matrix labeled against 40 kD non-collagenous protein at the IM periphery and mesoglea. scale bar = 1 µm. Fig. 6a. Winter spicule matrix with little labels against 140 kD protein. scale bar = 1 µm. Fig. 6b. Winter matrix and mesoglea labeled against 40 kD protein. scale bar = 1 µm. Fig. 7. Periphery of summer IM labeled against 88 kD soluble protein. scale bar = 1 µm.

fractions contain Hyp, Hyl, and Gly indicating their collagenous nature. The 500 kD protein is not likely to be a phosphoprotein since it does not stain by the method of Hegenau et al.[16]. The 40 kD protein may correspond to that of the pepsin insoluble protein shown in Table 1 F, since both proteins are present in the insoluble matrix regardless of the seasons, although a direct comparison is not possible.

The amino acid compositions of the whole lyophilized soluble matrix of the July and December spicules are similar and show very high content of aspartate (Asp) which is over 50 % of the total residues, and Gly and alanine (Table 1). A highly acidic calcium-binding glycoprotein is present in the soluble matrix [17], and SDS-PAGE indicates that the apparent molecular weight of this protein is 88,000 (Fig. 4b). In addition, approximately 10 minor protein bands ranging from 100 kD to 12 kD are present in 12 % gels of both the summer and winter samples.

Clearly, the organic matrix of the Leptogorgia spicules consists of season- dependent and -independent components. The composition of the soluble matrix is independent of the season and consists mainly of a calcium-binding acidic glycoprotein. On the other hand, the major proteins of the insoluble matrix are collagenous which occur mostly in the summer but very little in the winter and spring. Non-collagenous components are present in the insoluble regardless of the season.

The findings mentioned above are further substantiated by immunocytochemistry at the ultrastructural level. Polyclonal antibodies directed against the 140 kD collagenous proteins (insoluble), the 40 kD non-collagenous proteins (insoluble), and 88 kD calcium-binding protein (soluble) were raised in male New Zealand white rabbits. The antigens were prepared by SDS-PAGE (for the insoluble), or by DEAE ion exchange chromatography (for the soluble). For immunocytochemistry, undecalcified thin sections of branch tips of Leptogorgia were first treated with 0.1 % cetylpyridinium chloride (CPC) in 2 % (para)formaldehyde to decalcify spicules and to retain soluble and insoluble matrix as much as possible. Immunolabeling the antigens was carried out by incubation of the sections with each antibody and staining with 15 nm gold conjugated with goat anti-rabbit IgG.

In the summer spicules, the immunogold labels with antibodies against the collagenous 140 kD proteins are localized within the insoluble matrix framework (Fig.5a). The mesogleal fibers surrounding the scleroblasts and spicules are labeled very little. On the other hand, the labels against the non-collagenous 40 kD insoluble proteins decorate the periphery of the insoluble matrix framework; and the mesogleal fibers are heavily labeled (Fig. 5b). In the winter/spring spicules, both the insoluble matrix framework and the mesogleal fibers are labeled very little with the 140 kD collagen antibodies (Fig.6a), whereas the periphery of the insoluble matrix framework and the mesogleal fibers are labeled heavily for the non-collagenous 40 kD proteins (Fig.6b). However, the 40 kD labels are not as heavy as in the summer. More often than not, the insoluble matrix framework is less dense and the mesogleal fibers are fewer in the winter/spring than in the summer [18]. Similar to the labels for the 40 kD insoluble proteins, the immunogold marker for the soluble 88 kD glycoproteins decorates the periphery of the insoluble matrix framework (Fig. 7). However, the spatial relationship between the 88 kD soluble and 40 kD insoluble proteins has not been resolved.

In good agreement with the biochemical results, the immunocytochemical investigation indicates that the spicule collagen component is very much reduced in the winter/spring animals but not in the non-collagenous components. This implies that the degradation of the collagen component occurs within the spicules during the fall/winter months, preceded by (partial) spicule decalcification. It also implies that remineralization of the old spicules should occur in addition to new spicule formation towards summer. At present, no information is available concerning the process of degradation of this collagen. It may involve activation of some enzymes such as a specific collagenase by certain seasonal cues. Spicule decalcification may be resulted from shifts in the activities of some enzymes, e.g., lysosomal enzymes, carbonic anhydrase (CA) etc. In fact, these enzymes have been cytochemically localized in the scleroblasts [8, 19], and inhibition of CA activities by Diamox affected the spicule formation in rather complex ways [19]. It is known that lysosomal enzymes are affected by thyroid hormone [20] and substances which immunologically cross-react with thyroxine (T4) has been localized in Leptogorgia [21]. Furthermore, an inhibition of T4 activities has been shown to affect the calcification in this organism [22]. Thus, it is conceivable that season-dependent activities of endocrine or endocrine-like substances and enzymes regulate organic matrix formation and degradation as well as spicule calcification and decalcification in Leptogorgia.

The immunocytochemical investigation shows that the collagen is the backbone of the organic matrix framework of the summer spicule, and the non-collagenous insoluble and soluble fractions are attached to it. Functions of the organic matrix in spicule formation are not clear. It is generally accepted that acidic glycoprotein macromolecules influence calcium carbonate crystal nucleation and growth in vitro [23, 24]. At present, we have no information concerning the molecular organization of various protein fractions of the spicule matrix. However, considering the fact that the collagenous component occurs only in the summer when the spicule calcification is presumably very active, this spicule collagen may be essential for the initial calcite nucleation. In fact, the insoluble matrix obtained after decalcification of spicules of Leptogorgia has been shown to recalcify in vitro [1]. The Leptogorgia spicule collagen is different from the vertebrate collagens. It is also different from any of the collagens reported in the cnidaria [25] for its lack of periodic banding and/or its difference in the amino acid composition. If indeed the spicule collagen is the nucleator for calcite, this association of $CaCO_3$ crystals with a collagenous matrix is different from that in another cnidaria Veretillum cynomorium. In this organism, the axial rod contains calcite crystals through which collagen fibers penetrate. Ledger and Franc [26] report that the collagen fibers are never impregnated with the calcite and are not responsible for the initial nucleation.

Another candidate for the spicule calcite nucleator in Leptogorgia is the non-collagenous protein(s). The 40 kD insoluble and 88 kD soluble proteins are localized at the periphery of the matrix framework and presumably more accessible for Ca and CO_3 ions. Further immunocytochemical investigation is in progress to identify the type of proteins at the site of initial calcification of spicules.

ACKNOWLEDGEMENTS: Much of the work mentioned here has been supported by funds from the National Science Foundation grant #PCM-8201389, DCB-8502689, DCB-880109, and DIR-8805143. We thank Betty Bynum and Robert Tye Whitaker for assistance in the preparation for the manuscript.

REFERENCES

[1] Watabe N, Bernhardt AM, Wilbur KM, Kingsley RJ (1987) Trans Amer Microsc Soc 105: 311-318
[2] Weiner S, Eraz J (1984) J Foraminif Res 14: 206-212
[3] Weiner S, Traub W, Lowenstam HA (1983) In: Westbroek P, deJong LW (eds) Biomineralization and Biological Metal Accumulation. D Reidel Publishing, Dordrecht, Boston, London, pp. 205-224
[4] Swift DM, Sikes CS, Wheeler AP (1986) J Exp Zool 240: 65-73
[5] Crenshaw MA (1972) Biomineral Res 6: 6-11
[6] Kasai H, Ohta N (1981) In: Habe T, Omori M (eds) Study of Molluscan Paleobiology. Niigata University, Japan, pp. 107-123
[7] Wheeler AP, Rusenko KW, Sikes CS (1988) In: Sikes CS, Wheeler AP (eds) Chemical Aspects of Mineralization. University of South Alabama, Mobile, AL, pp. 3-13
[8] Kingsley RJ, Watabe N (1983) Comp Biochem Physiol 76B: 443-447
[9] Watabe N (1981) Prog Crystal Growth Charact 4: 99-147
[10] Kingsley RJ, Watabe N (1982) Cell Tissue Res 223, 325-334
[11] Watabe N, Kingsley RJ (1989) In: Crick RE (ed) Origin, Evolution, and Modern Aspects of Biomineralization in Plants and Animals. Plenum Press, New York, pp. 209-223
[12] Kingsley RJ, Tsuzaki M, Watabe N, Mechanic GL (1990) Biol Bull (In press)
[13] Oishi M, Kingsley RJ, Watabe N (Unpublished)
[14] Kingsley RJ, Watabe N (1984) Cell Tissue Res 235: 533-538
[15] Laemmli UK, Favre M (1973) J Mol Biol 80: 575-599
[16] Hegenau J, Ripley L, Nace G (1977) Anal Biochem 78: 308-311
[17] Samata T, Kingsley RJ, Watabe N (1989) Comp Biochem Physiol 94B: 651-654
[18] Dupree J, Kingsley RJ (Unpublished)
[19] Kingsley RJ, Watabe N (1987) J Expt Zool 241: 171-180
[20] Gaton, DD, Gaton E, Wolman M (1987) Cell Molec Biol 33: 615-624
[21] Kingsley RJ, Corcoran M (Unpublished)
[22] Kingsley RJ, Watabe N (Unpublished)
[23] Addadi A, Berman A, Oldak JM, Weiner S (1989) Conn Tiss Res 21: 127-135
[24] Mann S, Didmus JM, Sanderson NP, Heywood BR (1990) J Chem Soc Faraday Trans 86: 1873-1880
[25] Franc S (1985) In: Bairati A, Garrone R (eds) Biology of Invertebrate and Lower Vertebrate Collagens. Plenum Press, New York, pp.197-210
[26] Ledger PW, Franc S (1978) Cell Tiss Res 192: 249-266
[27] Nowack H, Nordwig A (1974) Eur J Biochem 45: 333-342
[28] Herring GM (1972) In: Bourne GH (ed) The Biochemistry and Physiology of Bone, Structure/Vol 1. Academic Press, New York, pp. 127-189

CHAPTER 1.3

Mineral-Associated Proteins from Modern Planktonic Foraminifera

L. STATHOPLOS and N. TUROSS

Smithsonian Institution, Conservation Analytical Laboratory, Washington, DC 20560, USA

Key words: planktonic foraminifera, organic matrix proteins, calcite, biomineralization

INTRODUCTION

Planktonic foraminifera are unicellular protists that live in open ocean surface waters. They make calcite tests (shells) by a complex series of events that may include biologically induced and controlled calcification [1, 2]. The specific function of the organic matrix in foraminiferal calcification is unknown [2, 3]. In this paper we describe chemical dissection techniques that improve our ability to isolate and characterize putative proteins involved in foraminiferal calcification.

About half of the extant foraminiferal species are spinose, such as *Globigerinoides ruber* (d'Orbigny), possessing mineralized elongate spines radiating from the shell wall which can serve as supports for rhizopodia. Other species, such as *Globorotalia menardii* (d'Orbigny), possess no spines. Calcification occurs during discrete chamber formation events [2]. A cytoplasmic bulge emerges from the existing shell, then the scaffold upon which subsequent calcification occurs is constructed. The site of shell wall bilamellar calcification is the primary organic membrane (POM). No proteins have ever been isolated and sequenced from the POM of modern planktonic foraminifera.

MATERIAL AND METHODS

Procedures used have been described in detail elsewhere [4, 5], and are only summarized here.

Collection. Live planktonic foraminifera were collected from surface waters (0-10 m) of the North Atlantic near the Gulf Stream (37°-38°N, 69°-72°W) by a small 150μ mesh plankton net. Bulk plankton samples rinsed into collection jars with sea water were frozen to about -40°C by immersion in refrigerated ethylene glycol. Samples were stored at -80°C. Foraminifera from bulk plankton thawed on ice were identified under a stereomicroscope, removed by pipette, and pooled in monospecific samples for analysis. Fossil foraminifera were separated with a vacuum picking device from the >125μ fraction of DSDP (Deep Sea Drilling Project) Site 586-1, core catcher sample (280 ky; [6]), and cleaned by repeated ultrasonication in distilled water.

Protein extraction. A modification of the dissociative demineralizing protocol commonly used in isolation of noncollagenous proteins from vertebrate tissue [4] was used. Briefly, whole foraminifera were extracted at 4°C first in "G" solution: 4 M guanidine HCl (Sigma), 0.05 M Tris (BioRad), 0.03 M $CaCl_2$ (Fisher), pH 7.4, in the presence of three protease inhibitors, 0.1 M 6-aminocaproic acid (Sigma), 5 mM benzamidine HCl (Kodak), and 1 mM phenylmethylsulfonyl flouride (Sigma). Then, shells were demineralized with "E" solution: "G" solution without $CaCl_2$, plus 0.5 M tetrasodium EDTA (ethylenediaminetetraacetic acid). "G" and "E" extracts were desalted by exchange into urea on Centricon microconcentrators (Amicon Co.; mw cutoff, 10,000). The insoluble residue remaining after demineralization was washed exhaustively with distilled water and freeze dried.

PAGE and Electroblotting. Proteins were separated on polyacrylamide gradient SDS mini-gels (4-20% or

4-12%; Novex) in Laemmli buffer [7]. Proteins were electroeluted onto 0.45 μ nitrocellulose and visualized with colloidal gold (Janssen; [8]).
Amino acid analysis. Samples were hydrolyzed under nitrogen in excess 6 N HCl at 110°C for 22 hours or 150°C for 20 minutes [9]. Bovine serum albumin (Sigma) was used to estimate the yield of hydrolyzable amino acids by weight from a simple protein with minimal post-translational modification. Amino acids were separated by ion-exchange high-performance liquid chromatography (St. John Assoc.) and detected by fluorescence after post-column reaction with OPA (*o*-phthalaldehyde; [10]).

RESULTS

Amino acid compositions. Proteins in each of the three separated fractions, the "G" soluble, the "E" soluble, and the insoluble, have different amino acid compositions for the spinose species *Globigerinoides ruber* (Fig. 1). This suggests that the sequential extraction procedure segregates unique or enriched components from the foraminifera. The amount of protein extracted in both the "G" (nondemineralizing) and "E" (demineralizing) extracts is small, probably on the order of low nanogram amounts per test. The mineral-associated "E" fraction is about 10% hydrolyzable amino acids by weight.

Components of the "E" extract from both the spinose species *G. ruber* and from the nonspinose species *Globorotalia menardii* are rich in glycine. Unextracted *G. menardii*, however, have an amino acid composition very similar to the insoluble residue (Fig. 2), further evidence that the "E" fraction proteins represent only a small proportion of the total hydrolyzable amino acids recovered from the organism (<1%). About a microgram of insoluble material was recovered per test from *G. menardii*.

The bulk of the proteinaceous material in four species of planktonic foraminifera is found in the insoluble fraction, which is at least 20% hydrolyzable amino acids by weight. The amino acid compositions of the insoluble fractions from two nonspinose *(Globorotalia truncatulinoides* and *G. menardii)* and two spinose *(G. ruber* and *Globigerinoides sacculifer)* species, are nearly identical, with maximal differences of only 1-2% in Asx and Ala (Fig. 3a). The overall distribution of amino acids in the insoluble replicas obtained upon decalcification is fairly even; no fraction dominated by acidic residues is observed. This contrasts with the amino acid compositions of fossil tests of three species (Fig. 3b). The compositions of 280 ky fossil tests show differences in relative amounts of Asx and Ala of 5-10% between species (Fig. 3b).

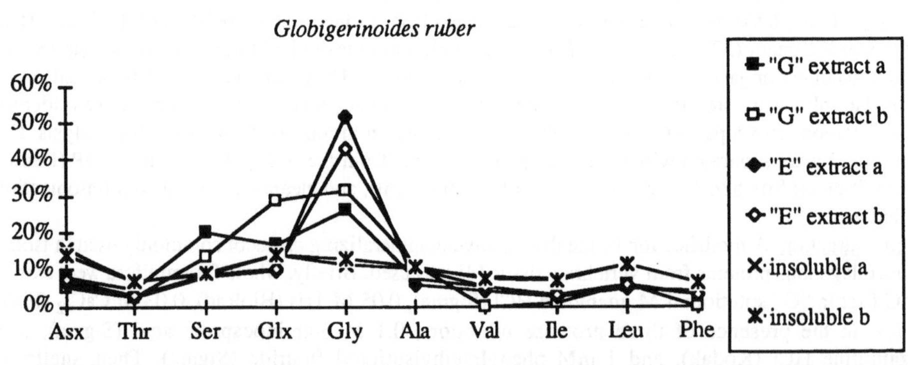

Fig. 1. Partial amino acid compositions, as percent of the sum of the amino acids shown, for two samples each of the "G", "E" and insoluble extracts from modern spinose *Globigerinoides ruber*. The different amino acid compositions suggest each fraction contains different cellular components. The "E" extracts are glycine rich, but neither the "G" nor the insoluble fractions are dominated by one amino acid.

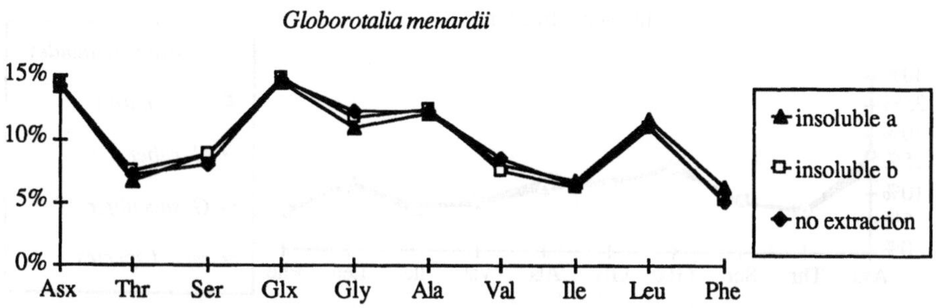

Fig. 2. Partial amino acid compositions, as percent of the sum of the amino acids shown, for two insoluble samples (after "G" and "E" extraction) and unextracted individuals (no "G" and "E" extraction) from the nonspinose species *Globorotalia menardii*. The glycine-rich "E" component found in both this species and in spinose *Globigerinoides ruber* cannot be a large percentage of the hydrolyzable amino acids recovered because of the similarity in composition between extracted and unextracted samples.

Gel electrophoresis. The insoluble fraction from a mixed species sample was composed of many proteins, susceptible to proteinase K digestion, spanning a wide range of apparent molecular weights. Some banding was observed, suggesting the presence of a high molecular weight component of greater than 110 kd (Fig. 4). Based on the broad banding behavior observed, and the relatively low recovery of hydrolyzable amino acids by weight from the insoluble fraction (20%), this component may be a proteoglycan.

DISCUSSION AND CONCLUSIONS

The extraction of freshly collected planktonic foraminifera with the dissociative agent guanidine HCl appears to solubilize the ectoplasm (the cytoplasm outside the shell; [11]). Components from the primary organic membrane (POM) and other shell organic layers are then solubilized by demineralization. The condition of these extracted shell proteins is unknown. In vertebrates, extracellular biomineralization proteins degrade and get trapped in the mineralization space as a normal consequence of mineralization [12]. Degraded cellular products likewise may be trapped during planktonic foraminiferal chamber formation.

None of the protein fractions recovered from core-top fossil *Globorotalia menardii* or *Globigerinoides ruber* tests [13] have amino acid compositions resembling the compositions reported here. The compositions of total soluble and insoluble mineral-associated fractions of the modern benthic foraminiferal species *Heterostegina depressa* also do not correspond to compositions of fractions extracted from modern planktonic tests [14]. It is not known if taxonomic differences, differences in separation protocols, or differences in starting material (modern vs. fossil, lab-reared vs. field-collected) are responsible for the discrepancy.

The insoluble fraction holds the gross morphology of the calcite test even after demineralization. Its amino acid composition resembles fossil test compositions much more closely that either the "G" or "E" fraction compositions, suggesting the preservation of endoplasmic components in fossil tests (Figs. 1, 3). Even small amounts of endoplasm would overwhelm the mineral-associated amino acid composition. The composition of fossil tests cannot be accounted for solely by mineral-associated components released during "E" solution demineralization. The enrichment of Asx in fossil shells compared to insoluble modern fractions (Fig. 3) could reflect either preferential retention of as yet uncharacterized high-Asx components by the shell during diagenetic loss [3], or adsorption of Asx-rich components from the surroundings [15 -

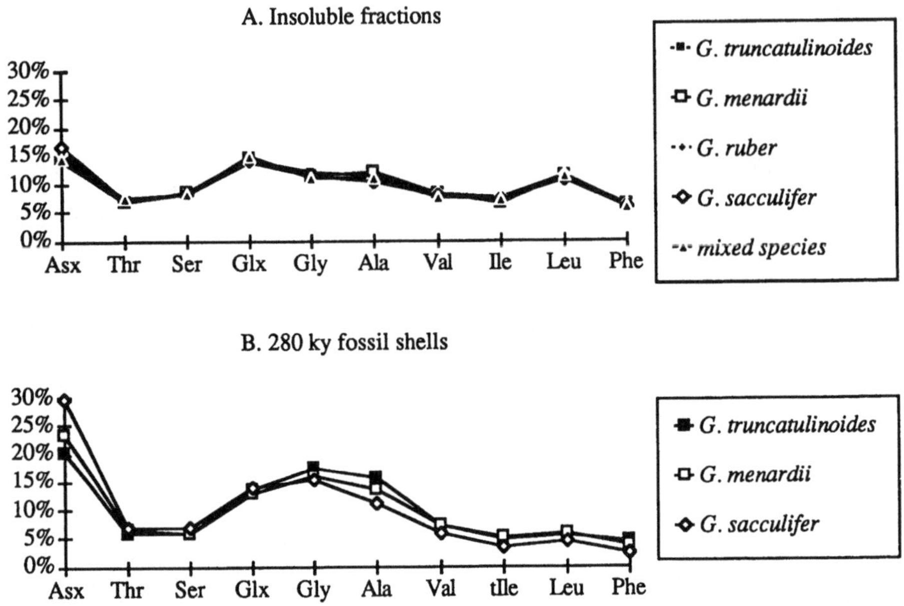

Fig. 3. (A) Partial amino acid compositions of insoluble fractions from two nonspinose species, *Globorotalia truncatulinoides* and *Globorotalia menardii*, two spinose species, *Globigerinoides ruber* and *Globigerinoides sacculifer*, and a mixed species sample. (B) Partial amino acid compositions of fossil shells (280 ky, from DSDP 586-1, core catcher) of three species, *G. truncatulinoides, G. menardii* and *G. sacculifer*. For fossil compositions, isoleucine and *allo*-isoleucine have been combined to give total isoleucine (tIle). Note that species-specific differences, especially in proportions of Asx and Ala, are far more pronounced in fossil than in modern insoluble compositions.

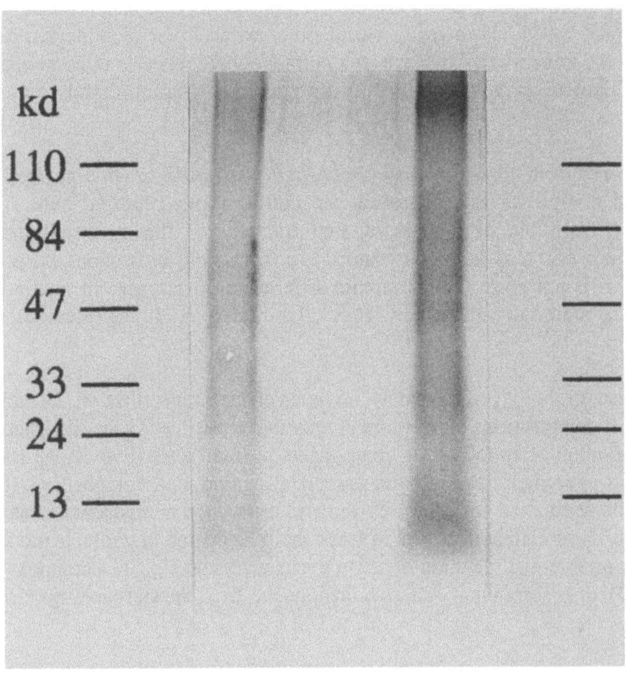

Fig. 4. Gold-stained nitrocellulose blot of a gradient (4-20%) polyacrylamide gel electrophoretic separation of two subsamples of the insoluble fraction from a mixed species sample (right lane 2x loading of left lane). Proteins of a nearly continuous range of apparent molecular weights (indicated as kilodaltons, kd) are observed. Staining disappears if samples are pretreated with proteinase K (not shown). A broad band is visible at greater than 110 kd, which may be a proteoglycan.

17]. The enrichment of Gly in fossil compared to modern insoluble compositions might reflect preferential retention of Gly-rich mineral-associated components compared to insoluble endoplasm.

ACKNOWLEDGEMENTS

This work was partially supported by a Smithsonian postdoctoral fellowship to LS. We thank P. Ortner and T. Rossby for allowing LS to participate in two BIOSYNOP cruises. J. P. Kennett and the Ocean Drilling Program (formerly DSDP) generously supplied fossil foraminiferal samples.

REFERENCES

[1] Towe K, Cifelli R (1967) J Paleont 41: 742-762.

[2] Reviewed in Hemleben C, Spindler M, Anderson OR (1989) Modern Planktonic Foraminifera. Springer-Verlag. New York, pp. 187-219.

[3] Lowenstam HA, Weiner S (1989) On Biomineralization. Oxford Univ Press. New York, 324 pp.

[4] Termine JD, Belcourt AB, Christner PJ, Conn KM, Nylen MU (1980) J Biol Chem 255: 9760-9768.

[5] Stathoplos L (1989) Amino acids in planktonic foraminiferal tests. PhD thesis, URI.

[6] Kennett JP, von der Borch CC, et al. (eds) (1986) Init Repts DSDP 90. US Gov't Printing Office. Washington, DC.

[7] Laemmli UK (1970) Nature 227: 680-685.

[8] Moeremans M, Daneels G, DeMey J (1985) Anal Biochem 145: 315-321.

[9] Hare PE (1977) Methods Enzymol 47: 3-18.

[10] Benson JR, Hare PE (1975) Proc Natl Acad Sci USA 72: 619-622.

[11] Hottinger L (1986) In: Leadbeater BSC, Riding R (eds) Biomineralization in Lower Plants and Animals. Clarendon Press, Oxford, pp. 219-235.

[12] Fisher LW, Termine JD (1985) In: Ornoy A, Harell A, Sela J (eds) Current Advances in Skeletogenesis. Elsevier, New York, NY, pp. 467-472.

[13] Robbins LL, Brew K (1990) Geochim Cosmochim Acta 54: 2285-2292.

[14] Weiner S, Erez J (1984) J Foram Res 14: 206-212.

[15] Carter PW (1978) Geochim Cosmochim Acta 42: 1239-1242.

[16] Carter PW, Mitterer RM (1978) Geochim Cosmochim Acta 42: 1231-1238.

[17] Müller PJ, Suess E (1977) Geochim Cosmochim Acta 41: 941-949.

CHAPTER 1.4

Structure and Function of the Organic Matrix in the Nacreous Layer of *Pinctada fucata*

T. SAMATA
Azabu University, Sagamihara, 229 Japan

Key words: Amino acid analysis, Electrophoresis, Ca-binding, Acidic proteins, Nacreous layer.

INTRODUCTION

The organic matrix in molluscan shells can be roughly divided into a water-soluble and a water-insoluble matrix, those demarcation between which is sometimes not quite clear. Ca-binding components have been isolated from the water-soluble matrix [1, 2]. Because of the capacity of such components for binding of Ca^{2+}, many authors have postulated reactions whereby these molecules might promote nucleation of $CaCO_3$ crystals [3, 4]. Although diversity of matrices has been reported by many authors [5, 6], the complete fractionation of such a matrix and the identificaiton of each component, contained in the matrix have not yet been achieved.

In the present study, each component in the organic matrix of Pinctada fucata was purified and its structure was determined in order to estimate the role of each component in the process of formation of the shell. The organic matrix in the nacreous layer was used for this study because it is known to be composed of several components which may be related to different functions in the control of mineral growth.

MATERIALS AND METHODS

Material Fresh shells of the Japanese pearl oyster, Pinctada fucata, were collected at Nakatsuhamaura, Mie, Japan.

Extraction The nacreous layer was separated from the outer prismatic layer with a dental drill and then dipped in 1 N HCl for 1 min. and incubated in 1% NaClO for 30 min. to remove the rest of the outer layer and some organic contaminants. Cleaned shells were dried, powdered and decalcified. Decalcification was carried out in 10% EDTA at pH 7.0, by the method of Wheeler et al. [7]. Salts were removed by dialysis against distilled water in Spectrapore no. 3 tubing (Spectrum Medical Ind., USA). Each dialyzed sample was centrifuged (10,000 x g, 30 min.). The supernatant was the water-soluble matrix (W-SM) and the precipitate was the water-insoluble matrix (W-ISM). The water-insoluble matrix was further extracted with a mixture of chroloform and methanol (2:1, v/v) for 24 hours and then extracted matrix was centrifuged (10,000 x g, 30 min.). The chloroform phase was analyzed by

HPLC and infrared spectroscopy for identification of its lipid components. After repeated washing on Whatman no. 3 filter paper, the insoluble portion was collected from the methanol phase. It was extracted with dilute NH_4OH at pH 8.4 at room temperature for 12 hours, and the extracted mixture was centrifuged under the same conditions as described above. The supernatant was the alkali-soluble matrix (AL-SM).

Labelling with $^{45}CaCl_2$

An aliquot of 0.1 ml of radioactive $^{45}CaCl_2$ (0.05 μCi) was added to 2 ml aliquots of the W-SM, AL-SM and the lipid fractions before each was fractionated on a column of Bio-Gel P2. Radioactivity of each eluted fraction was measured in a liquid scintillaiton counter of Packard Tricarb Model 4530 liquid scintillation counter (Packard Instrument Co., New York, USA).

Gel electrophoresis SDS-polyacrylamide gel electrophoresis (SDS-PAGE) was carried out in 8.5% gels as discribed by Anderson et al. [8]. Gels were stained with 0.4% Coomassie brilliant blue R-250 (C.B.B.) and silver-stained by the method of Morrisey [9] with a silver-staining kit (Wako Pure Chemicals Co., Osaka, Japan).

Amino acid analysis Samples for amino acid analysis were hydrolyzed in individual ampoules under vacuum at 110°C for 24 hours, in 6 N HCl. The hydrolysate was then analyzed on an Atto MLC-703S automatic amino acid analyser (Atto Co., Tokyo, Japan).

RESULTS

Extraction of the components from the bulk matrix

After decalcification with EDTA, the bulk matrix was roughly separated into a water-soluble matrix (W-SM) and a water-insoluble matrix (W-ISM). Extraction with chloroform and methanol revealed the presence of a definite lipid components in the W-ISM. Moreover, only a tiny portion of the insoluble material after extraction with chloroform and methanol was released in dilute alkali and in b-mercaptoethanol under boiling for more than 5 min. (AL-SM). Further attempts to solubilize the remaining insoluble material were successful only when pure HCOOH was used. This result will be discussed elsewhere.

SDS-PAGE of the soluble components

After electrophoresis of the W-SM, one clear band was visualized with a mobility between BSA and chick egg albumin by the staining of C.B.B. and silver. The molecular weight of the peptide in this band was estimated to be 58 kD. Two additional faint bands were recognized with a slightly lower mobility than CA and almost the same mobility with trypsin inhibitor. A wide, faint smear was observed to overlap these two bands. By contrast, after electrophoresis of the AL-SM one clear band was visualized that migrated slightly slower than α-lactalbumin from bovine milk. The molecular weight of the peptide in this band was estimated to be 17 kD.

Amino acid analysis of the W-SM, AL-SM and W-ISM

The amino acid compositions of the W-SM, AL-SM and W-ISM are shown in Table 1. The W-ISM contains a large amounts of Ala and Gly, which accounted for 49% of the total amount of amino acids. Moreover, Leu and Ser were also present at relatively high levels, while the acidic amino acids were present at relatively low levels. By contrast, the compositions of the W-SM and the AL-SM were characterized by a large amounts of Asx. These fractions were rich in the acidic

Table 1: Amino acid compositions of the water-soluble, alkali-soluble and water-insoluble organic matrix in the nacreous layer of <u>Pinctada fucata</u> (in molar percent).

	W-SM	AL-SM	W-ISM
Asx	23.68	16.34	7.66
Thr	2.57	1.35	1.74
Ser	4.92	4.64	6.91
Glx	8.83	6.38	3.36
Pro	3.97	2.37	2.95
Gly	18.62	25.99	27.62
Ala	8.61	14.62	21.33
Cys	-	0.58	-
Val	3.29	2.65	4.13
Met	1.20	1.54	0.93
Ile	2.25	1.63	2.97
Leu	5.63	6.85	7.84
Tyr	2.30	4.87	3.06
Phe	2.73	3.21	2.80
Lys	5.22	3.01	2.69
His	2.46	0.40	0.43
Arg	3.73	3.57	3.57
Acidic A.A.	32.51	22.72	11.02
Basic A.A.	11.41	6.98	6.69
Acidic/Basic	2.85	3.26	1.65
Polar A.A.	72.33	67.13	57.04

- ; not detected, A.A. ; Amino Acids

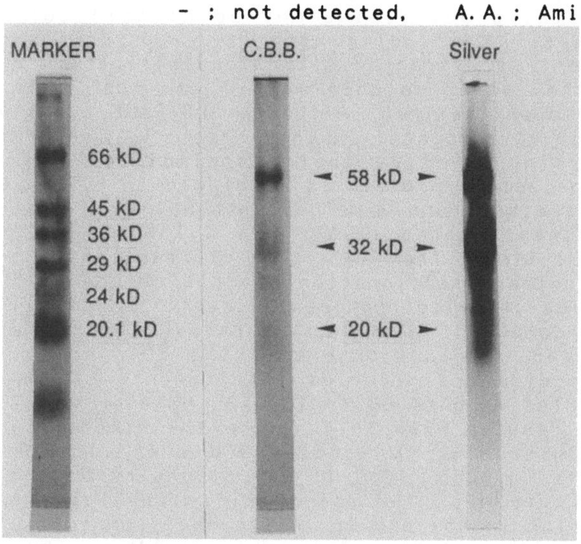

Figure 1: SDS-polyacrylamide gel electrophoresis of the water-soluble organic matrix in the nacreous layer of <u>Pinctada fucata</u>.
(left; stained with C.B.B., right; stained with silver)

amino acids, and the basic amino acids were present at fairly lower levels, with a resultant very high ratio of acidic to basic amino acids. While Gly was also the main component in both fractions, it was present at lower levels than in the W-ISM. Moreover, among the acidic components, Ala and Ser were minor amino acids, presented a levels more than 50% lower than those in the W-ISM. As a whole, the proportion of the polar amino acids in the W-SM and AL-SM was 10% greater than that in the W-ISM. Small differences could also be recognized between the two acidic components. In the W-SM, Asx was present at the highest levels followed by Gly , while Gly was present at the highest levels followed by Asx, in the AL-SM. In addition, Ala was more concentrated and the ratio of acidic to basic amino acids was slightly higher in the former than the latter

DISCUSSION

In the nacreous layer of molluscan shells, the organic matrix is thought to be present between the lamellar and crystal layers as an "interlamellar matrix" and as an "intercrystalline matrix", together with an "intracrystalline matrix" within the crystals [10]. Even though the diversity of the matrix has been clarified both biochemically and microscopically, the precise explanation of the function of each component of the matrix still remains to be clarified. For a more precise understanding of the ways in which the matrix can control formation of crystals, it is necessary to analyze each component in the bulk matrix.

In the present study, the author succeeded in the complete fractionation of the matrix, which is composed of five components. In these components, two acidic proteins and lipids may be of particular importance since they possess Ca-binding capacity. The dominant acidic proteinaceous component is extracted after decalcification by EDTA and is analogous to the "water-soluble matrix" or the "EDTA-soluble matrix" as most obviously defined. Although SDS-PAGE clearly shows the presence of the 58 kD component in this fraction, sevaral more components can be distinguished by their isoelectric points [Samata, in press]. The other acidic protein is insoluble in EDTA-H_2O and only dissolves in dilute alkali and β-mercaptoethanol with boiling. SDS-PAGE of this component showed it to have a low molecular weight of 17 kD and to be different from any other proteins extracted from the hard tissues. Further fractionation of this component by two-dimensional polyacrylamide gel electrophoresis identified it as a highly acidic, homogeneous component [Samata, in press]. Weiner et al. [11] have already analyzed by electrophoresis the EDTA-soluble matrix from 10 molluscan species and reported the presence of discrete components with molecular weights up to 100 kD. However, it is impossible to compare their results with this, since the composition of the matrix may vary from species to species and even, to some extent, within closely related groups [12]. It may also vary according to the shell ultrastructure [2]. The amino acid composition of the AL-SM shows that it is rich in Asx and is different from that of the W-ISM, which is rich in Ala and Gly. Thus, the AL-SM may be a different unit from the main component of the W-ISM. Disulfide bonds may serve to connect the AL-SM and W-ISM. Although the W-SM is also Asx-rich and acidic, a slight difference can be recognized in terms of composition between the two acidic proteins. As judged from the distinct differences in their molecular weights, the AL-SM can also be considered to be a component different from the W-SM. Although several hypotheses related to the relationship between the structure of the components of the matrix and their functions in calcification have been proposed, none has succeeded in offering an explanation of the primary mechanism by which matrix initiates, regulates and limits mineral growth. In particular, interactions between the soluble matrix and the insoluble matrix, which may be one of the key features, still remain speculative. Numerous authors have considered the Ca-binding properties of the matrix components in relation to the promotion of nucleation [3, 13]. By contrast, Wheeler [7] demonstrated that the fundamental process by which the matrix controls crystal growth is through adsorption to growing crystal surfaces. Instead of occurring via the binding of Ca^{2+} to the anionic matrix molecule, which were thought to be efficient inhibitors of crystal growth, nucleation was shown to be the result of the stabilization by the matrix of crystal nuclei. Although the idea of inhibition by the matrix of crystallization, based mainly on so-called " pH-stat assay " using W-SM from scallop shells, has some credence, several authors also reported that some specific protein, when free in solution, may indeed inhibit crystallization while the same molecule, when attached to an insoluble support, may promote crystallization [14, 15]. Therefore, the author is considering to apply the " pH- stat assay " using different combinations of the components that were isolated from the nacreous layer of <u>Pinctada</u>.

REFERENCES

[1] Crenshaw MA (1972) Biomineralization 6:6-11

[2] Samata T (1990) The Veliger 33-2:190-201

[3] Weiner S, Traub K (1981) In:Balaban M, Sussman JL, Traub W, Yonath A (eds) Structural aspects of recognition and assembly in biological macromolecules. Balaban I SS, Rehovot, Philaderphia, 467-482

[4] Addadi L, Moradian J, Shay E, Maroudas NG, Weiner S (1987) Proc Nat Acad Sci (USA) 84:2732-2736

[5] Krampitz G, Engels, J, Cazaux C (1976) In: Watabe N, Wilbur KM (eds) The mechanism of mineralization in the invertebrates and plants. Univ South carolina Press, Columbia, pp. 155-173

[6] Weiner S (1983) Biochemistry 22:4139-4144

[7] Wheeler AP, Rusenko KW, Sikes CS (1988) In:Sikes CS, Wheeler AP (eds) Chemical aspects of regulation of mineralization. Univ South Alabama, Alabama, pp. 9-13

[8] Anderson BL, Berry RW, Telser A (1983) Anal Biochem 132:365-375

[9] Morrisey j (1981) Anal Biochem 117:307-310

[10] Watabe N (1965) Jour Ultrastr Res 12:351-370

[11] Weiner S, Lowenstam HA, Hood L (1977) Jour Exp Mar Ecol 30:45-51

[12] Degens ET, Spencer DW, Parker RH (1967) Comp Biochem Physiol 20: 533-579

[13] Weiner S, Hood L (1975) Science 190:887-898

[14 Greenfield EM, Wilson DC, Crenshaw MA (1984) Amer Zool 24:925-932

[15] Addadi L, Weiner S (1985) Proc Natl Acad Sci (USA) 82:4110-4114

CHAPTER 1.5
Intracrystalline Proteins from a Sea Urchin and a Mollusk: A Comparison

L. ADDADI[1], A. BERMAN[2], and S. WEINER[2]

Departments of Structural Chemistry[1] and Isotope Research[2], The Weizmann Institute of Science, Rehovot 76100, Israel

Key words: Sea urchin, Mollusk shell, Intracrystalline proteins, Calcite, Mineralization.

INTRODUCTION

Control over mineral formation by organisms is achieved in a variety of ways. The unique feature in almost all of these processes is the intimate contact between the mineral and the assembly of organic macromolecules secreted by specialized cells at the site of crystallization [1,2]. This results in the formation of composite phases of mineral and organic material displaying physical and mechanical properties characteristic of the mixed phase, that are also presumably optimally adapted to fulfill specific mechanical functions. The crystals of the composite often display a unique orientation, or set of orientations of at least one of their crystallographic axes, as well as a unique morphology. We consider these properties as indications of involvement at the molecular level of the organic components in the processes of nucleation and growth of the crystals [3]. Less well understood, although recognized for a long time [4], is the ability of organisms to introduce organic molecules and macromolecules into the crystal itself, thus possibly altering the nature of the mineral phase. Here we compare the manner in which such processes occur in an echinoderm and a mollusk, and derive information on the mechanisms of intercalation and on the possible function of crystal-occluded proteins.

The crystallization of sea urchin spines and tests is essentially an intracellular process [5]. The calcite crystals, associated with some organic material, grow within a vacuolar cavity enclosed by a continuous cytoplasmic sheath (syncytium) [5,6]. Whole mature skeletal elements behave optically and diffract X-rays as good quality single crystals of calcite [7], notwithstanding their fenestrated spongy ultrastructure. Fracture of these elements does not follow the cleavage planes of calcite, but produces curved irregular surfaces, reminiscent of the fracture of glasses and amorphous materials [8] (Fig. 1a). The idea that organic material occluded inside the crystals may be responsible for the lower refractive index and specific gravity of echinoderm calcite was first proposed by Merker [9]. The presence of protein within the mineral is now well documented [10-14].

In contrast to the sea urchin, mollusk shell formation is the result of an extracellular process, where crystals of either calcite or aragonite grow inside a preformed matrix composed of (glyco)proteins. The crystals are separated by sheaths of organic material and are almost always oriented in a preferred direction. Clearly each crystal forms as a result of a separate act of nucleation and growth. The presence of organic material inside mollusk crystals was first reported by Watabe [4,15] and confirmed by Mutvei [16], Crenshaw [17] and Towe and

Thompson [18]. The question of whether this intracrystalline material is ordered or disordered [18] is still under dispute, and the functions, if any, of the macromolecules inside the crystals remain unknown. We have observed that upon fracture of shells of Mytilus californianus and Atrina serrata individual calcite prisms preferentially cleave along the smooth cleavage planes characteristic of pure calcite single crystals (Fig. 1b).

Here we review some recent data on the effects of intracrystalline proteins extracted from the prismatic layer of the mollusk Atrina serrata and from the skeletal parts of the sea urchin Paracentrotus lividus, on the texture and fracture behaviour of synthetic calcite crystals. We compare these synthetic crystals with those observed in vivo and present the hypothesis that the sea urchin intracrystalline proteins have the specific task of affecting the cleavage behaviour of these crystals, by being selectively adsorbed at and occluded through specific lattice planes of the growing crystals. In contrast, the mollusk intracrystalline proteins appear to be non-specifically trapped inside the crystals.

The bivalve Atrina has unusually large calcitic prisms, up to 500 μm long and 30 μm thick, greatly facilitating their isolation as single elements. The intracrystalline fraction, in this case, was operationally defined as the macromolecules obtained after dissolution of thoroughly hypochlorite treated and sonicated shell prismatic layer, that under examination in the SEM appeared to be totally disaggregated down to the level of single prisms. The sea urchin macromolecules were also prepared by first treating the ground test with sodium hypochlorite and then dissolving the mineral phase (0.5M EDTA, 0.05M TRIS pH 8.0) and removing the salt by dialysis at $4^{O}C$ (Spectrapor 3).

Table 1 shows the partial amino acid compositions of the soluble fractions of the glycoproteins isolated from the mollusk and sea urchin. The Atrina intracrystalline proteins have the same amino acid composition as the total soluble fraction, but constitute only about 15% of the latter. This represents about 0.1% by weight of the calcite. In addition some of the intracrystalline protein is insoluble (<0.05 percent by weight) and has an amino acid composition different from that of the soluble fraction and of the total insoluble fraction. In Paracentrotus the intracrystalline fraction constitutes almost all the soluble protein, and accounts for approximately 0.02% by weight of the mineral.

Table 1: Partial amino acid composition of the soluble (Sol) and insoluble (Ins) fractions of the total and intracrystalline (intracryst) proteins isolated from the mollusk A. serrata and the sea urchin P. lividus.

		ASP	THR	SER	GLU	PRO	GLY	ALA	VAL	LEU	TYR	protein/ mineral (%)
A. serrata												
INTRACRYS.	SOL.	53.0	2.5	6.3	11.3	2.5	8.7	8.3	2.7	0.8	0.9	0.1
	INS.	16.0	4.0	9.0	14.0	4.0	16.3	6.6	4.5	6.5	1.9	<0.05
TOTAL	SOL.	52.0	2.3	6.3	11.3	2.1	9.2	8.3	3.0	1.0	0.6	0.66
	INS.	1.2	0.8	5.5	0.6	n.d.	55.0	3.5	8.6	8.6	8.1	1.5-1.8
P. lividus												
TOTAL	SOL.	16.3	4.6	5.0	14.2	8.2	23.7	11.6	4.3	3.1	n.d.	0.02-0.03

When crystals of calcite were grown in vitro in the presence of 2 μg/ml of sea urchin or mollusk intracrystalline soluble proteins, their growth was affected in substantially different ways. The sea urchin proteins induced a change in morphology, characterized by the development of a new set of {1$\bar{1}$0} faces oblique to the {104} cleavage planes of calcite (Fig. 2a). Based on previous studies of stereospecific crystal growth inhibitors [3,20], we know that this phenomenon

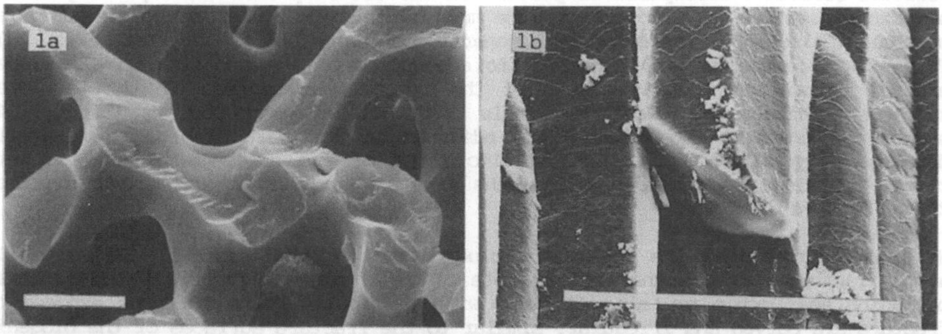

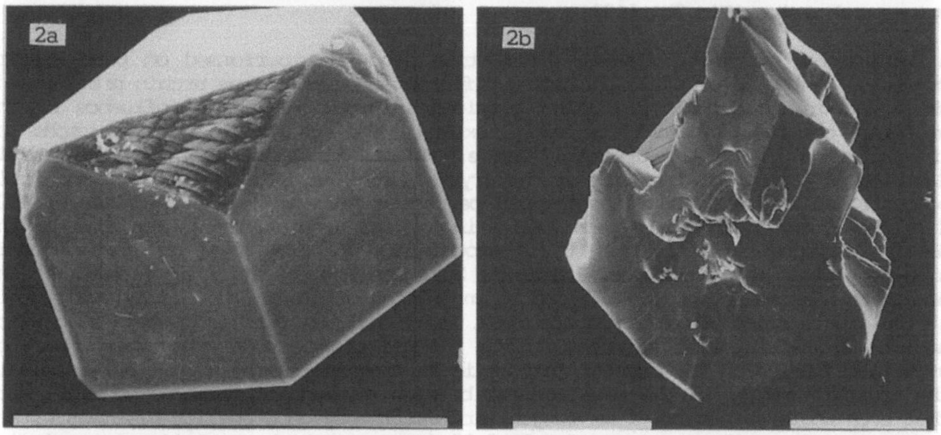

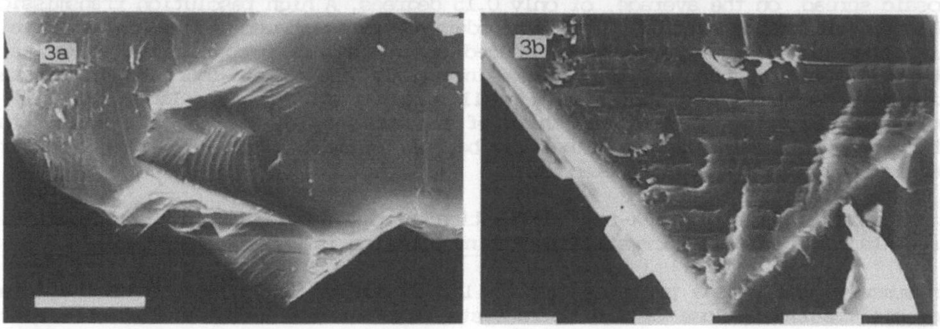

Fig.1 (a) Fractured surface of a spine of the sea urchin Paracentrotus lividus; the fracture is conchoidal. Scale bar 10μm. (b) Fractured prism from the shell of the mollusk Atrina serrata. The inclination of the cleavage plane corresponds to the {104} cleavage plane of pure calcite. Scale bar 0.1mm.

Fig.2 (a) Calcite crystal grown in the presence of sea urchin proteins. The rough faces are $(1\bar{1}0)$ and the smooth ones are {104}. Scale bar 0.1μm. (b) Calcite crystal grown in the presence of mollusk intracrystalline proteins. Scale bar 0.1mm.

Fig.3 Cleaved calcite crystals. (a) Calcite crystal grown in the presence of sea urchin proteins. The fracture is conchoidal. Scale bar 10μm. (b) Calcite crystal grown in the presence of mollusk shell intracrystalline proteins. Note the regular parallel sets of cleavage planes. Scale bar 10μm.

reflects selective adsorption of protein on the set of planes corresponding to the new faces developed. In contrast, the mollusk proteins only induce an irregular morphological change, reflected by a general 'corrosion' of the crystals (Fig. 2b). This effect corresponds to non-specific adsorption of the proteins on all planes exposed to solution during growth of the crystal. Inspection of the developed surfaces at higher magnification reveals that these are composed of sets of {104} steps. The proteins thus seem to be adsorbed on the most stable and naturally developed crystal faces, which are also those of the cleavage planes. In both cases, proteins are occluded inside the crystals in concentrations varying from 200-500 ppm [19].

We also observed that the crystals occluding mollusk shell protein cleave easily along fracture surfaces that follow sets of {104} planes (Fig. 3a). The crystals occluding sea urchin proteins fracture with difficulty. The surfaces exposed by fracture are curved and irregular, similar to the fracture surfaces of the biogenic spines and plates (Fig.3b) [19].

A comparative synchrotron X-ray diffraction study was performed on pure calcite crystals, synthetic crystals occluding different amounts of sea urchin protein, and whole sea urchin spines [21]. This provided information on the influence of the proteins on the internal crystal texture, and thereby on their mode of intercalation inside the crystal lattice. It was concluded that the protein is distributed throughout the bulk of the crystal, but not in a random fashion; it is concentrated at the boundaries of quasi-perfect domains with average sizes of 4500 Å in the synthetic crystals, and 2000 Å in the sea urchin spines. The size of the domains in pure synthetic calcite is around 5000 Å. The coverage of the domain area by protein was roughly evaluated as 30%. The concentration of protein at the domain boundaries is conceivably due to inhibition of growth induced by the protein at the growth sites where it is adsorbed, which induces further protein adsorption on the exposed surfaces adjacent to the blocked growth site. The main effect of protein occlusion on the crystal texture is an increase in the mosaic spread within the crystal, namely in the misalignment between perfect crystalline domains. Note that even whole spines of ~1 mm length, are still very good quality crystals with a mosaic spread, on the average, of only 0.15 degrees. A high resolution transmission electron microscope study [22] of echinoderm ossicles showed a much larger mosaic misalignment. This may have been induced during sample preparation, as we have observed that fracture of the spines inevitably causes an increase in crystal disorder. We also noted that a very small single tooth element (150x10 μm) had a mosaic spread value of about one third of the whole spine (800x80 μm), suggesting that the misalignment in the spines may be due to smaller local effects which are cumulative through the whole element.

A comparable synchrotron X-ray diffraction study using biogenic and synthetic crystals containing intracrystalline Atrina proteins is under way.

In summary, the Atrina proteins are similar in composition to those of the total soluble fraction. Some of the intracrystalline material remains insoluble after dissolution of the mineral, suggesting primary local aggregation. During growth of synthetic calcite crystals, these proteins adsorb on the normally produced crystal faces, and although occluded inside the crystal, they do still fracture along the normal cleavage planes of calcite. Individual prisms of Atrina also cleave along the normal calcite cleavage planes. These observations are all consistent with the proteins being non-specifically adsorbed at growth sites on the growing crystal surfaces, being then overgrown and essentially trapped inside the crystal, sometimes in local pockets. This scenario also conforms well to the direct observations of Towe and Thompson [18], of the organic matter occluded in Mytilus calcitic prisms and aragonitic nacreous tablets.

In contrast, the sea urchin proteins are almost all intracrystalline in location and almost all soluble. They appear to be targeted to specific crystal planes during crystal growth, and are eventually evenly distributed within the crystal bulk, although segregated at the boundaries of microscopic domains. Their presence causes in vitro and, we suggest in vivo as well, specific interference with the smooth propagation of cleavage planes within the single crystals, thus eventually reinforcing them against fracture. We also note that as the sea urchin proteins are less acidic than those of the mollusk, they are better candidates for specific interactions with charged surfaces, that require delicate structural recognition at the molecular level.

REFERENCES

[1] Lowenstam HA, Weiner S (1989) On Biomineralization. Oxford University Press. New York, Chapters 2-3.

[2] Simkiss K, Wilbur KM (1989) Biomineralization. Cell Biology and Mineral Deposition, Academic Press. San Diego.

[3] Addadi L, Weiner S (1989) Biomineralization. Chemical and Biochemical Perspectives. Mann S, Webb J, Williams RJP eds. VCH Press. Weinheim pp 133-156.

[4] Watabe N (1963) J Cell Biol 18: 701

[5] Okazaki K (1960) Embryologia 5: 283-320

[6] Märkel K, Röser U, Makenstedt U, Klosterman M (1986) Zoomorphology 106: 232-243

[7] West CD (1937) J Paleontol 11: 458-459

[8] Currey JD, Nichols D (1967) Nature 214: 81-83

[9] Merker E (1916) Zool Jahrb Abt Allg Physiol Tiere 36: 25-205

[10] Pilkington JB (1969 J Mar Biol Assoc UK 49: 857-877

[11] Klein L, Currey JD (1970) Science 169: 1209-1210

[12] Weiner S (1985) J Exp Zool 234: 7-15

[13] Swift DM, Sikes, S, Wheeler AP (1986) J Exp Zool 240: 65-73

[14] Benson SC, Benson NC, Wilt F (1980) J Cell Biol 102: 1878-1886

[15] Watabe N (1965) J Ultrastruct Res 12: 351-370

[16] Mutvei H (1970) Biomineralization Res Repts 2: 49-72

[17] Crenshaw MA (1972) Biomineralization Res Repts 6: 6-11

[18] Towe KM, Thompson GR (1972) Calc Tiss Res 10: 38-48

[19] Berman A, Addadi L, Weiner S (1988) Nature 331: 546-548

[20] Addadi L, Weiner S (1986) Mol Cryst Liq Cryst 134: 305-322

[21] Berman A, Addadi L, Leiserowitz L, Weiner S, Nelson M, Kvick A, Science, in press.

[22] Blake DF, Peacor DR, Allard LF (1984) Micron Microscopia Acta 15: 85-90

CHAPTER 1.6

Intracrystalline Molecules from Brachiopod Shells

G.B. Curry, M. Cusack, K. Endo, D. Walton, and R. Quinn

Department of Geology and Applied Geology, University of Glasgow, Glasgow, G12 8QQ Scotland UK

Key words: Brachiopod, intracrystalline, protein primary sequence, amino acid analysis.

INTRODUCTION

Shells are composed of both organic and inorganic constituents. It is believed that the organic compounds have important functions at several stages during the formation of biominerals. In brachiopod shells the disposition of inorganic biominerals and their enclosing organic sheaths have been thoroughly investigated using both scanning and transmission electron microscopy but little is known about the biochemistry of the intracrystalline molecules i.e. those enclosed within the inorganic portion. Such information is crucial for an understanding of biominerals if, as has been suggested, these compounds (i) induce crystal nucleation by providing a surface for precipitation, (ii) form compartments that determine the shape and volume of the biocrystal and (iii) determine the pattern of growth in the mineral phase in what is termed 'matrix mediated mineralisation' [1].

This study presents the first details of the organic intracrystalline components from the shell of the articulate brachiopod, *Neothyris lenticularis* (Deshayes). The shell of *N. lenticularis* is composed of numerous long calcite fibres from which the proteins studied here have been extracted. Although this protocol has disadvantages in that it is very time-consuming and the great proportion of the shell is discarded, the major advantage of such a strategy is that it avoids the possibility of including extraneous molecules both from contaminating organisms, such as bacteria, which may infest the organic sheaths of shell calcite fibres and from human finger tips during collection and preparation.

Partial N-terminal sequence and amino acid analyses of two shell proteins are presented here alongside SDS PAGE and hplc analyses of the intracrystalline molecules of *N. lenticularis*.

MATERIALS AND METHODS

Extraction of Shell Proteins

Shells of living *N. lenticularis* were collected from Stewart Island, New Zealand and killed by dehydration. The shells were cleaned thoroughly and incubated for 2 h at 22°C in an aqueous solution of bleach (5% v/v) to destroy the organic sheath and any possible bacterial contamination. The shells were then powdered in a ceramic pestle and mortar before incubating overnight at 4°C in an aqueous solution of bleach (1% v/v). The bleach was removed by repeated washes with Milli Q™ water followed by centrifugation (8 g.h). The precipitate was washed until no bleach could be detected (typically ten 2 l washes) and then lyophilised and EDTA (20% w/v), pH 11 added in the ratio of 23 ml to 1 g shell. The entire mixture was agitated at 4°C for 72 h or until the inorganic phase had dissolved. Following centrifugation (20 g.h) the supernatant was concentrated and the EDTA removed using the Millipore Minitan™ tangential flow system. The preparation was further concentrated in a minicon static concentrator (Amicon) with a 10 kDa cut-off membrane.

Separation of Proteins by hplc

An aliquot of concentrated shell extract was applied to a reverse-phase Aquapore™ RP-300 narrow bore (2.1 mm diameter x 30 mm length) column in trifluroacetic acid (0.1% v/v) at a flow rate of 0.1 ml / min. After 5 min, a 40 min linear gradient of 0 to 70 % (v/v) acetonitrile, in 0.1% (v/v) trifluroacetic acid, was applied to fractionate the shell proteins. The eluate was monitored at 280 and 214 nm.

Separation of Proteins by SDS PAGE

Small gels (9 cm x 7 cm) of 0.75 mm thickness containing 15% polyacrylamide were prepared according to the method of Schagger and Van Jagow [2]. Glycine, which is used in most SDS PAGE systems is here replaced by tricine. Samples for electrophoresis were heated at 100°C for 4 min in an equal volume of sample buffer containing final concentrations of 0.15 M Tris / HCl, pH 6.8, 0.2 M 2-mercaptoethanol, 0.1% (w/v) SDS, 30% (v/v) glycerol and 0.0002% (w/v) of the tracking dye, bromophenol blue. Molecular weight standards; bovine albumin (66 kDa), egg albumin (45 kDa), glyceraldehyde-3-phosphate dehydrogenase (36 kDa), carbonic anhydrase (29 kDa), trypsinogen (24 kDa), trypsin inhibitor (20.1 kDa) and α-lactalbumin (14.2 kDa) were included on every gel. Electrophoresis of samples in the small gel system required a constant voltage of 100 V for 2 h. Following electrophoresis, proteins were either fixed in the gel and visualised using Coomassie Brilliant Blue-R or electroblotted onto ProBlott™ membrane (Applied Biosystems).

Electroblotting of Proteins

Following SDS PAGE, the proteins in the gel were transferred to ProBlott™ membrane. The transfer was performed in transfer buffer (10 mM CAPS buffer, pH 11, 10% (v/v) methanol) in a Bio-Rad Trans Blot cell. A constant voltage of 50 V for 0.5 h moved the proteins from the gel towards the membrane. Coomassie Blue staining was used to reveal the protein bands on the ProBlott™ membrane.

Amino-Terminal Sequence Determination and Amino Acid Analysis

Automatic Edman degradation was carried out on the stained bands using a pulsed liquid protein sequencer (Applied Biosystems 477A). Bands were also loaded onto the 420-H amino acid analyser with automatic hydrolysis (Applied Biosystems) to determine the overall amino acid composition. Amino acid analysis of the hplc fractions was employed to identify the major proteins by comparing with the analyses from the homogeneous electroblotted proteins. Stained ProBlott™ membrane with no protein attached and all buffer solutions employed were analysed to determine the background level of amino acids present.

RESULTS AND DISCUSSION

The shell of *N. lenticularis* contains a mixture of proteins of different molecular weight, as determined by SDS PAGE (Figure 1). The major proteins are of molecular weight 10.5, 13 and 47 kDa. An identical copy of the the protein pattern in an SDS gel is obtained on a membrane, such as ProBlott™, using electroblotting. This process, first described by Towbin *et al* in 1979 [3], is ideal for the isolation of single proteins from crude mixtures for N-terminal sequence analysis.

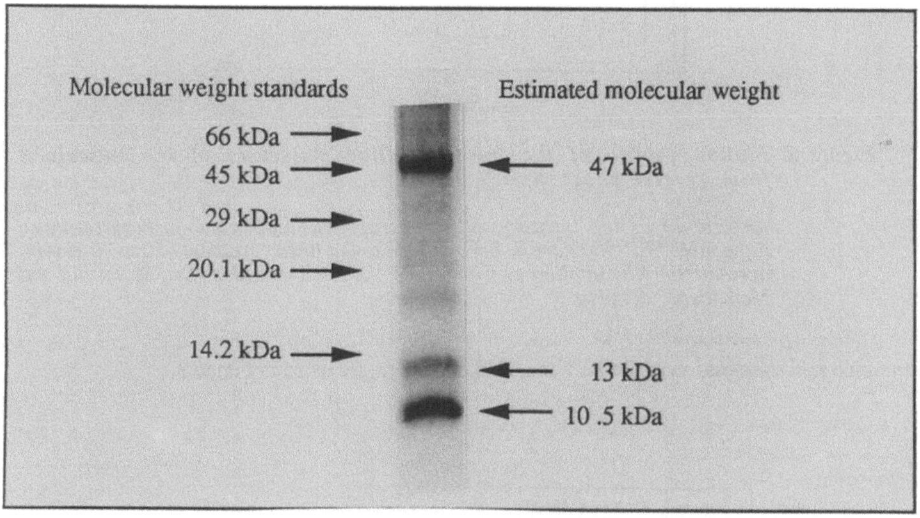

Figure 1 Intracrystalline Proteins of N. lenticularis shell

The proteins were separated by SDS PAGE (see Materials and Methods for details) and then revealed using Coomassie Brilliant Blue staining.

The elution profile of the shell proteins from the hplc system is presented in Figure 2. The light pink colour of Recent *N. lenticularis* becomes deep red as the shell extract is concentrated. The red pigment elutes from the hplc system in the elution volume 42 to 50 ml.(Figure 2). This corresponds with the elution of the 10.5 kDa protein. One possible explanation for the elution of the 10.5 kDa protein as several different peaks is that the molecule may in fact be a glycoprotein and the various peaks represent various degrees of glycosylation. The harsh conditions employed in SDS PAGE causes the protein(s) and pigment to dissociate.

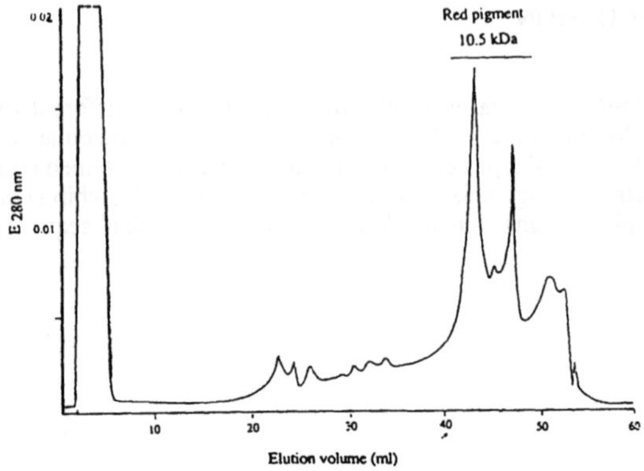

***Figure 2* Elution profile of the intracrystalline molecules of *N. lenticularis* from reverse-phase hplc**

An aliquot of the intracrystalline extract was applied to a reverse-phase Aquapore™ RP-300 narrow bore column and a linear gradient of 0 to 70 % (v/v) acetonitrile was applied to fractionate the shell proteins (See Materials and Methods for details).

The amino acid composition of the 10.5 and 47 kDa proteins are listed in Figure 3.

	47 kDa	10.5 kDa
Residue	mole %	
D/N	8.53	10.33
E/Q	8.18	9.06
S	5.57	5.65
G	8.89	11.50
H	1.10	0.76
R	0.10	3.21
T	6.12	6.21
A	13.02	9.21
P	6.47	16.23
Y	4.27	1.58
V	3.97	7.02
C	1.10	0.66
M	1.81	1.58
I	8.13	2.39
L	10.49	10.28
F	3.21	2.29
K	9.04	2.03

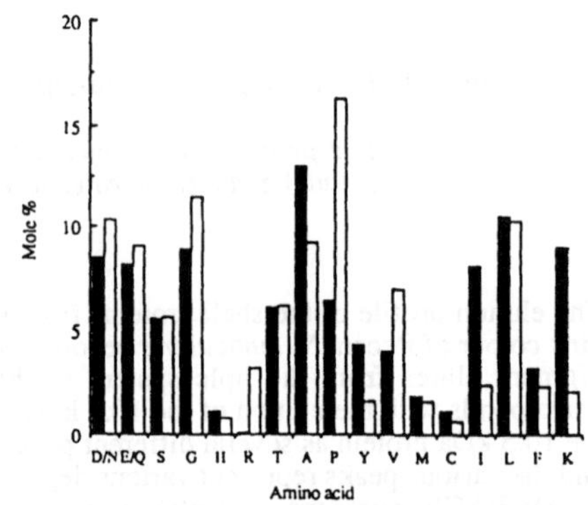

***Figure 3* Amino acid composition of two of the major intracrystalline proteins of *N. lenticularis* shell.**

Proteins were extracted and purified to homogeneity using SDS PAGE and electroblotted onto ProBlott™ and the amino acid composition determined using the 420-H amino acid analyser (Applied Biosystems). Values are presented for both the 10.5 (■) and 47 kDa (□) proteins.

Amino acid concentrations are presented as mole percentages to enable direct comparison of the two proteins. Actual yields of intracrystalline amino acids from the bulk mixture extracted from *N. lenticularis* indicate that intracrystalline amino acids occur at about 80 nmole/g shell material.

The conditions employed for the hydrolysis of the peptide bonds also destroys 100 and up to 80% of tryptophan and cysteine residues respectively. However,the hydrolysis reaction is automated and highly reproducible and thus each protein should lose the same proportion of these vulnerable residues allowing direct comparison of the compositions.

The N-terminal amino acid sequence of the 10.5 and 47 kDa proteins are presented in Figure 4.

	1	6	11	16
10.5 kDa	GPEQL	PYATM	ISKTS	NATKP
	1	6		
47 kDa	ANLVL	AGRGD		

Figure 4 ***Amino-terminal sequence of 10.5 and 47 kDa protein from Neothyris lenticularis shell.***

Proteins were extracted from N lenticularis and then fractionated by SDS PAGE before being transferred to ProBlott™ for sequence determination (see Materials and Methods).

These protein sequences show no significant similarity to any protein sequence listed in the NBFR or EMBL data base or to any implied peptide sequence in the EMBL or GenBank DNA sequence data base on the basis of searches conducted by the 'FastA' and 'TFastA' programs in the GCG sequence analysis package (Version 6) [4]. Elucidation of the entire primary sequence of these intracrystalline proteins should make sequence comparisons more meaningful and may enhance our understanding of the possible role these proteins play in the process of biomineralisation.

REFERENCES

[1] Lowenstam, H.A. (1981) Science **211** 1126 - 1131

[2] Schagger, H. and Von Jagow, G. (1987) Anal. Biochem. **166** 368 - 379

[3] Towbin, H., Staehelin, T. and Gordon, J. (1979) Proc. Natl. Acad. Sci. **76** 4350 - 4354

[4] Devereux, J., Haeberli, P. and Smthies, O. (1984) Nucleic Acids Research **12**(1) 387 - 395

CHAPTER 1.7

Morphological Evidence of Coherent Organic Material within the Stereom of Postmetamorphic Echinoderms

P. Dubois

Laboratoire de Biologie marine (CP 160), Université Libre de Bruxelles, B-1050 Bruxelles, Belgium

Key words: Skeleton, Echinoderms, Organic matrix, Crystallography.

SUMMARY. The occurrence of a coherent organic material within the mineral phase (i.e. the stereom) of the skeleton of postmetamorphic echinoderms is demonstrated in both light and scanning electron microscopies. This material is structured in concentric laminae bridged by radial threads. It most probably introduces extensive discontinuities in the crystalline network of the high-magnesium calcite which forms the skeleton. Consequently, echinoderm ossicles should presumably be true polycrystalline aggregates.

INTRODUCTION

Echinoderms have a high-magnesium calcite skeleton of mesodermal origin. This skeleton is observed in most postmetamorphic individuals as well as in echinoid and ophiuroid larvae. Postmetamorphic ossicles consist of a rounded tridimensional network of trabeculae (the so-called stereom), whereas spicules correspond to isolated -possibly branched- trabeculae. Most postmetamorphic ossicles and larval spicules have the particularity to behave like monocrystals in polarization microscopy and X-ray diffraction (1,2,3).

It is now well-established that both postmetamorphic ossicles and larval spicules contain intratrabecular organic material (IOM) which is supposed to govern calcification (4,5,6). However, the tridimensional arrangement of this material within the spicules and trabeculae remains unclear, making the comprehension of both the stereom structure (i.e. monocrystalline vs polycrystalline status) and the mineralization strategy of echinoderms uneasy. Indeed, if the IOM is structured in a coherent framework, it should presumably induce discontinuities in the crystalline network, making echinoderm spicules and ossicles true polycrystalline aggregates (7) (and this could imply multiple -simultaneous or successive nucleation events). On the contrary, if the macromolecules which build up the IOM are scattered in the mineral phase, the stereom elements probably have a mosaic block structure (8,9) (and this could imply a unique nucleation site). So far, the in situ morphology of the IOM was only successfully investigated in the spicules of an echinoid larva. This IOM is a coherent fibrous material formed by concentric laminae interconnected by radial threads (10). The present paper demonstrates that the IOM of postmetamorphic echinoderms is also structured in a coherent framework made of concentric and radial components.

MATERIAL AND METHODS

Adult Asterias rubens L. were collected by SCUBA diving at Scharendijk (Zealand, Netherlands). They were kept in a closed-circuit aquarium (30% , 12°C) in the laboratory until required.

Investigations were done on adambulacral spines. Spines were sampled on living individuals and preserved in 70% ethanol. Associated tissues -viz. epidermis,

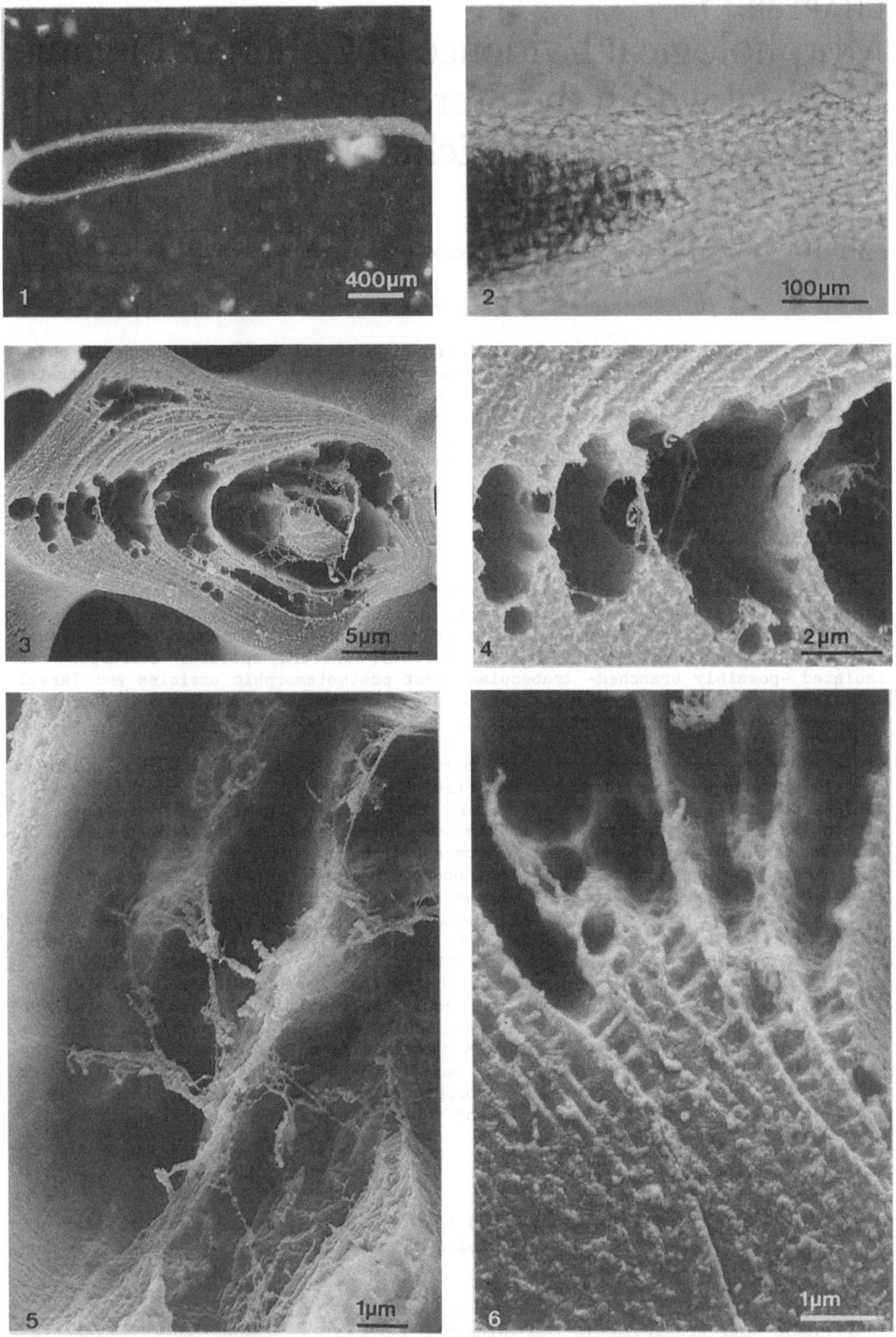
1
400μm
2
100μm
3
5μm
4
2μm
5
1μm
6
1μm

dermis, and ossicle stroma- were digested in incubating the spines for 24h at 37°C in a solution of 0.1% proteinase N (Serva) in 0.1M Tris-HCl buffer (pH 7.2) (11). The spine skeletons were then washed three times in Milli-Q desionized water (Millipore) and preserved in 96% ethanol until investigated. (Spines were air-dried immediately before experiments.) Each batch of cleaned spines was controlled for effective removing of the soft tissues by observation of three spines in scanning electron microscopy (SEM) (11).

For light microscopy (LM), the spine skeleton was partly dissolved on a microscopical slide by 0.01M or 0.1M trichloroacetic acid (TCA). The stereom dissolution is continuously observed by either phase-contrast microscopy or Nomarsky interference microscopy.

For SEM, the spine skeletons were either kept intact or cut transversely with a razor blade. Some intact spines were imbedded in Spurr's resin, ground using abrasive papers of incrasing gradations (250-1200), and etched for 15 or 30sec in 0.1M TCA (12). Others were submitted to partial dissolution through incubation for 2 to 5min in either 0.1M TCA or Bouin's fluid (with TCA in place of acetic acid). Partly dissolved spines were then fixed for 1h in 4% glutaraldehyde in 0.1M cacodylate buffer (pH 7.8) and post-fixed for 1h in 1% OsO_4 in the same buffer. The basal parts of the transversely cut spines were etched through incubation in different fixatives:

- 4% glutaraldehyde in 0.1M cacodylate buffer (pH 7.8) for 18h followed by 1% OsO_4 in the same buffer for 1h
- 4% glutaraldehyde, 1% tetrasodic EDTA in 0.1M cacodylate buffer (pH 7.8) for 1 or 3h followed by 1% OsO_4 in the same buffer for 1h
- 1% unbuffered OsO_4 (pH 6) for 1 or 3h followed by 4% glutaraldehyde in 0.1M cacodylate buffer (pH 7.8) for 1h
- 1% OsO_4 in 0.1M cacodylate buffer (pH 7.8) for 18h followed by 4% glutaraldehyde in the same buffer for 1h.

After incubation in the etching and/or fixing solutions, all spines were dehydrated in graded ethanols and dried by the critical point method using CO_2 as transition fluid. The specimens were then mounted on aluminium stubs, coated in a sputter coater and observed with an ISI-DS 130 SEM. At least three spines were prepared according to each method. The more relevant results were obtained using the glutaraldehyde-EDTA solution.

RESULTS AND DISCUSSION

SEM observation shows that proteinase incubation effectively removes all stereom-associated soft tissues and results in perfectly cleaned ossicles (11). When such proteinase-cleaned ossicles are decalcified on a microscopical slide, both phase-contrast and Nomarsky interference microscopies reveal that the dissolving stereom leaves a "ghost" which is insoluble in TCA (Fig.1) and not birefringent between crossed nicols. This ghost makes a coherent structure all along the decalcified ossicle and reproduces the architecture of the dissolved stereom (Fig.2). It is fibrillar in structure and may be stained by methylene blue/ azure II. Its properties indicate that the ghost is organic in nature. As the proteinase incubation which precedes the decalcification of the ossicles

---

Figs 1-2: Proteinase-cleaned adambulacral spines of Asterias rubens in the process of decalcification in 0.01M TCA (phase-contrast microscopy).

Figs 3-6: Sectionned proteinase-cleaned adambulacral spines of A.rubens etched by glutaraldehyde-EDTA (scanning electron microscopy).

Figs 3-4: General and detailed views of an etched fracture face of a sectionned trabecula.

Fig.5: Deeply excavated trabecular fracture face showing a sheet and radial threads.

Fig.6: Slightly etched trabecular fracture face with sheets and threads appearing in high-relief.

removes all extrastereomic organic structures, it is highly probable that the ghost is an intrastereomic organic structure which runs all along the ossicle.

Incubation of cleaned ossicles in weak acids, ultrastructural fixatives, or chelating agent results in the etching of the stereom. Particularly, when the ossicles are sectionned before etching, the fracture faces of the sectionned trabeculae show characteristic etching patterns made of concentric mineral layers (11). In some cases (especially when etching is performed by a glutaraldehyde-EDTA solution), some of these mineral layers are deeply excavated (Fig.3). In such places, where most of the mineral was dissolved, the trace of the mineral layers is preserved as finely granular concentric sheets (Fig.4). These sheets are bridged by fibrillar threads, giving to the whole structure the aspect of a tridimensional spider web which pervades the whole trabecula. At higher magnification, both sheets and threads appear as fibrillar structures decorated by small granules (Fig.5). In less deeply etched places, the sheets appear as concentric ridges and the threads as short high-relief radial segments (Fig.6).

Both the morphology and the low susceptibility to EDTA of the sheets and threads strongly suggest that these are organic structures. They are obviously intrastereomic and should correspond to the ghost observed in LM. SEM observations show that this material is organized in concentric laminae linked by radial bridges. That organization is strongly reminiscent of that reported for the intraspicular organic material of larvae (although the latter was demonstrated by transmission electron microscopy -TEM) (10).

That mineral layers and organic laminae alternate within echinoderm trabeculae suggests that the successive pairs made of a mineral layer and an organic lamina could represent successive growth episodes of the trabecula, and consequently that multiple nucleation events could occur during stereom formation. Indeed, delayed thickening of trabeculae has been reported (13,14,15).

Observations in LM and SEM (this study) and in TEM (10) show that the IOM is a coherent structure. Consequently, it is most likely that this material produces extensive discontinuities in the crystalline network of the magnesium calcite which forms the trabeculae (see Towe -7- for discussion). Such discontinuities would mean that echinoderm ossicles and spicules are true polycrystalline aggregates (whose constituting microcrystals must be highly ordered to explain their single crystal behaviour). This hypothesis contrasts with that of Berman et al. (8) who considered that the IOM is quantitatively too scarce to build a "continuous phase" within the stereom and suggested that stereom elements are single crystals made of mosaic blocks. However, these authors only considered the EDTA-soluble constituents of the IOM and overlooked the insoluble constituents which are more probably responsible for the coherent framework (16,17). It is here demonstrated that echinoderm IOM is actually a coherent structure.

ACKNOWLEDGMENTS

The author thanks Prof. M.Jangoux for critical reading of the manuscript and J.Harray for technical assistance. Work supported by FRFC contract n°2.4527.89.

REFERENCES

(1) Raup DM (1959) J Geol 67: 661-674
(2) Donnay G, Pawson DL (1969) Science 166: 1147-1150
(3) Okazaki K, Inoué S (1976) Dev Growth Differ 18: 413-434
(4) Benson S, Benson N, Wilt F (1986) J Cell Biol 102: 1878-1886
(5) Weiner S (1985) J exp Zool 234: 7-15
(6) Dubois Ph, Chen CP (1989) Echinoderm Studies 3: 109-178
(7) Towe KM (1972) Biomineralization 4: 1-14
(8) Berman A, Addadi L, Weiner S (1988) Nature, Lond 331: 546-548

(9) Addadi L, Weiner S (1989) In: Mann S, Webb J, Williams RJP (eds) Biomineralization: chemical and biochemical perspectives. VCH Verlaggesellschaft, Weinheim, BRD, pp.133-156.
(10) Benson S, Jones EME, Crise-Benson N, Wilt F (1983) Exp Cell Res 148: 249-253
(11) Dubois Ph, Jangoux M (1985) In: Keegan BF, O'Connor BDS (eds) Proceedings of the fifth international echinoderm conference, Galway. AA Balkema, Rotterdam, pp. 507-512.
(12) Clark II GR (1980) In: Rhoads DC, Lutz RA (eds) Skeletal growth of aquatic organisms. Plenum Press, New York, pp.607-612.
(13) Heatfield BM (1971) J Morph 134: 57-90
(14) Mischor B (1975) Zoomorphologie 82: 243-258
(15) Dubois Ph, Jangoux M (1990) Zoomorphology 109: 263-272
(16) Degens ET (1976) Top Curr Chem 64: 1-112
(17) Weiner S (1984) Amer Zool 24: 945-951

CHAPTER 1.8

Molecular Recognition in Biomineralization

S. MANN, B.R. HEYWOOD, S. RAJAM, and V.J. WADE

School of Chemistry, University of Bath, Bath BA2 7AY, UK

Key words: Oriented nucleation, Matrix-mediated biomineralization, Ferritin, Site-directed mutagenesis, Langmuir monolayers, Modelling biomininineralization.

INTRODUCTION

Molecular recognition is a term used widely in many branches of chemistry and biochemistry. It encompasses a multitude of molecular systems ranging from the small molecule-small molecule interactions of macrocyclic ligands with metal cations, through macromolecular processes involving enyzme-substrate reactions, protein-protein coupling and antibody-antigen specificity to supramolecular events involving cell-cell communication. In each case, the specificity of interaction is determined by complementarity in size, charge and molecular shape, often generated by the interplay of structural, stereochemical and dynamical relationships. The fact that these processes are intrinsic to biochemical reactions in solution and within membrane assemblies suggests that all functional systems evolved from chemical recognition.

Many features of controlled biomineralization indicate that such processes are genetically controlled and subject to the requirements of molecular recognition. One is continually astonished by the specificity in crystal chemical design of minerals optimised to structural, storage and receptor functions. Furthermore, the precise replication of biominerals of controlled structure, morphology, size, orientation and texture is not confined to higher organisms; it is as much a feature of magnetic crystals formed within membrane-bound vesicles in the cells of magnetotactic bacteria [15,16] as it is of bone crystals located within the intrafibrillar spaces of turkey tendon [18] (figure 1). Thus, the fundamental strategies of controlled biomineralization were established early in the evolution of organisms and we must therefore search for common mechanisms within the divergences of different biological systems.

It has long been recognised that the pivotal issue which separates biomineralization from mineralization is the involvement of organic macromolecules and the elucidation of the role(s) of the organic matrix remains a central issue to the field. (Note, however, that many geological processes involve interactions with organic molecules, eg. calcitic ooids and aspartic acid [1]. Furthermore, many theories of the origin of life invoke organic synthesis on inorganic minerals [2,3] suggesting that mineral-matrix relationships may have undergone a "functional reversal" during the early evolution of prokaryotic life). In this paper, we discuss some of the potential molecular recognition processes that may occur at mineral-matrix interfaces and highlight several of these features using model systems involving (a) the iron storage protein, ferritin, and (b) compressed Langmuir monolayers as organized organic substrates for mineralization *in vitro*. We show that systematic changes in the nature of the organic matrix, either by site-directed mutagenesis or the selection of an appropriate headgroup moiety, result in specific changes in mineralization. Electrostatic, geometric and stereochemical factors have been determined as important factors responsible for oriented nucleation by organic surfaces.

MATRIX-MEDIATED NUCLEATION

The nucleation of an inorganic mineral at the surface of an organic matrix can be considered as a phase transformation reaction involving surface and bulk processes. Here we will not consider the influence of solution thermodynamics and kinetics (see [4] for details) but will limit our discussion to the interfacial processes that can modify the reaction profile. In many ways we can liken the role of the matrix to an enzyme with the incipient nuclei acting as the substrate. A difficulty arises in the choice of description with regard to the forming clusters:- in no cases do we have knowledge of their structure (periodic, amorphous, polyhedral), size or composition and we do not know whether the initial interactions involve ion-binding or larger scale polynuclear events. Whatever the details, the role of the matrix is primarily to lower the activation energy of nucleation by increasing the

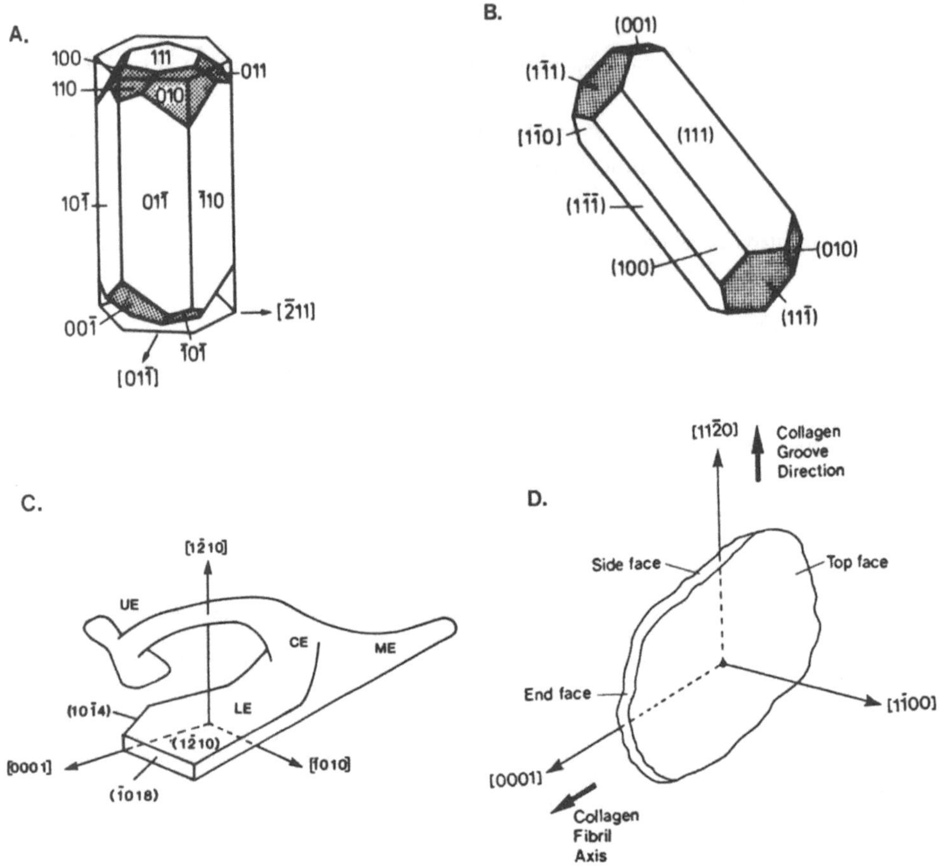

Fig.1. Examples of controlled biomineralization. (a) and (b), magnetite (Fe_3O_4) single crystals in magnetotactic bacteria [15,16]; (c) calcite single crystal in coccolith of *E. huxleyi* [17]; (d) hydroxyapatite single crystals in calcified turkey tendon [18].

encounter time of ionic collisions thereby stabilizing and perhaps catalysing the transition state of the phase transformation reaction. We know several general aspects about the matrix;

Primary Structures.
The matrix is matched to the coordination chemistry of the ionic species comprising the mineral through the choice of ligands exposed at the interface. For example, calcified invertebrate tissues contain macromolecules rich in carboxylate residues [5] such that Ca binding mimics Ca-CO_3 interactions, Ca-PO_4 motifs in bone, dentine and bacterial deposits can be established through $ROPO_3$-Ca binding (phosphoproteins in bone [6] and dentine [7], proteolipids in calcifying bacteria [8]) and ice nucleation is activated [9] and suppressed [10] by hydroxy-rich macromolecules.

Secondary and Tertiary Structures.
There are only two options; nucleation at a planar organic surface or at one which is curved. The former can be generated by antiparallel β-pleated sheets and there is X-ray diffraction evidence [11] to support the role of such structures in biomineralization. Other planar surfaces could be derived by elongation of phospholipid membranes or by crystallization such as the proteinaceous S-layer of bacterial membranes. Curved surfaces are more common; localized protein pockets and grooves, membrane-bound vesicles, α-helical (antifreeze proteins) and triple-helical (collagen) conformations are all possible.

Quaternary Structures.
Quaternary structures may play a fundamental role in biomineralization. In evolutionary terms, ferritin (as bacterioferritin) is an ancient matrix and functions primarily through its ability to self-assembly into a 24-mer capable of sequestering and deposition Fe(III) oxide [12]. Similarly, it is known that the membrane aggregation of proteins are crucial to the efficacy of ice nucleating bacteria [13]. It seems feasible that the "two-component" model of shell mineralization may also rely on the ordered assembly of the nucleator macromolecules on the relatively inert framework proteins. For example, the EDTA soluble macromolecules are considered to adopt the β-sheet conformation of the underlying structural proteins [14].

In conclusion, there are two key structural factors in the use of organic matrices in controlled nucleation. Firstly, the matrix is preorganized with respect to nucleation through processes such as self-assembly, aggregation, membrane vesiculation and controlled polymerization (cross-linking) which impart spatial regulation of functional groups. Secondly, nucleation at the matrix surface is regiospecific with a limited number of sites being confined to discrete loci. These two factors may be temporally linked since ion-binding could result in specific conformational changes in the preorganised matrix such that nucleation is activated within localized domains.

MOLECULAR RECOGNITION AND ORIENTED NUCLEATION

The examples of oriented biominerals shown in figure 1 illustrate several features which reflect the degree of molecular control involved in nucleation. In each case, the crystals are oriented along preferred crystallographic axes. Bacterial magnetites (Fe_3O_4) are elongated along the [111] axis [15] and coccolith scales [17] and bone crystals [18] along the *c* axes of calcite ($CaCO_3$) and hydroxyapatite ($Ca_{10}(PO_4)_6(OH)_2$), respectively. Electron microscopy studies of immature crystals have shown that the {111} face of bacterial magnetite probably represents the nucleation face [16]. In coccoliths, the nucleation face has been identified as {110} [17] which comprises the basal element of these structures. Interestingly, the arrangement of side faces on the basal element exhibits a handedness (for example, the small truncated {104} face is always to the right of the element when viewed from above) implying that there is selectivity of symmetry equivalent crystal faces. In essence, the nucleation process appears to discriminate between the front and back of the same face! This is only possible if the face exhibits chirality and this is the case for {110} of calcite. Thus, if we accept that the arrangement of binding sites on the underlying organic matrix dictates the orientation of nuclei, then it follows that sites disposed in chiral arrangements will favour the mirror image crystal face rather than the related enatiomeric configuration.

Some of the possible modes of molecular complementarity existing at the mineral-matrix interface are shown in figure 2 [19,20]. It is likely that several of these factors act cooperatively in real systems. The most fundamental aspect of recognition involves the matching of charge and polarity distributions and it seems feasible that the earliest biological approach to regulate nucleation was based on the clustering of charged centres, particularly if these sites were also redox active. The localization of Mn-oxidizing proteins in the cell walls of bacteria such as *Bacillus* [21] and ferroxidase centres in ferritin [22] are typical examples. The primary role of charge matching is to favour electrostatic accumulation and hence increase the encounter times for nucleation. In this regard, the topography of the matrix surface may play a significant role (figure 3). For example, it is clear that localized pockets or grooves can give rise to high spatial charge densities over dimensions commensurate with stable nuclei (1-5 nm). Note that high affinity binding may be disadvantageous since it restricts the structural rearrangement of the dispersed hydrated ionic clusters to the stable nuclei. Thus ion-pair accumulation may be important.

The curvature of molecular cavities provides three-dimensional control over nucleation and a limit on the size of the nucleation site. Planar surfaces, in contrast, provide a nucleation site only in two dimensions and, like convex surfaces, in the absence of other factors, would be less active since the binding sites are not constrained in close proximity to each other as for concave localities (figure 3). The advantage of planar surfaces, however, is that long range structural matches can be readily established in principle. This leads to the concept of epitaxy which has been a central hypothesis of controlled biomineralization [23,24].

It is important to make a distinction between inorganic epitaxy and "biological epitaxy". The former is well documented and in many cases, geometric matching at the unit cell level is apparent. However, large degrees of mismatch can occur and ordered aggregation of non-oriented nuclei to give oriented overgrowths has been observed at the post-nucleation stage. Biological substrates

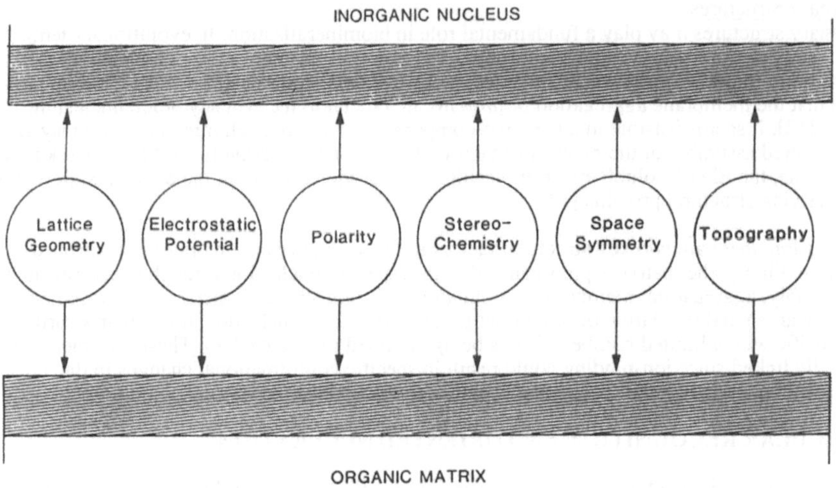

Fig.2. Possible aspects of molecular recognition at the interface between an inorganic nucleus and organic matrix.

differ from analogous inorganic templates in that they do not show the molecular smoothness or rigidity implicit in epitaxial mechanisms. On the other hand, they exhibit surface stereochemistry due to exposed functional groups and this may be advantageous since it provides for variability in surface conformation depending on the binding/ionization state of the residues. Thus, there may be several potential geometric matches for a given crystal face.

An additional possibility is that geometric epitaxy is superceded by stereochemical correspondence as determined in model systems [25-27]. This is a crucial concept since it relegates the need for matrix periodicity and is applicable to both planar and non-planar surfaces. A corollary of a stereochemically-driven mechanism is that it may select general rather than specific features of a crystal face. This is important because the putative crystal faces of nuclei are likely to be extensively reconstructed, non-stoichiometric and high in defect sites and general features such as the orientation of carbonate groups may be the only discernable property. In crystallographic terms, nucleation on organic surfaces may be selective for specific zone axes rather than particular crystal faces.

In summary, electrostatic, geometric and stereochemical recognition processes can occur at mineral-matrix interfaces. Understanding these interactions depends very much on determining the nature of assembled organic surfaces and in elucidating the surface structure of inorganic clusters. There is much greater crystal chemical flexibility through the use of organic surfaces although this may be at the expense of non-specificity. Hence the need for precise chemical regulation of mineralizing fluids in biomineralization.

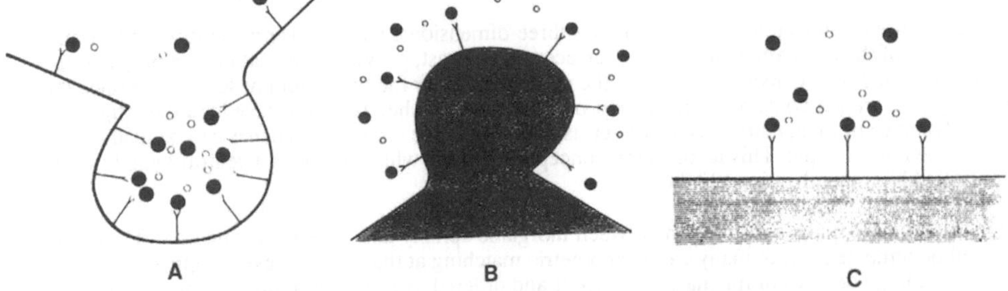

Fig.3. Influence of matrix topography on spatial charge distribution; (a) concave, (b) convex and (c) planar matrix surfaces.

MODELLING BIOMINERALIZATION

There have been several approaches directed at modelling the chemical role of organic macromolecules in biomineralization. These include (a) the extraction of matrix macrmolecules and the investigation of their nucleation efficacy when adsorbed onto solid substrates [25-28], and ability in solution to inhibit crystal growth [29-32], (b) use of phospholipid vesicles as constrained reaction environments [33-35], (c) functionalization of polymer substrates in nucleation studies [26,36,37] and (d) molecular design of synthetic peptides in crystal growth [38]. Recently we have explored two further possibilities, *viz.* site-directed mutagenesis [39] and controlled nucleation under compressed Langmuir monolayers [40-42]. We summarize our results below.

Ferritin.

The iron storage protein, ferritin, is found in bacteria, animals and plants and comprises a self-assembled protein shell of 24 polypeptide subunits which surround a hydrated Fe(III) oxide core of *ca.* 5 nm diameter [12]. Iron enters the cavity by means of symmetry-related molecular channels formed at the interfaces of individual polypeptide subunits. The remarkable property of this molecule is its ability to preferentially sequester and deposit Fe within the cavity in competition with non-specific bulk precipitation. Although native ferritins isolated from human spleen, limpet hemolymph and bacterial cells are structurally, magnetically and compositionally different [43], and *in vitro* reconstitutions of apoprotein in Fe(II) solutions indicate that structural variations can be induced in the iron oxide cores [44], the degree of structural change is limited by the organic matrix to a sequence of defect ferrihydrite ($Fe_2O_3.nH_2O$) phases. The corresponding product formed in free solution is microcrystalline lepidocrocite (γ-FeOOH).

Ferritin is the best characterized biomineralization matrix to date. The subunits are of two types, designated H and L which show only 55% homology in amino acid sequences [45]. The crystal structure of ferritin has been determined at high resolution [12] and has identified two important features of the molecule; (a) the channels are of two types, short (6 Å) 3-fold hydrophilic channels lined with conserved aspartate and glutamate residues and long (12 Å) 4-fold hydrophobic channels of variable composition but generally lined with leucine residues; (b) the inner cavity surface has a cluster of conserved glutamate residues which could be involved in Fe(II) oxidation and nucleation. The recent success in expressing human recombinant H-chain [46] and L-chain ferritins [47] as well as H-chain mutants [48] has shown that only the H-chain protein has ferroxidase activity and that depletion of the inner surface residues Glu 62, His 65 resulted in loss of this activity in mutant H-chain ferritins [22].

In collaboration with Professors P. M. Harrison (Sheffield) and P. Arosio (Milan) we have used electron microscopy to study the biomineralization of Fe in recombinant homopolymer and mutant ferritins reconstituted to 2000 Fe atoms from Fe(II) solutions [39]. Interestingly, no structural changes in the mineral cores were observed for H-chain, L-chain or mutants modified in the 3-fold (Asp 131→His, Glu 134→Ala) or 4-fold channels (Leu169→Arg, His 173→Leu), ferroxidase centre (Glu 62→Lys and His 65→Gly) and inner surface carboxylates (Glu61,64,67→Ala). Thus, the stabilization of the ferrihydrite structure is a general effect of molecular architecture and not related to specific amino acid residues.

Superimposed on the structural invariance observed in these experiments were specific effects related to amino acid modifications. These are summarized diagrammatically in figure 4. Whereas the H-chain protein exhibited rapid Fe uptake kinetics compared with the ferroxidase-inactive L-chain homopolymer [47], the mean size of the mineral cores of the former was significantly reduced. A similar effect was observed for the H-chain mutant depleted of the ferroxidase centre which showed larger cores than the unmodified protein. In comparison, depletion of the inner surface glutamate residues, and to a lesser extent, modification of the 3-fold channel, resulted in mineral cores of smaller mean diameter.

These observations can be attributed to changes in the rate determining mechanisms of mineralization. In particular, nucleation of the Fe cores can be driven by rapid increases in supersaturation within the protein cavity via rapid oxidation of Fe(II) entering through the molecular channels. This is the case in the presence of the ferroxidase centre and results in essentially uniform core nucleation in all the ferritin molecules. Alternatively, nucleation can be driven by the lowering of interfacial energy through interactions at the protein surface. This process is relatively slow compared with the oxidation/supersaturation route. Thus, discrimination between ferritin molecules will arise since statistically some molecules will contain nuclei prior to others and these will develop

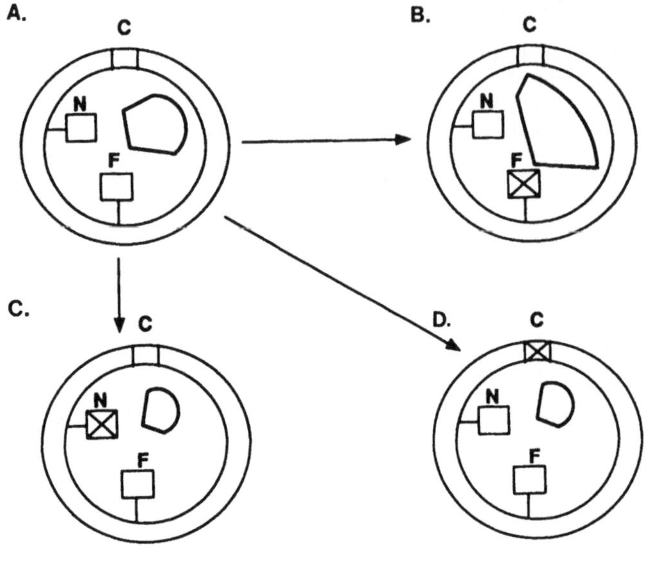

Fig.4. Scheme showing effects of site-directed depletions in ferritin. (a) molecule reconstituted with unmodified channel (C), ferroxidase centre (F) and putative nucleation (N) sites; (b) depletion of ferroxidase centre results in larger Fe cores; (c) depletion of putative nucleation sites results in smaller Fe cores; (d) depletion of three-fold channel sites results in partially smaller Fe cores.

at the expense of molecules without nuclei by autocatalytic processes involving Fe(II) oxidation at the surface of the growing core. Consequently, the available Fe is distributed amongst fewer ferritin molecules with some cavities remaining devoid of mineral. Those cores that do develop attain a larger diameter than those formed from rapid precipitation induced by ferroxidase activity.

Depletion of the inner surface glutamate clusters results in a reduced core size because the matrix-mediated process becomes less competitive with regard to bulk non-specific oxidation. The ferroxidase centre is present in this mutant suggesting that the nucleation of the initially formed Fe(III) oligomers is retarded and that the Glu 61,64,67 residues comprise the nucleation site in ferritin. Significantly, whereas the ferroxidase centre is not conserved in H-chain and L-chain homopolymers, the above glutamate residues are conserved in both subunit types.

Langmuir Monolayers.
Compressed Langmuir monolayers of amphiphilic molecules provide an excellent experimental system to study the role of two-dimensional organized organic surfaces in the controlled nucleation of inorganic solids. They were originally used in the study of organic crystallization [49-50]. Inorganic materials such as NaCl [50], $CaCO_3$ [40-42] and $BaSO_4$ (B.R.Heywood, unpublished data) have been investigated. In general, all these systems show oriented nucleation at the organic surface which can be rationalized in terms of geometric, electrostatic or stereochemical interactions.

Studies of of $CaCO_3$ crystallization from supersaturated Ca bicarbonate solutions by passive loss of CO_2 gas under carboxylate, amine and alcohol headgroups [42,51,52] indicate that charged headgroups are required for oriented nucleation. In the absence of a monolayer or in the presence of neutral alcohol monolayers, crystallization from a solution of total [Ca] = 9.5 mM gave randomly intergrown rhombohedral calcite crystals at the air/water interface (figure 5a). In the presence of a stearate monolayer, the calcite crystals were, by contrast, discrete and crystallographically oriented (figure 5b). The crystal surface apposed to the monolayer surface was roughened and contained an elevated central region that represented the point of contact with the air/water interface. Transmission electron microscopy and electron diffraction studies of crystals collected 15 minutes after film compresssion showed plate-like calcite crystals oriented with the crystallographic $[1\bar{1}0]$ direction normal to the monolayer surface. Selective nucleation of the $(1\bar{1}0)$ face can be explained by electrostatic, structural and stereochemical relationships at the organic surface. Firstly, Ca binding is a prerequisite for oriented calcite nucleation. For example, the number of oriented calcite crystals

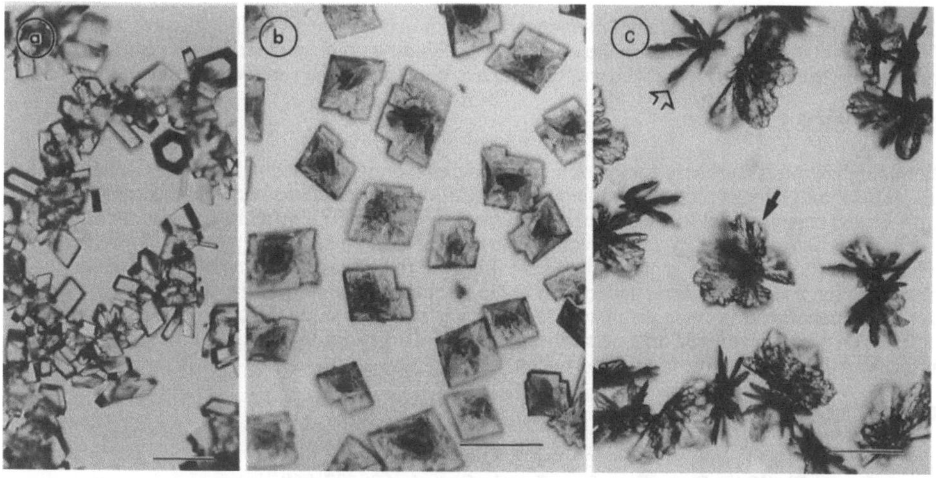

Fig.5. Oriented nucleation under compressed Langmuir monolayers. Optical micrographs of (a) control calcite crystals (no monolayer), (b) calcite crystals under stearate monolayers, and (c) vaterite crystals under octadecylamine monolayers; arrow, crystal oriented with the *c* axis normal to the monolayer; open arrow, crystal oriented with the *a* axis normal to the monolayer. Scale bar = 100 μm in all micrographs.

rapidly diminishes in the presence of competitive ion-binding of Na at Ca:Na ratios of 1:10 (S. Rajam, unpublished data), oriented vaterite is nucleated under stearate films at total [Ca]<4.5 mM [40-42] and nucleation on positively charged amine monolayers also results in oriented vaterite [42] (figure 5c, see below). Secondly, there is a close structural match in two dimensions between the hexagonal net of the stearate molecules (a = 5 Å) and Ca-Ca distances on the ($1\bar{1}0$) face. Ca atoms coplanar in this face form a rectangular lattice, 4.99 x 8.6 Å, which matches the orthogonal 5 x 8.7 Å spacings of stearate molecules in the hexagonally packed monolayer. Modifications in the interheadgroup spacing, for example, by using isostearic acid (a = 5.8 Å) results in non-oriented nucleation at the carboxylate surface (S. Rajam, unpublished data). Thirdly, there is a close stereochemical correspondence between the orientation of the carbonates in the ($1\bar{1}0$) face and that of the headgroup carboxylates; in both cases they lie perpendicular to the monolayer/solution interface. This is an important factor because there are other crystal faces besides ($1\bar{1}0$), eg.(001), which exhibit a close geometric match with the stearate lattice.

Nucleation under octadecylamine monolayers gave two types of oriented vaterite crystals (figure 5c). Floret-shaped and butterfly-shaped crystals were aligned with their *c* and *a* axes normal to the monolayer surface respectively. We propose that these orientations arise from a kinetic effect due to accumulation of HCO_3^- ions at the amine headgroups. Both the (001) and (110) faces have carbonates perpendicular to the crystal surface and this could be simulated by bidentate binding of bicarbonate under the monolayer. Thus, stereochemical ordering in the boundary layers adjacent to organic surfaces may be important in oriented nucleation.

These experiments clearly indicate that the proposal that organic surfaces can mediate inorganic nucleation is justified. Furthermore, they show that two-dimensional organic surfaces can recognize inorganic nuclei through epitaxial and stereochemical mechanisms and these may be cooperative. Electrostatic accumulation is a primary factor; strong ion-binding of Ca is important in oriented calcite nucleation but diffuse anionic interactions at the organic surface can give rise to metastable phases (vaterite). Significantly, the degree of monolayer compression was not critical in determining oriented nucleation provided that Ca-binding could induce localized ordered domains of appropriate interheadgroup spacing. Indeed, nucleation under liquid phase monolayers tended to produce more uniform crystals compared with fully compressed surfaces. The latter remain catalytic over a relatively long time course resulting in episodic nucleation as the supersaturation rises and falls whilst the former require activation by localized film condensation such that the initial burst of nuclei are sustained by autocatalysis. One can envisage a similar effect occurring in biomineralization where a degree of dynamical freedom may be beneficial in regulating nucleation. The process could be synergistic in the sense that ion-binding to the organic matrix results in a specific conformational change which in turn drives more ions to the surface resulting in nucleation.

CONCLUSIONS

Biomineralization is a process comparable in complexity to many biological reactions. It has attained this status through the evolution of molecular recognition events between organic macromolecular surfaces and crystal nuclei formed from supersaturated solution. The matching of spatial charge distributions, lattice geometry and stereochemistry are possible modes of interaction. Site-directed mutagenesis is one possible approach in determining the role of primary structure functionality on mineralization and this has been undertaken in the Fe storage protein, ferritin. Alternatively, Langmuir monolayers provide viable model systems in which to study the controlled nucleation of minerals on planar organized organic surfaces. Such investigations indicate that charged membrane surfaces are highly effective in oriented nucleation.

REFERENCES

1. Mitterer RM (1968) Science 162: 1498-1499.
2. Bernal JD (1949) Proc. Phys. Soc. (Lond.) Sect. A. 62: 537-558
3. Wächtershäuser G (1988) Microbiol. Rev. 52: 452-484.
4. Mann S (1983) Structure and Bonding. 54: 125-174.
5. Weiner S (1986) CRC Crit. Rev. Biochem. 20: 365-408.
6. Glimcher MJ (1984) Phil. Trans. R. Soc. Lond. B. 304: 479-508.
7. Lee SL, Veis A, Glonek T (1977) Biochemistry 16: 2971-2979.
8. Boyan-Salyers B, Boskey A (1980) Calc. Tiss. Intern. 30: 167-174.
9. Wolber PK, Warren GJ (1989) Trends Biol. Sci. 14: 179-182.
10. DeVries AL (1984) Phil. Trans. R. Soc. Lond. B. 304: 575-588.
11. Weiner S, Traub W (1980) FEBS Letts. 111: 311-316.
12. Ford GC, Harrison PM, Rice DW, Smith JMA, Treffry A, White JL Yariv J (1984) Phil. Trans. R. Soc. Lond. B 304: 551-565.
13. Warren GJ, Wolber PK (1987) Cryo-Lett. 8: 204-215.
14. Weiner S, Traub W (1984) Phil. Trans. R. Soc. Lond. B. 304: 425-434.
15. Mann S, Moench T, Williams RJP (1984) Proc. R. Soc. London B 221: 385-393.
16. Mann S, Sparks NHC, Blakemore RP (1987) Proc. R. Soc. London B 231: 477-487.
17. Mann S, Sparks NHC (1988) Proc. R. Soc. Lond. B 234: 441-453.
18. Heywood BR, Sparks NHC, Shellis RP, Weiner S, Mann S (1990) Connective Tiss. Res. 25: 1-17.
19. Mann S (1988) Nature 332: 119-124.
20. Mann S (1989) In "Biomineralization: Chemical and Biochemical Perspectives" eds. S Mann, J Webb and R J P Williams, VCH Publishers, Weinheim, pp 35-62.
21. de Vrind JPM, de Vrind-de Jong EW, de Voogt JW, Westbroek P, Boogerd FC, An Rosson RA (1986) Appl. Envir. Microbiol. 52: 1096-1100.
22. Lawson DM, Treffry A, Artymiuk PJ, Harrison PM, Yewdall, SJ, Luzzago A, Cesareni G, Levi S, Arosio P (1989) FEBS Letts. 254:207-210.
23. Neuman WF, Neuman MW (1958) "The Chemical Dynamics of Bone Mineralization" University of Chicago Press, Illinois.
24. Bachra BN (1973) In "Biological Mineralization" ed. I. Zipkin, J. Wiley & Sons, pp. 845-881.
25. Addadi L, Weiner S (1985) Proc. Natn. Acad. Sci. USA 82: 4110-4114.
26. Addadi L, Moradian J, Shay E, Maroudas NG, Weiner S (1987) Proc. Natn. Acad. Sci. USA 84: 2732-2736.
27. Addadi L, Weiner S (1989) In "Biomineralization: Chemical and Biochemical Perspectives", eds. S Mann, J Webb and RJP Williams, (VCH Publishers, Weinheim) pp. 134-156.
28. Greenfield EM, Wilson DC, Crenshaw MA (1984) Amer. Zool. 24: 925-932.
29. Wheeler AP, George W, Evans CA (1981) Science, 212: 1397-1398.
30. Berman A, Addadi L, Weiner S (1988) Nature, 331: 546-548.
31. Borman AH, de Jong EW, Huizinga M, Kok DJ, Westbroek P, Bosch L (1982) Eur. J. Biochem. 129: 179-183.
32. Campbell AA, Ebrahimpour A, Perez L, Smesko SA, Nancollas GH (1989) Calc. Tiss. Intern. 45: 122-128.

33. Mann S, Williams RJP (1983) J. Chem. Soc. Dalton Trans. 311-316.
34. Mann S, Hannington JP, Williams RJP (1986) Nature 324: 565-567.
35. Eanes ED, Hailer AW, Calcif. Tiss. Int. (1987) 40: 43-48.
36. Kallitsis J, Koumanakos E, Dalas E, Sakkopoulos S, Koutsoukos PG (1989) J. Chem. Soc. Chem. Commun. 1146-1147.
37. Tarasevich BJ, Reike PC (1990) Mat. Res. Symp. 174: 51-60.
38. Wheeler AP, Sikes CS , (1989) In "Biomineralization: Chemical and Biochemical Perspectives", eds. S Mann, J Webb and R J P Williams, (VCH Publishers, Weinheim) pp 95-132.
39. Wade VJ, Mann S, Harrison PM, Treffry A, Arosio P (1990) submitted to J. Mol. Biol.
40. Mann S, Heywood BR, Rajam S, Birchall JD (1988) Nature 334: 692-695.
41. Mann S, Heywood BR, Rajam S, Birchall JD (1989) Proc. R. Soc. Lond. A. 423: 457-471.
42. Mann S, Heywood BR, Rajam S, Walker JBA, Davey RJ Birchall JD (1990) Adv. Materials 2: 257-261.
43. Mann S, Bannister JV, Williams RJP (1986) J. Mol. Biol. 188: 225-232.
44. Mann S, Williams JM, Treffry A, Harrison PM (1987) J. Mol. Biol. 198: 405-416.
45. Boyd D, Vecoli C, Belcher DM, Jain SK, Drysdale JW (1985) J. Biol Chem. 260: 11755-11761.
46. Levi S, Cesareni G, Arosio P, Lorenzetti R, Sollazzo M, Cortese R (1987) Gene 51: 267-272.
47. Levi S, Salfield J, Franceschinelli F, Cozzi A, Dorner M, Arosio P (1989) Biochemistry, in press.
48. Levi S, Luzzago A, Franceschinelli F, Santambrogio P, Cesareni G, Arosio P (1989) Biochem. J. 264: 381-388.
49. Landau EM, Levanon M, Leiserowitz L, Lahav M, Sagiv J (1985) Nature 318: 353-356.
50. Landau EM, Popovitz-Bior R, Levanon M, Leiserowitz L, Lahav M, Sagiv J (1986) Molec. Cryst. liq. Cryst. 134: 323-335.
51. Rajam S, Heywood BR, Walker JBA, Davey RJ, Birchall JD, Mann S (1990) submited to J. Chem. Soc. Farad. Trans.
52. Heywood BR, Rajam S, Mann S (1990) submitted to J. Chem. Soc. Farad. Trans.

CHAPTER 1.9

Otolith Matrix Proteins of Walleye Pollock; Biochemical Properties and Immunohisto-chemical Localization in the Saccular Tissue

K. BABA, M. SHIMIZU, Y. MUGIYA, and J. YAMADA

Faculty of Fishery, Hokkaido University, Hakodate, 041 Japan

Key Words: Fish otolith, EDTA-soluble proteins, Calcium binding proteins, Immunoblotting, Teleost.

INTRODUCTION

In teleosts, the sacculus of the inner ear contains a solid calcareous otolith, sagitta. The otolith is fixed over a region of the sensory epithelium (macula), intervened by a membrane called the otolithic membrane [1, 2]. The otolith grows in the endolymph of the saccular lumen without attachment of cells. It consists of aragonitic crystals of calcium carbonate [3], deposited on a small amount of an organic matrix consisting of proteins, glycosaminoglycans and lipids [4, 5]. The organic matrix is suggested to be produced at the macula region and diffuses into the endolymph fluid [4]. This study was performed to clarify the biochemical properties and the origin of this organic matrix. This investigation particularly focused on the calcium binding proteins extracted from walleye pollock otoliths. Biochemical properties of matrix proteins were examined by electrophoresis and ^{45}Ca autoradiography. The origin of these matrix proteins was studied by applying immunohistochemical techniques to the saccular epithelium.

MATERIALS AND METHODS

Otoliths of walleye pollock, Theragra chalcogramma (Pallas), were cleaned and soaked with 1% sodium hypochlorite, 5% NaOH and distilled water, and then air dried. The finely ground and powdered otoliths were made into a slurry with a 10% EDTA solution, pH 7.0, and dialyzed against 10% EDTA in dialysis tubing (size 27/32, 12000-14000 MW cutoff) until all otolith materials were dissolved (48-60h). The entire extract was separated into soluble and insoluble fractions by centrifugation at 15000g for 30 min. The soluble matrix was then lyophilized and stored at -40°C until use.
Total otolith matrix proteins were measured by the Kjeldahl method. Soluble matrix proteins were measured by Lowry's method as modified by Bensadoun and Weinstein [6]. Amino acid analyses of the soluble matrix were performed with an automatic amino acid analyzer (Hitachi-835). Electrophoretic analyses of soluble matrix proteins were carried out by SDS-PAGE using the discontinuous buffer system [7]. Dimercaptoethanol (2ME) reduced and non-reduced samples were prepared. Gels were stained with Coomasie brilliant blue R-250 for proteins, toluidine blue for glycosaminoglycans and proteoglycans [8], alcian blue for acidic groups of glycosaminoglycans [9], Sudan black B for lipoproteins [10] and periodic acid Schiff's reagent (PAS) for glycoproteins [11]. Calcium binding proteins (CaBP) in otolith soluble matrix separated by SDS-PAGE were detected by ^{45}Ca autoradiography[12].
Antisera against the whole otolith matrix (a mixture of insoluble and lyophilized soluble matrices) and CaBPs (anti-OM and anti-CaBPs) were obtained by injecting them subcutaneously into a rabbit.
Immunochemical analyses of the soluble matrix were performed by the immunoblotting technique of Towbin et al.,[13]. Anti-OM or anti-CaBPs

was used as the primary antibody. Horseradish peroxidase (HRP) conjugated with goat IgG against rabbit IgG and diluted to a ratio of 1: 2000 with Tris-buffered saline was used as the secondary antibody. Otolith organic matrix producing cells were identified by the peroxidase-antiperoxidase (PAP) technique of Sternberger et al., [14].

RESULTS

Proteins contained in the otolith were 0.40±0.05% (w/w, n=7) and soluble proteins were 0.06±0.006% (w/w, n=4). The amino acid composition showed characteristically high contents of aspartic and glutamic acids and a low content of proline (Table 1). The total amount of acidic amino acids was at a level 2.5 times that of basic amino acids.

Table 1. Amino acid composition of EDTA-soluble matrix from otoliths of walleye pollock, *Theragra chalcogramma*

Amino acid	residues / 1000 residues
Asp	115.7
Thr	79.9
Ser	102.6
Glu	135.0
Pro	nd
Gly	110.3
Ala	99.7
Cys/2	13.3
Val	70.2
Met	16.7
Ileu	20.9
Leu	82.8
Tyr	23.4
Phe	27.1
Lys	36.1
His	19.0
Arg	45.8
acidic amino acid	250.7
basic amino acid	100.8
acidic / basic	2.49

nd: not determined

SDS-PAGE patterns of non-reduced and reduced proteins from the otolith soluble matrix were shown in Fig. 1A and 1B, respectively. Four major bands, several minor bands ranging from 10 to 100 KDa, and a macromolecule band beyond that range were evident in both samples. Molecular weights of the major proteins were determined to be 95, 56, 42 and 25 KDa in the non-reduced sample and 95, 80, 66, 56, 52, 42 and 25 KDa in the reduced sample. Three protein bands of 80, 66 and 52 KDa in the reduced sample were not observed in the non-reduced sample. This suggests these three proteins were subunits of the macromolecule in the non-reduced sample. Staining properties of the major bands in the non-reduced and reduced samples were shown in Fig. 1A, 1B and Table 2. Protein bands of 95, 42 and 25 KDa in the non-reduced sample and those of 95, 52 and 25 KDa in the reduced sample were found to have calcium binding ability (Table 2). The calcium binding ability of the non-reduced 42-KDa protein disappeared by the reduction with 2ME. Calcium binding proteins of 42-KDa (HCaBP) and 25-KDa (LCaBP) were used as antigens to raise rabbit antisera.

All the non-reduced and reduced otolith soluble protein bands separated by SDS-PAGE were detected by immunoblotting with anti-OM. Almost all serum protein bands of walleye pollock also reacted with anti-OM.

The reactions between serum proteins and anti-OM vanished when anti-OM was previously absorbed by serum. A single protein band of 25-KDa was detected by immunoblotting with anti-LCaBP. When the anti-LCaBP was absorbed by serum, this band vanished. Several bands of otolith soluble matrix were detected by immunoblotting with anti-HCaBP. Only a single protein band of 42 KDa was detected by immunoblotting with anti-HCaBPab.

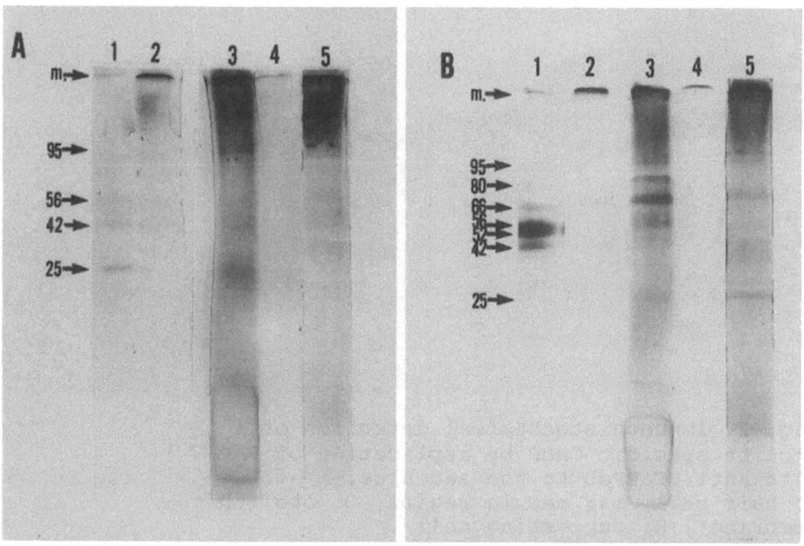

Fig. 1. SDS-PAGE patterns of non-reduced (A) and reduced (B) samples. Left side figures represent molecular weights estimated in KDa. m: macromolecule Stained with 1. Coomasie brilliant blue; 2. PAS; 3. alcian blue; 4. toluidine blue; 5. Sudan black B.

Table 2. Staining properties of major EDTA-soluble otolith proteins, after non-reduced or reduced with 2-mercaptoethanol.

	Moecular weight (KDa)	TB	AB	PAS	SB	^{45}Ca
Non-reduced	macromolecule	+	+	+	+	+
	95	-	+	+	+	+
	56	-	-	+	-	-
	42 (HCaBP)	-	+	+	-	+
	25 (LCaBP)	-	+	+	+	+
Reduced	macromolecule	+	+	+	+	+
	95	-	+	+	+	+
	80	-	+	+	-	-
	66	-	+	+	+	-
	56	-	-	+	-	-
	52	-	-	+	-	+
	42 (HCaBP)	-	+	+	-	-
	25 (LCaBP)	-	+	+	+	+

TB: toluidine blue, AB: alcian blue, SB: Sudan black B

The saccular wall was composed of three types of epithelia: sensory epithelium, transitional epithelium and squamous epithelium. Almost all saccular cells, except hair cells, were deeply stained with anti-OMab, but not with anti-HCaBPab. Hair cells were specifically stained with anti-HCaBPab.

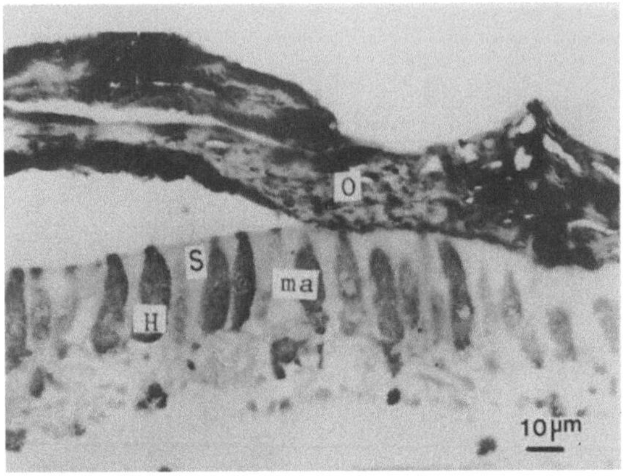

Fig. 2. Immunohistochemical detection of an otolith specific CaBP by application of PAP with anti-HCaBPab to the sacculus.
H: hair cell, ma: macula region, O: otolithic membrane , S: supporting cell

DISCUSSION

The organic matrix of fish otoliths was reported to be characterized by abundant aspartic and glutamic acids with a low contents of aromatic and basic amino acids [5]. In walleye pollock otoliths, the amino acid analysis of soluble proteins showed a similar composition except that only a trace level of proline was observed. Asano et al., [15] also reported a similar result in EDTA-soluble otolith proteins of tilapia. Similar patterns of amino acid composition have been observed in various calcified tissues which deposit calcium carbonate [16]. Degens et al., [5] reported that the molecular weight of major matrix proteins in otolith, or otolin, is greater than 150 KDa as ascertained by a molecular sieving technique. In SDS-PAGE of soluble proteins of walleye pollock otoliths, a macromolecule over 200 KDa was observed. This macromolecule may be referred to as an otolin. At least three proteins of 80, 66 and 52 KDa were identified as subunits of the macromolecule. These three proteins showed different staining and Ca-binding properties; 80-KDa protein was a glycoprotein, 66-KDa protein was a lipoprotein and 52-KDa was a glycoprotein and had a calcium binding ability. These different properties of the subunits of otolin may play some role in otolith calcification.
Four protein bands of 95, 56, 42 and 25 KDa other than the above three subunits of macromolecule were observed as major proteins in SDS-PAGE. Except for the 56-KDa protein, the other three proteins, and the macromolecule, all showed calcium binding ability by ^{45}Ca autoradiography. The 42-KDa protein lost its calcium binding ability by the 2ME treatment. This suggests that the calcium binding ability of this CaBP depends on its high dimensional structure which is retained by disulfide bonds.

All otolith soluble proteins separated by SDS-PAGE were detected by immunoblotting with anti-OM. Furthermore, almost all serum proteins separated by SDS-PAGE were detected by immunoblotting with anti-OM. It is suggested, therefore, that serum proteins are incorporated into the otolith organic matrix, as similarly reported to occur in human bone [17]. A single protein band of 25 KDa was detected by immunoblotting with anti-LCaBP. However, this band vanished when the antiserum was absorbed by walleye pollock serum. This indicates that LCaBP (25-KDa protein) must be a serum protein. The other single protein band (HCaBP, 42-KDa protein) detected by immunoblotting with anti-HCaBPab is certainly an otolith specific Ca-binding protein. The SDS-PAGE band of this protein was positively stained with PAS and alcian blue, indicating that it is a glycoprotein containing acidic groups of glycosaminoglycans.
Almost all cells of the saccular wall and the otolithic membrane reacted with anti-OMab. This means that these cells produce the otolith matrix. On the other hand, only hair cells of the macula were immunohistochemically identified as HCaBP producing cells. It is highly possible that a factor causing otolith daily increments is a diurnal rhythm of secretion of calcium binding glycoprotein (HCaBP) from the hair cells.

REFERENCES

1. Dunkerberger DG, Dean JM, Watabe N (1980) J Morph 163: 367-377.
2. Saitoh S, Yamada J (1989) Trans Am Microsc Soc 108: 223-238.
3. Morales-Nin B (1987) In: Summerfelt RC, Hall GE (eds.) The Age and Growth of Fish, Iowa State Univ Press, Ames, pp. 331-343.
4. Mugiya Y (1968) Bull Jap Soc Sci Fish 34: 1096-1106.
5. Degens ET, Deuser WG, Haedrich RL (1969) Mar Biol 2: 105-113.
6. Bensadoun A, Weinstein D (1976) Anal Biochem 70: 241-251.
7. Lammeli UK (1970) Nature 227: 680-685.
8. Heinegård D, Sommsrin Y, Hedbom E, Weislander J, Larsson B (1985) Anal Biochem 151: 41-48.
9. Hata R (1973) Protein Nucleic acid Enzyme 18: 843-857.
10. Prat JP, Lamy JN, Weill JD (1969) Bull Soc Chem Biol 51: 1367.
11. Zacharius RM, Zell TE, Morrison JH, Woodlock JJ (1969) Anal Biochem 30: 148-152.
12. Maruyama K, Mikawa T, Ebashi S (1984) J Biochem (Tokyo) 95: 511-519.
13. Towbin H, Staehelin T, Gordon J (1979) Proc Natl Acad Sci USA76: 4350-4354.
14. Sternberger LA, Hardy PH, Cuculis JJ, Meyer HG (1970) J Histochem Cytochem 18: 315-333.
15. Asano M, Mugiya Y, Yamada J (1988) Report Sci Res Fund Monbusho, No. 61480065.
16. Samata T (1988) In: Omori M, Suga S, Goto M (eds.) Tokai Univ Press, Tokyo, pp. 182-202.
17. Mbuyi JM, Dequeker J, Bloemmen F, Stevens E (1982) Calcif Tissue Int 34: 229-231.

CHAPTER 1.10

The Water-Soluble Organic Matrix in the Egg Shells of the Extinct Moa and the Extant Rhea

H. SOMIYA[1], V.B. MEYER-ROCHOW[2], and T. SAMATA[1]

Azabu University, Sagamihara, 229 Japan[1] and University of Waikato, Hamilton, New Zealand[2]

Key words: Egg Shell, Organic Matrix , Amino acid composition, Moa, Rhea.

INTRODUCTION

Moas (Dinornithiformes) were New Zealand's unique flightless birds which comprise about nineteen species. Most species of moas became extinct between the tenth and seventeenth centuries. A few researchers have analyzed the moa egg shells structurally and histochemically [1-3], but as far as we know, there have been no published details of the chemical analysis of the organic matrix of moa egg shells. The present work was performed to provide information on the chemical nature of the water-soluble organic matrix of moa egg shells and compare them with those from egg shells of the extant south American rhea.

MATERIAL AND METHODS

Material. Fragments of moa egg shells were made available from the Auckland Institute and Museum, Auckland. They were about 300 - 400 years old and collected from sand dunes, Ocean Beach, Whangarei, New Zealand. Egg shells of the extant rhea (*Rhea americana*) were sampled at Ueno Zoo, Tokyo, Japan. Only the true egg shells, remaining after removal of the cuticle and membranes, were used as the material in both ratite birds.

Extraction. These true egg shell fragment were dipped in 1 N HCl for 1 min. and in 1% NaClO for 30 min. to remove organic contaminants. Cleaned egg shells (10 g)were dried, ground into powder and decalcified in 10 % EDTA under pH 7.0. Salts were removed by dialysis against distilled water in Spectapore No.3 tubing (Spectrum Medical Ind. USA). The dialysate was centrifuged at 10,000 x g and separated into supernatant (water-soluble) and precipitated (water-insoluble) fractions. This procedure is essentially the same as that of Wheeler et al. (1988) [4].

SDS polyacrylamide gel electrophoresis (SDS-PAGE). SDS-PAGE was carried out in 8.5 % gel as described by Anderson et al (1983) [5]. Gels were stained with 0.4 % Coomassie Brilliant Blue R-250 (CBB).

Isoelectric focusing (IEF). IEF was carried out in ampholine (pH 3.0 to 10.0) as described by O'Farrell (l975) [6]. Gels were packed in glass tubings (130 x 2.5 mm inside diameter) and run at 400 volts for 7 hours, using disk gel electrophoresis system of Marysol KS-8110 (Marysol Co., Japan). Gels were stained with 0.4 % CBB.

Two-dimensional polyacrylamide gel electrophoresis (2D-PAGE)
By the method of O'Farrell (l975), 2D-PAGE was carried out and the gels were stained with 0.4 % CBB.
Amino acid analysis.
Samples for the amino acid analysis were hydrolyzed in an ampoule under vacuum at 110 ^{0}C for 24 hours in 6N HCl. The hydrolysate was then analyzed on an Atto MLC-703S automatic amino acid analyzer (Atto Co., Japan).

RESULTS

SDS polyacrylamide gel electrophoresis .
The results of the SDS-PAGE of the water-soluble organic matrix in moa and rhea egg shells are shown in Fig.1. In the rhea egg shells, a thick main band was observed at the position of molecular weight of 18 kD. A faint band was also observed at the position of a marker protein, CA (bovine erythrocyte, 29 kD). Two thin bands were observed at the high molecular weight region (greater than 66 kD). By contrast, in the moa egg shells, bands were not recognized clearly but a faintly-stained smear was observed at the low molecular weight position between 10 kD to 40 kD.
Isoelectric focusing.
The results of the IEF of the water-soluble matrix in the moa and the rhea egg shells are shown in Fig 2. In the rhea, the most prominent thick-stained band was seen at the position of pH 5-6, and other bands and smears were observed at the position of around pH 5. In the moa, the organic components were separated and located only in the acidic side. Three obscure bands with smear were observed at the positions of around pH 4.2, 4.5 and 5.0, respectively.
Two-dimensional polyacrylamide gel electrophoresis.
In the rhea, only one thick stained band was recognized (Fig. 3). In the moa, a clear band was not recognized, except for a faintly-stained smear.
Amino acid composition.
Amino acid composition of the water-soluble shell matrix in the moa and the rhea is shown in Table 1 and Fig. 4. There is considerable similarity between the amino acid composition of the matrix in the moa and the rhea egg shells. The most dominant amino acid is glutamate and rather a high content of glycine, alanine, aspartate and leucine was observed in both egg shells. Total contents of acidic amino acid is more dominant than that of basic amino acid. Total amounts of polar amino acids are also higly concentrated with more than 60 % in both egg shells.

DISCUSSION

In both egg shells, that of the extinct moa and the extant rhea, the water-soluble organic matrix was extracted in the same order of amounts. Electrophoretic experiments revealed that the main protein of the rhea egg shell has relatively low molecular weight (18 kD) with acidic pI (about 5.5). This is clearly demonstrated in the 2D-PAGE experiments (Fig. 3). On the other hand, in the moa egg shell, only a broad smear is observed and isolated bands are not detected on the gels of the SDS- and 2D-PAGE experiments. But at least three acidic components are separated as rather obscure bands in the IEF-

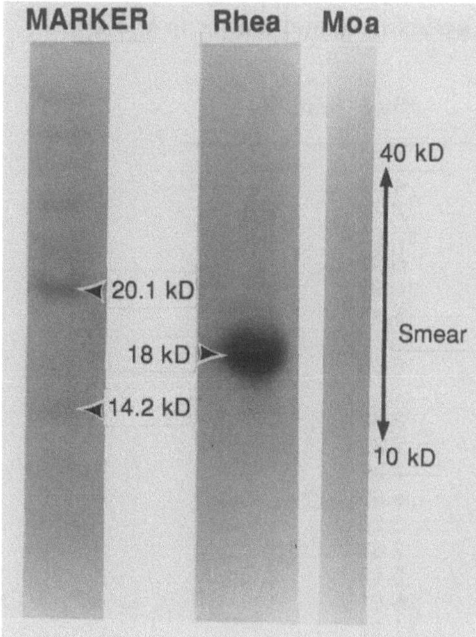

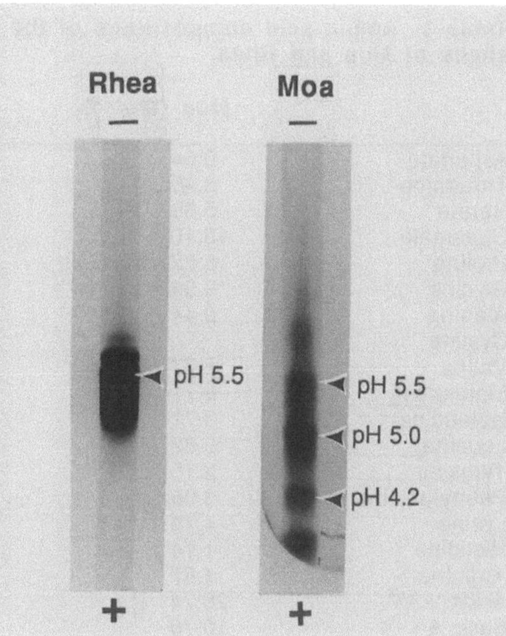

Fig.1 SDS-polyacrylamide gel electrophoresis of the organic matrix in rhea egg shell (left) and moa (right).

Fig.2 Isoelectric focusing of the organic matrix in rhea egg shell (left) and moa (right).

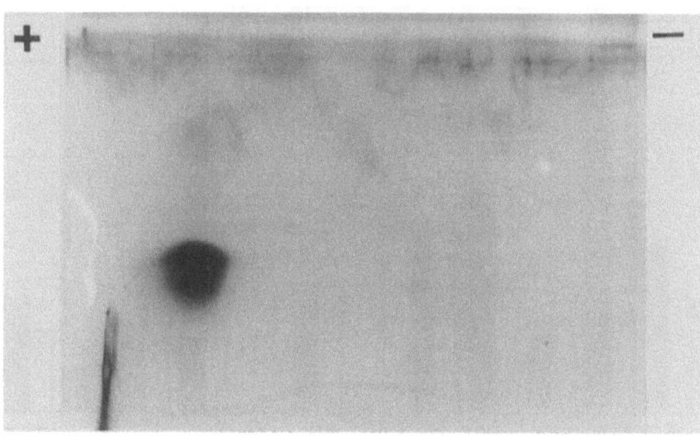

Fig.3 Two dimensional gel electrophoresis of the organic matrix in rhea egg shell.

Table 1 Amino acid compositions of the soluble organic matrix in the egg shells of Moa and Rhea.

	Moa (Mol %)	Rhea (Mol %)
Aspartate	9.64	8.32
Threonine	3.40	4.46
Serine	5.69	8.24
Glutamate	19.10	17.69
Proline	6.22	5.01
Glycine	9.99	10.97
Alanine	9.44	9.45
Cystine	-*	-
Valine	5.37	5.06
Methionine	0.71	-
Isoleucine	3.71	3.82
Leucine	9.82	7.89
Tyrosine	2.15	2.22
Phenylalanine	3.96	5.65
Lysine	4.78	3.25
Histidine	1.14	2.86
Arginine	4.87	5.12
Acidic AA**	28.74	26.01
Basic AA	10.79	11.23
Acidic AA/Basic AA	2.66	2.32
Polar AA	60.76	63.13

* -; not detected ** AA; Amino Acid

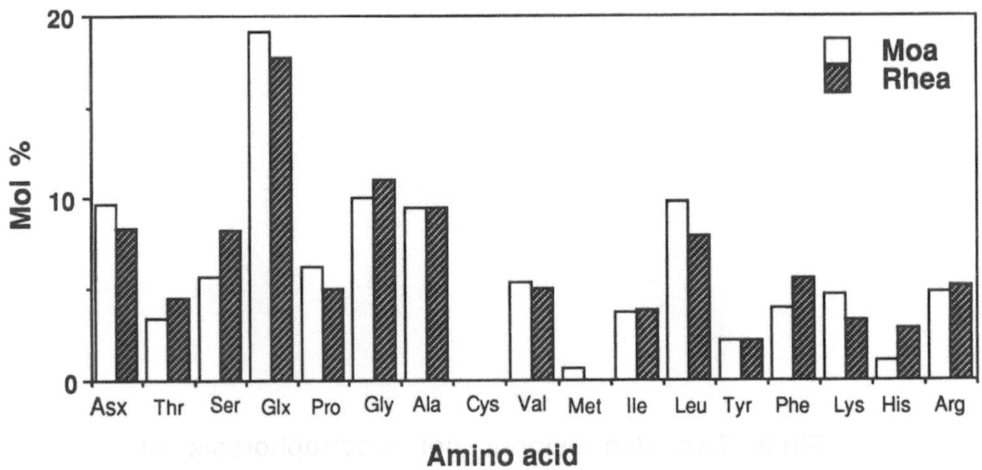

Fig.4 Diagram of the amino acid compositions of the soluble organic matrix in the egg shells of Moa and Rhea.

experiments. Thus, the electrophoretic profiles of the matrix protein of the moa were apparently different from that of the rhea. However, it was found clearly that the amino acid compositions of the egg shell matrix are quite similar in both birds. If we combine the results of the electrophoretic experiments with that of the amino acid analysis, the present phenomenon can be explained as follows: "fresh" egg shells of the moa may have the acidic protein just like that of the rhea and "erosion" of over 300 years may degrade the acidic protein into various kinds of acidic peptides. It is also possible that polymerization of small peptides may have happened. But the side chains of each amino acids in the organic matrix did not breake in the present moa egg shells. It is known that the most heat-unstable amino acids are methionine, cystein, arginine, threonine, histidine and so on [7]. Existence of methionine in the moa egg shells indicates that the present specimen has been preserved in a good condition over about several hundred years.

It is found that the main protein of the water-soluble organic matrix in the rhea egg shells is acidic with low molecular weight (18 kD). Such a low molecular protein with high percentage of glutamate has never been reported as the chief component in any of the organic matrices of hard tissues . "Amelogenins" which are the main matrix protein in the developing mammalian tooth, are relatively low molecular but contain high percentage of proline [8]. Recently, low molecular matrix protein (about 17 kD) has been separated in the molluscan shells of Pinctada fucata [9]. However, it contains a high percentage (about 20 %) of aspartate. The present type of acidic protein with low molecular weight found in the rhea egg shells may be specific in bird egg shells. This point will be the subject of future research.

Acknowledgments- We wish to thank Dr.E. Narushima, Ueno Zoo for his gift of the rhea egg shells.

REFERENCES

[1] Tyler C (1957) J Polynes Soc 66:110-130
[2] Tyler C, Simkiss K (1959) Proc Zool Soc Lond 133:201-243
[3] Meyer-Rochow VB (l982) Kosmos 78:26-31
[4] Wheeler AP, Rusenko KW, Sikes CS (1988) In: Sikes CS, Wheelar AP (eds) Chemical aspects of regulation of mineralization. Univ South Alabama, Alabama, pp.9-13.
[5] Anderson BL, Berry RW, Telser A (1983) Anal Biochem 132:365-375
[6] O'Farrell PH (1975) J Biol Chem 250:4007-4021
[7] Vallentyne JR (1964) Geochim Cosmochim Acta 28:157-188
[8] Takagi T, Suzuki M, Baba T, Minegishi K, Sasaki S (1984) Biochem Biophys Res Commun 121:592-597
[9] Samata T , in this Proceedings.

CHAPTER 1.11

Morphological Manifestations in Hypocalcemic Conditions — Their Relation to Molecular Structrual Changes of Glycosaminoglycans in Basement Membrane and Matrix

H. YAMAGUCHI and M. MORISADA

Department of Pathology, School of Medicine Keio University, Tokyo, Japan

Key words: Calcium ion, Matrix, Gycosaminoglycans, Molecular structures, Electron charges.

INTRODUCTION

Glycosaminoglycans are widely distributed throughout the animal body; they are in basement membrane (BM) which regulates the exudation across the vessel wall, and in the matrix of connective tissue, which provides the site for cellular movement and other cellular functions. In some organs, such as glomerulus or vessel, they may also maintain the organ's structures. Their functions are characterized by their high viscosity. Glycosaminoglycans in general have a lot of negative charges in one molecule. When these negative charges are freed, the molecular structure becomes elongated and loosened. On the other hand, when these negative charges are conjugated with cationic ions, their molecular structures become shrunken and condensed(1). Therefore, the molecular structure of glycosaminoglycans, namely colloid-chemical or high viscous properties, is changed by the presence of cationic materials conjugated to their negative charges. Among the many cationic materials, calcium is the most common and widely distributed bivalent salt, which is easily ionized, and is related to both the functions and structures of living animals.

In the present series of experiments, a low-calcium condition was induced by the administration of Na2EDTA. Under this condition, calcium ions, incorporated in the structures, shifted into the blood. It is possible to suggest that calcium conjugated to glycosaminoglycans, having strong ionization activities, is released into the blood more easily than calcium of the bone. Therefore, the results obtained demonstrate the remarkable morphological distortion in the various organs.

MATERIALS AND METHODS

Guinea pigs of both sexes, weighing 250-300 gm, were used for this series of experiments.

Experimental 1: Four ml of 6% Na2EDTA saline solution was administered intraperitoneally. The average serum calcium level was dropped down to 2/3, compared to normal. All animals treated in this way were suffered from tetanic shock. About 60% of the animals died immediately after administration and the remaining ones recovered within several hours. The surviving animals were sacrificed following schedule. The various organs were removed.

Experimental 2: Two weeks after Na2EDTA administration when angiolytic changes in the mesenteric arteries could be expected, 4 ml of 6% calcium lactate solution was administered intraperitoneally. One week after the above procedures, the animals were sacrificed.

The light and electron microscopic specimens were made following the routine techniques.

RESULTS

In kidney, light microscopic observation showed the cystic dilatation of the glomerular tufts with edematous loosening of the mesangial tissue. Glomerular capillary lumina were two or three times larger than those of normal ones (Fig. 1). An EM study showed the decreased density of the mesangial matrix. The remarkable laminification and reticulation, and the subepithelial protrusion of GBM was also demonstrated (Fig. 2). (2)

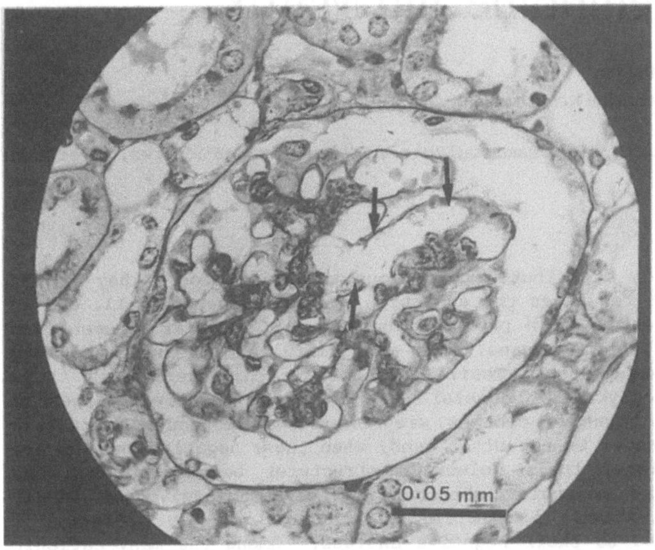

Fig.1 In kidney, mesangiolytic changes with cystic dilatation of the glomerular tuft were demonstrated.

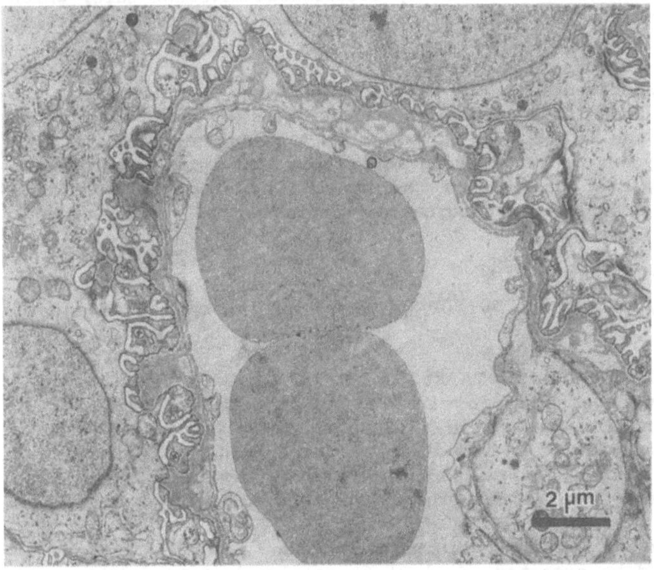

Fig.2 An EM studies showed laminification of glomerular basement membrane with edematous loosening of the mesangial matrix.

In the mesenteric arteries, with the administration of Na2EDTA, vascular wall revealed the loosening and dissociation of the muscular layer with transformation of muscle cells into a spindle shape (Fig. 3). A few leucocytic reactions were noted. Endothelial desquamation and the lifting up phenomenon were also noted (3). With EM, endothelial lift-up with fragmentation of internal elastic lamellae was

noted. In the media, intermuscular spaces were expanded remarkably with desquamation of BM from spindle-shaped muscle cells. No degenerative changes in muscle cells were seen (Fig. 4). No morphological changes were demonstrated in the vein. These vascular changes should be called "angiolysis" (4).
Administration of calcium lactate after Na2EDTA induced the aneurysmal changes in the arterial wall. With Kossa staining, in injured areas, where the vascular wall became aneurysmal in shape, were remarkably stained calcium positive (Fig. 5) (5).

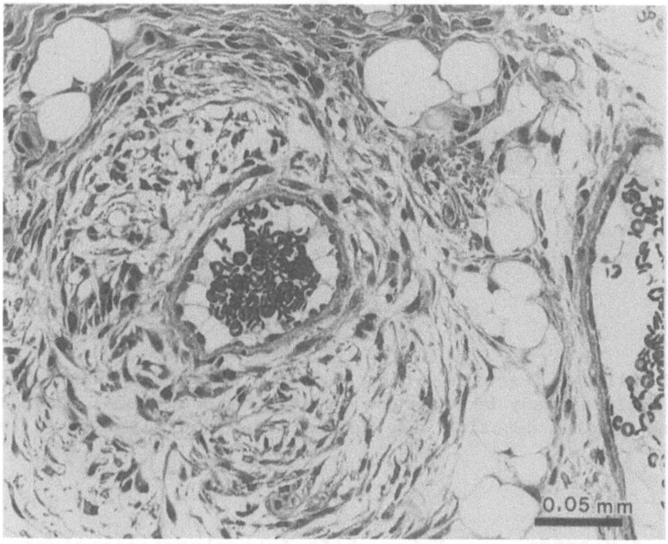

Fig.3 Five days after administration of Na2EDTA, angiolytic changes of mesenteric arteries with expansion of intermuscular spaces, accompanying by transformation of muscle cells into spindle shape were seen.

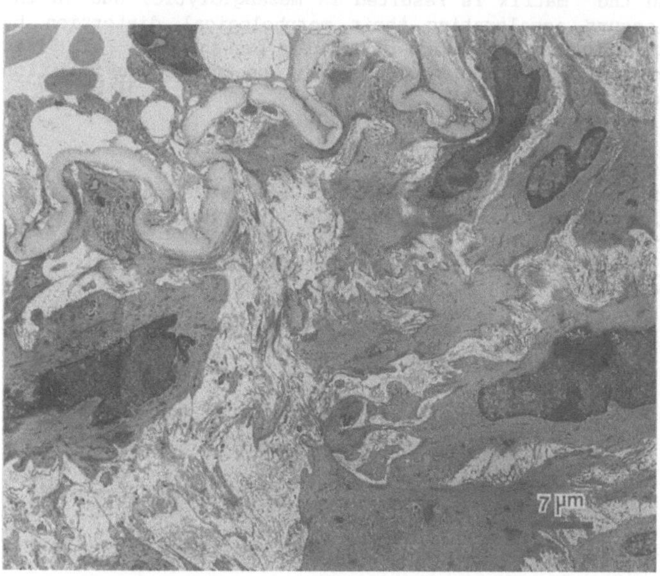

Fig.4 With EM, lift up phenomenon of the endothelial cells with fragmentation of internal elastic lamellae was noted. In the media, intermuscular spaces were expanded remarkably with desquamation of BM from spindle shaped muscle cells.

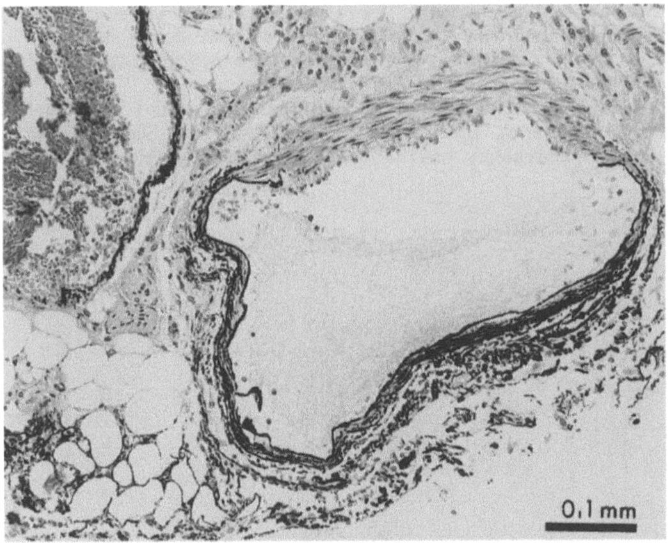

Fig.5 Administration of calcium lactate after Na2EDTA injection when the angiolytic changes would already be expected, induced the aneurysmal changes of arterial wall and calcium deposits were demonstrated strictly limited in the injurious areas.

DISCUSSION

With the administration of Na2EDTA, a calcium shift from glycosaminoglycans molecules into the blood is induced. Therefore, the negative charges of their molecules are exposed, resulting in the Sol diversion of the pasty matrix which is located intercellularly. As a result, the renal mesangium where there exist no fixing component other than the matrix is resulted in mesangiolytic, and in the vascular wall angiolysis occurs accelerating their morphological distortion by arterial vasomotion. The conspicuous deposition of calcium (as shown in Fig. 5) in the aneurysmally injured areas supports that the necessary and sufficient condition for the provocation of above phenomena.

REFERENCES

1) Pearse AGE (1960) Histochemistry: Theoretical and Applied. JA Churchill, London
2) Yamaguchi H, Takeuchi H, Tovikata C, Sakaguchi H (1978) Exp Path, Bd 15: 153-160
3) Yamaguchi H, Usui H, Tajima T (1981) Exp Path, 20:26-30
4) Yamaguchi H, Usui H, Tajima T Sakaguchi H (1981) Exp Path, 20:3-40
5) Yamaguchi H, Tajima T, Morisada M (1983) Exp Path, 23: 1987-200

CHAPTER 1.12
Enamelin and Enameloid

D. Deutsch[1], A. Palmon[1], L. Dafni[1], A. Shenkman[1], J. Sherman[1], L. Fisher[2], J.D. Termine[2], and M. Young[2]

[1]Dental Research Unit, Hebrew University-Hadassah Faculty of Dental Medicine, Jerusalem, Israel and [2]Bone Research Branch, N.I.D.R., National Institutes of Health, Bethesda, USA

Key words: Enameloid, Enamelin, ELISA, Western blot, Indirect Immunohistochemistry

INTRODUCTION

Immunological studies have indicated that enamel proteins have common antigenic determinants across a wide range of vertebrates suggesting the structure of some of these enamel proteins to be highly conserved during the 450 million years of vertebrate evolution [1]. These and other biochemical studies [2, 3, 4, 5, 6] have further shown that the acidic glycoprotein enamelins are predominant in certain aquatic vertebrates such as fishes and sharks, whereas terrestrial vertebrates enamelins are detected in lower proportion relative to the predominant amelogenins [7, 8, 9]. This finding, that enameloid mineralization occurs in the presence of enamelins but does not require the presence of amelogenins, is indicative of the important biological role of these proteins.

The elucidation of the primary structure of enamelin would provide a base line for better understanding of the role of enamelin in enamel and enameloid mineralization. Recently [10] we have screened and identified two different enamelin cDNA clones (2.8 Kb and 1.8 Kb) [14], from bovine enamel cDNA library using affinity purified antibodies prepared against bovine enamelin [11]. This antibody did not cross react with amelogenin, bovine serum albumin or dentin. The 2.8 Kb cDNA enamelin clone was sequenced, and the results revealed the deduced protein (named tuftelin), of the sequenced cDNA, to be a novel hydrophilic and acidic protein. Its composition was similar to tuft proteins (proteins remaining in mature enamel [12] some of which have recently been indicated to originate from enamelin proteins secreted in early stages of enamel formation [13]).

The identity and localization of the deduced protein was confirmed by ELISA Western-blotting, indirect immunohistochemistry and high resolution protein A-gold immunochemistry employing antibodies directed against synthetic peptides corresponding to the deduced cloned cDNA sequences [14]. The synthetic peptide antisera reacted with the purified denatured enamelin proteins and with proteins found in enamel cells during secretion and in extracellular matrix. They also reacted with the tuft proteins in mature erupted bovine and human enamel. They did not react, however, with the underlying dentin and its cells odontoblasts. Using the cloned cDNA probes and the synthetic peptide antisera, we have begun to characterize the structure and expression of enamelin in shark enameloid.

MATERIAL AND METHODS

Source of Material A series of freeze dried developing cat shark teeth of increasing age obtained from a single jaw (Fig. 1) was used.

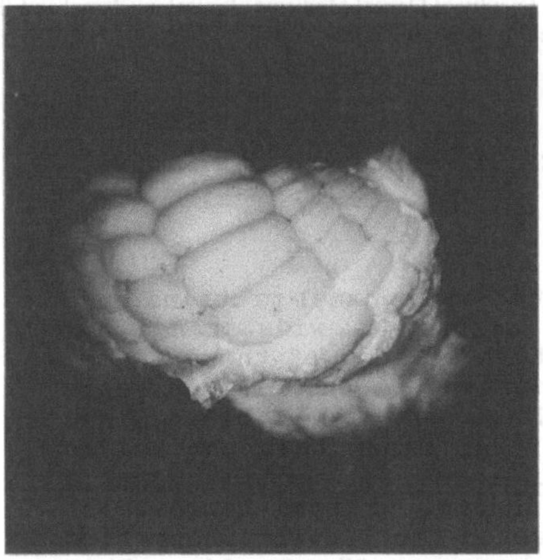

Fig. 1

ELISA Relative titers of synthetic peptide antisera (LF-73, LF-74, LF-75) against enameloid protein were determined by coating Dynatech micro-titre plates with protein extracts and proceeding with ELISA using peroxidase-conjugated second antibody.

Indirect Immunofluorescence Indirect immunofluorescence using enamelin (tuftelin) synthetic peptide antisera [14] as first antibody and fluorescene-conjugated antirabbit (FITC) as second antibody, was performed on longitudinal cryostat sections (5um thick) of non-decalcified young cat shark teeth (containing forming enameloid) and on 30-40 um thick longitudinal ground sections of erupted cat shark teeth containing mature enameloid. Dilution of 1:300 was used for the first antibody and 1:150 for the second antibody respectively. PBS was used as washing solution and 1% fetal calf serum (FCS) to block unspecific sites. The tissue specimens were incubated with the first antibody for 1 hr and with the second antibody for 7 mins. Longitudinal tooth sections treated with corresponding pre-immune sera served as a control.

Western-Blotting and Immunodetection Forming and more mature enameloid scraped from the cat shark tooth was extracted with urea in phosphate buffer at 4 C for 48 hours. The enameloid extracts were electrophoresed on SDS PAGE 10-18%, and stained with Coomassie blue. Electroblotting was according to the method of Towbin et al [15]. Nitrocellulose electrotransfers were stained with amido-black and others were processed for immunodetection using the conjugated alkaline phosphatase second antibodies. Synthetic peptide antisera [14] served as first antibodies, and the respective pre-immune sera as control.

RESULTS

The results showed that tuftelin synthetic peptide antisera reacted with the extracted denatured shark enameloid proteins, both in forming and more mature tissue (Fig. 2). They also reacted with protein both in unerupted forming enameloid (Fig. 3) and mature erupted enameloid (Fig. 4). No reaction was observed with pre-immune (control) sera.

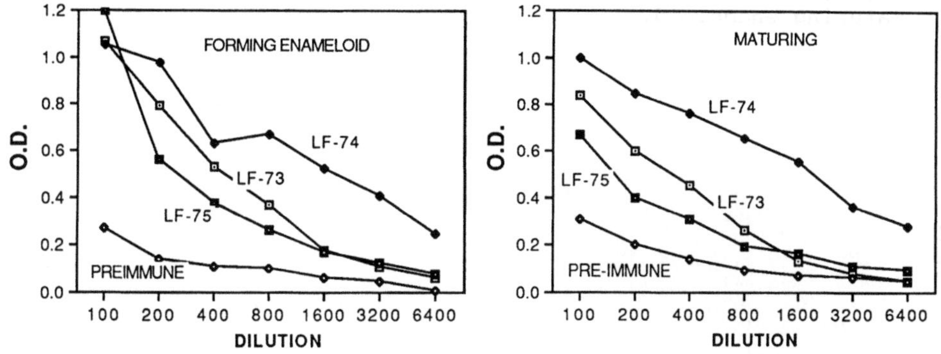

Fig. 2

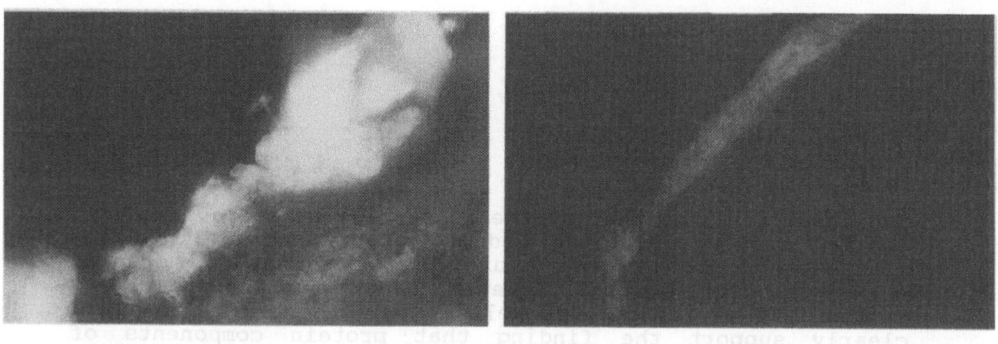

Fig. 3

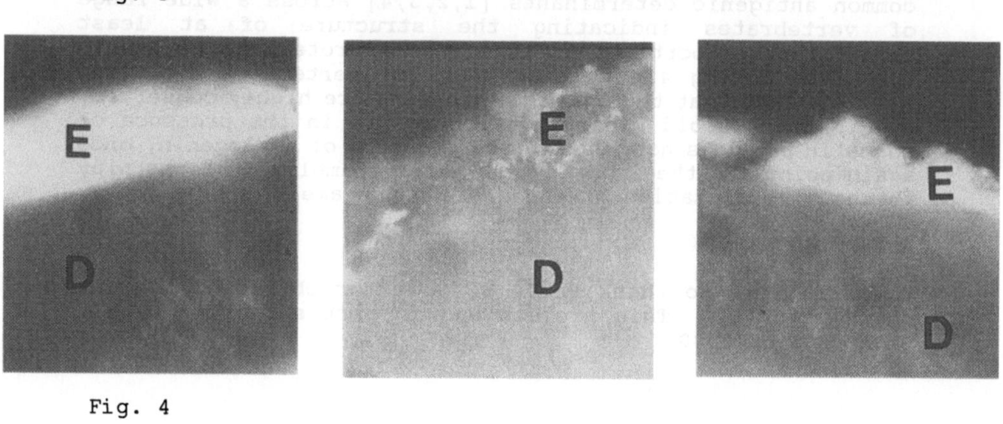

Fig. 4

Immunoblotting of forming and more maturing denatured enameloid proteins, separated by SDS PAGE, electroblotted on nitrocellulose and stained with amido-black (Fig. 5a) revealed protein bands from 10 kD to > 70 kD. The distribution and presence of proteins varied between forming and more maturing enameloid, indicating biochemical changes in extracellular matrix as the enameloid matures and mineralizes. Western-blotting using bovine enamelin synthetic peptide antisera as first antibodies (Fig. 5b) reveal a strong reaction with 66 Kd protein band both in forming and more maturing enameloid.

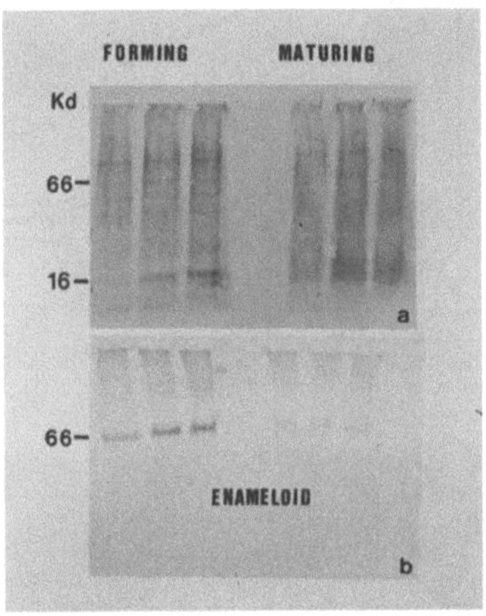

Fig. 5

DISCUSSION

The results reveal that synthetic peptide antisera, produced against sequences of the deduced protein, tuftelin [14], reacted with both forming unerupted and mature, erupted enameloid. They did not, however, react with the underlying dentin. These results, as well as our recent results [14], clearly support the finding that protein components of extracellular enamel matrix - the enamelin proteins - have common antigenic determinants [1,2,3,4] across a wide range of vertebrates indicating the structure of at least significant proportions of this enamel protein to be highly conserved during 450 million years of vertebrate evolution [1]. The fact that the enamelin proteins are highly conserved, and that enameloid mineralization occurs in the presence of enamelin but does not require the presence of amelogenin, once again points to the important role the enamelin proteins play in the mineralization of enameloid and enamel.

ACKNOWLEDGEMENT

We would like to thank Prof. S. Suga for obtaining the cat shark material. This project was in part supported by NIH Grant RO1 DE05780.

REFERENCES

1. Slavkin HC, Zeichner-David M, Sneed ML, Bringas P, Bessem C, Santos V (1984) In: Fearnhead RW, Suga S (eds) Tooth Enamel IV. Elsevier Science Publishers, Amsterdam, pp. 167-171.

2. Slavkin HC, Graham E, Zeichner-David M, and Hildemann W (1983) Evolution 37:404-412.

3. Herold R, Graver H, Christner PJ (1980) Science 207: 1357-1358.

4. Herold R, Rosenbloom J, Granovsky M (1989) Calcif Tiss Int 45:88-94.

5. Clement JG (1984) In: Fearnhead RW, Suga S (eds) Tooth Enamel IV. Elsevier Science Publishers, Amsterdam, pp. 422-426.

6. Graham EE (1985) J Exp Zool 234: 185-191.

7. Termine JD, Belcourt AB, Christner PJ, Conn KM, Nylen MU (1980) J Biol Chem 225: 9760-9768.

8. Fincham AG, Belcourt AB (1985) In: Butler WT (ed) The Chemistry and Biology of Mineralized Tissues. EBSCO Media Inc. Birmingham, Alabama, pp. 240.

9. Deutsch D (1989) Anatomical Record 224: 189-210.

10. Deutsch D, Palmon A, Fisher L, Termine JD, Young M (1989) In: Fearnhead RW (ed) Tooth Enamel V, Florence Publishers, pp. 308-312.

11. Deutsch D, Palmon A, Catalano-Sherman J, Laskov R (1987) Adv Dent Res 1(2): 282-288.

12. Robinson C, Lowe NR, Weatherell JA (1975) Archs Oral Biol 20:29-42.

13. Amizuka N, Ozawa H (1989) In: Fearnhead RW (ed) Tooth Enamel V. Florence Publishers, pp. 410-416.

14. Deutsch D, Palmon A, Fisher LW, Kolodny N, Termine JD, Young M. Sequencing of Bovine Enamelin ("Tuftelin") - A Novel Acidic Enamel Protein, Submitted.

15. Towbin H, Staehelin T, Gordon J (1979) Proc Natl Acad Sci USA 76:4350-4354.

CHAPTER 1.13

Amelogenin Degradation by an Enzyme Having Acidic pH Optimum and the Presence of Acidic Zones in Developing Bovine Enamel

S. SASAKI, T. TAKAGI, and M. SUZUKI

Department of Biochemistry, School of Dentistry, Tokyo Medical and Dental University, Tokyo, 113 Japan

INTRODUCTION

Protein content in immature enamel decreases dramatically during development. Disappearance of high-molecular weight proteins and relative increase in low molecular weight fractions are observed at the later stage of the development. This fact suggests degradation of matrix components during the process of enamel development and maturation. We purified and characterized an enzyme responsible for amelogenin degradation in developing enamel. Interestingly, the optimum pH of the enzyme was found to be in acidic pH, approximately at 6. This result suggested importance of acidic conditions in degradation process of the amelogenin matrix during enamel maturation. The authors attempted to stain developing teeth with pH indicators and found the presence of alternating acidic and neutral zones in developing enamel.

MATERIALS AND METHODS

Developing bovine enamel was homogenized with 0.05M phosphate buffered saline at pH 7.4 and extracted at 4°C. Enzyme activity was recovered in 45 to 65% ammonium sulfate precipitates and this fraction was concentrated through a Diaflow PM-10 membrane. The enzyme was purified by filtration through molecular cut membrane (PM-10), DEAE-Sephacel ion exchange and molecular sieve HPLC. Unextractable fraction from the enamel was further solubilized by 0.5M acetic acid and amelogenin and enamelin were each separated and purified through a series of chromatographic procedures and were utilized as substrates.

Enzyme activity was assayed by mixing the enzyme solution with substrate, 0.01% amelogenin or other proteins, at various pH values and incubating at 37°C. After the digestion, the mixture was precipitated with 15% ammonium sulfate and the absorbance of the supernatant at 280 nm was determined. In addition, SDS-polyacrylamide gel electrophoresis with 15 to 30% gradient gel was carried out after lyophilization of the reaction mixture.

The developing enamel, free of ameloblasts and other adhering tissues, was stained with pH indicators; methyl red, bromothymol blue and a commercial preparation of pH indicator mixture (Universal indicator, BDH Ltd.).

After staining the bovine enamel with the indicator, each acid or neutral staining zone was scraped separated with a spatula and suspended in distilled-deionized water at a concentration of 50 mg/ml. The respective pH values of the supernatants of the suspensions were measured with a glass-electrode pH meter.

Glyoxal bis(2-hydroxyanil) solution (20 mg GBHA dissolved in 2 ml 0.1 M NaOH) was used for staining for 30 to 60 seconds.

RESULTS

In gel filtration HPLC using Asahipac GS-510, amelogenin-degrading enzyme activity was eluted separately from gelatinolytic activity. The amelogenin-hydrolyzing enzyme activity was purified by ion-exchange chromatography with a DEAE-Sephacel column. The active fraction was collected and examined by 12% SDS-polyacrylamide gel electrophoresis. It was a single band at a position of about 30 KD by silver staining.
The enzyme activity after purification was found almost 1,000 times as high as the starting extract. It is clearly shown that the enzyme digested 28 KD amelogenin into two fragments after the treatment for 20 hours. However, it did not hydrolyze purified 70 KD enamelin at all.
The band of the smaller fragment was recovered from the gel and purified by HPLC with a C-18 column and its amino acid composition and sequence were determined.
The amino acid sequence of recovered peptide corresponded well to that of so-called "tyrosine-rich amelogenin peptide", TRAP, reported by Fincham et al.(1).
As acidic conditions in degradation process of the amelogenin matris were suggested, the pH values in developing teeth were measured with color indicators and the presence of acidic and neutral zones was found in developing enamel.
When the forming enamel was stained with methyl red solution, stripes of orange staining indicating acidic pH were clearly shown. When stained with bromothymol blue solution, acidic (yellow) and neutral (blue) bands were observed. The enamel was also stained with a preparation of pH indicator mixture (Universal indicator). The enamel manifested several alternative stripes of orange color corresponding to pH 5.5-6.0 and of green color pH 7.0. When incisors of Sprague-Dawley rats were stained with the universal indicator, almost the same pattern of alternating stripes of orange and green were observed.
After staining the bovine enamel with the universal indicator, each zone stained in orange or in green was scraped separately with a spatula and suspended in water.
Respective pH values of the supernatants of the suspensions after standing for 48 hours at 4°C were measured with a glass-electrode pH meter. The average pH of the orange-stained enamel sample was 6.2 and that of the green-stained 7.2, indicating the pH difference even though a little higher than when measured by staining.
A bovine tooth preparation was sagittally cut in half with a dental diamond disc. While half of the preparation was stained with GBHA solution, the other half was stained with the universal pH indicator and the staining patterns were compared. Red stripes of GBHA staining corresponded to neutral bands of green staining with the pH indicator and unstained white zones to acidic orange zones of pH staining. The pH staining in orange and green with the universal indicator solution of a bovine enamel section sliced with a diamond disc was found not only on the forming surface but also in depth and in some areas throughout the whole enamel thickness.

DISCUSSION

The purified enzyme from the developing bovine enamel digested 28 KD amelogenin into two fragments and the smaller one was identified to be TRAP from its amino acid composition and sequence. From this result, it is suggested that this enzyme has an ability to hydrolyze amelogenin molecule at the position

between 45th tryptophan and 46th leucine, producing TRAP. The most important finding on this enzyme was its acidic pH optimum, between 5.5 and 6.0. This result suggested importance of acidic conditions in maturation enamel.
We discovered in the first time the banded staining pattern of the enamel in terms of differences in pH; the maturation enamel consisting of alternating striations of acidic and neutral zones. The pH values of the enamel samples scraped from respective bands were confirmed by measuring them in suspension with a pH-meter. Neutral bands corresponded to red stripes of glyoxal bis(2-hydroxyanil) staining and acidic bands to unstained ones, suggesting a correlation of the acidic and neutral stripes with the zones of ruffle-ended and smooth-ended ameloblasts (2, 3). Mechanisms and physiological significance of the cyclical change in pH in the developing enamel is most interesting. During the growth of inorganic phase in the enamel, eight H^+ ions are released for the formation of every unit cell of hydroxyapatite crystal (4). Removal of the H^+ ions released by the formation of hydroxyapatite is a basic requirement for the progress of mineralization (4). Continuous or frequent neutralization by adding alkali into the reaction mixture to eliminate excess H^+ ions is usually conducted in an _in vitro_ hydroxyapatite synthesis system (5, 6).
In the case of enamel mineralization _in vivo_, we propose a theory of the presence of a periodic physiological neutralization mechanism by which alternative acidic and neutral zones are modulated in relation to ruffle-ended and smooth-ended ameloblasts.
The importance of acidic conditions in order to activate the enzyme for amelogenin matrix degradation during enamel maturation is confirmed in this study.
It has been believed that biological calcification during tooth formation proceeds under neutral conditions. Discovery of repeated acidic environments in the forming enamel by the present study may compel a critical revision of current theories on the mechanism of biological mineralization.

REFERENCES

1. Fincham A G, Belcourt A B, Termine J D, Butler W T, Cothran W C (1981) Biosci Rep **1**: 771-778
2. Takano Y, Crenshaw M A, Bawden J W, Hammarstrom L, Lindskog S (1983) J Dent Res **61**: 1580-1586
3. McKee M D, Martin J R, Landis W J (1989) J Dent Res **68**: 101-106
4. Samachson J (1968, 1969) Nature **218**: 1262-1263, **221**: 1247-1248
5. Nancollas G H, Mohan M S (1970) Archs Oral Biol **15**: 731-745
6. Eanes E D, Meyer J L (1977) Calcif Tissue Res **23**: 259-269

CHAPTER 1.14

On the Physical Role of Amelogenin in the Biomineralisation of Tooth Enamel

R.W. FEARNHEAD and K. KAWASAKI
Tsurumi School of Dental Medicine, Yokohama, 230 Japan

INTRODUCTION.

Vertebrate tooth enamel and enameloid is a secretory product of ameloblast cells which are of epithelial origin. These cells first produce an organic matrix (amelogenin) on the outermost surface of the dentine, in intimate association with the mineral and collagen fibres which are oriented perpendicular to that surface. Growth of hydroxyapatite crystals into the newly secreted organic matrix from this collagenous boundary quickly follows, thus establishing the position of the enamel/dentine junction.

The principal obstacle to a better understanding of the role of amelogenin is the lack of knowledge about it`s physico-chemical properties. Some data however is available for example, according to the description by Nikiforuk and Simmonds in 1965 (1) and later by Nikiforuk and Sognnaes in 1966 (2), they noticed that when developing bovine enamel protein was solubilised in EDTA or dilute formic acid pH. 6.8- 7.0, it changes it`s state from a clear gel at 4°C., to an opalescent solution at a temperature of 18°C this change of state being reversible. Many biochemists have also made similar observations referring to the phenomenon as a precipitation or aggregation during the preparation of amelogenin samples, although giving no further details. Optical activity, temperature, and concentration sensitivity is one of the characteristics of the "so-called liquid crystal mesophase states" which so far as the author is aware is a physico-chemical phenomenon that up to the present has not been considered in relation to amelogenin although it is a property which could possibly explain the enigmatic behaviour of amelogenin during amelogenesis.

Many compounds and mixtures have been found to have the property of entering these `ordered-liquid` states and it would not be an exaggeration to say that it is rapidly becoming apparent that liquid state mesophases may be involved at some point in most biological processes under cellular control, since it provides the cell with a rapid and sensitive reversible mechanism with which to respond to local environmental changes. Transitions from the isotropic liquid, through ordered liquid mesophase states, fall into two main types, (nematic and smectic). Since a large number of other physico-chemical processes have already been considered in relation to the role that amelogenin may have in the secretory phase of amelogenesis without success, it was decided that it would be worthwhile trying to find out whether amelogenin might exhibit self orienting properties characteristic of liquid crystal mesophases.

MATERIALS and METHODS.

One 150 mg. sample of the total enamel protein extracted from pooled developing pig molars, and one 50 mg. sample of an extract separated by electrophoresis and characterised as a single band with a molecular weight of 21 kDa were generously given to us by Dr.'s Shimizu and Fukae. Both samples were readily soluble in distilled water, and were studied in this state progressively as they evaporated to the dry state using polarised light microscopy, x-ray diffraction and electron microscopy. Small vials made by cutting an open ended cavity in a sheet of 0.5mm silicone rubber sandwiched between two glass microscope slides were prepared and a small aliquot (0.05cc) of each solution was then placed in separate vials and allowed to evaporate at room temperature (22° - 23°C) and examined daily for possible change in optical behaviour.

RESULTS.

In the case of the total protein extract a change to a positive birefringence occurred abruptly after ten days when the material was then already in a gel like state Fig 1. The 21 kDa sample was started a day later and was placed in a vial of different shape having less surface exposed to evaporation, the room temperature had risen slightly to between 23° - 24°C, the first change occurred after seven days by the appearance of a small opaque mass on the side of the vial Fig. 2a & b, which grew by the extension of tiny finger-like rows of small globules of material very rapidly indeed so that within two minutes the whole of the contents of the vial was filled with the opaque mass which was isotropic. This change occurred in the evening when the room temperature had risen to 25.5°C. On cooling to 24.5°C the new phase completely cleared within one minute. The temperature was then raised and lowered many times within a range of 24° - 40°C the reversible transition occurring between 25.5° and 26°C. In an attempt to stop or at least slow the evaporation a small amount of liquid paraffin was then layered over the sample. Following the application of the liquid paraffin the rows of tiny globules became joined into threads radiating from the main mass and directed towards the interface, some threads developed side branches Fig. 3. When the temperature was lowered below 25°C again the whole sample became a clear isotropic liquid once more, quickly returning to the opaque mass with threads when the temperature was raised above 25.5°C. The temperature was then raised to 37°C and maintained at this temperature for several weeks. During this period the opaque mass developed many large globules some of which were weakly birefringent, and some of the threads separated into short birefringent rods, meanwhile extremely small crystal nuclei appeared in association with both the threads and some of the globules Figs. 4a & b. After two months the liquid adjacent to some of the threads developed sheet-like birefringent planes Figs. 5a & b. At this stage the whole contents of the vial were in a viscous fluid state.

The first total enamel protein sample Fig. 1, however, developed steadily into a birefringent gel without exhibiting the globular and thread-like stages. This sample was not covered by liquid paraffin, and the vial had a large mouth, thus evaporation was faster than from the 21 kDa sample. X-ray diffraction from this total enamel protein gel, revealed a broad diffuse ring with a dimension of 4.4 Å closely similar to the 4.48Å obtained from developing human foetal, and adult "tuft" enamel (Fearnhead 1964)[(3)], and foetal goat enamel (unpublished). When this experiment was repeated using a flask of a different design which resulted in a slower evapouration rate, thread-like Schlieren planes developed as the solution became more viscous Fig. 6 but the globular mesophase did not develop at all. However when a small drop of the second sample of the total enamel protein was evaporated to dryness onto glass or on a carbon coated E. M. grid, a specimen consisting of an amorphous phase and two crystalline species were obtained, the latter having a morphology similar to those found developing in the 21 kDa sample, Fig.7a, b.

CONCLUSIONS.

It can be deduced from these observations :

1) That the 21 kDa porcine amelogenin sample under the conditions imposed by the experiment, is capable of undergoing temperature and concentration dependent mesophase transitions most probably of a nematic type.

2) That the changes after covering the 21 kDa sample with liquid paraffin may well be related to a lipophilic property of this fraction. This is of particular importance since the extraction procedures used for electrophoresis studies most probably remove the various lipids which are known to be present in developing enamel (Shapiro, Wuthier and Irving 1966) [(4)], and their role in amelogenesis still remains unsolved.

3) That the total enamel protein complex is also capable of developing mesophase transition states although in the present experiments the mesophases did not produce such dramatic birefringent phases in the absence of the hydrocarbons which were provided in the 21kDa experiment by covering the solution with liquid paraffin.

ACKNOWLEDGEMENTS

We wish to thank Y. Shimanuki and S. Nagasaka for their generous help, and M. Shimizu, M. Fukae and T. Tanabe for the supply of protein samples.

REFERENCES.

1) Nikiforuk G, Simmonds N.S. (1965) No. 441 Abstr. p. 148, Int. Assoc. Dent .Res.
2) Nikiforuk G., Sognaes R.F. (1966) Clin. Orthopaed. 47: 229 - 248.
3) Fearnhead R.W. (1964) Tooth Enamel John Wright & Sons Ltd. Bristol.
4) Shapiro I. M., Wuthier R.E., Irving J.T. (1966) Arch. oral Biol. 11: 501 - 512.

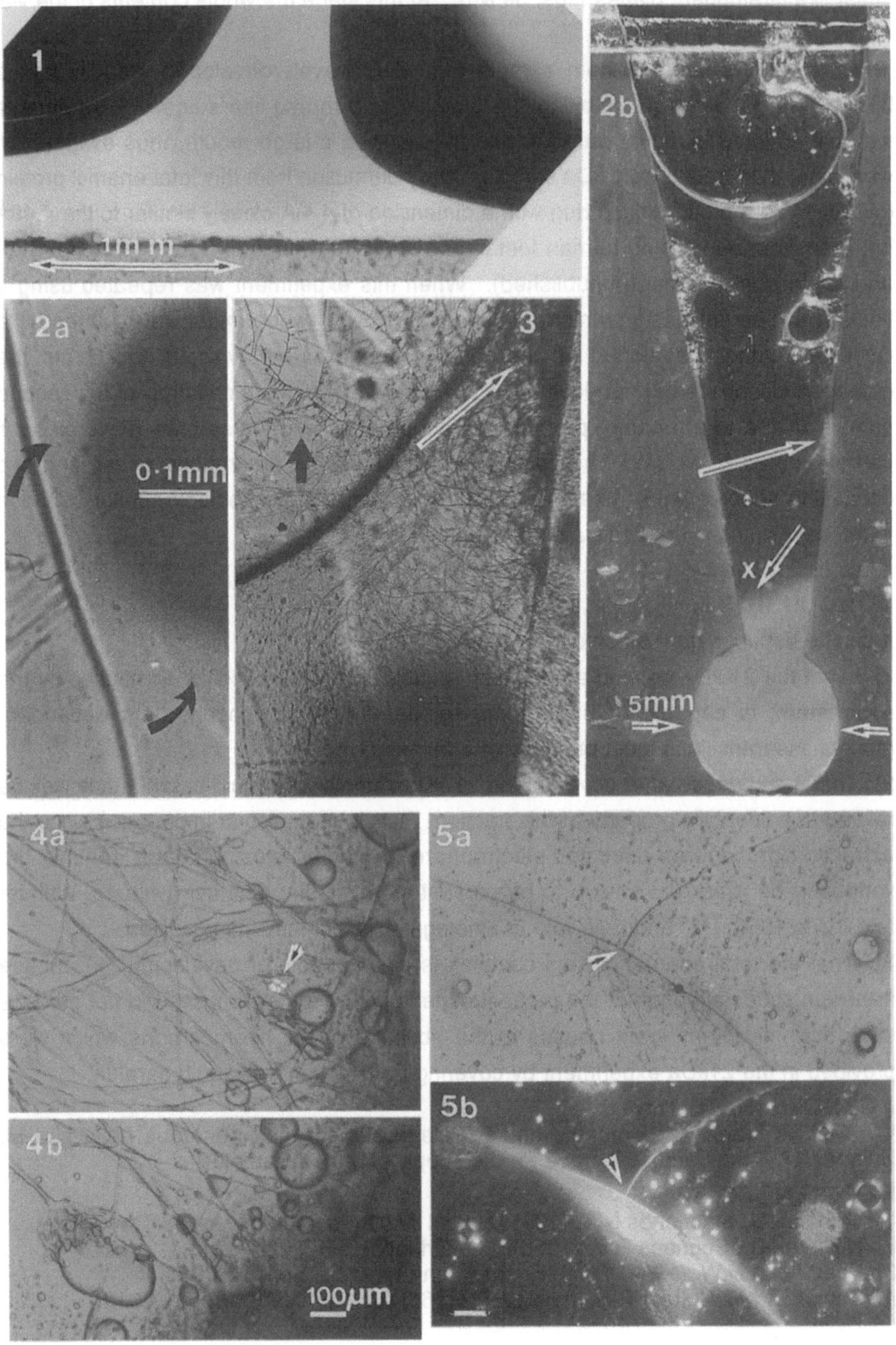
1
1mm
2a
0·1mm
3
2b
x
5mm
4a
4b
100μm
5a
5b

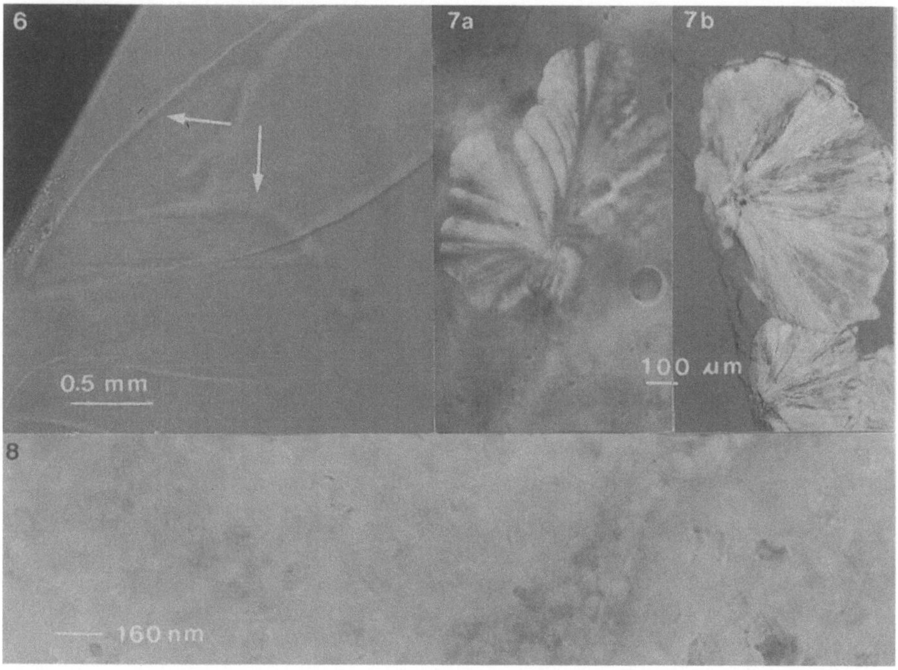

Figure Legends.

1) Total enamel protein in the birefringent gel stage, polarised light.

2a) The 21 kDa sample at 25.5°C, showing the opaque mass with streams of tiny globules spreading into the clear solution.

2b) The small vial formed between two glass slides separated by a 0.5mm sheet of silicone rubber, containing the 21 kDa sample covered with liquid paraffin. The opaque region arrowed corresponds with the field illustrated in Fig. 3.

3) The junction between the 21 kDa sample and the liquid paraffin showing the branched nematic fibre-like transition.

4a&b) A later stage of the 21 kDa sample from the region labelled (x) in Fig. 2b note tubular-like extensions into the isotropic part of the sample, and the increase in size of the globular structures. Also a small birefringent `battonette` arrowed has developed in relation to one of the `fibres`. Polarised light with gypsum plate at 45°.

5a&b) A later stage than Fig. 4 showing in (a) with the 45° gypsum plate the nematic threads, and in (b) with the plate removed, the sheet-like birefrigent planes.

6) The second sample of total enamel protein showing the development of a nematic threaded Schlieren texture.

7a) Fan shaped crystal which developed as the 21 kDa sample became very viscous.

7b) Similar fan shaped crystals which formed in the second total enamel protein sample when evaporated to dryness.

8) A thin film of the 2nd. total enamel sample formed on a carbon support film by evaporation from the solution, (JEM-200 CX.,200 kV -158°C). The polycrystalline sheet appears to have an hexagonal habit, and where two crystals overlap moiré patterns can be detected.

CHAPTER 1.15
Noncollagenous Proteins of Sea Bream Dentine

M. FUKAE, T. TANABE, and K. KAWASAKI[1]
Departments of Biochemistry and Anatomy[1], School of Dental Medicine, Tsurumi University, Yokohama, Japan

Key words: Noncollagenous proteins, Proteinases, Sea Bream dentine, Porcine dentine.

INTRODUCTION

The most interesting noncollagenous proteins for mineralization in mammalian dentine are at present thought to be the highly phosphorylated protein, phosphophoryns which may be involved in regulating the ordered deposition of apatite crystals[1]. Lussi et al[2] have shown that phosphophoryn induce mineralization in an 'in vitro' system using a phosphophoryn agarose complex. However, Takagi and Sasaki have reported that phosphophoryn is absent in the dentine of dentinogenesis imperfecta type II[3], thus the function of phosphophoryn is still at present obscure.

The bulk of organic matrix of fish dentine consists of collagen the same as that of mammalian dentine. However, there is little information about the noncollagenous proteins in fish dentine.

Our report is concerned with the acidic noncollagenous proteins and proteinase of Sea Bream and porcine dentines, to make a comparison between fish and mammalian dentine.

MATERIALS AND METHODS

Tissue Preparation and Extraction Sea Bream teeth were obtained from the living fish and porcine teeth from the permanent tooth germs of 6 month old pig's mandibles obtained from an abattoir. After the surrounding tissues were removed using a dental burr, both dentines (1g) were crushed in liquid nitrogen and then washed out with 4M guanidine HCl solution according to the method of Termine et al[4] to get the mineralized phase. Both mineralized phases were demineralized by stirring at 4°C in 100ml of 0.5M acetic acid for 6 days exchanging with a new solution after centrifugation every 2 days. The supernatants were recovered as an acetic acid extract(A-extract). The residues were extracted by stirring in 50ml of 4M guanidine HCl, 50mM Tris, pH 7.4 for one day. The extractants of both dentines (G-extract) were concentrated by ultrafiltration with YM-5 membrane(Amicon), equilibrated and recovered with 2.5ml of 50mM Tris, pH 7.4 and stored in a freezer(-25°C).

Chromatography Anion exchange high performance liquid chromatography(HPLC) were carried out on a column (7mmIDx 5cm) of TSK-GEL DEAE-5PW(TOSOH). The column was equilibrated with 6M urea, 30mM Tris, pH 7.0 at room temp. and eluted with a starting buffer for 10 min, followed with a 0-1.2 M NaCl linear gradient. The flow rate was 0.4ml/min and the gradient rate was 1%/min. The eluted protein was detected at 280nm. Samples for chromatography were equilibrated with a starting buffer using a YM-5 membrane.

Detection of Proteinase Activity Proteinase activities were detected by substrate-gel electrophoresis using gelatin according to the method of Heussen and Dowdle[5]. Samples for this method were equilibrated with 1% SDS containing 1% sucrose. After electrophoresis, the gel was shaken in 2.5% Triton X-100 for 1h and then incubated in 0.02% NaN_3, 50mM Tris, pH 7.5 for 16 h at 37°C. In some cases, 1mM Ca was added in the incubation buffer. The gels were stained with Coomassie Brilliant blue(CBB). The gelatinase activities were detected as unstained bands.

Incubation of G-extract Both dentine G-extracts were incubated with or without 1mM Ca at 37°C for various intervals. The reaction was terminated by adding an equal volume of 2% SDS, 2mM EDTA, 50% glycerin, 20mM Tris, pH 8.0 and analyzed on SDS electrophoreses [6]. Gels were stained with CBB and / or Stains-all.

Histological Study Sea Breams were killed by decapitation and the jaws were dissected from the head and immersed in 10% neutral formol saline. Ground longitudinal sections approx. 80μm thick were stained with haematoxylin and examined in light microscopy.

Analytical Procedure The samples were hydrolyzed with 6N HCl in an evacuated sealed tube at 110°C for 24 h. Amino acid analyses were performed using a JEOL JLC 300 automatic amino acid analyzer. In some cases, after electrophoreses, the proteins were recovered by the improved electrodyalysis system from the gels which contained the Stains-all positive bands and then hydrolyzed[7]. The protein contents were calculated using the results of amino acid analyses. The quantification of calcium and phosphate were carried out by colorimetric methods[8,9].

RESULTS

It was observed on the histological sections that the fine structure of Sea Bream dentine was similar that of porcine dentine(Fig. 1).

Chemical compositions of both dentines were almost similar(Sea Bream dentine: 17.1% water, 23.0% Ca, 35.6% PO_4 and 18.1% protein; porcine dentine: 15.6% water, 21.9% Ca, 34.2% PO_4 and 19.4% protein). A similar amount of proteins(0.6% per total weight) were solubilized from both dentines by washing with 4M guanidine solution. The extractable protein obtained from the Sea Bream and porcine mineralized phases during and after demineralization were approx. 0.8% and 1.6% per total weight respectively.

On SDS electrophoresis, it was observed that phosphophoryns(Mr=86,000 and 92,000) which were not stained with CBB, but stained blue with Stains-all, were found only in the G-extract of porcine dentine(Fig. 2). The Stains-all positive bands(Mr=66,000 and 147,000) were also found in the G-extract of Sea Bream dentine and these were eluted at a more acidic position on DEAE 5PW ion exchange HPLC than that of the proteoglycan of porcine dentine(Fig. 3). Their amino acid compositions were different from those of phosphophoryn and proteoglycan of porcine dentine(Table 1). These 66 and 147kD proteins were thought to be acidic glycoproteins, since these were stained blue with Stains-all. However it is not yet sure whether or not these are really glycoproteins, because uronic acid was detected in the fractions which were extracted from the electrophoresed gels containing the Stains-all positive 66 and 147kD bands.

On searching by substrate-gel electrophoresis in the porcine dentine G-extract, distinct proteinase activities which were activated by Ca ions, were detected at the molecular weight of 56, 61 and 117kD, although weak and different molecular weight activities were also found in the incubation without Ca ions(Fig. 4). They were completely inhibited by 10mM EDTA and 2mM 1,10 phenanthroline(data not shown). On the other hand, the main band of Sea Bream dentine proteinases migrated with a molecular weight of 97kD(Fig. 4). It was partially inhibited by 10mM EDTA, but not by 2mM 1,10 phenanthroline(data not shown). Both dentine proteinases were active over the pH range 6-9 against gelatin as a substrate.

When the G-extract of porcine dentine was incubated, as it was, after adding 1mM Ca, the degradation of some noncollagenous proteins except phosphophoryn was observed during the increase of the reaction time(Fig. 5-b), but this degradation was not observed in the incubation samples without Ca ions. In the case of Sea Bream dentine G-extract, although Ca ions were not added in the incubation buffer, the noncollagenous proteins which stained with CBB were degraded acutely, while the Stains-all positive broad 66kD protein appeared to be unchanged during incubation(Fig. 5-a).

DISCUSSION

The fine structure and the degree of mineralization of Sea Bream dentine are similar to those of porcine dentine. However phosphophoryn which may be related to the induction of mineralization in mammals, was not found in Sea Bream dentine. It is possible therefore that the inductive mineralization mechanism in the Sea Bream uses a different protein. Another possibility is that it is not possible to discover the protein related to the induction of mineralization in the matrix of the already mineralized dentine, since it may have been degraded at an early stage of development by the proteinases which exist in the mineralized matrix of Sea Bream and porcine dentines.

REFERENCES

1. Veis A(1984) In:Butler WT(ed) The Chemistry and Biology of Mineralized Tissues. Ebsco Media, Inc., Birmingham, Alabama, pp. 170-182.
2. Lussi A, Crenshaw MA, Linde A (1988) Archs Oral Biol 33: 685-691
3. Takagi Y, Sasaki S (1986) J Oral Pathol 15: 463-467
4. Termine JD, Belcourt AB, Miyamoto MS, Conn KM (1980) J Biol Chem 255: 9769-9772
5. Heussen C, Dowdle EB(1980) Anal Biochem 10: 196-202
6. Laemmli UK (1970) Nature 227: 680-685

7. Fukae M, Tanabe T (1987) Calcif Tissue Int 40: 286-293
8. Connerty HT, Briggs AR (1966) J Clin Pathol 45: 290-296
9. Drewes PA (1972) Clin Chim Acta 39: 81-88

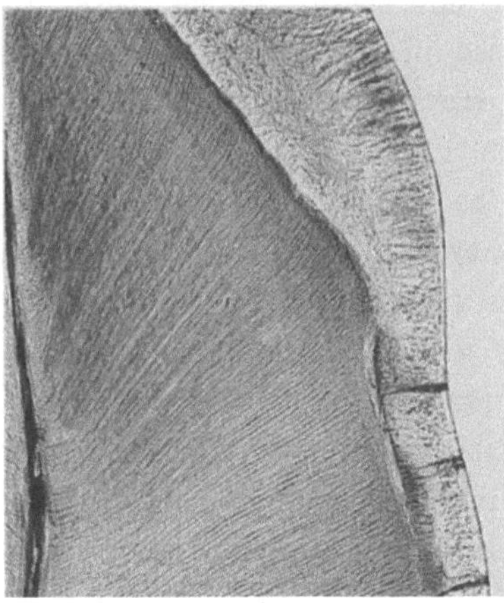

Fig. 1. Longitudinal ground section through the distal surface of a functinal Sea Bream tooth. The tubular structure in the Sea Bream dentine is similar to mammalian dentine. x50

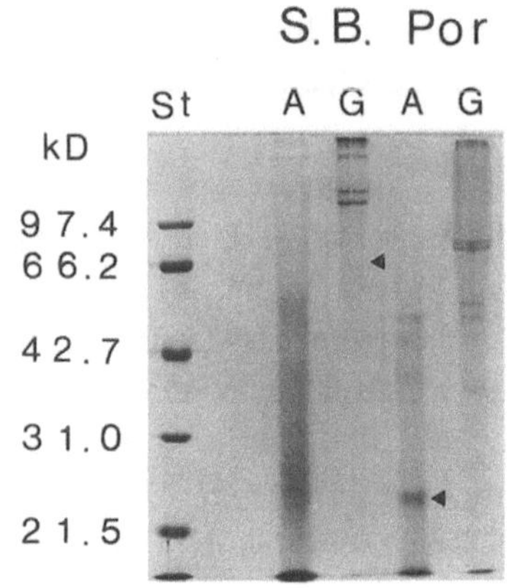

Fig. 2. SDS electrophoresis of A-extract(A) and G-extract(G) of Sea Bream and porcine dentines. S.B., Sea Bream; Por, porcine. Marks show the Stains-all positive band(◄◄, phosphophoryn).

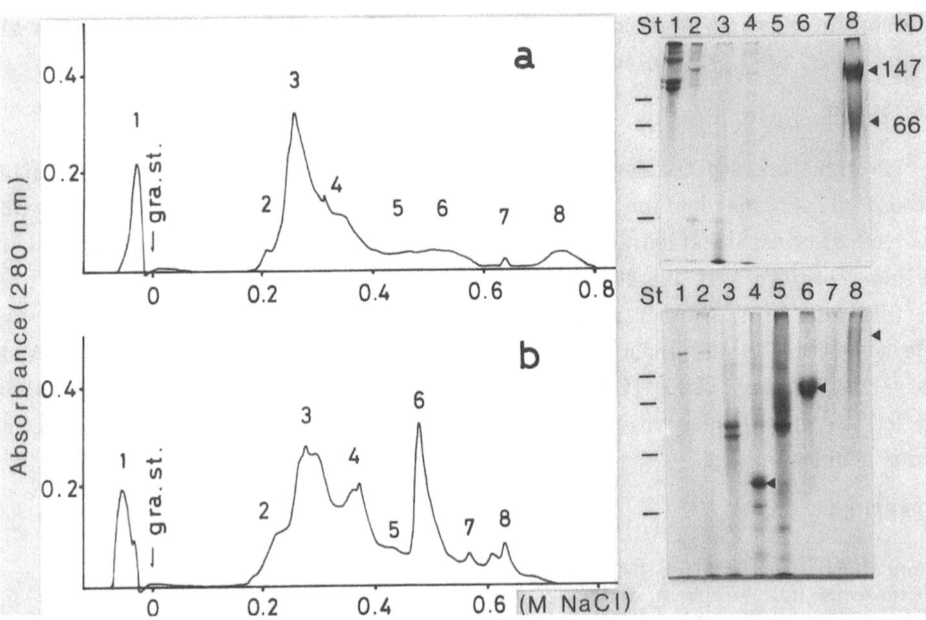

Fig. 3. Elution profiles of Sea Bream(a) and porcine(b) dentine G-extracts on DEAE 5PW ion exchange HPLC. Each fraction indicated by a number was analyzed on SDS electrophoresis. Stains-all positive bands are shown by marks. St, Bio-Rad LMW protein standard(from top to bottom 97.4, 66.2, 42.7 and 31.0 kD).

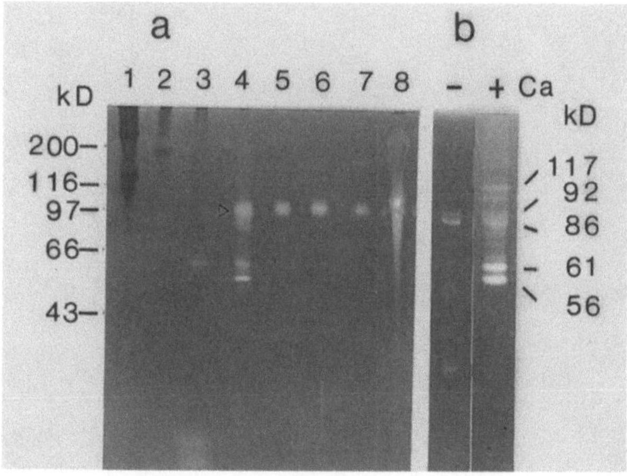

Fig. 4. Substrate-gel electrophoresis of Sea Bream and porcine dentine G-extracts. Each fraction of Sea Bream dentine G-extract indicated by numbers in Fig. 3-a was examined and their incubation buffer contained no Ca ions(a). In the case of porcine dentine, the incubation was carried out with or without 1mM Ca (b). Mark shows 97kD. Although the unstained band shows usually protainase activity, since 86 and 92kD unstained bands found in b were stained blue with Stains-all and equal to the migration position of phosphophoryn, these unstained bands were thought to be phosphophoryn, but not proteinase activity. The unstained band found in the column of a-8 were also not proteinase activity, since these were stained blue with Stains-all.

Table 1. Amino acid compositions of phos phophoryn(PP) and proteoglycan (PG) of porcine dentine shown in Fig. 3-b and the Stains-all positive 147 and 66kD bands of Sea Bream dentine (Fig. 3-a-8).

	PP	PG	147kD	66kD
Asp	359	166	119	93
Thr	12	45	40	86
Ser	406	100	62	127
Glu	59	162	116	180
Pro	9	41	63	74
Gly	40	240	97	118
Ala	19	57	68	63
Cys	2	3	6	-
Val	6	26	55	55
Met	1	2	16	7
Ile	3	24	43	32
Leu	7	25	105	55
Tyr	4	7	30	11
Phe	3	10	42	22
His	8	25	32	23
Lys	54	36	51	30
Arg	9	32	56	24
Hex	-	+	+	+

Hex, Hexosaamine.

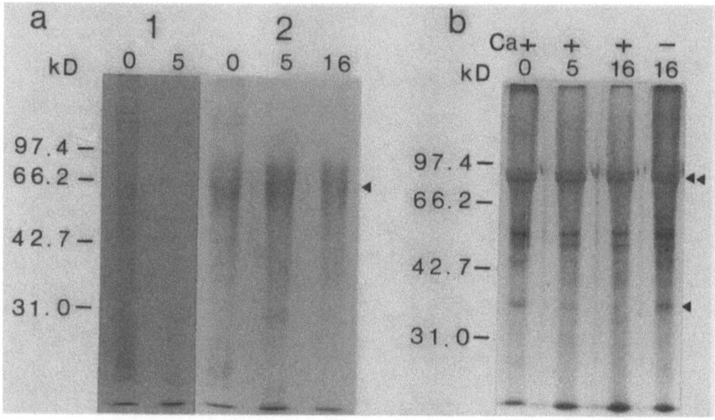

Fig. 5. The changes of components in Sea Bream and porcine dentine G-extracts during incubation. The incubation of Sea Bream dentine was carried out without Ca ions(a), and the incubation of porcine dentine with 1mM Ca or without Ca(b). a-1, CBB staining; a-2 and b, Stains-all staining after CBB staining. 0: 0 h, 5: 5 h, 16: 16 h. Marks show the Stains-all positive band(◂◂, phosphophoryn).

ACKNOWLEDGMENT

We are grateful to Prof. R. W. Fearnhead, Department of Anatomy, Tsurumi University, for helpful comments.

CHAPTER 1.16

Immunoblotting Study of Developing Enamel Matrix Proteins

M. KAKEI, H. NAKAHARA, and H. TAKEYAMA[1]

Departments of Oral Anatomy and [1]Orthodontics, Meikai University School of Dentistry, Saitama, Japan

Key words: Enamel matrix protein, Organic envelope, Crystal growth, Calcifying hard tissues, Salivary gland.

INTRODUCTION

It is generally thought that the calcification mechanism of epidermally derived enamel is strikingly different from that of dentin and bone, both of which are of mesenchymal origin. One of the reasons for this is that developing enamel contains specific matrix proteins, namely, "Amelogenins and Enamelins" [1,2]. With regard to enamel matrix proteins, we have demonstrated the presence of some common functional proteins such as carbonic anhydrase (CA), troponin subunit, and calmodulin in the extracellular matrix of developing enamel of the rat incisor by means of an immunological technique[3-5]. These findings gave us the idea that other matrix proteins of developing enamel, which have not yet been identified, may not be specific to enamel. Therefore, using polyclonal antibody against 25K enamel matrix protein, we conducted experiments to demonstrate the presence of immunoreactive protein in other calcifying and non-calcifying tissues. If the immunoreactive proteins are really present in other calcifying and non-calcifying tissues, we can assume that the calcification mechanism among vertebrate hard tissues is controlled by a similar basic mechanism. The present work is the initial step for an understanding of the exact functions of these substances.

MATERIALS AND METHODS

Tissue preparation for immunological detection: Developing enamel, dentin, and bone were obtained from Sprague-Dawley rats. Dentin material was dissected from the lingual side of lower incisor to avoid contamination of enamel tissue. Calvaria was treated with collagenase digestion; then the precipitated materials, rich in crystals, were used for analysis. These were briefly treated with a SDS sample buffer containing protease inhibitors, then subjected to SDS-PAGE on 12% slab gel. Non-calcifying tissues examined in this study were also prepared in the same manner.

Preparation of polyclonal antibody: Enamel matrix protein used for antibody production was prepared from the developing enamel of rat incisors. After two-dimensional electrophoresis, one principal enamel protein, showing a pI value of 6.6 and a molecular weight of approximately 25K daltons, was collected. The protein was eluted from the gel, dialyzed against distilled water, freeze-dried, and used as the immunogen. New Zealand white rabbits were injected subcutaneously with 300μg of the immunogen emulsified with Freund's complete adjuvant. Rabbits were bled and immunogloblin G (IgG) was fractionated from whole antiserum by ammonium sulfate precipitation and protein A-Sepharose chromatography [6].

Electrophoresis and immunological detection of protein on nitrocellulose filter: SDS-polyacrylamide gel electrophoresis was carried out on 12% gels according to the method of Laemmlie[7]. Immunoblotting analysis of proteins on nitrocellulose membranes was performed according to the method of Towbin *et al.* [8]. Two-dimensional electrophoresis was carried out according to the modified method of O'Farrell[9].

Immunocytochemical procedures: Immunoelectron microscopy was performed by a double immunogold technique. Sections were treated with 1% H_2O_2 solution for 3 minutes, rinsed with PBS, and preincubated for 30 minutes in PBS containing 1% BSA to reduce the nonspecific reaction. Then, the sections were incubated with the polyclonal antibody against 25K enamel protein for one hour in a moisture chamber at room temperature. Control sections were incubated with 1% preimmune serum. Next, the sections were rinsed by immersion for 15 minutes in PBS and subsequently treated with anti rabbit-IgG-gold complex reagent. Finally they were stained with uranyl acetate and lead citrate and observed under the electron microscope.

RESULTS

Immunoblotting analysis on calcifying and non-calcifying tissues. The specificity of antibody used in this study was tested by immunoblotting. Immunoblotting showed that most of enamel matrix proteins share the same epitopes against this antibody. In other calcifying tissues, the immunoreactive proteins were detected in both matrix of dentin and bone (Fig. 1-a and -b). These positively reacting proteins were observed over the molecular weight range between 23K and 30K, and at approximately 44K daltons. Single-dimensional patterns of immunoreactive proteins detected in matrix of both dentin and bone were exactly the same as the pattern seen with enamel.

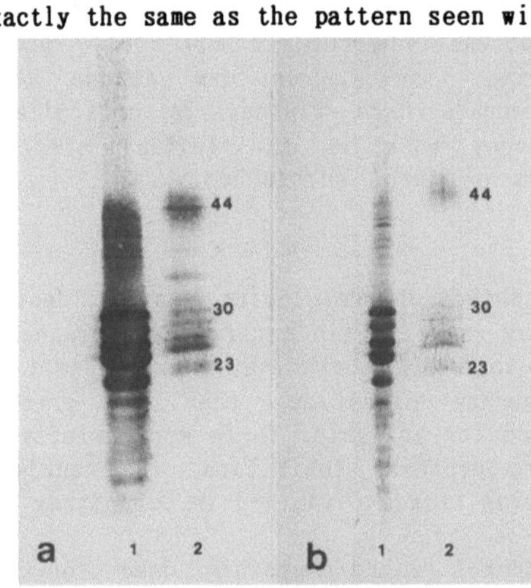

Fig. 1

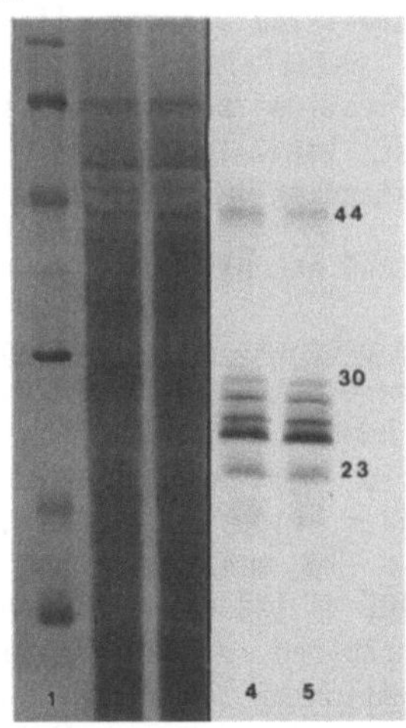

Fig. 2

Fig. 1 (a and b): Immunoblotting analysis of enamel, dentin, and bone matrix using antibody against 25K enamel protein after electrophoresis in a single dimension. Similarity of immunoreaction pattern was observed over the range of molecular sizes from 23K to 30K daltons, and at 44K daltons. (a): Developing enamel matrix (lane 1) and dentin matrix (lane 2). (b): Developing enamel (lane 1) and bone (lane 2). Fig. 2 (a and b): Electrophoretic and immunoblotting analyses of the submandibular gland. Immunoreactive proteins were demonstrated in the matrix of the submandibular gland. (a): Molecular weight markers (lane 1) and matrix proteins of the submandibular gland stained with Amido black (lanes 2 and 3). (b): Immunoblot with antibody against 25K enamel protein (lanes 4 and 5).

Immunoreactive proteins were clearly detected in the sample of submandibular gland, a non-calcifying tissue, as shown in Fig 2. The sublingulal gland also presented a similar pattern of reaction (data not shown). Single-dimensional analysis revealed that the positions of molecular weight of each reactive protein was similar to those of enamel, dentin, and bone. A 44K dalton protein occasionally exhibited a positive reaction against the antibody. When immunoreactive proteins from 23K to 30K daltons were removed from the gel and subjected to two-dimensional analysis, most of the positions of the major spots of immunoreactive proteins also remarkably coincided with those of enamel (Fig. 3). Furthermore, a small amount of proteins giving a positive reaction appeared at a relatively high molecular weight position (see arrow in Fig. 3 b-2).

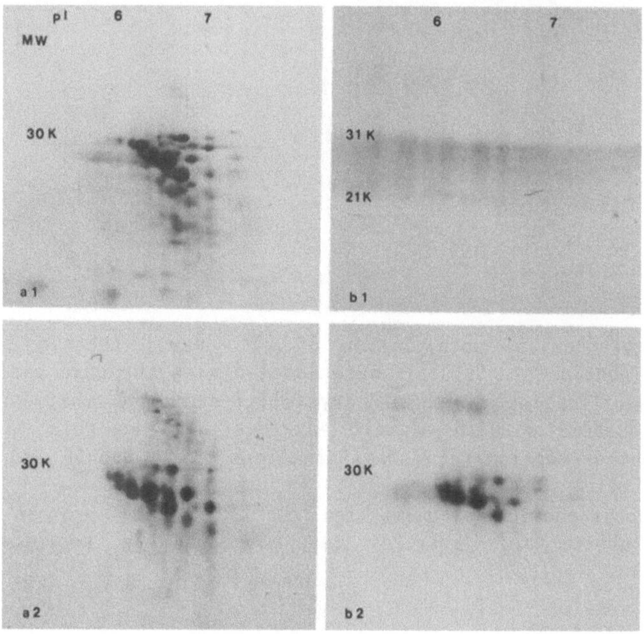

Fig. 3: Two-dimensional electrophoretic and immunoblotting analyses of the matrix proteins of developing enamel and submandibular gland. Major enamel proteins had pI values of 6-7.4 and molecular weights of between 20K and 30K. When proteins ranging from 21K to 31K daltons, were removed from the single-dimension gel and applied to two-dimensional analysis, smeared protein bands were obtained with the sabmandibular gland sample. A number of major immunoreactive proteins were detected with similar positions in the two samples. (a-1 and 2): Developing enamel sample. (b-1 and 2): Sample of submandibular gland. 1: Amido black staining. 2: Antibody staining.

Immunoelectron microscopic observations: Immunocytochemical observations revealed that immunolabeled-gold particles, indicating the presence of 25K enamel protein, were observed in the extracellular matrix of both enamel and dentin. Specific immunolabeled-gold particles were localized along the crystals in enamel and dentin (Fig. 4). Higher magnification showed that particles were closely associated with organic substances (organic envelope). In dentin, the region of the calcification front was relatively densely stained with the gold particles. Immunocytochemical study also showed the immunolabeled-gold particles to be localized exclusively in secretory granules of the salivary gland, as shown in Fig. 5. The Golgi apparatus was poorly labeled. Negligible gold particles were observed in control sections.

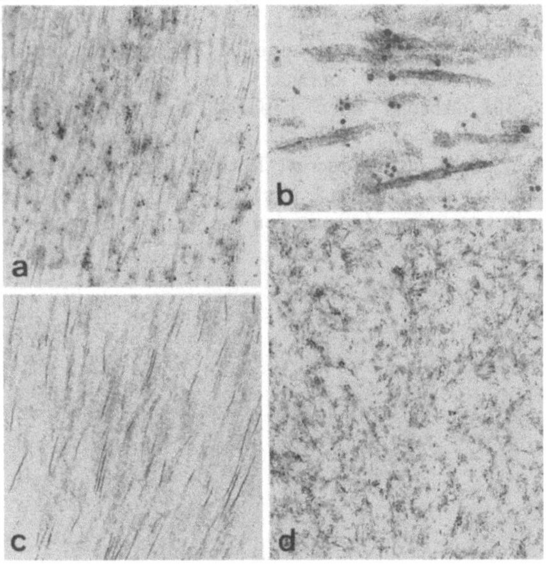

← Fig. 4

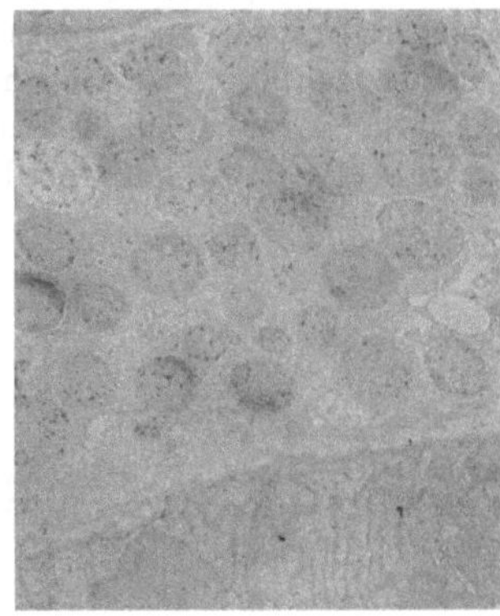

Fig. 4. Immunocytochemical localization of 25K enamel protein in developing enamel (a) and dentin (d). Crystals were labeled with the gold particles (a and d). Higher magnification of enamel crystallites showed that gold particles were closely associated with organic envelope (b). No gold particles were observed in control section (c). Magnifications (a):× 30,000 (b):× 90,000 (c): × 35,000 (d):× 6,000

Fig. 5. Immunocytochemical localization of 25K enamel protein in salivary gland. Immunolabeled-gold particles were observed in secretory granules. Magnification × 3,500

DISCUSSION

Recently, ultrastructural observations have provided some evidence to show a similarity in the basic mechanism of mineralization among calcifying hard tissues in which calcium phosphate minerals are formed. One is the presence of the central dark line of hydroxyapatite crystallite, which indicates the site of nucleation in enamel, dentin, and bone[10-12]. Another is that crystal formation always takes place within the organic envelope, which structure may control the crystal growth in all calcifying tissues[13,14]. Furthermore, we have proposed that carbonic anhydrase functions to supply carbonate ions to the crystal nucleation site at the calcification front[15-17]. These findings have led to a single hypothesis for the calcification mechanism. However, the nature of "Amelogenins and Enamelins" of developing enamel has remained largely obscure so far[1,2]. The polyclonal antibody used in this study reacted with most of the enamel matrix proteins, which ranged from low to high in molecular weight. This reaction pattern on single-dimension electrophoresis was similar to that of stained with a polyclonal antibody against amelogenin reported previously [18]. From the present studies, we demonstrated as expected, that dentin and bone matrix contain proteins immunoreactive against anti 25K enamel protein antibody. Moreover, immunocytochemical observation showed the close association between organic substance coated crystal and immunolabeled-gold particle in both enamel and dentin, suggesting the proteins detected here might be components of the organic envelope as a group of common proteins. These findings further confirmed that the calcification mechanism among calcifying hard tissues including enamel, dentin, and bone is controlled by a similar group of proteins. The present study also indicated that these proteins are not specific to calcifying tissues, because immunoblotting clearly revealed that the salivary gland contains these immunoreactive proteins as well. In addition, two-dimensional analysis revealed that the major spots of immunoreactive proteins obtained from the salivary gland were coincident with those of developing enamel matrix. This suggests that matrix proteins relating to biomineralization are phylogenetically identical with those of the submandibular gland. When the proteins from 21K to 31K daltons, were subjected to two-dimensional analysis, high molecular weight proteins showing a positive reaction were detected. This indicates that aggregation may have taken place during electrophoresis. Although further investigations are necessary to discuss their functions in terms of the calcification process, we predict that these proteins might be associated with the supply of calcium and/or phosphate ions during growth of the apatite crystal among calcifying hard tissues.

In conclusion, the present study supports the view that the calcification mechanism among vertebrate hard tissues is governed by a similar basic mechanism, and demonstrates that proteins responsible for calcification processes are not specific to calcifying hard tissues.

REFERENCES

1. Eastoe, J.E. (1960) Nature 187, 411-412.
2. Termine, J.D., Belcourt, A.B., Christner, P.J., Kathleen, M., Conn, K.M. and Nylen, M.U. (1980) J. Biol. Chem. 255, 9760-9768.
3. Kakei, M. and Nakahara, H. (1985) Jpn. J. Oral Biol. 27, 357-361.
4. Kakei, M. and Nakahara, H. (1985) Jpn. J. Oral Biol. 27, 1001-1005.
5. Kakei, M. and Nakahara, H. (1986) Bull. Josai Dent. Univ. 309-312.
6. Prouese, P.L.EY,S.J. and Jenkin, C.R. (1978) Immunochemistry 15, 429-436.
7. Laemmlie, U.K. (1970) Nature 227, 680-685.
8. Towbin, H., Staehelin, T. and Gorolon, J. (1979) Proc. Natl. Acad. Sci. USA. 79, 4350-4354.
9. O'Farrell, P.H. (1975) J. Biol. Chem. 250, 4007-4021.
10. Marshall, A.F. and Lowless, K.R. (1981) J. Dent, Res. 60, 1773-1782.
11. Nakahara, H. (1982) Bull. Josai Dent. Univ. 11, 209-215.
12. Nakahara, H. and Kakei, M. (1984) Bull. Josai Dent. Univ. 13, 259-263.
13. Nakahara, H. and Kakei, M. (1984) in Tooth EnamelⅣ (eds. Fearnhead, R.W. and Suga, S.) Elsevier, Amsterdom, 42-46.
14. Nakahara, H. and Kakei, M. (1983) Bull. Josai Dent. Univ. 12, 1-7.
15. Kakei, M. and Nakahara, H. (1983) Jpn. J. Oral Biol. 25, 374-377.
16. Kakei, M. and Nakahara, H. (1983) Jpn. J. Oral Biol. 25, 1129-1133.
17. Kakei, M. and Nakahara, H. (1984) Jpn. J. Oral Biol. 26, 554-558.
18. Shimokawa, H.,Wassmer, P., Sobel, M.E. and Termine, J.D. (1984) in Tooth EnamelⅣ (eds. Fearnhead, R.W. and Suga, S.) Elsevier, Amsterdom, 161-166.

CHAPTER 1.17

Mineral Induction by Immobilized Polyanions

M.A. Crenshaw

University of North Carolina Chapel Hill, NC 27599-7455 USA

Key words: Apatite, Clam soluble matrix, Phosphoryn, Phosvitin, Mineral induction.

INTRODUCTION

I shall briefly trace the developments leading to the hypothesis that polyanionic proteins from molluscan shell and dentin play a predominate role in initial mineral formation and present our recent experiments that support this hypothesis. I shall concentrate on the developments related to the role of the soluble matrix from molluscan shell because more progress has been made with the system.

For some time, it was thought the insoluble matrix proteins obtained from mineralized tissues nucleated the mineral and controlled its subsequent growth. Collagen of bone and dentin and conchiolin of molluscan shells were the two matrices that received the most attention. More recently, collagen and conchiolin have been regarded as the structural frameworks for the respective mineralized tissues, and acidic proteins associated with these networks are thought to be the functional elements in nucleation and control [1, 2].

An EDTA-soluble fraction of molluscan shell matrix was first isolated from the shell of the clam, *Mercenaria mercenaria..* Amino acid analysis showed that about 30% of the total residues were aspartic acid [3]. More thorough analyses of the soluble matrices from other molluscs showed that the soluble fraction was a common constituent of shell matrix and that it contained several macromolecules. Further, a high aspartic acid content was a distinguishing feature [4, 5. 6, 7, 8, 9, 10]. Using histochemical methods, we localized the soluble matrix at the site of initial crystal formation and concluded that the soluble matrix nucleated the mineral [11, 12]. However, soluble matrix in solution inhibits mineral precipitation from spontaneously precipitating solutions [7, 13].

Recently, we described the induction of mineral by whole molluscan matrix from solutions having a low degree of supersaturation [(ion activity product)/ Ksp = 3.7 for calcite]. Mineral formation occurred only when the soluble matrix remained fixed to the insoluble matrix by using a quarternary ammonium salt during decalcification. Under these conditions, an induction time of six to eight hours was required for the deposition of mineral on the whole matrix. Insoluble matrix prepared without a quarternary ammonium salt or at high ionic strengths, conditions which dissolve the soluble matrix from the insoluble matrix, did not induce mineral formation even when the calcium activity was increased eight-fold [14, 15, 16].

In solution, dentin phosphoprotein (phosphoryn) inhibits apatite precipitation and modifies the habit of crystals formed from spontaneously precipitating solutions [17]. Phosphoryn may induce hydroxyapatite formation in a similar manner when bound to dentin collagen [18].

To determine the effect of immobilizing dentin phosphoprotein on its ability to induce mineral, we covalently attached phosphoryn to agarose beads [19, 20]. We found that when phosphoryn was attached to agarose beads and incubated in metastable calcium phosphate solutions, apatitic mineral formed on the agarose beads. Bovine serum albumin did not induce mineral formation. Mineral formation by immobilized phosphoryn was inhibited by phosphoryn in solution, but the concentrations required for this inhibition were high relative to the amount of immobilized phosphoryn. Complete inhibition was noted with free phosphoprotein concentrations greater than 160 μg/ml. This is 2,000 times the amount of immobilized phosphoryn.

MATERIALS AND METHODS

More recently, we attached unfractionated clam soluble matrix on agarose beads by carbodiimide condensation [21]. Bovine serum albumin was used as a control. Because the compositions were easier to control, metastable calcium phosphate rather than calcium carbonate solutions were used as the calcifying media. The media were prepared from stock solutions at pH 7.4 as previously described [22], except that 10 mM Hepes was the buffer. Each stock solution was filtered with a 0.22 μm filter before the preparation of the calcifying solutions, which had an ionic strength of 0.165 a molar Ca/P of 1.67 and calcium phosphate concentration products of 1.4 to 1.8 mM^2. For incubation, 10 ml of medium was added to 10 mg beads in an acid-washed polyethylene vial. Incubation was carried out at 37±0.1º C in a water bath equipped with an oscillating platform operated at 100 Hz. Samples were withdrawn and filtered through a 0.22 μm filter. The filter was washed with water made alkaline (pH ca 8) with dilute ammonium hydroxide, air dried, mounted on a specimen holder, sputtered with gold/palladium (80/20) and examined with a scanning electron microscope. The presence of calcium phosphate mineral was confirmed by energy dispersive X-ray analyses.

RESULTS

Within 2 hours, calcium phosphate deposits were evident on about 25% of the agarose beads. With time, the size of the mineral deposits and the fraction of beads mineralized increased. The fraction of beads bearing mineral and the amount of mineral per bead generally increased with the calcium-phosphate product. By five days, large deposits were evident (Fig. 1). X-ray diffraction analysis showed the mineral was apatitic. No mineral deposits were detected on the beads carrying serum albumin at any concentration product used even when the incubation was extended to 10 days.

Phosphoryn [19, 20] (Fig 2) and the egg yolk phosphoprotein [23], phosvitin (30 wt. % phosphate, 65 residues asp/1000) (Fig 3), similarly immobilized also induced apatitic mineral formation over the same time course. Because phosvitin is a much smaller protein, it was coupled in the interior of the beads and induced mineral below the surface of the bead (Fig 4).

CONCLUSIONS

We found that apatite was induced from metastable calcium phosphate solutions, that do not spontaneously precipitate, by molluscan soluble matrix, phosphoryn and phosvitin when they were attached to an insoluble substrate, agarose. Another calcium-binding protein, serum albumin, did not induce mineral formation under the same conditions.

When immobilized, molluscan soluble matrix, phosphoryn and phosvitin would present a relatively fixed polyanionic surface to the solution. This surface may induce mineral formation by effecting an organization of the Stern layer [24]. Post-nucleation growth and control of the type and habit of the mineral deposited may be determined by the same polyanions in solution [13, 25, 26, 27] or attached to an oriented substrate [28]. The subsequent overgrowth of mineral onto an ordered, insoluble substrate also influence the microarchitecture of the deposited mineral.

Acknowledgements: I thank Professor K.M. Wilbur for reviewing this manuscript. This work was supported, in part, by NIH grant # DE08885.

REFERENCES

Veis A (1978) In: Everett DH, Vincent G (eds) Ions in Macromolecular and Biological Systems. Scientechnia, Bristol, pp. 259-267

Weiner S (1986) CRC Crit Rev Biochem 20: 365-408

Crenshaw MA (1972) Biomineralization 6: 6-11

Weiner S, Hood L (1975) Science 190: 987-989

Krampitz G, Engels J, Cazaux C (1976) In: Watabe N, Wilbur KM (eds) The Mechanisms of Mineralization in the Invertebrates and Plants. University of South Carolina Press, Columbia, pp. 155-173

Weiner S, Lowenstam H (1977) J Exp Mar Biol Ecol 30: 45-51

Wheeler A, George J, Evans C (1981) Science 212: 1397-1398

Weiner S, Traub W, Lowenstam HA. (1983) In: Westbroek P, de Jong EW (eds) Biomineralization and Biological Metal Accumulation. D. Reidel, Dodrecht, pp. 205-224

Worms D, Weiner S (1983) J Exper Zool 237:11-20

Lowenstam H, Weiner S (1989) On biomineralization. Oxford University Press, New York, 324 pp.

Crenshaw MA, Ristedt H (1975) Biomineralization 8: 1-8

Crenshaw MA, Ristedt H (1976) In: Watabe N, Wilbur KM (eds) The Mechanisms of Mineralization in the Invertebrates and Plants. University of South Carolina Press, Columbia, pp. 36-368

Wheeler AP, Sikes CS (1984) Am Zool 24: 933-944

Greenfield EM, Wilson DC, Crenshaw MA (1984) Am Zool 24: 925-932

Greenfield EM. (1987) In vitro mineral induction by soluble matrix from molluscan shells. Ph D thesis University of North Carolina at Chapel Hill

Greenfield EM, Crenshaw MA (1990) In: Crick RE (ed) Origin, Evolution, and Modern Aspects of Biomineralization in Plants and Animals. Plenum Press, New York, pp. 303-308

Termine JD, Eanes ED, Conn KM (1980) Calcif Tiss Int 31: 247-251

Stetler-Stevenson WG, Veis A (1986) Calcif Tiss Int 38: 135-141

Crenshaw MA, Linde A, Lussi A (1988) In: Aksay IA, McVay GL, Stoebe TG, Wager JF (eds) Atomic and Molecular Processing of Electronic and Ceramic Materials: Preparation, Characterization and Properties. Materials Research Society, Pittsburgh, pp. 99-107

Lussi A, Crenshaw MA, Linde A (1988) Arch Oral Biol 33: 685-691

Parikh I, March S, Cuatrecasas P (1974) In: Jakoby W, Wilchek M (eds) Affinity Techniques. Academic Press, New York, pp. 77-103

Crenshaw MA, Ramp WK, Gonnerman WA, Toverud SU (1974) Proc Soc Exptl Biol. Med 146: 488-493

Linde A, Lussi A, Crenshaw MA (1989) Calcif Tissue Int 44: 286-295

Mann S, Heywood BR, Rajam S, Birchall JD (1988) Nature 334: 692-695

Nawrot CF, Campbell DJ, Schroeder JK, Valkenburg MV (1976) Biochem 15: 3445-3449

Addadi L, Weiner S (1985) Proc Nat Acad Sci US 82: 4110-4114

Mann S, Didymus N, Sanderson N, Heywoos B (1990) J Chem Soc Frarday Trans 86: 1873-1880

Addadi L, Berman A, Oldak JM, Weiner S (1989) Conn Tiss Res 21: 457-465

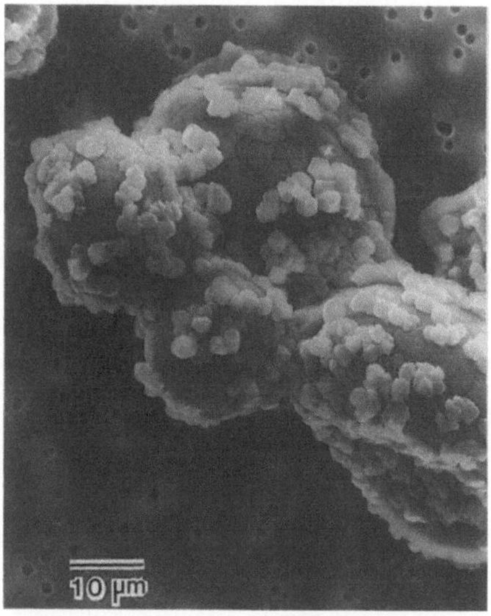

Fig.1 Mineral induced by soluble clam matrix attached to argarose beads. Incubated for 5 days in Ca x P = 1.8 mM2. Secondary electron image.

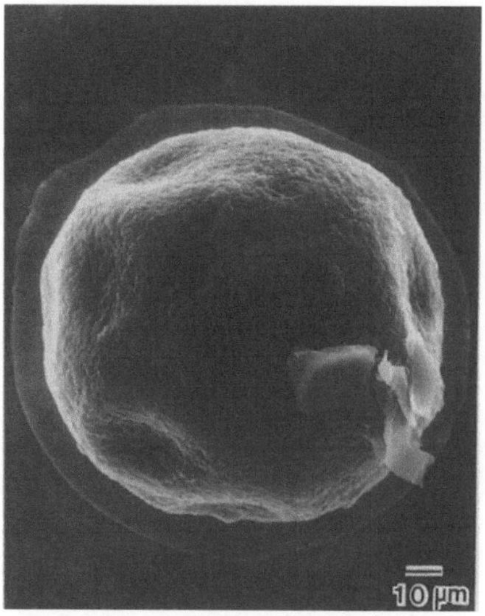

Fig.2 Mineral induced by phosphoryn attached to argarose beads. Incubated for 5 days in Ca x P = 1.8 mM2. Secondary electron image.

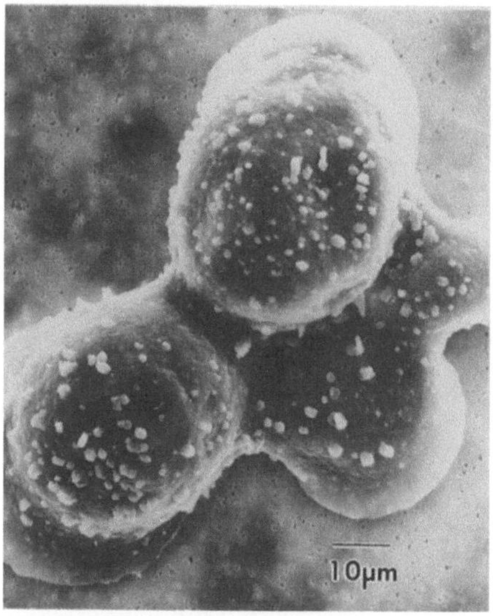

Fig.3 Mineral induced by phosvitin attached to argarose beads. Incubated for 5 days in Ca x P = 1.8 mM2. Secondary electron image [23].

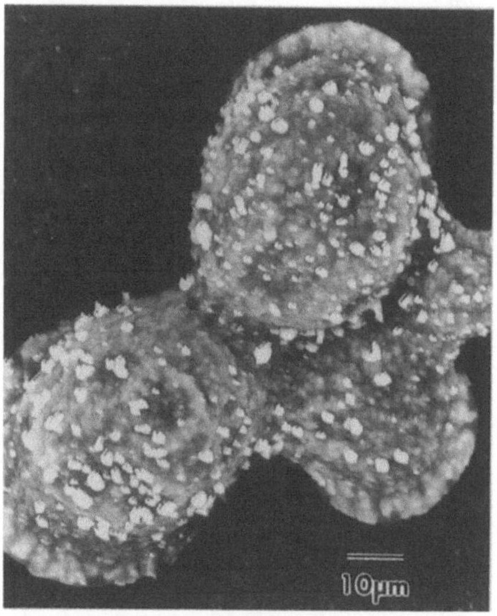

Fig.4 Backscattered electron image of beads in Fig 3 showing that mineral was deposited within the agarose beads [23].

CHAPTER 1.18

Interactions of Bovine Dentin Phosphophoryn with Calcium Phosphates

R. FUJISAWA and Y. KUBOKI

Department of Biochemistry, School of Dentistry, Hokkaido University, Sapporo, 060 Japan

Key words: Phosphoprotein, Dentin, Hydroxyapatite, Mineralization, Surface adsorption.

INTRODUCTION

Mineralization of dentin is a highly organized process that is controlled by many kinds of modulators. These modulators affect the nucleation and crystal growth of the mineral phases. Phosphophoryn, a highly phosphorylated protein of dentin [1], is a typical example of such modulators. Phosphophoryn binds calcium ions with modest affinity [2-6] and changes its conformation and molecular dynamics [2, 7]. Binding a large number of calcium ions, this protein aggregates into granules [8,9] and finally precipitates [3].
As expected from its polyanionic character, phosphophoryn has an affinity to hydroxyapatite crystals [10], and this affinity is dependent on the phosphate groups. The molecule in solution inhibits the nucleation and crystal growth of hydroxyapatite [11, 12]. This protein also reduces the amounts of calcium phosphate precipitates in a gel diffusion system of in vitro mineralization [13], although local density of nucleation was elevated. Contrasting results were obtained in an in vitro mineralization system using phosphophoryn immobilized on a solid matrix. Linde and colleagues demonstrated the induction of hydroxyapatite by phosphophoryn immobilized on agarose beads [14, 15].
Both in inhibitory and inductive actions, the elementary process is an interaction between phosphophoryn and hydroxyapatite. Such interaction is well illustrated in calcium carbonate formation in a system containing mollusc shell acidic proteins [16]. In the present study, the authors attempted to demonstrate the interaction between phosphophoryn and hydroxyapatite using fluorescence-labeled phosphophoryn. The labeled protein was adsorbed on the crystal in a surface-specific manner.

MATERIALS AND METHODS

Preparation of Phosphophoryn and Labeling. Phosphophoryn was extracted from bovine incisor and was purified chromatographically as described in a previous report [17]. The purified phosphophoryn was labeled with fluorescein isothiocyanate (Wako Chemical Co.) in borate buffer, pH 9.3, at 4°C for 24 hours. Unreacted fluorophore was removed completely by gel filtration.

Hydroxyapatite Crystals. Synthetic hydroxyapatite powder and single crystals were generous gifts from Dr. H. Aoki, Tokyo Medical and Dental University [18]. Apatitic character was confirmed by X-ray diffraction. The hydroxyapatite powder had a specific surface of 61.5 m^2/g. Plate-like hydroxyapatite crystals (Bio Gel HTP) were purchased from Bio-Rad Laboratories.

Binding Experiments. The hydroxyapatite powder was equilibrated with TBS (150 mM NaCl, 50 mM Tris/HCl, pH 7.4) overnight, and degassed. One hundred micrograms of the hydroxyapatite powder was incubated with 1-100 µg/ml of fluorescein-labeled phosphophoryn in the TBS for 1 hour with gentle agitation at 25°C. The mixture was then centrifuged and the supernatant was recovered as the unbound fraction. The hydroxyapatite was washed again with TBS and the supernatant was combined with the unbound fraction. The residual hydroxyapatite was decalcified with TBS containing 0.1 M EDTA, and then the decalcified extracts were collected as the bound fraction. The protein in the bound and unbound fractions was quantified by fluorometry at an excitation wave length of 494 nm and an emission wave length of 525 nm using a Hitachi F-3000 spectrofluorometer. From these data, dissociation constants were calculated with double reciprocal plots. The labeled phosphophoryn was incubated with hydroxyapatite single crystals in TBS as described above. After the incubation, the crystals were washed with TBS and observed with a fluorescence microscope.

RESULTS AND DISCUSSION

Phosphophoryn purified from bovine dentin was stained purple with Stains-All and migrated at an apparent molecular mass of 130-150 kD on SDS gel electrophoresis using 5-15% gradient gel. One molecule of fluorescein was conjugated per one molecule of phosphophoryn. A Langmuir-type binding curve was obtained for the binding of phosphophoryn (Fig. 1). The binding site had a dissociation constant of 19 µg/ml or 0.13 µM. This value is consistent with our previous report [10]. The maximum binding capacity of the hydroxyapatite was not as high for phosphophoryn as for bone Gla protein or serum albumin. A small number of phosphophoryn molecules can cover the crystal surface.

Although the whole surface of the hydroxyapatite crystal could adsorb the phosphophoryn, there was a preference in the adsorption on each surface. The fluorescence was more prominent on the (100) face than on the (001) face, when observed with the fluorescence microscope (Fig. 2). This preferential adsorption was observed both on plate-like crystals and on rod-like crystals. Surface adsorption on hydroxyapatite has been investigated by Kawasaki et al. [19, 20] in reference to hydroxyapatite chromatography. They proposed that acidic proteins had a tendency to be adsorbed on the (100) face of the hydroxyapatite crystal [20]. This is consistent with the results in this study. Calcium ions are exposed on the (100) surface of the crystal and can interact with acidic proteins.

Adsorption of a crystal growth inhibitor such as phosphophoryn on a specific crystal surface may result in a change in the crystal

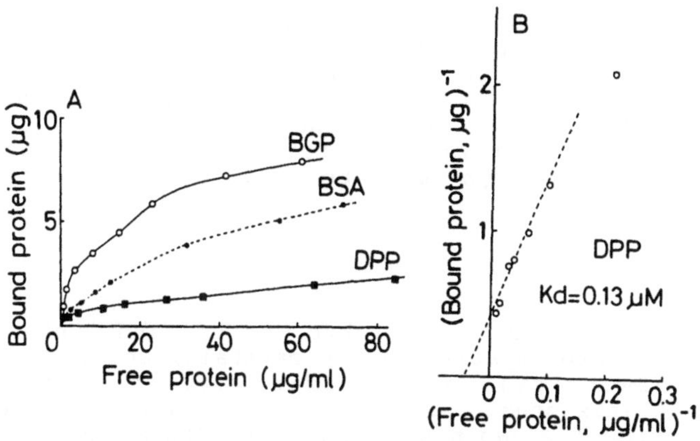

Fig. 1 Binding of phosphophoryn on hydroxyapatite. Fluorescence-labeled phosphophoryn (DPP), bone Gla protein (BGP) and bovine serum albumin (BSA) were incubated with 100μg of synthetic hydroxyapatite powder, and bound and unbound fractions were recovered (A). The right figure is double reciprocal plot of the binding data of phosphophoryn (B).

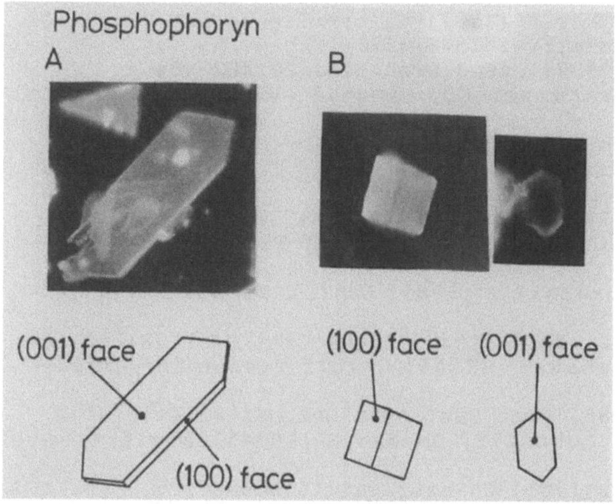

Fig. 2 Adsorption of phosphophoryn on each crystal face of hydroxyapatite. Plate-like (A) or rod-like (B) crystals of hydroxyapatite were incubated with fluorescence-labeled phosphophoryn, and observed with fluorescence microscope. Phosphophoryn was adsorbed preferentially on the (100) face of the crystal.

shape. As in the case of mollusc aspartate-rich protein [16], adsorption of phosphophoryn on the (100) face inhibits crystal growth perpendicular to this face. The crystal, thus, has an elongated shape along the c-axis as a result of this inhibition. During mineralization of bone and dentin, the first crystals to appear are long platelet-like crystals with their c-axis along the collagen fiber [21]. Phosphophoryn could be responsible for the shaping of such crystals.

ACKNOWLEDGEMENTS

The authors are grateful to Dr. H. Aoki (Tokyo Medical and Dental University) for providing hydroxyapatite single crystals, and to Dr. T. Domon (Hokkaido University) for providing access to the fluorescence microscope. This study was supported by a Grant-in-Aid for Scientific Research from the Japanese Ministry of Education.

REFERENCES

1. Butler WT (1984) Collagen Rel Res 4:297-307
2. Lee SL, Veis A, Glonek T (1977) Biochemistry 16:2971-2979
3. Kuboki Y, Fujisawa R, Aoyama K, Sasaki S (1979) J Dent Res 58: 1926-1932
4. Zanetti M, de Bernard B, Jontell M, Linde A (1981) Eur J Biochem 113:541-545
5. Stetler-Stevenson WG, Veis A (1987) Calcif Tissue Int 40:97-102
6. Marsh ME (1989) Biochemistry 28:346-352
7. Fujisawa R, Kuboki Y (1987) Jpn J Oral Biol 29:482-484
8. Cocking-Johnson D, Van Kampen CL, Sauk JJ (1983) Collagen Rel Res 3:505-510
9. Marsh ME (1989) Biochemistry 28:339-345
10. Fujisawa R, kuboki Y, Sasaki S (1986) Calcif Tissue Int 39:248-251
11. Termine JD, Conn KM (1976) Calcif Tissue Res 22:149-157
12. Termine JD, Eanes ED, Conn KM (1980) Calcif Tissue Int 31:247-251
13. Fujisawa R, Kuboki Y, Sasaki S (1987) Calcif Tissue Int 41:44-47
14. Lussi A, Crenshaw MA, Linde A (1988) Archs Oral Biol 33:685-691
15. Linde A, Lussi A, Crenshaw MA (1989) Calcif Tissue Int 44:286-295
16. Addadi L, Weiner S (1985) Proc Natl Acad Sci USA 82:4110-4114
17. Fujisawa R, Takagi T, Kuboki Y, Sasaki S (1984) Calcif Tissue Int 36:239-242
18. Akao M, Aoki H, Kato K (1981) J Mater Sci 16:809-812
19. Kawasaki T, Takahashi S, Ikeda K (1985) Eur J Biochem 152:361-371
20. Kawasaki T, Ikeda K, Takahashi S, Kuboki Y (1986) Eur J Biochem 153:249-257
21. Landis WJ, Paine MC, Glimcher MJ (1977) J Ultrastruct Res 59:1-30

CHAPTER 1.19

Immunohistochemical and Enzyme Histochemical Studies in Bone Formation Induced by Bone Morphogenetic Protein (BMP) in Mouse Muscle Tissue

A. KAMEGAI, T. TANABE, N. SHIMAMURA, S. KUMASA, K. YAMADA, and M. MORI
Department of Oral and Maxillofacial Surgery, Asahi University School of Dentistry, Gifu, Japan

Key wards: Bone morphogenetic protein, Chondro-osteogenesis, Histochemistry

INTRODUCTION

It has been a well-known phenomenon that bone morphogenetic protein (BMP) heterotopically induces bone formation by causing the differentiation of osteoprogenitor cells from mesenchymal cells [1,2]. Two hypotheses of the cell origination and mechanism of bone formation by BMP have been proposed: BMP irreversibly induces the differentiation of perivascular mesenchymal cells into osteoprogenitor cells [3,4]; BMP is capable of stimulating the differentiation of skeletal muscle into cartilage [5-7]. The biological roles of bone formation induced by BMP still remain unclear. In the presnt study, the process of chondro-osteogenesis and cell differentiation for endochondral ossification by BMP implanted into mouse skeletal muscle tissue, were evaluated in terms of the immunohistochemical method for laminin, fibronectin, glycosaminoglycan, and S-100 protein, and the enzyme histochemical method of phosphatase activity with wide range of buffers (pH 4.0 to pH 9.2).

MATERIALS AND METHODS

Preparation of bovine BMP Semipurified BMP was prepared from bovine bones [7]. These bone particles (1-2 mm), decalcifed with 0.6 N HCl, were extracted with 4 M guanidine hydrochloride for 48 hours at 4°C, and the extract was dialyzed against 0.5 M guanidine hydrochloride at the final concentration. The resulting precipitate was collected and used as the BMP preparation for this study.

Chondro-osteoidal tissue induced by BMP Five-week-old male ICR mouse was employed for the histologic assay. Five mg of semipurified BMP in a gelatin capsule were implanted in the thigh muscle. The post-implant survival times were: 3, 5, 7, 14, 21, 28 and 56 days. All mice were maintained under ordinary nutitional control, and given free access to food and water.

Materials Muscle tissue containing well developed and radio-opaque masses were fixed in 10 % formaline, decalcified with EDTA-4Na (pH 7.2), and embedded in paraffin. Materials including muscle tissue and bone were frozen in isopentane, precooled to -160° C in liquid nitrogen. Sections were cut at 6 μm in a -20° C cryostat. Serial sections were stained for phosphate activity.

Immunohistochemical methods The PAP method, using anti-human fibronectin antiserum and anti-laminin antiserum were used. Chondroitin-4-sulfate (C4S), chondroitin-6-sulfate (C6S), dermatan sulfate (DS), and keratan sulfate (KS) expression were examined; S-100 protein was demonstrated using polyclonal antiserum (S-100) and monoclonal antibodies to S-100 protein subunit alpha and beta.

Enzyme histochemical methods Cryostat sections were fixed in 70 % aceton for 30 min. and incubated in the appropriate pH-buffered solution. The alkaline phosphatase incubation medium consisted of Palitisch-buffer (pH 8.0 to pH 9.2) or Tris-maleate buffer (pH 7.0 to pH 7.8) containing AS-MX phosphate, and Fast Red Violet LB. For the acid phosphatase medium, Tris-maleate buffer (pH 5.8 to pH 6.8) containing AS-TR phosphate and Fast Red Violet LB.

RESULTS

Chondro-osteogenesis induced by BMP began with proliferation of undifferentiated mesenchymal cells which were transformed and undergone metaplasia from satellite cells of muscle fibers. Tissue specimens taken from 7 days after BMP implantation contained fibro-cartilage or hyaline cartilage tissue but no calcified bone tissue. Such cartilage forming tissue were composed mainly of well-differentiated chondrocytes, cartilage matrix, fibrillar cells and proliferating capillary vessels; and occasionally remains of the BMP implant were noticed. In the specimens taken from 14 days', chondroidal tissue was transformed to osteoidal and calcified bones with numerous osteoclastic cells. On day 21, osteoidal and well-calcified bones were gradually reduced and simultaneously bone marrow formation was focused, and afterwards induced bone structure disappeared.

Laminin and fibronectin (Table 1 and 2; Fig 1A and 1B) On day 3 post implantation of BMP, laminin and fibronectin stainings were strongly expressed in the basement membrane and fibronectin expressed in the ground substance of neighboring connective tissue. On day 7, laminin staining continued in the muscle sheath and reacted in proliferating capillary vessels. Positive fibronectin staining continued in sarcolemma. Fibronectin was confined to the cartilage matrix, and became negative in the newly formed bone matrix.

Glycosaminoglycan (Table 3) C4S showed strong staining in BMP implanted area through chondro-osteogenic process. The formed cartilage matrix began to show strong staining for C6S and DS, and slight staining for KS. KS also showed positive staining in chondrocytes. In the endochondrial ossification stage, KS staining showed negative, and DS staining showed strong in the formed matriax.

Phosphatase activities (Table 4, 5 and 6) The phosphatase staining was divided into 3 groups according to the expression by pH levels: alkaline phosphatase (pH 7.6 to pH 9.2), neutral phosphatase (pH 6.8 to pH 7.4), and acid phosphatase (pH 4.0 to pH 6.6). In 7 days' specimens, strong phosphatase activity was seen in mesenchymal and osteoprogenitor cells at pH 7.0 to pH 8.6, in chondroidal cells at pH 7.0 to pH 7.6, and in chondroclastic cells at pH 4.0 to pH 5.8. In 14 and 21 days' specimens, alkaline phosphatase staining was concentrated in both the outer and inner layers adjacent to newly formed bone. Neutral phosphatase activity was concentrated in chondrocytes. Macrophage lineage cells exhibited abundant acid phosphatase activity at pH 4.0 to pH 5.6 during chondro-osteogenesis induced by BMP.

S-100 protein (Fig 1C) All types of cartilage cells expressed strong staining for S-100 and S-100 alpha subunit, and slight staining for S-100 beta; and the S-100 alpha staining was more intense than that of S-100. Cartilage cells were stained positively for S-100 protein subunit alpha in their cytoplasms and nuclei, but the surrounding cartilage matrix was unstained. In some instances, cartilage cells around the BMP implant showed a very strong reaction for S-100 and S-100 alpha; and intermingled round-shaped cartilage cells and fibrillar osteoprogenitor cells were also highly reactive to antibodies specific for S-100 and S-100 alpha.

DISCUSSION

In the initial phase of BMP bone formation, myoblastic cells appeared in the muscle sheath at adjacent BMP implanted areas, and electron microscopic view showed satellite cell existed outside of the muscle fiber surrounded by plasma membrane [8]. Fibronectin is a marker of sarcolemma and cartilage matrix [9-11], and laminin is also a marker of muscle sheath, basement membrane, and blood vessels [12,13]. As in the initial phase of chondro-osteogenesis, immunohistochemical stainings of laminin and fibronectin were strongly expressed in the basement membrane and the ground substance of neighboring connective tissue. These results suggested that progenitor cells of chondroidal tissue originated from myogenic cells. Satellite cells of the muscle sheath in the BMP-implanted area should migrate, transform, and finally differentiate to chondrocytes. Simultaneously, capillary endothelial cells proliferated from peripheral regions of BMP implanted differentiated, and formed blood vessels channels into osteo-chondrial tissue.

From the evaluation of chondro-osteogenesis induced by BMP in terms of alkaline and acid phosphatase activities as a marker of bone formation and reduction, cells undergoing chondrogenic changes showed strong phosphatase reactions over a very wide pH range, from alkaline to acid. Strong alkaline phosphatase reactions in osteoprogenitor cells resembled that of cells in healing fractures [2] or abnormal bone proliferation as occurs in osteogenic sarcoma. Also, osteoclasts showing acid phosphatase activity were found in the cartilage tissue. This occurs during endochondrial ossification and normally exhibits numerous chondroclastic giant cells which characteristically exhibit strong activities for acid phosphatase. The giant cells found in this tissue were seen in the absorption region of chondroid or osteoid matrices, as found in normal ossification. The acidic environment enhanced new bone formation induced by BMP [14]. The acidic environment caused by inflammation and direct phagocytotic action on BMP by macrophage lineage cells seemed to be a trigger for new bone formation. In the process of BMP-induced new bone formation, BMP regulates differentiation of mesenchymal cell into cartilage and bone [1]. Macrophage lineage cells invade inward and differentiate into chondroclast-like or osteoclast-like cells. During this biological process, BMP should induce undifferentiate dmesenchymal cells to form a chondroid matrix and endochondrial ossification.

Proteoglycans are involved in mineralization process in vivo [15,16]. Bone lineage cells synthesized proteoglycans are included in the matrix formed. These proteoglycans were found in the calcified and uncalcified matrix both in vovo and in vitro [16]. In the present examination, the presence of C4S in the early stage of chondro-osteogenesis is seemed to have already provided a favorable environment for further mineralization, which is available to form chondroid matrix and ossification.

Immunoreaction for S-100 protein is a marker of chondrocytes, and precalcified chondrocytes showed the highest staining for S-100 protein reaction [17-19]. It is possible that the calcium-binding proteins including S-100 protein are related to the mechanisms of calcium signaling in mineralized tissue and calcium regulation in hard tissue. This relationship is suggested by the localization of more intensely reactive chondrocytes for S-100 protein in the vicinity of foci of cartilaginous matrix mineralization. Skin mixed tumors and salivery pleomorphic adenomas occasionally exhibited S-100 protein in chondroidal-changed cells of hyalinous and myxomatous tissues [20], and these chondroid cells are originated from myoepithelial cells or epimyoepithelial cells as well as the laminin expression.

Bone mineralization is a complex process under the control of bone cells involving complex factors. From the present study, osteoclastic cells originate from histiocytes and are involved in resorption mechanisms of formed chondro-osseous tissue. Progenitor cells relate to satellite cells in muscle fiber. Simultaneously, capillary endotherial cells proliferate and form blood vessels channels into osteo-chondroidal tissue. In the chondro-osteogenesis induced by BMP, there exists controlling substances for matrix forming or mineralization.

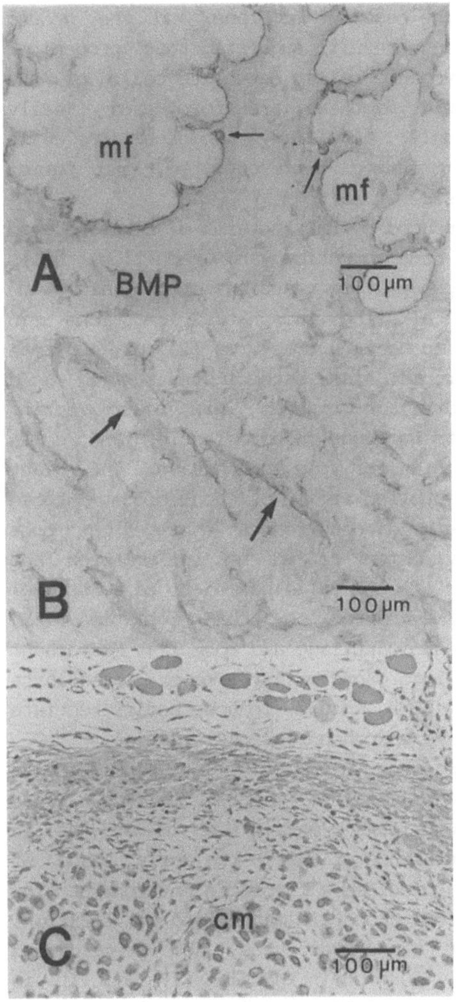

Table 1 Laminin expression

days	normal muscle fiber sheath	muscle fiber sheath adjacent to BMP	satellite cells	basement membrane of blood vessels	chondro-osseous matrix
3	+1	+3	+3		
7	+1	+3	+3	+2	
14	+1	+2	+2	+2	+2
21	+1	+1	+1	+2	+2
28	+1	+1		+2	+2
56	+1	+1			

− negative. ± trace. +1 slight. +2 moderate. +3 strong

Table 2 Fibronectin expression

days	normal muscle fiber sheath	muscle fiber sheath adjacent to BMP	ground substances of connective tissue	cartilage or bone matrix
3	+1	+3	+1	
7	+1	+1	+1	
14	+1	±	+1	+2
21	+1	−	+1	−
28	+1	−	+1	
56	+1	−	+1	

Fig. 1A: Laminin staining in muscle fiber (mf) sheath and satellite cells (arrows) of 3 days after BMP implantation.

Fig. 1B: Laminin staining in proliferating blood capillaries (arrows) of 7 days after BMP implantation.

Fig. 1C: S-100 alpha staining in chondrocytes of 7 days after BMP implantaion.

Table 3 Expression of glycosaminoglycan

days	chondroitin -4-sulfate			chondroitin -6-sulfate			dermatan sulfate			keratan sulfate		
	A	B	C	A	B	C	A	B	C	A	B	C
3	+3	+3	±	−	+1	−	−	−	−	±	+1	−
7	+1	+3	−	±	±	−	+2	−	−	+1	+1	−
14	+2	+3	−	+1	+3	−	+1	±/+3	−	+1	±/+1	−
21	+2	+3	−	−	±	−	−	−/+3	−	+1	±/+2	−

A, cell proliferating layer; B, BMP implanted area;
C, neighboring connective tissue

Table 4 Phosphatase activities on day 7

pH	mesenchymal cells chondroblasts osteoblasts	chondroid cells	chondroclasts osteoclasts
9.2-8.8	+2	−	−
8.6-8.4	+2〜+3	−	−
8.2-7.8	+3	−	−
7.6-7.4	+3	+1	−
7.2-7.0	+2〜+3	+1	−
6.8	+1〜+2	−〜±	−
6.6	±〜+1	−〜±	−
6.4-6.2	±〜+1	−	−
6.0	−〜±	−	−
5.8-5.6	−〜±	−	+2〜+3
5.4-4.0	−〜±	−	+3

Table 5 Phosphatase activities on day 14

pH	chondroblasts osteoblasts	chondrocytes	osteoclasts
9.2-8.8	+1〜+2	−	−
8.6-7.8	+2〜+3	−	−
7.6-6.8	+2〜+3	+1	−
6.6	+1〜+2	+1	−
6.4	+1〜+2	±	+1
6.2	±	±	+1
6.0-5.8	−	−	+2
5.6-4.0	−	−	+3

Table 6 Phosphatase activities on day 21

pH	chondroblasts osteoblasts	osteoclasts
9.2-8.8	+2	−
8.6-7.6	+3	−
7.4	+2	−
7.2	+1	−
7.0	±	−
6.8-5.8	−	−
5.6-4.0	−	+3

REFERENCES

1. Syftestad GT, Urist MR (1982) Clin Orthop 162:288-297
2. Jee WSS (1983) In Weiss L (ed) Cell and Tissue Biology. MaCmillan Press, London, pp200-255.
3. Urist MR, DeLange RL, Finerman GAM (1983) Science 220:680-686
4. Urist MR (1968) Clin Orthop 56:37-50
5. Nathanson MA (1985) Clin Orthop 200:142-158
6. Mahy PR, Urist MR (1988) Clin Orthop 237:236-244
7. Kamegai A, Muramatsu Y, Tanabe T, Mori H, Mori M, Inoue S (1990) Acta histochem cytochem 23:209-218
8. Kamegai A, Tanabe T, Shimamura N, Shibata K, Yamada K, Mori M (1990) JBMM (in press)
9. Weiss RE, Reddi AH (1981) J Cell Biol 88:630-636
10. Reddi AH (1985) J Biomed Mater Res 19:233-239
11. Maor G, von der Mark K, Reddi AH, Heinegård D, Franzen A, Silbermann M (1987) Coll Relat Res 7:351-370
12. Foidart JM, Bere EW Jr, Year M, Rennard SI, Gullino M, Martin GR, Katz SI (1980) Lab Invest 42:336-342
13. Briggaman RA (1982) J Invest Dermatol 78:1-6
14. Yoshikawa H, Takaoka K, Shimizu N, Ono K (1986) Bone 7:125-128
15. Nefussi JR, Septier D, Collin P, Goldberg M, Forest N (1989) Calcif Tissue Int 44:11-19
16. Anderson HC (1989) Lab Invest Mar 60:320-330
17. Stefansson K, Wollmann RL, Moore BW, Arnason BGW (1982) Nature 295:63-64
18. Weiss APC, Dorfman HD (1986) J Bone Joint Surg 68A:521-526
19. Okajima K, Honda I, Kitagawa T (1988) Cancer 61:792-799
20. Noda Y, Horike H, Tanimura T, Tsujimura T, Mori M (1988) Virchows Archiv B, 54:371-380

CHAPTER 1.20

Analysis of the Organic Matrix Associated with Calcium Oxalate Crystals in *Vitis mustangensis*

M.A. Webb

Department of Botany and Plant Pathology, Purdue University, West Lafayette, IN 47907, USA

Key Words: Calcium oxalate, *Vitis mustangensis*, Organic matrix, Glycoprotein, Raphides

INTRODUCTION

In plants large amounts of calcium are sequestered as crystalline calcium oxalate, which commonly occurs within specialized cells in anatomically and developmentally specific patterns [for review see 1, 2]. As is common in other systems of biomineralization, calcium oxalate deposition within plant cells is preceded by formation of an organic matrix within which crystallization occurs [3]. However, very little is known about the biochemical nature of that matrix or the factors inducing its formation within plant cells.

Two types of calcium oxalate crystals are found in leaves of *Vitis*. Bundles of raphides, needle-like crystals, occur in large cells called idioblasts that are distributed throughout the leaf. In addition, spherical crystal aggregates known as druses are also common. In this study crystals isolated from leaves of *Vitis mustangensis* have been treated in a variety of ways to dissolve their matrix components for biochemical analysis.

MATERIALS AND METHODS

Crystal isolation. Crystals isolated from leaves of *Vitis mustangensis*, collected in Arlington, Texas, were kindly provided by H. J. Arnott. Methods used were similar to those reported for isolation of cystoliths [4]; absolute ethanol was used in all procedures rather than aqueous solutions. Crystals were stored at 4°C in ethanol in a capped vial until use.

Extraction of organic matrix. For each experiment approximately 200 µl of crystals in suspension were removed from the vial. A crystal pellet was collected by briefly spinning the suspension in an Eppendorff microfuge. The pellet was washed by resuspending it in cold ethanol, and the suspension was again centrifuged to pellet the crystals. The resulting crystal pellet was then treated to obtain extracts of matrix components. 1) Guanidine-HCl extract: The pellet was resupended in 6M guanidine-hydrochloride, pH 5.3, and incubated at 4°C overnight or at room temperature for 1-2 hr. Crystals were pelleted out of the suspension in the microfuge by centrifuging at 16,000 x g for 5 min, and the supernatant was removed for analysis. Proteins were precipitated from the extract by adding six volumes cold ethanol. The precipitate was redissolved in 0.1M $LiNO_3$ in 0.1M TES, pH 7.5, for column chromatography or directly in SDS sample buffer consisting of 2.3% sodium dodecyl sulfate (SDS), 10% glycerol, 5% β-mercaptoethanol in 0.063 M Tris, pH 6.8, for gel electrophoresis. 2) TES buffer extract: Alternatively, the pellet was resuspended in 0.1M TES buffer with or without 0.4 M $LiNO_3$. Crystals were immediately pelleted out of the solution and the supernatant was collected for analysis. Resuspension and pelleting of crystals was repeated four times in succession. Supernatants from each resuspension were saved for analysis by SDS-polyacrylamide gel electrophoresis (SDS-PAGE).

3) SDS buffer extract : To extract additional components SDS sample buffer, described above, was added directly to the crystal pellet left after TES buffer washes.

Electrophoresis of matrix components. SDS-PAGE was performed by standard methods [5] with acrylamide concentration of 12.5% or with a linear gradient of acrylamide from 7.5 to 17.5%. Gels were stained with silver or with the periodic acid-Schiff (PAS) procedure for carbohydrate localization [6].

Column chromatography. Gel filtration was performed on a Pharmacia FPLC automated protein chromatography system using a Superose 12 column. The buffer used for chromatography was 0.1M $LiNO_3$ in 0.1M TES, pH 7.5, and samples were applied in the same buffer. Flow rate was 0.5 ml/min, and fractions of 1 ml were collected.

RESULTS

The crystal fraction isolated from *Vitis mustangensis* leaves contained both raphides and druses. Raphide bundles isolated in ethanol were intact, consisting of a central packet of many needle-like crystals surrounded by a non-mineralized region with a granular appearance (Figs. 1-3). Addition of aqueous solution to the bundles caused dissolution of material surrounding the raphides, resulting in a thick mucus-like solution, and release and separation of raphides from the bundles (Fig. 4). No change in druse structure was noted after addition of aqueous solutions.
Treatment with aqueous guanidine-hydrochloride extracted proteineaceous components from the matrix. These were analyzed by gel filtration chromatography (Superose 12) and SDS-PAGE. Superose 12 chromatography showed the presence of several distinct species with absorbance at 280 nm (Fig 5). Approximate molecular weights of the first three peaks eluted were 100, 47, and 15 kD; the remaining peaks corresponded to very low molecular weights estimated to be less than 5 kD. Analysis of the extract by SDS-PAGE with silver-stained gels exhibited high background staining, which obscured stained bands (Fig 6A). However, two bands which stained a golden-brown color could be detected at

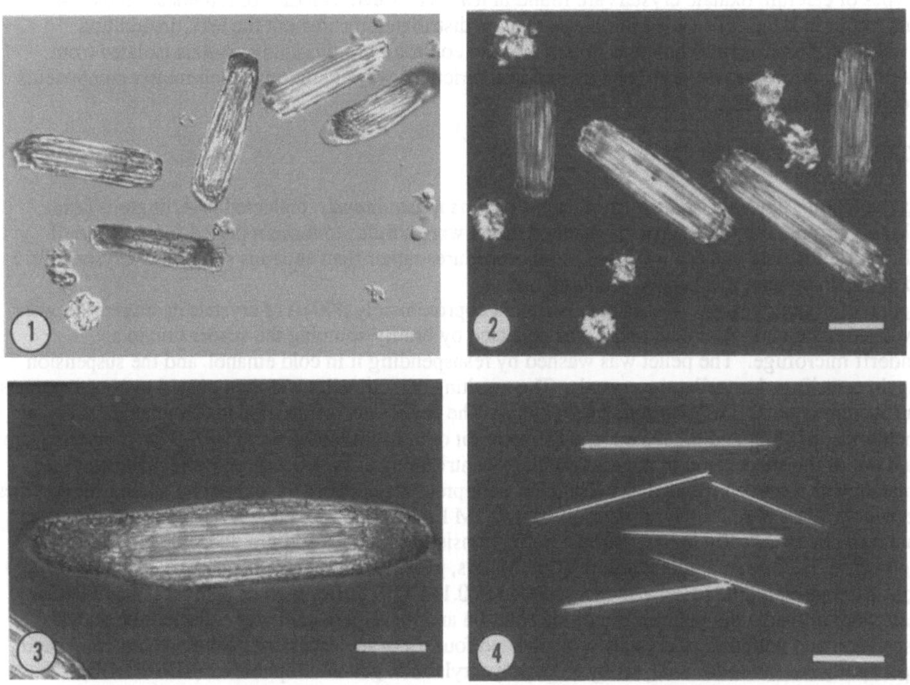

Figs. 1-4. Light microscopy of isolated crystals. Fig. 1. A portion of the crystal fraction isolated from *Vitis mustangensis*, showing intact raphide bundles (R) and a druse (D). Fig. 2. A similar field shown under polarized light with crossed nicols. Fig. 3. A raphide bundle under partially crossed nicols, illustrating the birefringent raphides, surrounded by granular matrix. Fig. 4. Individual raphides released from the bundle by treatment with aqueous solution. Bars = 20 μm.

Fig. 5. Profile of absorbance at 280 nm for fractions eluted from Superose 12 column (gel filtration). Sample was prepared by incubating the crystal fraction in 6 M guanidine-HCl, pH 5.3, and precipitating extracted protein with cold ethanol. 250 µl of sample was applied to the column in 0.1 M $LiNO_3$ in 0.1M TES buffer, pH 7.5.

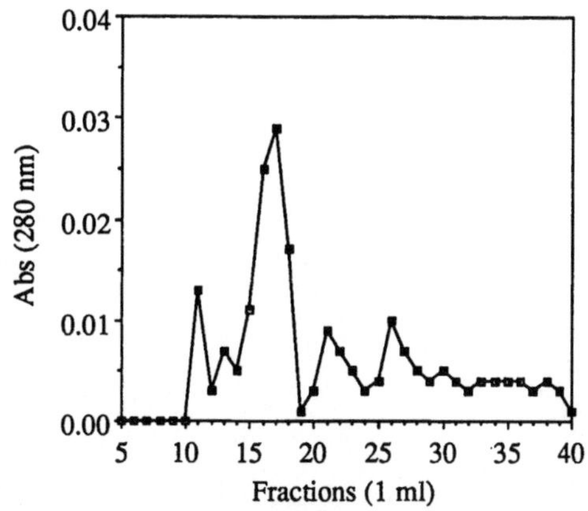

the top of the gel. These bands were strongly PAS-positive, indicating presence of glycoprotein (Fig. 7A). The mobility of the glycoprotein in acrylamide gels was very limited and corresponded to species with molecular weights greater than 200 kD; however, it is well known that glycoproteins may form aggregates and that their molecular weight cannot be accurately estimated by SDS-PAGE. Aggregate formation may also have prevented entry of the glycoprotein into the Superose 12 column. The first few fractions eluted from an inverted column following separation of the guanidine-HCl extract had strong absorbance at 280 nm, however, the content of those fractions has not been analyzed.

In other experiments the crystal fraction was washed with 0.1M TES, pH 7.5, with or without 0.4M $LiNO_3$. This treatment also resulted in dissolution of the material surrounding the raphide bundles and release of raphides from the bundles. The resulting solution gave a positive reaction with the Molisch test for carbohydrates. Equal aliquots from four sequential washes of the raphides were analyzed by SDS-PAGE (Fig. 8). A number of bands were detected in silver-stained gels, having approximate molecular weights of 83, 76, 70, 64, 58, and 47 kD. The intensity of bands diminished with sequential washes, as did background staining. A component with limited mobility similar to that in the guanidine-HCl extract was also present in washes and was shown to be PAS-positive (Fig 7C).

The crystal pellet remaining after four washes in buffer was then treated directly with sample buffer containing SDS. This resulted in release of additional proteins, presumably those closely associated with the individual crystals. Analysis of this solution by SDS-PAGE showed a number of bands, with prominent bands at mobilities corresponding to molecular weights of approximately 70, 66, 58, and 55 kD (Fig. 6B) and a large number of weakly-staining bands. A weak reaction was observed at the top of the gel after PAS staining (Fig 7B).

DISCUSSION

Formation of calcium oxalate crystals with a raphide morphology is unique to plants and has not been duplicated in artificial systems [7]. It has been known for some time that raphide formation in plants occurs within membrane-like matrices, such as crystal chambers described in *Eichhornia* or membrane sheets in *Lemna* [4]. Organic matrices have also been found in association with formation of other types of crystals in plants, for example, paracrystalline structures observed at the center of developing druses in *Capsicum annum* [8]. Despite many structural descriptions of these matrices, little has previously been known about their biochemical composition.

The results of this study show that the organic matrix associated with crystals isolated from *Vitis mustangensis* contains an assortment of proteins, and that a glycoprotein is a major component of the matrix. Glycoproteins are commonly associated with a variety of different biomineralization systems, such as bone, tooth, spine, and shell formation [see Table 2.3 in ref. 9]. Thus it appears likely that at the biochemical level calcium oxalate formation in plants has features in common with biomineralization in animal systems. Additional research will be necessary to gain information about both the protein and carbohydrate moieties of the glycoprotein present in *Vitis* for comparison with these systems.

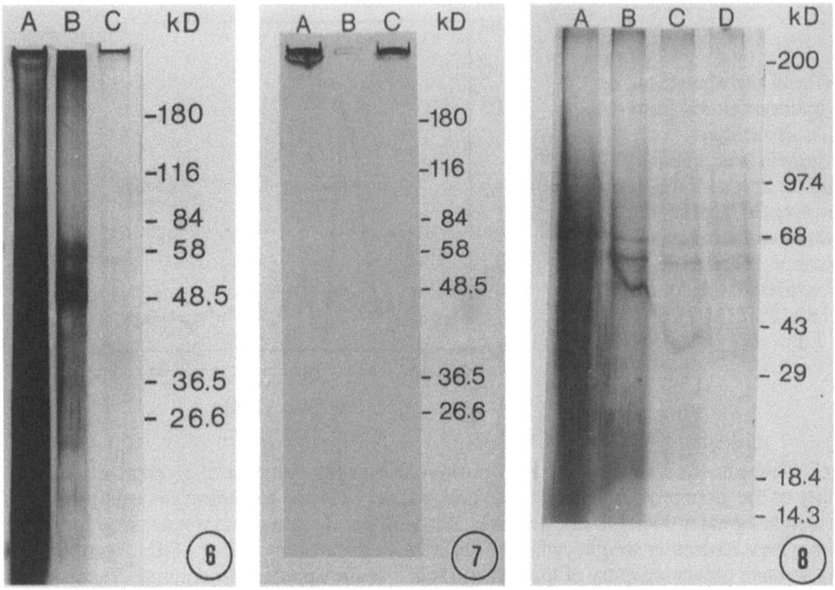

Figs. 6-8. SDS-PAGE analysis of matrix components extracted from the crystal fraction. Figs. 6 and 7. Duplicate gels stained with silver (Fig. 6) or with the PAS procedure (Fig. 7); gradient gel of 7.5 to 17.5 % acrylamide without stacking gel. Lanes correspond to: A) Guanidine-HCl extract, B) SDS buffer extract, C) TES buffer extract. Lanes A) and C) both exhibited strong staining with PAS of a component with very limited mobility, corresponding to faint bands in the silver stained gel. When a stacking gel of 5% acrylamide was used, the PAS-positive component also failed to move through the stacking gel (not shown). Fig. 8. Equal aliquots from sequential washes of crystal fraction with 0.4M $LiNO_3$ in 0.1M TES buffer, showing diminishing extraction with each wash.

More than one compartment is present in isolated raphide bundles, and it appears from this study that protein profiles are distinct for each compartment. Current evidence suggests that the major glycoprotein component is present in material around and/or between the individual raphide crystals, since it is present in buffer washes that dissolve that material. The glycoprotein present in the SDS extract may be the same as that present in buffer and guanidine-HCl extracts, or it may be a distinct protein. Some bands detected on SDS gels, which had similar apparent molecular weights in the TES buffer extracts and the SDS buffer extract, may represent proteins loosely bound to the crystal chambers. Analagous compartmentation of matrix components has also been shown in other biomineralization systems, for example, the soluble and insoluble fractions of molluscan shell matrices [10].

Although druses that were present in the crystal fraction appeared intact following all treatments, some proteins detected in the extracts may have come from organic cores or sheaths of druses. Non-crystalline cores can be seen in druses from *Vitis* leaf under crossed polars, and druses in *Vitis* endosperm have been shown to contain proteinaceous cores [11]. Methods are being refined to separate raphides and druses into discrete fractions for analysis.

Proteins released only by treatment with SDS-containing buffer or guanidine-HCl may represent proteins tightly associated with the crystals, such as integral proteins of crystal chambers. These proteins potentially have important roles in crystal formation, either directly by contributing to a template or nucleation center or indirectly by controlling the influx of calcium or oxalate into the chambers.

ACKNOWLEDGEMENTS

The author wishes to thank H. J. Arnott for providing crystals from *Vitis mustangensis* and for helpful discussion, J. Lindell for technical assistance, and D. Bentlage for assistance in preparation of the manuscript.

REFERENCES

1. Arnott HJ, Pautard FG (1970) In: Schraer H (ed) Biological Calcification. Appleton-Century-Crofts, New York, pp.375-446.
2. Franceschi VR, Horner HT (1980) Bot Rev 46:361-427.
3. Arnott HJ (1966) In: Fleisch H, Blackwood HJ, Owen M (eds) Third European Symposium on Calcified Tissues. Springer-Verlag, Berlin, pp. 152-157.
4. Arnott HJ (1980) In: Omori M, Watabe N (eds) The Mechanisms of Biomineralization in Animals and Plants. Tokai University Press, Tokyo, pp. 211-218.
5. Laemmli UK (1970) Nature 227:680-685.
6. Gerard C (1990) In: Deutscher MP (ed) Methods in Enzymology, Vol 182. Academic Press, New York, pp. 529-538.
7. Cody AM, Horner HT, Cody RD (1982) Scanning Electron Microsc I:185-197.
8. Horner HT and Wagner BL (1980) Amer J Bot 67:1347-1360.
9. Lowenstam HA, Weiner S (1989) On Biomineralization. Oxford University Press, New York, p. 22.
10. Simkiss K, Wilbur KM (1989) Biomineralization. Academic Press, San Diego, pp. 237-238.
11. Webb MA, Arnott HJ (1983) Scanning Electron Microsc. IV:1759-1770.

2 Mineralization

CHAPTER 2.1

From Calcium in Molluscs to Coral Grafted into Bone: an Example of Basic to Applied Research Transfer

M. CHETAIL and J. FOURNIE
Dynamique des Tissus Calcifiés, Université Paris VII, 75251 Paris Cedex 05, France

Key words: Molluscs, Bone, Calcification-decalcification, Carbonic anhydrase.

This is an attempt to review our researches for 20 years on biomineralization. During this time, we have been focussing on cytological and biochemical study of the calcification-decalcification problems related with the intervention of carbonic anhydrase and ATPases in several biological systems.

Calcified tissues have to be considered as a reserve of Ca^{++} ions for animals in which a permanent and dynamic equilibrium exists between precipitation and dissolution of calcium salts to insure the homeostasis of their body fluids. Calcium is a very important element in living systems and particularly in animals where it is involved in cellular cohesion in embryos as well as in developped animals, muscular contraction, nervous functions, etc... The most important problems in calcium metabolism concern precipitation and liberation of this cation necessary for different physiological purposes such as, the maintenance of the acid-base and osmotic pressure balances, or the mobilization of this cation during reproduction. Since calcium is known to be a very active element , so its concentration in living cells is necessarily limited to about 10^{-6} M and consequently is carefully regulated to keep the properties of organelles and of various enzyme systems. So, the cell has to protect itself against the toxicity of this cation by inactivating it in 3 ways: intra or extracellular precipitation, linkage or elimination [1].

BORING MECHANISM AS AN EXAMPLE OF DEMINERALIZATION

Various predatory prosobranchs live on the flesh of other molluscs and to reach it they have to drill the calcareous shell of their prey. As explained by Carriker [2, 3], the boring mechanism consists in a chemo-mechanical process: the mechanical part is insured by the radula and the chemical part, mainly decalcification, by the boring organ, located in the foot in Thais lapillus as in the other Muricids. This mushroom-shaped organ (Fig. 1) is made of a simple epithelium. The main ultrastructural features of these cells are those found in epithelia implicated in active ionic transport and essentially consist in the presence of long and dense microvilli covered by a luxuriant cell-coat with a rapid turn over during cell activity; lacunar junctions (gap type) allow the immediate electrophysiological coupling between neighbour cells; dense mitochondrial population is located in the cells where the ionic transport takes place, just underneath the microvilli, remarkable amount of glycogenic inclusions able to relay the mitochondria in their role of ATP providers in case of metabolic emergency. The cells are arranged in clusters surrounded by large blood sinuses providing oxygen during the activity of the organ [4, 5].

Carbonic anhydrase (CA) has been demonstrated histochemically (HANSSON's method) and biochemically (manometric method) at the level of the boring organ. To demonstrate the main part of CA in the dissolution of $CaCO_3$ of the prey's shell of T. lapillus several experiments were carried out either to inhibit the CA activity by adding diamox, a specific inhibitor of this enzyme in the sea water, or to enhance the boring mechanism by adding CO_2, one of the two substrates of CA, in sea water. The results demonstrated that CA is responsible for demineralization of the valves of the preys: at low concentrations of diamox, partial inhibition of the enzyme occured, i.e. the number of complete holes decreased or disappeared and etchings increased; at higher concentrations, full inhibition took place. Inhibition is reversed when snails are replaced in normal sea water. The action of pure CO_2 or mixtures of CO_2 and O_2 accelerated boring; in the optimal mixture, 3 times more boreholes were produced by snails than in controls and in about half the time [6].

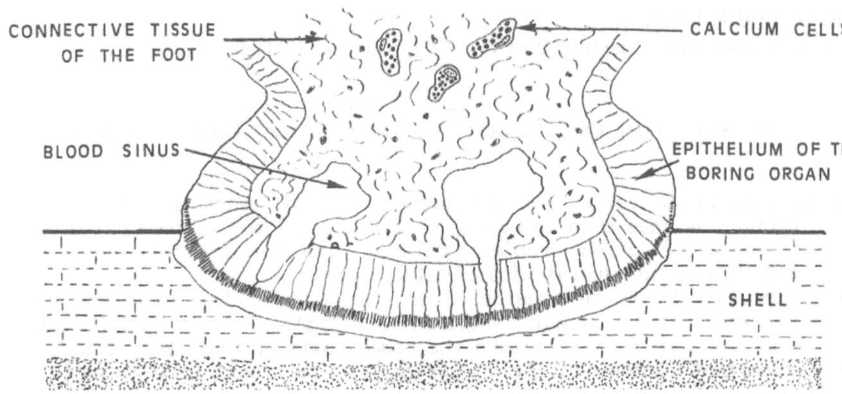

Fig. 1: Schematic view of the boring organ in position of shell drilling.

CA is well known to catalyze the following reversible reaction:

$$CO_2 + H_2O \rightleftharpoons H_2CO_3 \rightleftharpoons HCO_3^- + H^+$$

The stimulation of boring by CO_2 shows that in the case studied here the reversible reaction catalyzed by CA goes to the right in the sense of a release of H^+ ions; the resulting acidity is responsible for the dissolution of $CaCO_3$. This output of H^+ ions is balanced by a compensatory input of Ca^{++} ions, the concentration of which is particularly high at the level of the boring organ during activity. Then calcium passes through the blood sinuses and finally they enter the calcium cells located in the foot connective tissue where they are precipitated as calcium carbonate. ATPases intervene in these ionic movements [7].

CALCIUM CELLS : AN INTRACELLULAR SYSTEM OF CALCIFICATION-DECALCIFICATION

Four cell types are able to accumulate calcium intracellularly in molluscs under the form of microgranules in which calcium carbonate and/or phosphate are linked to a complex organic matrix [8]. These 4 cell types located at various anatomical levels in the molluscan body are:
- calcium cells located in the general connective tissue (foot, mantle, perivascular and perinervous) . They are generally widespread in gastropods.
- calcium cells inserted between the acinar cells of the digestive gland are observed in stylommatophora and also in some marine prosobranchs and in the land snail Pomatias.
- Amoebocytes observable at the hypostracal level of normal and regenerating shell elaborate minute calcium granules observed in stylommatophora and lamellibranchs. After bursting of amoebocyte, microgranules become free and able to grow, so forming the spherulitic hypostracal layer of the shell.
- excretory calcium cells located exclusively at the border of pedal and dorsal integument; they are known only in stylommatophora and their function is not yet clearly understood. The same excretory calcium cells are also found in the mantle

edge of exochocleates land pulmonates where they clearly intervene in the elaboration of the mineral part of the epiphragm.

All these calcium cells contain numerous vesicles in which calcium is stored. They represent an intracellular system for accumulation of calcium salts in molluscs added to the extracellular reserve constituted by the shell [9]. The organocalcic nature of these vesicles has been widely demonstrated. The mineral part is essentially made of calcium carbonate, with the exception of the calcium cells of the digestive gland where they essentially consist in calcium phosphate. The organic part of vesicles is complex and constituted by polysaccharides, amino acids and lipids. Two enzymes have been demonstrated at their level: CA and alkaline phosphatase, probably Ca-ATPase.

The fact that calcification is initiated in intracellular vesicles is widely recognized. This intravesicular calcification involves a calcium transit through the plasma membrane, the hyaloplasma and the membrane of the calcification vesicles with probably in addition a part played by the cisternae, ducts and vesicles of endoplasmic reticulum whose ability to sequestrate calcium is well known. It is well recognized that ionic calcium is noxious for the cell for levels arround 10^{-6} M for the hyaloplasma and 10^{-3} M for the extracellular medium. So, the main role of calcium cells is to sequestrate superfluous calcium ions and to precipitate them in closed compartments, that are the calcium vesicles. In the calcium vesicles, proteins, polysaccharides and lipids are packed and they play an important role to capture excess calcium in the hyaloplasma. The 2 well known enzymes which generally intervene in calcification systems, alkaline phosphatase and CA are present at the level of the calcification vesicle membrane, as previously described in several species. The presence of these 2 enzymes shows that these vesicles have to be considered as specific compartments of the intracellular calcification and that the most important calcium exchanges are those taking place at the level of the vesicle membrane, between the hyaloplasma and the intravesicular compartment.

Simkiss [10] interpretes the alkaline phosphatase observed in the calcification sites as a Ca-ATPase catalyzing the chelation of calcium by ATP. The resulting Ca-ATP chelate is hydrolyzed by carbonic acid resulting from the catalytic action of CA; this interpretation was the first to explain the simultaneous action of these 2 enzymes in biomineralizing systems, as well as the coexistence of calcium phosphate and carbonate in these systems. Moreover, this interpretation has 2 main advantages, first, it allows an efficient drainage of calcium towards the vesicles when the hyaloplasmic concentration is too high since the mechanism starts by intervention of a Ca-dependant ATPase and second it saves energy since the conversion of Ca-chelate into carbonate regenerates ATP.

Calcium vesicles are not only devoted to calcium precipitation; in fact, several observations on more than 10 species of gastropods, of these connective calcium cells showed different aspects: when the cells are numerous, generally they are completely filled by calcium loaded vesicles; sometimes, only a few calcium cells can be seen and they contain vesicles whose organic part alone may be observed. This indicates that their calcium contents is of a peculiar lability and is released when physiologically required. So, at the level of these vesicles, permanent and bidirectionnal movements of calcium take place through their membrane. As long as an equilibrium exists in the calcium cells between the amount of precipitated and released calcium, one can consider that they are still living; but when calcium precipitation exceeds calcium release, the cells are condemneded to death, as indicated by their pycnotic nucleus.

The connective origin of calcium cells has been clearly recognized by Prenant [11] and more recently their ultrastructural study [12, 13] has shown that they have to be considered as one path of differentiation of the pore cells or groove cells (Fig. 2). This kind of connective cells is widespread in molluscs and represents a pluripotent step of the connective cell which is able to express various functional and structural characters when they differentiate [14, 15]. Taking into account their plasticity, it looks reasonable to compare the pore cells of molluscs to the vertebrate mesenchymal cells.

The main ultrastructural device in pore cells is the presence at the level of their plasma membrane of several pores in communication with the cell by a group of ducts and cavities forming the so-called undergroove system. The initiation of calcium granules takes place in these ducts limited by a membrane.

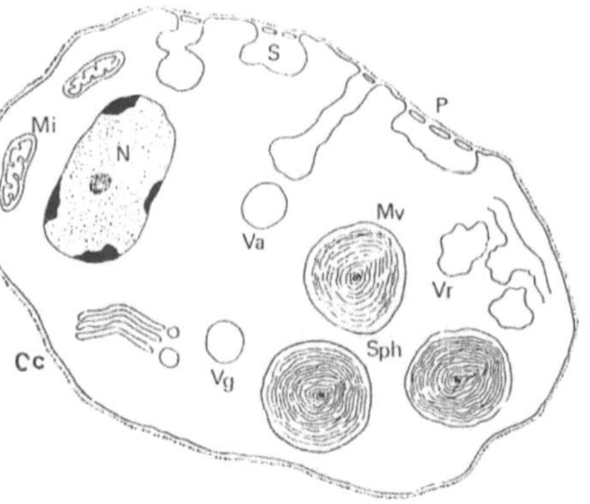

Fig. 2: Interpretation of the ultrastructure of connective pore cell in molluscs. Cc: cell coat; M: mitochondria; Mv: vesicle membrane; N: nucleus; P: pores; S: undergroove system; Vg: golgi vesicle; Vr: endoplasmic reticulum vesicle;

CALCIUM MOBILIZATION, A MUST FOR REPRODUCTION

It appeared to us that in molluscs, the calcium cells, the extracellular granules and the shell itself have to be considered as a stock of calcium useful for animals for different physiological purposes. This utilisation of calcium reserves during life and reproduction cycle is well illustrated by the case of Deroceras reticulatum [16]. To this end, calcium measurements and histological studies were carried out on the calcium storage compartments (shell, integument and digestive gland), on the genital duct (albumen gland and oviduct) at various stages of the reproductive cycle and also on the new laid eggs, embryos and egg albumen during the incubation period. The results were as follows: during growth period, animals store calcium at the level of their inner shell and of the integumental and digestive calcium cells. This storage reaches its maximal level at the end of the male period, during which the shell contains about 45% of the whole calcium stock while the calcium contents for the whole lot of calcium cells is 30%; the remaining 25% represent calcium bound to organic compounds and free calcium. From the beginning of oogenesis, a mobilization of calcium reserves takes place from the storage compartments towards the genital duct. This calcium mobilization is maximal at the oviposition period. At this time, we noticed a progressive drop of calcium level in the shell, the digestive and integument calcium cells, as shown by erosion shapes of calcite crystals in the shell and by a decrease of the number of calcium cells. Simultaneously, we noticed an increase of calcium rate at the level of the female genital duct (albumen gland and oviduct) which contains 15% of total calcium stock at the beginning of oogenesis and an average of 30% for slugs during oviposition. At each laying, animals lose about 20% of the whole calcium stock. During the interval between 2 layings, we noticed a new increase of calcium in the storage compartments followed by a sudden decrease, while in the albumen gland and oviduct we observed an appreciable increase of calcium rate just before each laying, followed by a calcium drop at the time of laying. Calcium stored in the eggs is used to form the shell embryo which is well calcified 10 days after laying and to be precipitated in the calcium cells which differentiate in the tissues of embryo before hatching [17].
These results indicate that the shell and the connective calcium cells are calcium reservoirs for reproduction requirements. The calcium dynamics, accumulation as well as release, appears to be very quick and precise. After each calcium release due to a laying, the storage compartments are immediately reloaded. All these observations suggest a precise correlation between calcium movements and reproduction physiology revealing a sophisticated dynamics of calcium as well as a tight linkage between the

regulation of calcium metabolism and the physiology of the female reproductive system, probably under the dependance of an hormonal factor as already known in birds and mammals.

ELIMINATION OF EXCESS CALCIUM IONS IN Pomatias elegans

Pomatias elegans is a prosobranch adapted to an exclusively land life which illustrates the fact that sometimes the animal receiving too much calcium from their environment have to eliminate it to avoid the noxious effect of this cation [18]. Pomatias lives only in calcareous soils covered by humus known to be acid; this acidity attacks the calcareous substrate and calcium ions are then liberated , trapped in the mucous sole film and absorbed by the epithelial cells of the sole through their apical microvilli. Such a percutaneous calcium uptake is known to take place in other land snails [19,20]. For land snails, the 2 main supplies of calcium come from the food and from the substratum. Pomatias is a species fasting for long periods during which the only supply of calcium is absorption through the pedal sole which is the only organ in close contact with the substratum. The aim of this input of Ca^{++} ions is not only to provide calcium but is also to allow a flux of water to enter the snail body. If too many calcium ions enter the animal, they can be precipitated into salts in the connective calcium cells located along the sole epithelium, just above the basal lamina or in the other storage compartments (shell, connective calcium cells, calcified operculum) or bound to organic compounds. In P. elegans, the need of calcium is high since the osmotic pressure reaches 385 milliosmoles against 200 to 250 for ordinary land snails. Moreover, in Pomatias, this osmolarity relies substantially on calcium, since the level of calcium ions in the haemolymph is 20 mM/l against 10 mM/l for Helix . It has been shown [21] that the osmotic pressure of haemolymph and urine is higher in Pomatias than in any other land gastropod and that both fluids are isoosmotic and isoionic, thus arguing against the intervention of the kidney in the hydromineral equilibrium which is realized by the tubulous gland located in the pedal sinus. This gland intervenes when the storage compartments are unable to store more calcium ions and functions like the salt glands of marine and desert vertebrates by eliminating excess calcium ions from the haemolymph. In fact, the ultrastructure of the cells of the tubulous gland reveals characteristic features of epithelial cells involved in active ionic transport, which are: lacunar junctions (septate desmosomes to realize the immediate physiological coupling of the cells as gap junctions do; infoldings of the basal plasma membrane enclosing rod-like mitochondria to provide energy just at the place where active ionic secretion takes place; finally glycogenic inclusions in the hyaloplasma as an alternative source of energy. Moreover these cells contain CA and alkaline phosphatases (ATPases?) both enzymes known to catalyze active transports. The non intervention of the kidney in the regulation of hydromineral equilibrium and the fact that the tubulous gland replaces it, means that in Pomatias, adaptation to a land life is minimal for this prosobranch in the conditions of its semi arid biotope.

CORAL GRAFTED INTO MAMMALIAN BONE: RESORPTION OF THE IMPLANT AND INDUCTION OF NEW BONE DEPOSITION

At this point of our researches, we were contacted by an orthopedist and a veterinary surgeon as consultants on the subject of calcium from a basic point of view. To enter into their subject, they gave us a paper by Roy and Linneham [22]. These authors were able by an hydrothermic substitution reaction to convert the carbonated coral microstructure into pure hydroxyapatite. They grafted these replicas into bone diaphysis of rabbits or dogs; later on, Holmes [23], and also ourselves did the same thing. The grafted replicas were analyzed by X rays from 1 week to 6 months and sometimes two years after implantation, with, for the 3 teams, the same following results: a quick ingrowth of host bone into the porous implant was observed when the bone defect does not exceed a length of 12 mm, while for larger defects, more than 20 mm, the implant is only invaded by non mineralized fibrous tissue. In both cases, the replicas keep about the same size and shape as

shown on the radiographies.
Only one paper by Chiroff [24] refers to the use of pieces of carbonated madrepore skeleton grafted into bone diaphysis of 12 dogs. These grafts may be observed on the radiographies up to the fourth month after intervention and later on they progressively diminished and more often they completely disappeared; this disappearance was not explained probably because these authors failed to study histologically what happens to the grafts and surrounding tissues.

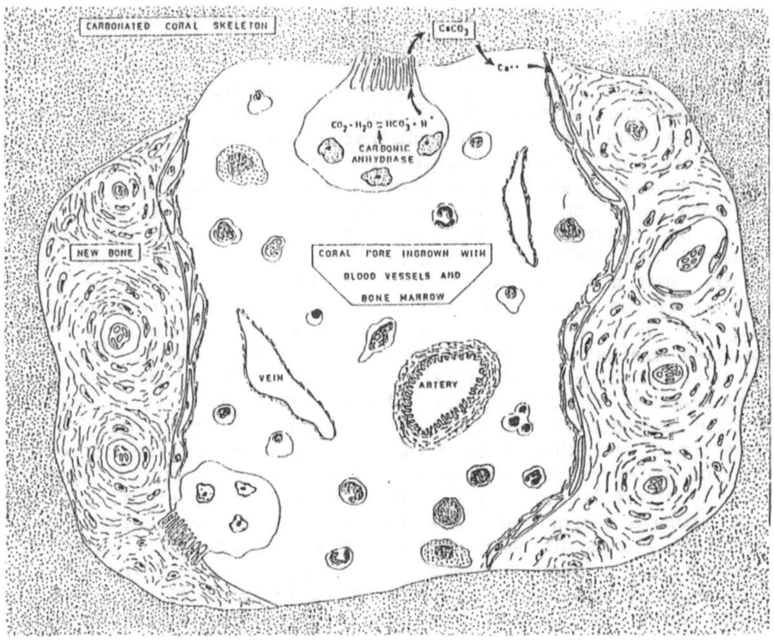

Fig. 3: Mechanism of coral resorption and bone apposition inside the coral pores.

Since for orthopedists the principal aim is to find out a favourable material to be grafted into bone clearly defined as: biocompatible, readily available, easily cut and shaped and at least invaded or better replaceable by host bone. So, in the light of Chiroff's preliminary results and of our own researches on calcium in molluscs, we felt that natural madrepore skeleton could certainly be a good candidate as a favourable material to be grafted into bone. So we advised the veterinary surgeon team to test pieces of natural madrepore skeleton rather than their hydroxyapatite replicas, as they had intended using and at the same time we recommended them to try, on some grafted animals, the effect of acetazolamide in convenient concentration to elucidate the resorption mechanism of the carbonated grafts.
Several genera of madrepore were tested: Porites, Favites, Goniopora and Lobophylla. 60 ulnar or femoral defects (from 8 mm to 28 mm in length) were grafted with fragments of Porites or Favites in dogs; all the grafts were stabilized with bone plate to allow weight bearing; among them, 10 dogs received daily an oral dose of 250 mg of acetazolamide, a potent inhibitor of CA, while in 20 other dogs, the bone defects were left empty. Radiographies were taken after operation and then every 2 weeks. Histological evolution was followed on undecalcified samples embedded in polyester resin and other samples were decalcified for routine histology.[25, 26, 27]. X ray study shows a progressive and centripetal resorption of the grafts. For small defects, the whole implant is resorbed after 8 weeks and replaced by woven bone. For the more important defects, the resorption process needs more time, up to

18 months and goes on until the setting up of a firm osseous union between the two diaphysis parts. Histological study has shown (Fig. 3) that a short time after the implantation of coral grafts, the whole porosity is invaded by bone marrow elements and by blood vessels. Simultaneously the coral skeleton is progressively resorbed by the numerous osteoclasts tightly applied on it by their ruffle border while osteoblasts begin to lay down new bone upon the partially resorbed coral grafts (Fig. 3).In the ungrafted control dogs, a fibrocartilagenous non-union (pseudarthrosis) is observed as well as in the dogs treated with acetazolamide in which a significant retardation of coral resorption is observed, preventing the two diaphysis parts to unite.

The presence of CA has been demonstrated in osteoclasts [28, 29]. We think that osteoclasic production and liberation of H^+ ions catalyzed by CA explain the dissolution of the carbonated skeleton as previously demonstrated by us in the boring mechanism in predatory prosobranch gastropods [6].

From a surgical point of view, the biodegradability of coral skeleton compared to other kinds of implants makes it a useful material for grafting certain bone defects in veterinary and human orthopedy, in maxillofacial reconstruction and also in dentistry. The first coral grafts in human bones were implanted in 1979 [30, 31] for orthopedic purposes. Up to now, this material has been used successfully to fill aseptic bone defects after trauma, tumor resection or removal of autogenic bone grafts from the iliac crest. It has also been used in grafting large bone defects like loose pseudarthrosis, bone lengthening, vertebral arthrodesis, cranio and maxillofacial surgery [30, 31, 32, 33, 34, 35]. In dentistry, it is indicated for parodontal purposes and for preprosthetic buccal surgery [36, 37] in order to fill the root dental hole after avulsion allowing a bone crest to grow. In this last case, the coral has to be powdered and mixed with the patient's blood before being set in place. Surgeons can now find this material, ready to use, under the name of Biocoral and in different shapes:pre-shaped blocks for orthopedic surgery, and for plastic surgery, and beads or granules of different sizes for dental surgery [38].

CONCLUSIONS

From a basic point of view, we begin to understand the general calcium dynamics during intracellular calcification realized into free connective cells in molluscs. But however, studies on transepithelial extracellular calcification-decalcification process through the molluscan mantle during shell formation and demineralization are still in progress. Both calcification processes despite their originality, show 3 common features :

- association of calcium to an organic matrix
- presence of two enzymes : Carbonic anhydrase and ATPases
- permanent remodelling by formation-destruction.

At last, it is interesting to notice that the same features characterize also vertebrate ossification. So, molluscs can be considered as a choice model to study the ability of initiation of intracellular calcification which is not yet recognized during vertebrate bone formation and also to understand the part of extracellular calcification during molluscan shell formation as well as during vertebrate ossification. Moreover, if we compare what occurs during the 2 calcification processes in both phylla, the same correlations become evident:

- pore cells and mesenchymal cells represent respectively in molluscan and vertebrate connective tissue a stock of stem cells able to differentiate in several directions outlining the great plasticity of connective cells.
- molluscan calcium cells and osteoblasts initiate calcium precipitation by their activity and both cell types represent the last step of these highly differentiated cells more or less rapidly condemned to death.
- epithelial cells of the boring organ of muricids and osteoclasts in vertebrate bone play the same part in the destruction of calcified substrates via intervention of CA.

The major advantage of coral grafted into bone is the fact that the coral skeleton is competent to allow apposition and differentiation of osteoblast and osteoclast precursor cells similarly to what happens during normal ossification. Consequently,

they have to be considered as a promising system to study osteoclast and osteoblast differentiation processes.
Finally, our own progress from molluscs to coral grafts showed clearly that, in the future, it may be fruitful for a better coordination between basic and applied researches, to promote multidisciplinary working groups, taking advantage of mutual and complementary approaches.

REFERENCES

[1] Chétail M, Krampitz G (1982) Malacologia 22: 1-2
[2] Carriker MR (1959) Proc. XVth Int. Congr. Zool. London: 373-376
[3] Carriker MR, Charlton G, Van Zandt D (1967) Science 158: 920-922
[4] Nylen UM, Provenza DV, Carriker, M.R. (1969) Am. Zool. 9: 935-965
[5] Chétail M, Derer M, Fournié J (1982) Malacologia 22 (1-2): 305-311
[6] Chétail M, Fournié J (1969) Am. Zool. 9: 983-990
[7] Chétail M, Fournié J (1970) C. R. Acad. Sci. Paris 271: 118-121
[8] Fournié J, Chétail M (1982) Malacologia 22 (1-2): 265-284
[9] Watabe N, Meenakshi VR, Blackwelder PL, Kurtz EM, Dunkelberger DG (1976) In: Watabe N, Wilbur K (eds) The mechanisms of mineralization in invertebrates and plants. University of South Carolina Press, pp 283-308.
[10] Simkiss K (1976) In: Watabe N, Wilbur K (eds) The mechanisms of mineralization in invertebrates and plants. University of South Carolina Press, pp1-31.
[11] Prenant M (1924) Bull. Biol. Fr et Belg. 58: 331-380
[12] Plummer JM (1966) Proc. Malac. Soc. Lond. 37: 189-197
[13] Nicaise G, Garrone R, Pavans de Ceccaty M (1966) C. R. Acad. Sci. Paris 262: 2248-2250
[14] Richardot M (1976) Thèse, Lyon, n° 76-6
[15] Vovelle J, Grasset M (1979) Malacologia 18: 557-560
[16] Fournié J, Chétail M (1982) Malacologia 22 (1-2): 285-291
[17] Fournié J, Chétail M (1984) Am. Zoologist 24: 857-870
[18] Bensalem M, Chétail M (1982) Malacologia 22 (1-2): 293-303
[19] Kado Y (1960) Journ. Sci. Hiroshima Univ. 19: 163-210
[20] Fournié J, Néauport C, Bizet C, Chétail M (1988) Haliotis 18: 311
[21] Rumsey T.J. (1972) Journ. Exp. Biol. 57: 205-215
[22] Roy DM, Linneham SK (1974) Nature 247: N. 5438: 220-222
[23] Holmes RE (1979) Plast. and Reconstr. Surg. 63: 626-633
[24] Chiroff RT, White EW, Weber JN, Roy DM (1975) J. Biomed. Mat. Res. 6: 29-45
[25] Guillemin G, Fournié J, Patat JL, Chétail M (1981) C.R. Acad. Sci. Paris 293: 371-376.
[26] Guillemin G, Patat JL, Fournié J, Chétail M (1987) Journ. Biomed. Mat. Res. 21: 557-567.
[27] Guillemin G (1981) Thèse Paris
[28] Simasaki M, Yagi T (1960) Dent. Bull. Osaka Univ. 1: 89-98
[29] Gay CV, Müller WJ (1974) Science 183: 432-434
[30] Patel A, Honnart F, Guillemin G, Patat JL, Chétail M, Fournié J (1980) Rev. Chir. Orthop. 66: 63-64
[31] Patel A, Honnart F, Guillemin G, Patat JL (1980) Chirurgie: 106: 199-205
[32] Levet Y, Jost G (1983) Ann. Chir. Plast. Esth. 28 (2): 180-181
[33] Levet Y, Guero S, Guillemin G, Jost G (1988) Ann. Chir. Plast. Esth. 33 (3): 279-282
[34] Souyris F, Chevalier JP, Payrot C, Pellequier C, Gary-Bobo A, Merlier C (1984) Ann. Chir. Plast. Esth. 29 (3): 256-260
[35] Souyris F, Pellequier C, Payrot C, Servera C (1985) J. Max. Fac. Surg. 13: 64-69
[36] Issahakian S, Ouhayoun JP (1988) J. Parodontol. 8 (3): 251-259
[37] Issahakian S, Ouhayoun JP, Shabana H, Sawaf H (1989) Journ. Dent. Res. 68 (4) Abstract N 274
[38] From coral to biocoral (1989) Inoteb, Saint Gonnery 56920 Noyal- Pontivy, France.

CHAPTER 2.2
Fine Structure of Skeletal Growth-Ridge in Some Scleractinian Corals

Y. Isa
Department of Biology, University of the Ryukyus, Okinawa, 903-01 Japan

Key words: Skeleton formation, Microstructure, Scleractinian, Coelenterata.

INTRODUCTION

To elucidate the skeleton formation mechanism in scleractinian corals, microstructures of the surface and inside of skeleton have been investigated using petrographic and scanning electron microscopic techniques (1,2,3,4). Studies showed the basic trabecular system of scledodermites were composed of several bundles of needle-like aragonite crystals. Development of the ridge and side surfaces of the actively growing portions of the coral skeleton remains to be studied.

Scanning electron microscopy of the growing tips of axial corallites in two species of *Acropora* indicated skeletal accretion by aggregation and coalescence of spherular and spindle-shaped crystal units (5,6). This study found the fine structure of the crystals exibited a hollow texture of peripheral tiny granular crystallites and that formation of the crystals occurred by a hollow organic precursor substance secreted by calicoblastic cells into the sub-epithelial space between calicoblastic tissue and $CaCO_3$ skeleton surface (7).

This article describes the appearance of the skeleton microstructure and the possible pattern of accretion of skeletal growth-ridges in some shallow water reef-building corals.

MATERIALS AND METHOD

Living scleractinian corals (Anthozoa: Coelenterata) were collected at a fringing reef, 2-3m deep, in the northern part of Okinawa, Japan. Species used were included *Stylophora pistillata, Acropora microphtalma, Acropora hebes, Montipora aequituberculata, Pavona decussata, Porites andrewsi, Hydnophora rigida, Galaxea fascicularis and Euphyllia grabrescence.* Small pieces of the growing marginal portiond of these colonies were individually immersed for 12 hrs into a 1N NaOH solution, and the skeletal samples were air-dried, coated with gold-paradium and then observed with a scanning electron microscope (HS-530) at 20 KV.

RESULTS

Figure 1 shows the framework of the skeleton margin from apical tips of branching species (A-D) and peripheries of foliated species(E,F). The thecal wall of a phaceloid corallite (A) is constructed with a thin skeletal plate, while that of axial corallite (B) is composed of 3-4 layers of annular thecal plates connected with

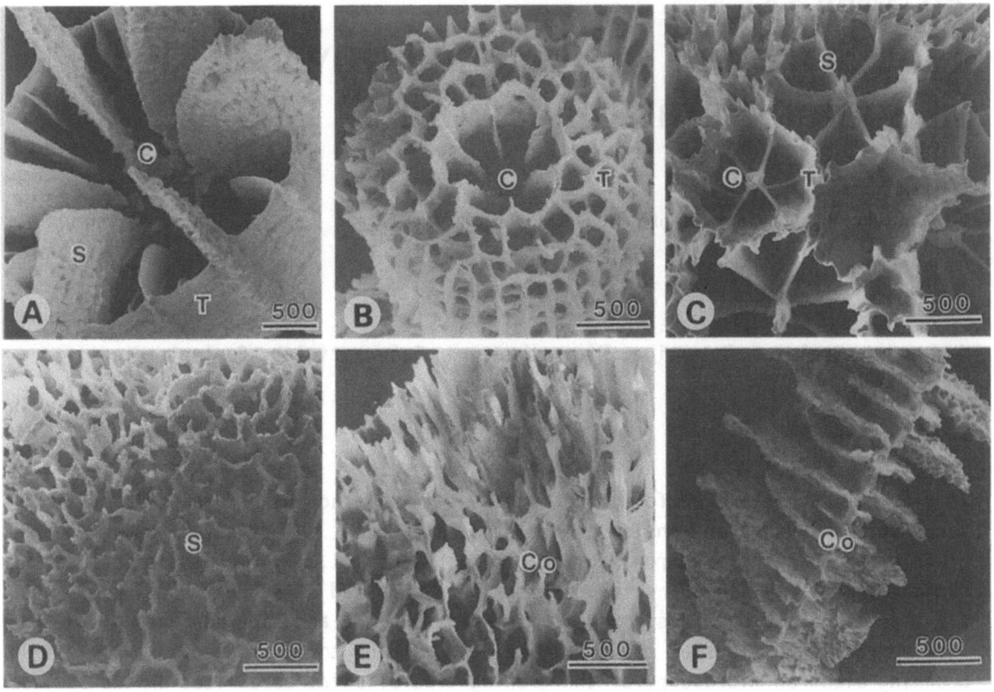

Fig.1 Top view of the marginal portion of some scleractinian skeletons. A, *Galaxea fascicularis;* B, *Acropora microphthalma;* C, *Stylophora pistillata;* D, *Porites andrewsi;* E, *Montipora aequituberculata,* F, *Pavona decussata.* *Abreviations*: C, calice; Co, coenosteum; S, septum; T, thecal wall. Bar scales are in microns.

lengthwise skeletal bars. In case of cerioid species, each corallite uses a thin layer of spinose thecal plates without any coenosteum. In another type of branching species (D), thamnasterioid, there exists no distinct thecal plates, hence, each corallite exhibits only delicate septal plates entering into a calice. Apecies of the two foliate species reveal the coenosteum texture of a number of skeletal plates are situated in random orientations (E) or are parallel to one another (F). Figure 2 shows the microstructure of the ridges of skeletal elements which constitute the corallites and coenosteum of a corallum (A-F). The main skeletal plates of the corallite, thecal wall (A) and septum (B), are composed of an aggregation of spindle-shaped and spherular crystals as well as the elements of the coenosteum (C) and the thecal bar and costae (D,E). A skeletal spine of the thecal wall also reveals a conglomeration of the crystal units (F). Most of the crystals are randomly oriented and are variable in size ranging from 0.5 µm to 15 µm in length. The crystals are arbitrarily divided into two classes in size, i.e., larger crystals of above 5 µm in length and smaller ones ranging from 0.5 µm to 5 µm long. In some cases, the larger spindle-shaped crystals are deposited with the long axis parallel to the marginal plane, forming the porous structures of a thecal bar (D). The majority of smaller crystals are distributed in random orientations, while several of them cluster in the form of a rosette on the ridges of skeletal plates (E). The clusters often exhibit a nodular structure on the skeletal margin as shown in Fig-

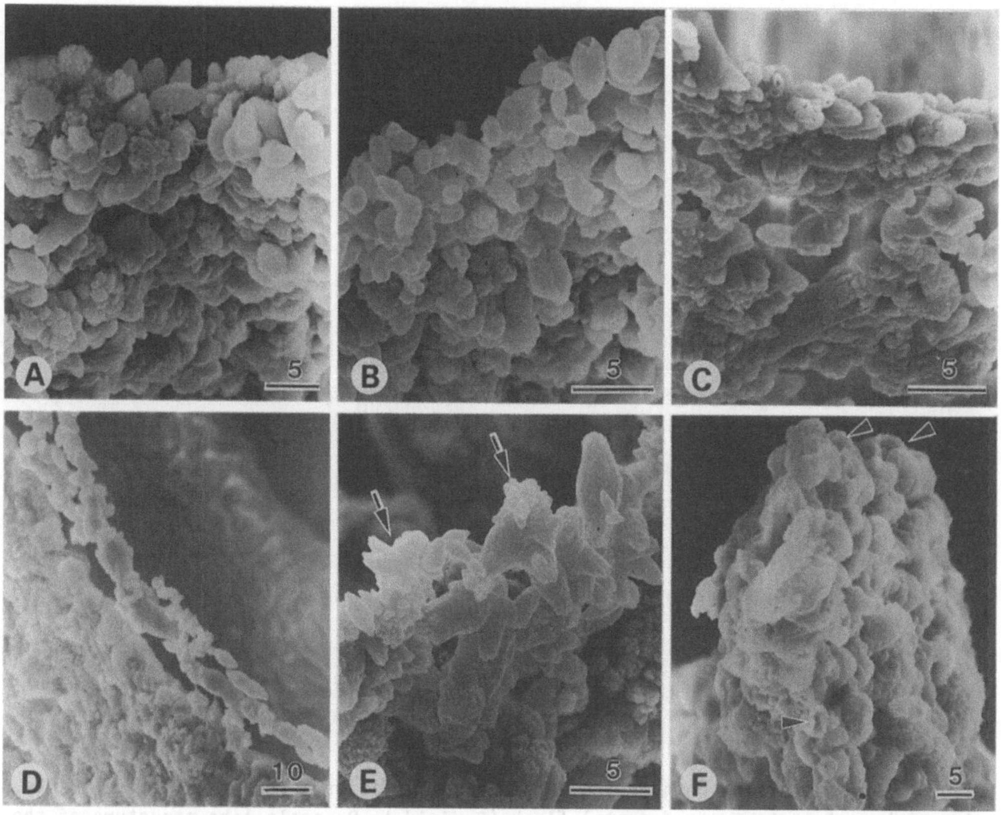

Fig.2 Enlarged ridge area of skeletal elements composing the coral corallum. A, theca (*G.fascicularis*); B, septum (*G.fascicularis*); C, coenosteum (*M.aequituberculata*); D, thecal bar (*A.hebes*); E, costae (*A.microphthalma*); F, thecal spine (*S. pistillata*). Arrows(E) indicate clustering of the crystals in a rosette texture. Bar scales are in microns.

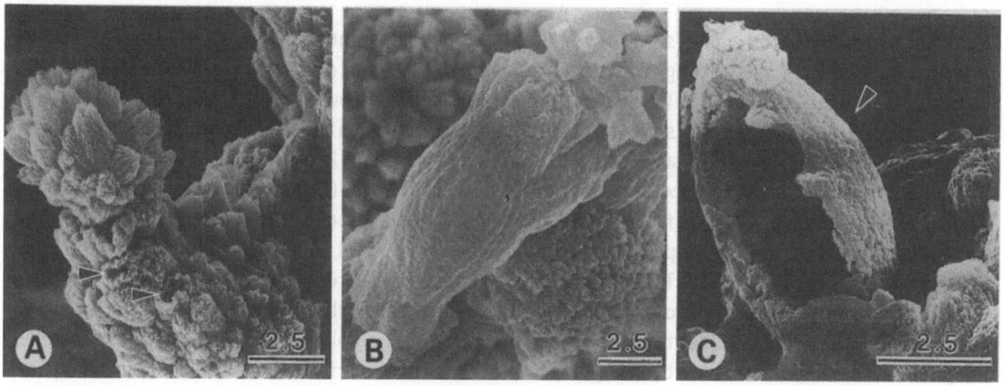

Fig.3 Characteristics of the crystals deposited on the margin of several skeletal plates. A, a nodular structure (*P.decussata*, coenosteum). B, spindle-shaped crystals showing irregular appearance (*A.microphthalma*, theca). C, partially broken spindle-shaped crystals of hollow structure (*E.grabrenscence*, septum). Bar scales are in microns.

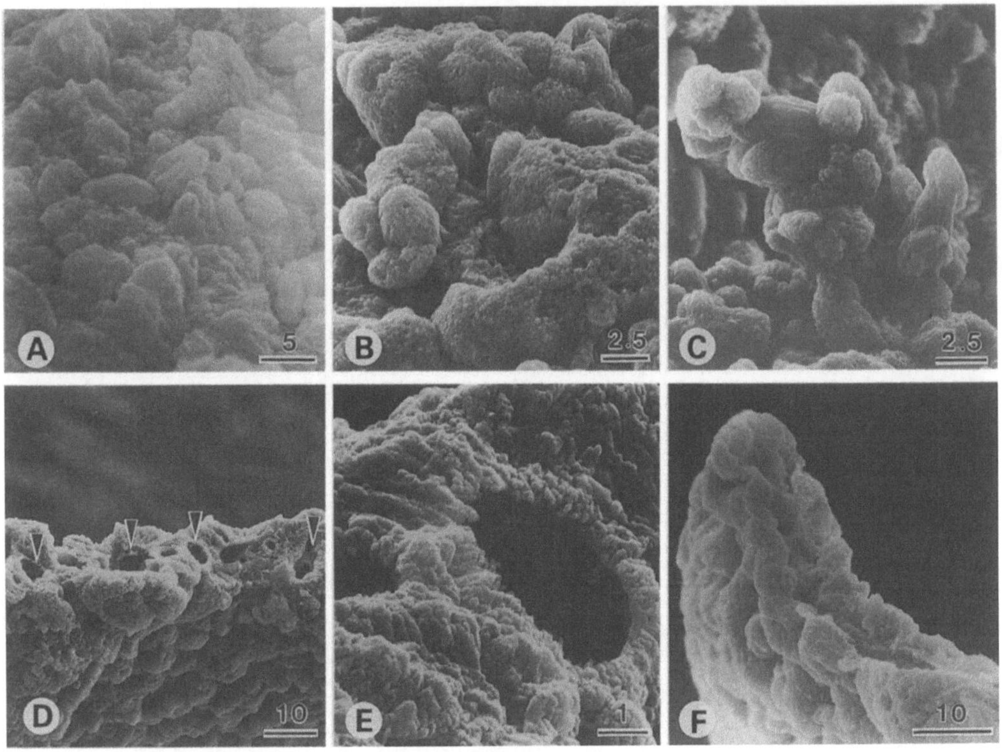

Fig.4 Surface texture both of the outside and inside of several skeletal elements. A, rough surface texture of septal flank (*H.rigida*). B, scale-like structure of the septal side surface (*E.grabrescence*). C, a small spine on the septal side surface (*E.grabrescence*). D, broken septum with many pores, arrowheads, in the plate (*E. grabrescence*). E, enlarged pores in D, showing tiny granular crystallite compositions. F, a broken thecal spine (*S.pistillata*). Bar scales are in microns.

ure 3A. In addition, the smaller crystals aggregate and coalesce in the pole of larger crystals, resulting in the formation of irregular spindle-shaped crystals (Fig.3B). When the crystals are partially broken (arrowheads in Fig.2F, Fig.3A,C), they reveal the characteristic hollow structure of a thin wall consisting of a number of tiny granular crystallites of about 0.1 μm to 0.2 μm diameter.

Figure 4 shows the fine texture of the side surface (A-C) and that of the septal and spine fracture (D-F) of skeletal plates. The skeletal flank has a rough surface structure caused by the depositions of spherular and unoriented spindle-shaped crystals (A). The scale-like texture of the septal side surface is due to the aggregation and coalescence of several crystal units of about 4 μm in diameter (B). The small spine on the septal flank is a clump of spherular and spindle-shaped crystals in random orientations (C). The fracture surface of the septum and spine indicates that the inside of the skeletal elements keeps the configuration of an aggregation of hollow crystals (D-F). The pores in the fracture are variable in diameter from 3 μm to 8 μm, and exhibit the crystal composition with the peripheral clustering of tiny granular crystallites of about 0.2 μm (E).

DISCUSSION

For the various growth forms of scleractinians, radioisotopic tracer experiments (^{45}Ca uptake) indicated to the apecies and peripheral regions of a colony to be the sites of active growth (8). It might therefore be advantageous to study the skeleton accretion at the growth-ridges of a corallum in clarifing the skeleton formation process in scleractinians. In addition to the general appearance of a porous framework structure of some skeletal plate components at the marginal portions of 9 species of shallow water coral skeletons used, it has been found that the periphery and side surface of these skeletal elements have a common configuration of aggregation of spindle-shaped and/or spherular crystal units, by scanning electron microscope. The crystals, varing in size from 0.5 μm to 15 μm long, characteristically showed a hollow texture and were mostly deposited in random orientations. Several crystals were responsible for the accretion of microarchitectures such as nodular, scale-like and spinose structures which were evident on the skeletal plate surface. These structures suggest that skeletal development in scleractinians occurs by formation and deposition of the crystal units, in agreement with the spontaneous and regenerating process of the accretion of the *Acropora* skeleton (5,6,9), and with larval skeleton development in *Pocillopora damicornis* (10). Although the continuous growth of crystal fibers from the calcification center had tentatively been proposed to in the formation of the scledodermite units of coral skeletons (1,11), the accretion of skeletal plates by an aggregation of hollow crystals is supported by the broken surface texture of porous crystal units of peripheral granular crystallites compositions, as the pores correspond to the center and the rod-like aragonitic crystals to the basic conglomeration of granular tiny crystallites of 0.1- 0.2 μm diameter. The formation of different shapes of crystals might be due to the activities of skeletogenic tissue, because there exists the intimate relationship between crystal development and cytoplasmic characters of calicoblastic cells (7).

ACKNOWLEDGEMENTS

The author is indebted to Dr. R. H. Richmond of University of Guam for his critical reading of the manuscript. This work is part of a doctorial study in the Faculty of Fisheries, Hokkaido University.

REFERENCES

1. Barnes DJ(1970) Science 17:1305-1308
2. Wise SW(1972) Biomineralization 6:160-175
3. Sorauf JE(1972) Paleontology 15:88-107
4. Jells JS(1974) Proc 2nd Int Coral Reef Symp 2:301-320
5. Isa Y,Yamazato K(1981) Proc 4th Int Coral Reef Symp 2:99-105
6. Gladfelter EH(1982) Coral Reefs 1:45-51
7. Isa Y(1986) Mar Biol 9:91-101
8. Goreau TF(1961) Endeavour 20:32-39
9. Isa Y(1987) Galaxea 6:153-162
10. Yamasu T,Isa Y(1987) Galaxea 6:125-151
11. Wainwright SA(1964) Exp Cell Res 34:213-230

CHAPTER 2.3

Fusiform and Needle-shaped Crystals Found on the Skeleton of a Coral, *Galaxea fascicularis*

M. HIDAKA
Department of Biology, Univeristy of the Ryukyus, Okinawa, 903-01 Japan

Key words: Coral, Skeleton, Calcification, Scanning electron microscope

INTRODUCTION

Two kinds of crystals, fusiform and needle-shaped crystals, have been observed on the surface of skeletons of hermatypic corals [1, 2, 3, 4]. Fusiform crystals are found only on the growing edges of distal tips of skeletons. It has been suggested that needle-shaped crystals are deposited on the surface of the fusiform crystals and that further growth and addition of needle-shaped crystals result in parallel bundles of needle-shaped crystals [2, 3]. Tips of these bundles protrude above the surface of the skeleton to form irregular-shaped, fish scale-like fasciculi. However, it is not clear whether the fasciculi are formed only in this manner and are always derived from fusiform crystals.

No fusiform crystals or needle-shaped crystals were found on the skeleton of polyps kept in darkness for a prolonged period. It has been suggested that, in darkness, the surface of skeletons becomes smooth because fine crystals are deposited to fill the spaces between previously deposited crystals [3]. However, the possibility that the flat surface of the skeleton is the result of skeletal dissolution in darkness has not been excluded.

In the present study, the skeletons of polyps and newly formed buds of a coral, Galaxea fascicularis were examined with a scanning electron microscope to examine the role of fusiform crystals in the formation of fasciculi and in the formation of species-specific architecture of the skeleton. In addition, growth rate and surface structure of skeletons of polyps kept in darkness for various lengths of time were examined in order to make sure that the smooth surface of skeletons of polyps kept in darkness is due to filling of gaps between previously deposited crystals by fine crystals and not due to skeletal dissolution.

MATERIALS AND METHODS

Colonies of Galaxea fascicularis were collected from the reef at the Sesoko Marine Science Center of the University of the Ryukyus in Okinawa. Individual polyps with their corallites were isolated from a colony by breaking off the basal portion of the corallite with forceps [5]. Isolated polyps were held in a vertical position by tying the basal skeletal portion to acrylic rods with thread. In some cases isolated polyps were tied to glass slides with thread and were held in a horizontal position.

Isolated polyps were maintained in a tank exposed to sun light through

a glass roof and windows. The skeletons were prepared for scanning electron microscopy by immersing the polyps in a 10% commercial hypochlorite solution in water for 6 hours. The specimens were rinsed in tap water and then in distilled water, and were air-dried. They were glued onto aluminum specimen stubs, sputter-coated in an Eiko Ioncoater and observed in an Hitachi S-530 scanning electron microscope.
To observe changes in the surface structure of skeletons of polyps which were kept in darkness for various lengths of time, isolated polyps were placed into a dark tank supplied with running sea water. Then polyps were collected every day for 7 days and their skeletons were prepared for scanning electron microscopy. Skeletal growth rates of isolated polyps kept in darkness or under natural light were measured using a buoyant weight technique [6]. Six isolated polyps were mounted on a plastic board. Twelve boards each with six isolated polyps were prepared. Six of them were maintained under natural light and other six were maintained in darkness. These sets were weighed in sea water every night for 6 or 7 days and increases in the weight of the skeleton of polyps were calculated.

RESULTS

Fusiform crystals 0.3-3 um wide and 0.5-5 um long were observed on growing edges of septa (Fig. 1a). Small fusiform crystals were observed only at the edge of the septa. Larger fusiform crystals were found usually just below the edge. These large fusiform crystals often fused with each other or were embedded among other fusiform crystals. Thin needle-shaped crystals were observed on the surface of these large fusiform crystals running in parallel with the long axis of the fusiform crystals. More or less fusiform tufts of needle-shaped crystals or small fasciculi were also observed (Fig. 1b). The more proximal region of the septa had a fasciculate surface.
When isolated polyps were reared, column wall tissues grew outwards and began to secrete an extramural skeleton, coenosteum. No fusiform crystals were observed on growing edges of coenosteum. At the growing edges of coenosteum, fine, needle-shaped crystals ran parallel to each other in the direction of skeletal extension (Fig. 2b). On the

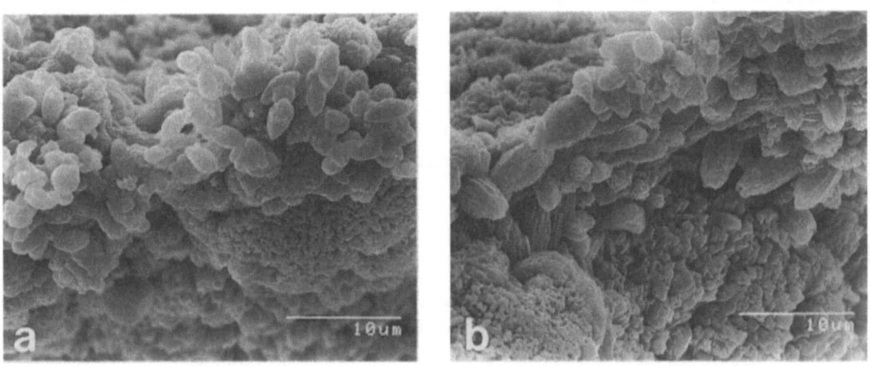

Fig. 1. Scanning electron micrographs of the growing edge of a septum. **a** Growing edge with many small fusiform crystals. **b** Adjacent area with larger fusiform crystals and fusiform tufts of needle-shaped crystals.

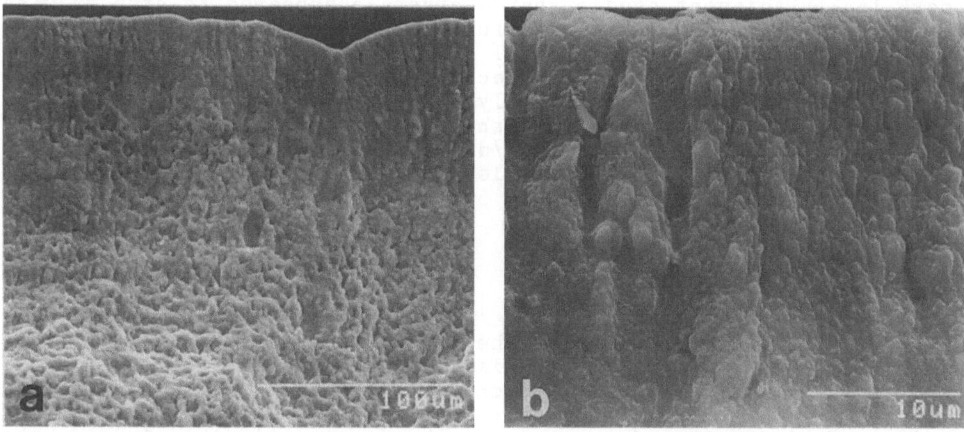

Fig. 2. Scanning electron micrographs of the coenosteum of an isolated polyp of <u>Galaxea fascicularis</u>. **a** Coenosteum with fasciculi on its proximal region. **b** Margin of coenosteum consisting of parallel running needle-shaped crystals.

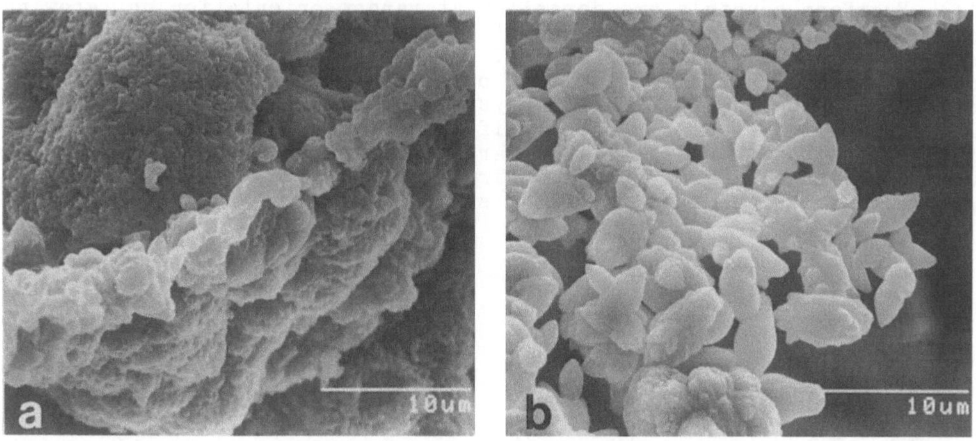

Fig. 3. Newly formed theca of a bud. **a** Rudiment of theca. Cluster of fusiform crystals were deposited on the fasciculate surface of the parent skeleton. **b** Theca of the bud at a more developed stage.

proximal region of coenosteum, needle-shaped crystals grew upwards from the surface of the previously secreted coenosteum and formed fasciculi (Fig. 2a).

Buds were formed on coenosteum as well as on outer column wall of parent polyps. When skeletons of buds were formed, fusiform crystals were first deposited on the fasciculate surface of theca or coenosteum of the parent polyp (Fig. 3a). Thus newly formed septa and theca of buds were composed of numerous fusiform crystals (Fig. 3b).

Fusiform crystals were absent on the septa of polyps kept in darkness for more than two days. The distal tips of the speta became almost completely smooth in four days after the polyps were transferred into

a dark tank. Adjacent needle-shaped crystals appeared to stop growing at the same plane and their tips fused with each other to form flat surfaces of irregular shape. Small dots of about 0.1 um diameter were sometimes observed on lateral surfaces of fasciculi and on apparently smooth surfaces of skeletons of polyps kept in darkness.
Measurements of daily growth rate showed that the skeleton of isolated polyps grew at a rate of 0.4-0.8%/day. Skeletons of polyps kept in darkness continued to grow for at least 7 days, though the growth rate decreased after 3-4 days to about 50% of those shown by polyps kept under natural light.

DISCUSSION

The present study showed that the septal margin had a different crystal structure from the proximal region of the septa. Small fusiform crystals were observed only at the edges of septa, larger fusiform crystals just below the edges, and fasciculi or parallel bundles of needle-shaped crystals in the more proximal region of the septa. These spatial differences in the crystal components of the skeleton may reflect the developmental process of fasciculi. The surface texture of large fusiform crystals also suggests that fusiform crystals grow by the deposition of needle-shaped crystals on their surface.
The present observations suggest the following scheme about the process of fasciculi formation on the septa of G. fascicularis (Fig. 4). Fusiform crystals are deposited at random orientation on growing edges of septa. The fusiform crystals grow by the addition of needle-shaped crystals on their surface. Further addition and growth of needle-shaped crystals result in parallel bundles of needle-shaped crystals. The needle-shaped crystals continue to grow in the longitudinal direction of the fusiform crystals. These bundles may fuse with other bundles to form larger bundles or may stop extending when other bundles grow across their path [7]. Tips of these bundles protruded above the surface of the skeleton make the fasciculi.

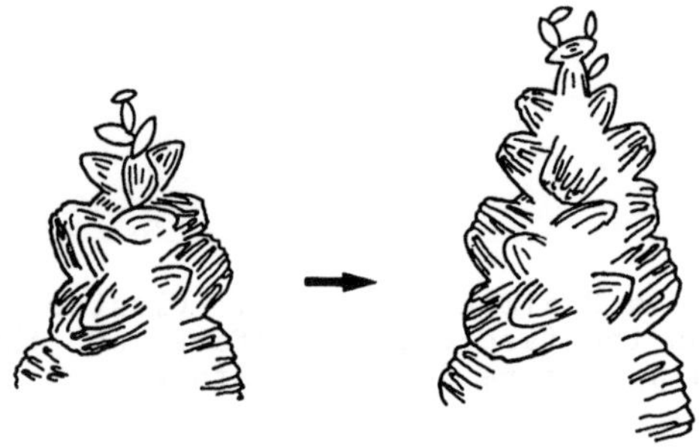

Fig. 4. Schematic diagram showing the process of fasciculi formation at the growing edges of the septa of Galaxea fascicularis.

Fasciculi are, however, not always formed in the manner described above, since fasciculi are formed on coenosteum where no fusiform crystals are deposited. In this case, needle-shaped crystals grow upwards from the surface of coenosteum, that is, perpendicularly to the calicoblastic tissue. These needle-shaped crystals somehow run in bundles, thus forming fasciculi.
Fusiform crystals often fuse with each other to form a meshwork, which provides a large surface area for needle-shaped crystals to be deposited upon. Thus the area where fusiform crystals are deposited extends faster than areas without them. When isolated polyps were

held in a horizontal position under natural light, fusiform crystals were deposited on the lateral margin of the septa which received most light rather than on distal edges of the septa [3]. This enables the coral skeleton to extend towards the direction of the maximum irradiance. Thus the phototropism displayed by the coral is at least partly due to displacement of actively calcifying site.

When bud skeletons were formed, fusiform crystals were first deposited. Since new fusiform crystals were deposited mainly on the previously deposited fusiform crystals, theca and septa of buds grew vertically to the parent skeleton. Species-specific architechture of coral skeletons is the result of different growth rates of various parts of the skeleton and thus depends on the pattern of deposition of fusiform crystals. Muscatine [8] questioned, "if the skeleton is deposited external to the living tissue, how is the species-specific architecture imposed?" One possible answer to this question is that corals make species-specific architecture of the skeleton by depositing the fusiform crystals in a specific pattern. The mechanism remains unknown how the deposition of fusiform crystals are regulated by cellular activity.

The measurements of growth rate of isolated polyps kept in darkness showed that the skeletal weight continued to increase in darkness at least for 7 days. Tips of septa became completely smooth within 7 days after the polyps were transferred into a dark tank. These observations strongly suggest that the smooth surface of the skeleton is not due to dissociation of skeletons in darkness but due to filling with fine crystals the spaces between needle-shaped crystals and channels between fasciculi.

Fusiform crystals are deposited at the growing edges of septa and also on newly formed septa and theca of buds. The deposition of fusiform crystals is not essential for the formation of fasciculi but increases the rate of extension of the area where the fusiform crystals are deposited. Thus the deposition of fusiform crystals might play an important role in determining the direction of skeletal growth and also in making the species-specific architecture of coral skeletons.

REFERENCES

[1] Gladfelter EH (1982) Coral Reefs 1: 45-51
[2] Gladfelter EH (1983) Coral Reefs 2: 91-100
[3] Hidaka M (1988) Proc Sixth Int Coral Reef Symp 3: 95-100
[4] Le Tissier MDA (1988) Coral Reefs 7: 81-88
[5] Hidaka M, Yamazato K (1982) Galaxea 1: 65-75
[6] Jokiel PL, Maragos JE, Franzisket L (1978) In: Stoddart DR, Johannes RE (eds) Coral reefs: research methods. UNESCO, pp.529-541
[7] Barnes DJ (1970) Science 170: 1305-1308
[8] Muscatine L (1971) In: Lenhoff HM, Muscatine L, Davies LV (eds) Experimental Coelenterate Biology. Universty of Hawaii Press, pp.227-238

CHAPTER 2.4

Fine Structure of the Regenerating Sites of the Stalk in a Solitary Coral *Fungia fungites* after Decalcification of the Skeleton

H. YAMASHIRO and K. YAMAZATO

Radioisotope Laboratory and Department of Biology, University of the Ryukyus, Okinawa, 903-01 Japan

Key words: Decalcification, Calcification, Regeneration, Coral Asexual reproduction

Introduction

A solitary coral Fungia fungites (Scleractinia, Fungiidae) reproduces by two ways. Sexual increase by planulation and asexual increase by detaching a disc from a stalk or budding [1]. It is known that the disc detachment is carried out by means of decalcification of the skeleton by the coral itself [2]. During the process of decalcification, the ordinary hard and translucent skeletons turn into weak, opaque and pulverized ones. After detachment, the stalk loses mouth and tentacles which are held on the disc. A new disc is formed not by budding but by direct re-growth of the corresponding structures in the stalk in Fungia sp. [3][4]. However, initial hard structural changes of the regenerating stalk of the coral have not been investigated.

In order to elucidate the sequential changes from skeletal dissolution to re-calcification, the microstructures of the regenerating sites of the stalk were observed by SEM.

Materials and Method

Coral samples were collected at Zanpa (127° 43'E, 26°26'N), transferred to Sesoko Marine Science Center (127° 52E', 26° 39'N) and reared in an aquarium supplied with running sea water. For the detached corals, the regenerating stalks were collected weekly, immersed in a 10% sodium hypochlorite solution in distilled water to remove the soft tissues, washed in distilled water, air-dried, coated with gold and observed under a scanning electron microscope (Hitachi S-530).

Results

After detachment, a detachment scar of a stalk was covered rapidly by the regenerating soft tissues almost within a day. One week later, in spite of still lacking both mouth and tentacles, such an incomplete polyp was observed to have already started to re-calcify. Figure 1 shows the site of initial deposition of a septum. The surface of a septum which is composed of pulverized crystal bundles has turned

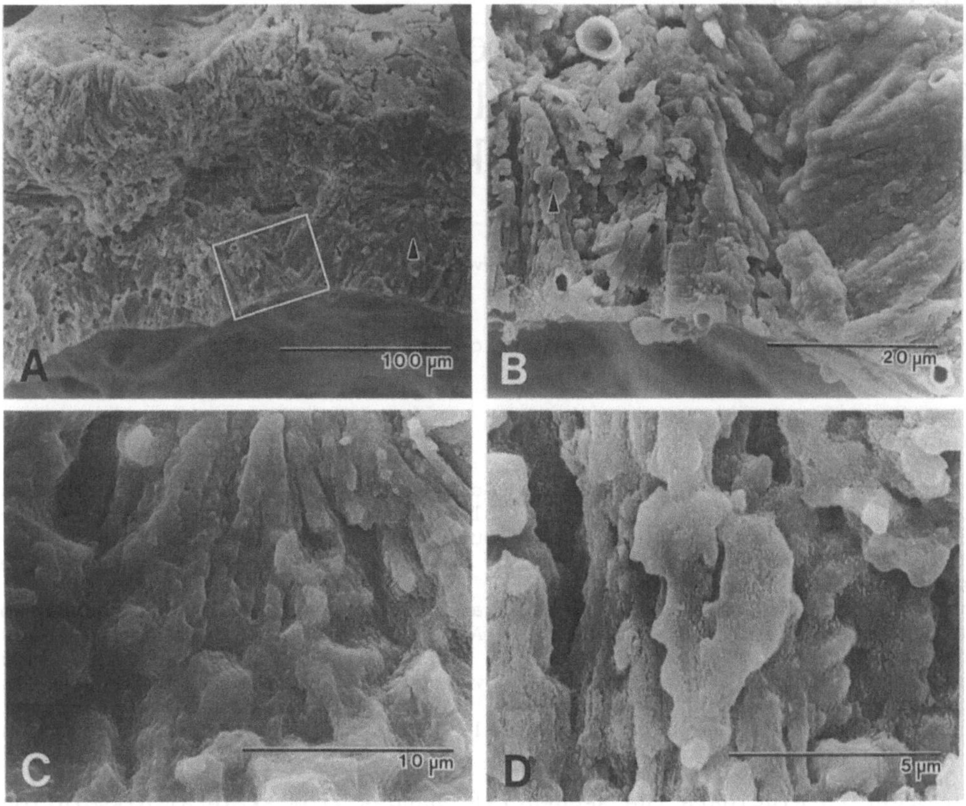

Fig.1. Fungia fungites. (A): Regenerating site of a septum of a stalk, one week after detachment. (B): Boundary region squared in A. Left half remains pulverized and right half shows the smooth surface, where the spaces between pulverized crystal bundles are filled up. Holes are made by boring algae. (C): Enlarged view of crystal bundles (arrow head in A) covered by fine crystals, less than 0.3 µm in diameter. (D): Enlargement of B (arrow head). Numerous fine crystals, less than 0.1 um in diameter, are being deposited around the surfaces of pulverized crystal bundles.

smooth and flat. The spaces between the bundles are cemented and filled with calcareous matter which is initially composed of very fine needle-shaped crystals less than 0.1 µm in diameter. These fine crystals were deposited on the crystal bundles which are exposed on the fractured surfaced and turned into round and smooth during crystal dissolution (Fig.1D). In the regenerating stalk of F. fungites, the degree of re-calcification greatly varied from place to place even in the same specimen.

Next step for the skeletal regeneration is upward growth. Figure 2 shows crystal particles arising from the smooth-surfaced septum which has been covered formerly by the needle-shaped fine crystals. These particles pile up, thus forming a cluster which will probably correspond to a trabecula, an element in structure of skeletal component, of a septum.

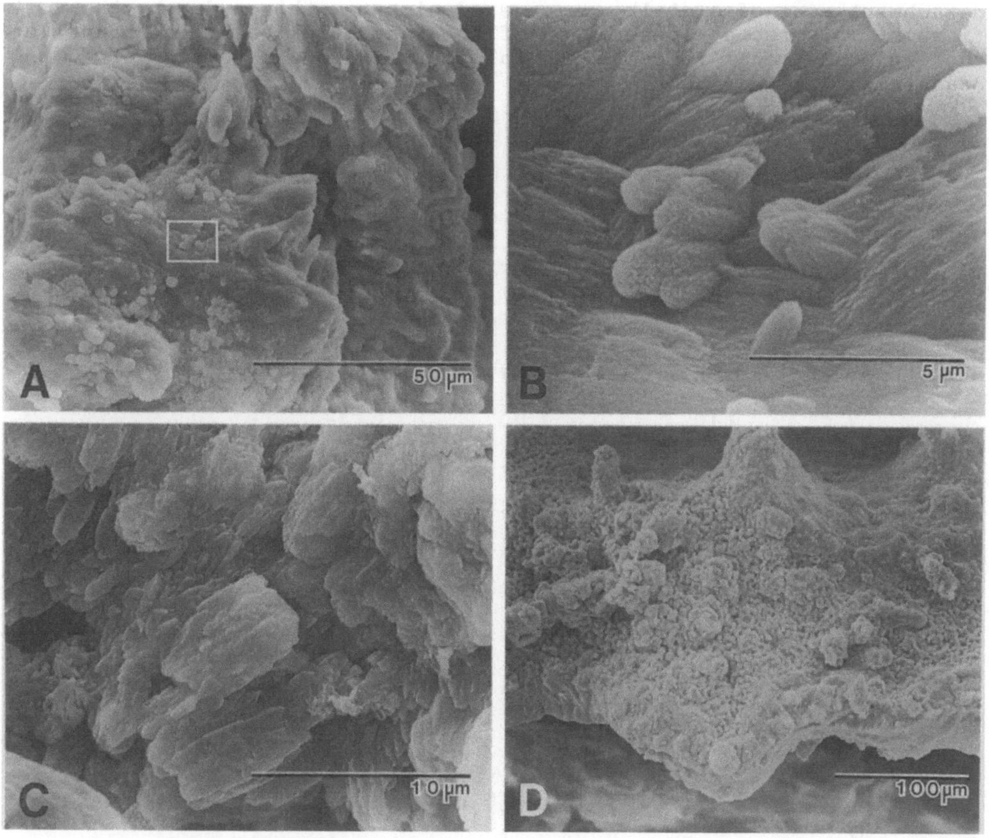

Fig.2. A regenerating site of a septum at the plane of detachment in Fungia fungites, 1 week after detachment. (A): Crystal particles being deposited on the smooth surface of a septum. (B): Enlarged view squared in A. Particles composed of fine crystals , less than 0.1 µm in diameter, growing upward. (C): Rugged crystals aggregate and pile up together. (D): Rugged crystals form clusters which make an initial septum.

Final stage of skeletal regeneration is accomplished by the reconstruction of skeletal components such as septum, wall and columella. Before or just after detachment the lateral surfaces of septum, wall and columella in a primary stalk are almost smooth and composed of fine needle-shaped crystals. However, after detachment, the surface became much more fasciculate towards the detachment plane (Fig.3). A regenerating septum was thin and was composed of a row of columnar structures. These regenerated skeletal components united together and make a secondary stalk which will grow upward and horizontally, thus forming a secondary disc afterwards.

Discussion

In damaged epitheca of some corals, the lappet which makes growth ridge was observed re-formed overnight [5]. Also in Fungia fungites, a detachment scar of a stalk was covered almost within a day by the soft tissues. This rapid regeneration of the scar is probably important for the injured coral to cope with falling silts and invasion of small organisms.

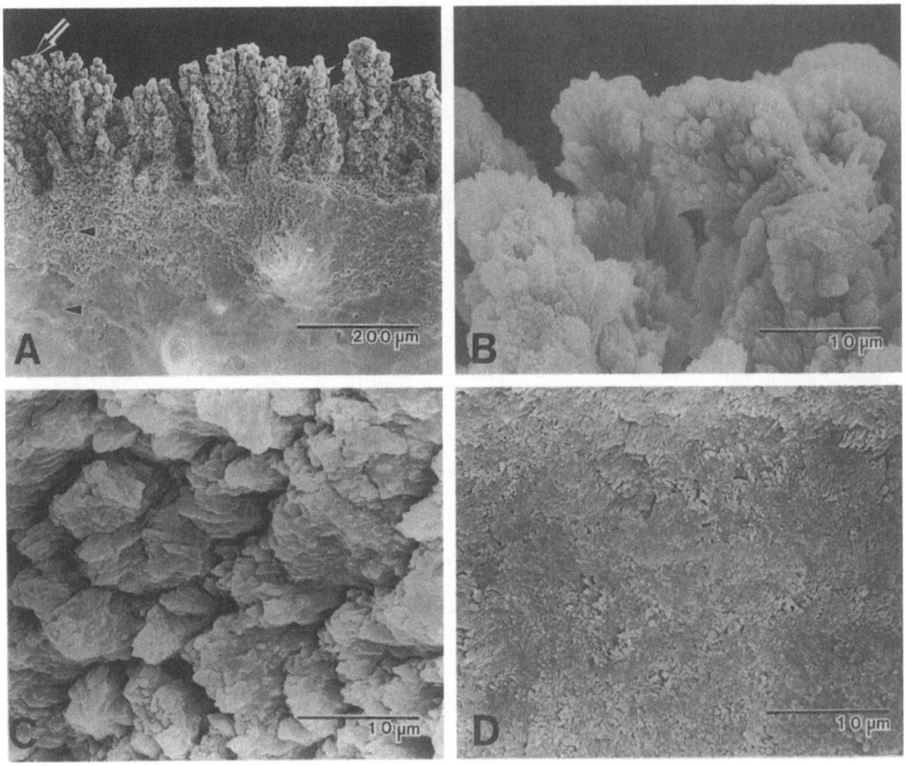

Fig.3. Fungia fungites, 3 weeks after detachment. (A): Lateral side of a regenerating septum. (B): Growing ridge (arrow in A) of the septum composed of blade-shaped crystals. (C): Much fasciculate surface observed in a primary septum, just beneath the detachment plane (upper arrow head in A). These crystals reinforce the detachment plane laterally. (D): Smooth surface of a primary septum (lower arrow head in A) which is composed of needle-shaped crystals.

Initially deposited crystals seem not only to strengthen the mechanically weak pulverized crystals [6] but to level the ground on which a secondary stalk will be built. Fusiform crystals observed at the growing ridges in some corals [7][8][9] were not found at the initial regenerating site in the present study. In a coral Acropora hebes 17 hours after fracture, polygonal crystal particles were deposited first on the fractured surface of the skeleton [10]. A comparative study of the artificially fracture stalk with that of naturally detached ones will yield a useful information on the regenerative processes.

Before detachment, lateral side of a primary stalk was almost smooth and composed of needle-shaped crystals (Fig.3D), which are similar to the crystals observed on a coral Galaxea fascicularis kept in continuous darkness [8]. On the other hand, the rugged crystals are deposited on the detachment plane of the primary stalk. This result suggests that laterally depositing skeletons beneath the detachment plane has a role of reinforcing the foundation for a coming secondary stalk. This suggests the low rate of calcification of a primary stalk before detachment.

It can be concluded that the drastic change from skeletal dissolution to re-calcification in Fungia fungites was supported by several steps of skeletal deposition which make rigid foundation required to construct sufficient mechanical strength for the next stalk. Firstly, the spaces between pulverized crystal bundles were cemented firmly and covered by fine crystals. Secondly, lateral side of the primary septum proximal to the detachment plane was reinforced by the deposition of crystals. In addition to cementation, septal thickening increase mechanical strength for the new generation.

References

[1] Wells, J.W. (1966) Symp.Zool.Soc.,London 16:223-246

[2] Yamashiro, H. & K. Yamazato (1987) Galaxea 6:163-175

[3] Lister, J.J. (1888) Quart.J.Micro.Sci. 29:359-363

[4] Bourne, G.C. (1893) Sci.Trans.Roy.Dublin Soc.5(Ser.II):205-238, pls.22-25

[5] Barnes, D.J. (1972) Proc.Roy.Soc.Lond.B. 182:331-350

[6] Motokawa, T., H.Yamashiro & K.Yamazato (1987) Galaxea 6:177-183

[7] Gladfelter, E.H. (1982) Coral Reefs 1:45-51

[8] Hidaka, M. (1988) Proc.6th Intern. Coral Reef Symp.3:95-100

[9] Isa, Y. (1986) Mar.Biol.93:91-101

[10] Isa, Y. (1987) Galaxea 6:153-162

CHAPTER 2.5

Larval Shell Formation and Mineralogy in *Neptunea arthritica* (Bernardi) (Neogastropoda: Buccinidae)

Y. Togo[1], S. Suzuki[2], K. Iwata[3], and S. Uozumi[3]

[1]Department of Earth Science, Hokkaido University of Education, Iwamizawa, 068 Japan; [2]Department of Earth Sciences and Astronomy, Fukuoka University of Education, Munakata, 811-41, Japan; and [3]Department of Geology and Mineralogy, Hokkaido University, Sapporo, 060 Japan

Key words: Larval shell, Mineralization, Microstructure, Ontogeny, Gastropoda.

INTRODUCTION

Internal structure of molluscan calcified shell, the shell structure, encompasses shell architecture, microstructure, shell layering and growth layering or growth structure [1] [2]. In gastropods, studies of completely calcified protoconchs sometimes preserved well on the teleoconch of adults indicate that no highly ordered microstructures, such as the crossed lamellar and nacreous, are formed in the protoconch shell, and that the protoconch calcification tends to produce not calcitic but aragonitic microstructures [3] [4] [5] [6]. Furthermore, studies of developing gastropod larvae show initial shell mineralization and the deposition of aragonite and trace calcite at the trochophore stage [7] [8]. Very little is, however, known about the detailed processes of the formation of shell microstructure during early developmental stages after the first precipitation of the mineral phase.
This paper presents descriptions of the sequential processes of microstructure formation of the pre-hatched developing larvae in the neogastropod *Neptunea arthritica*. Also reported are the relationship between two polymorphs of calcium carbonate, aragonite and calcite, both of which are precipitated during larval stages of this whelk.

MATERIALS AND METHODS

The prosobranch neogastropod *Neptunea arthritica* was used in this study. Adults of this species were collected from a rocky shore in Japan (41°42'N, 140°31.5'E) in every late June to early July of 1985, 1986 and 1987. Specimens were maintained in aquaria with flow-through natural sea water under natural conditions (20.3-23.2°C; pH: 6.8-7.2) in the laboratory and parents laid about 30 to 50 egg capsules in clusters. Capsules were marked and the time of deposition was recorded.
Encapsulated larvae were reared in aquaria, the conditions being the same as those in the adults. The larvae of various developmental stages were removed mechanically from capsules, placed in a shallow dish half filled with sea water, and were observed under a binocular microscope to estimate the approximate stage of development. They were then killed in distilled water or anesthetized in sea water with sodium sulfate solution. Specimens were fixed in 10% neutral formalin with or without sea water, 2% glutaraldehyde in 0.1M phosphate buffer, or in 70% ethanol. Larvae were dehydrated in a graded series of ethanols and were then air dried or critical-point dried and coated with gold for scanning electron microscopy (SEM), or embedded in paraffin and sectioned for light microscopy.
To obtain a clear outline of shell minerals, organic materials were removed from some calcified shells with 2.5% sodium hypochloride solution (NaClO).
Shell minerals were determined by X-ray powder diffractometry.

RESULTS AND DISCUSSION

Outline of Development

The fertilization of *N. arthritica* takes place internally, and parents lay egg capsules between late June and the middle of August. Each capsule contains many eggs, but most of these may be consumed as nurse eggs.
By about ten days after spawning larvae show an indefinite shape under the microscope. By 15 to 17 days larvae become spherical or ellipsoidal in shape with a diameter of 2.0 to 2.7 mm, having only the first

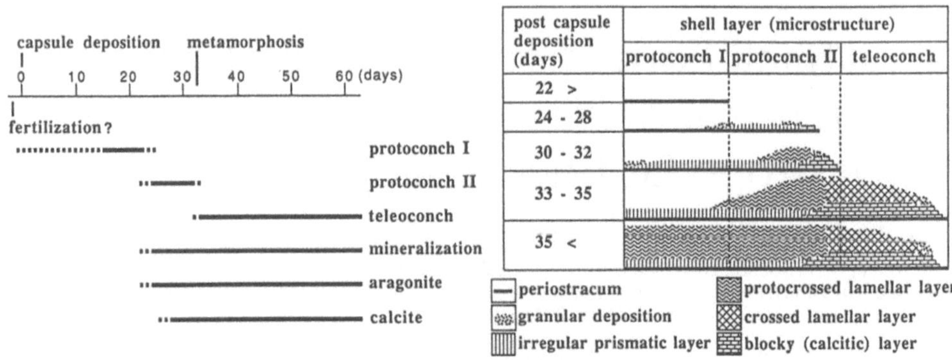

Fig. 1. Relationship of stages of development to mineralization and formation of shell whorl and shell layer (microstructure). Dotted lines indicate variability and uncertainty (left); shell length and thickness of shell layers not to scale (right).

whorl, provisionally named protoconch I (pc-I), consisting of an organic membranous shell layer. By about 23 to 24 days the subsequent whorl of the larval shell (protoconch II: pc-II) is formed. Incipient pc-II is a homogeneous membrane narrowly secreted around the mantle margin. Further development of the larvae leads to the increase of the diameter in both the pc-I and pc-II. Metamorphosis occurred in about 32±1 days post capsule deposition.

Adult ornament or teleoconch is formed during development within the capsule, and the increments of the teleoconch increase into 0.8 to 1.3 mm long by 39 days, 1.2 to 2.0 mm by 47 days and about 6.0 mm by 60 days. The growth of the larval shell (pr-I and II) appears to be completed by the time of hatching. The developing larvae are in the capsules for at least about 60 days, and only one (occasionally two or three) young survive to emerge.

Timing and Proceeding of Shell Mineralization

The larval shell (pc-I) consists of a homogeneous organic membrane of less than 1.0 μm thick by 17 to about 22 days, the pc-II not being formed by this time.

By about 24 days first deposition of minerals is observed on the periostracum of the pc-II; mineral depositions are dense on the major part of the pc-II, about 0.6 to 1.0 μm thick, but become thinner proximally, about 0.5 μm thick near the pc-I/II junction, and are not distributed on the major part of the pc-I. The shell edge remains organic through the developmental processes of mineralization. By 28 days mineralization proceeds toward both the proximal (apical) and distal (apertural) parts of the shell, but it is incomplete in the pc-I by this time. Although X-ray powder patterns have not been obtained clearly from the shell of the 24- to 28-day-larva, at about 28 days the larva appears to deposit both aragonite and calcite, because two different forms of deposition appear on the pc-II; one is granular aggregations and the other is the depositions with rhomboidal to hexagonal outline near the shell edge.

The pre-metamorphic 30- to 32-day-old larva apparently secretes aragonite and calcite. Both minerals are deposited on the pc-II, but only aragonite on the pc-I. By this stage mineralization of the pc-I and pc-II is completed (Fig. 1).

After the completion of mineralization, the thickness of the mineralized layers increases to 6 to 8 μm (pc-I) and 10 to 13 μm (pc-II) by 35 days, and 40 to 250 μm (pc-I) and 50 to 660 μm (pc-II) by the stage just before hatching.

There are two possibile interpretations for the observed phenomenon that mineralization does not begin stepwise from early formed shell; one is that the secretionary cells located in the pc-I region remain dormant or "semi-active" after they form the first organic shell and then become active in secretion of inorganic component; the other is that at least in the pc-I the mineralized shell is an inorganic residual not basically associated with the activities of the secretion cells.

Formation of Aragonitic Microstructure

Granular deposition. — While the initial state of aragonite deposition varied among the specimens examined, the deposition, in general, began as granules of less than about 1 μm in diameter (Fig. 2).

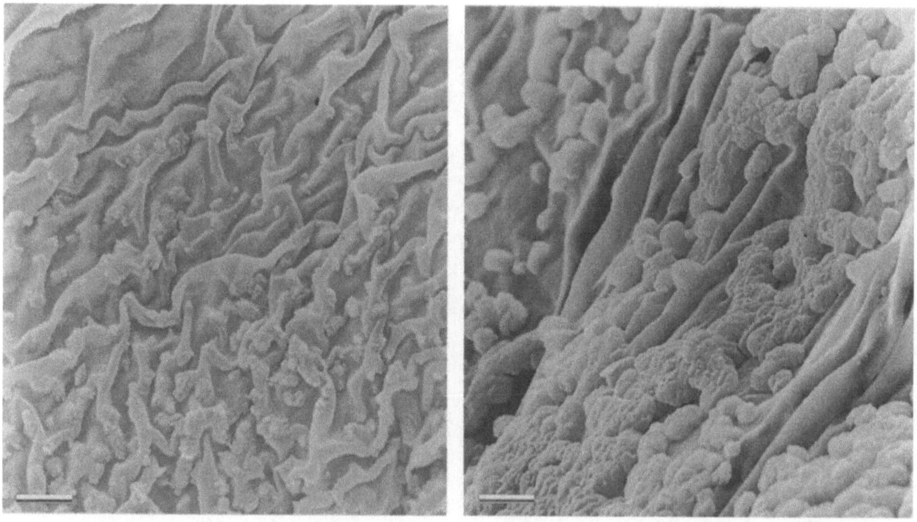

Fig.2. Inner surface views of granular depositions on the periostracum near pc-I/II boundary (left, bar = 2 μm) and early pc-II (right, bar = 2 μm); 24 days after spawning, air dried.

Although no detailed information has been obtained at the SEM level, from their smooth-faced appearance the granules may be "organic" initially and then crystallized, being the same as those in other molluscs previously examined [6] [9]. The granules grow into crystalline grains with rugged surfaces and are then aggregated to form a layer of about 0.6 to 1 μm thick on the periostracum. This layer is seen to be granular or homogeneous on the vertical and horizontal surfaces, appearing to contain an amount of organic materials.

Irregular prismatic structure (ipr). — The crystalline layer thickens to about 2.0 μm by 28 days and appears to be irregularly prismatic in the vertical view, the prisms arranged loosely and still having a granular aspect partly at this stage (Fig. 3). After this stage the layer becomes more rigid and compact by the growth of crystal faces of the prisms. The prisms, 0.5 to 2.0 μm in diameter, are arranged perpendicularly (or at high angle) to the shell surface so that the elongation of prisms produces parallel (or radial) features on the vertical surfaces of the shell. The prisms are composed of finer, basic units, the dimensions of which have not been confirmed in this study but appear to be about 0.2 μm in diameter.

The *ipr* layer is almost always formed on the periostracum of the pc-I and the major part of the pc-II as a substratum of the protocrossed lamellar layer described below and has a nearly uniform thickness of about 3.0 to 5.5 μm through the larval development of this species.

Characteristics of the *ipr*, simpler form and arrangement, considerably wide distribution [2] [10] [11] [12] [13], and occurrence as minor shell layer, suggest that the mineralogy of aragonite allows the *ipr* to grow automatically under conditions less influenced by organic materials.

Protocrossed lamellar structure (pcl). — At about 30 days more complicated but not so ordered arrangements of crystal aggregates are formed continuously on the *ipr*. The smallest unit of the structure, third order unit, is an acicular to prismatic crystal of aragonite, less than 0.2 to 0.4 μm in diameter. These acicular crystals aggregate to form a larger, second order structural unit which usually shows an irregular or indistinct outline but sometimes shows lamellar-like shape. The first order units, in which the second and/or third order units are organized radially or sometimes in parallel, are irregular and variable in shape, hemi-spherulitic, fan-shaped, bundle-like, spindle- or cylindrical-shaped, and so on, and usually show unsettled arrangements (Fig. 4). These aggregates grow into favorable orientations until they meet other aggregates and cease growing; however, the first order units occasionally and partially show the arrangement and shape similar to those in the crossed lamellar structure. These features indicate that the newly formed layer at this stage is constructed by the protocrossed lamellar structure which characterizes the larval shell microstructure in some meso- and neogastropods [2] [6].

Growth manner of the crystal aggregates suggests that the *pcl* is formed mainly by mineralogy like the low of geometric selection [14].

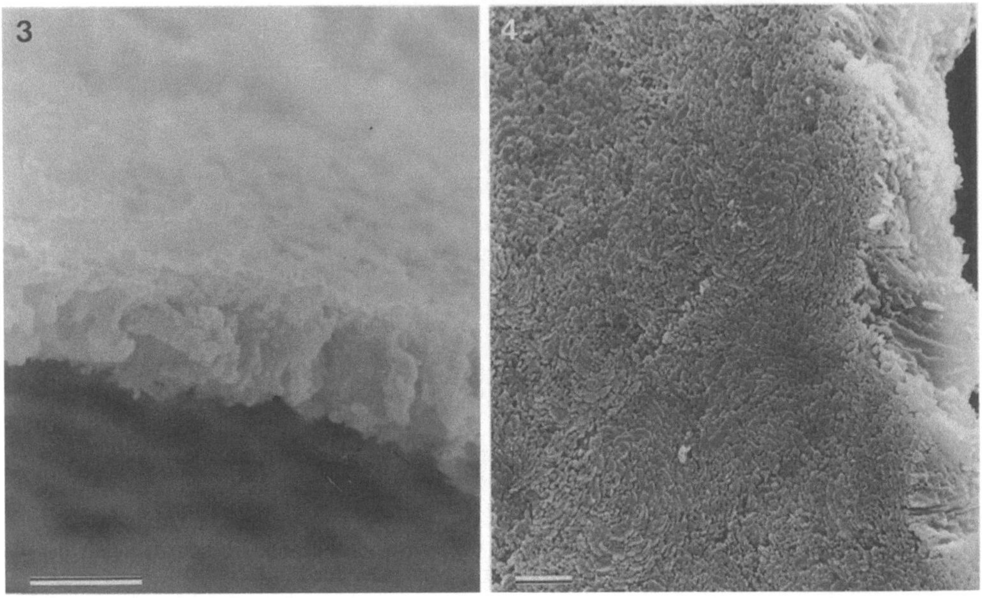

Figs. 3-4: 3, incipient irregular prismatic layer with granular appearance in part, late pc-II of 27-day-old larva (air dried: bar = 2 μm); 4, oblique view of protocrossed lamellar layer, middle pc-II of 32-day-old larva (organic materials removed: 2.5% NaClO, 20 hours, air dried, bar = 2 μm).

Crossed lamellar structure (cl). — By about 33 to 35 days the *pcl* changes into another structure with a more ordered arrangement, the crossed lamellar structure, in the pc-II. Vertical views of fractured or polished sections indicate that the structural change from the *pcl* to *cl* occurs gradually. The smallest structural units of the *cl* referred to as "third order lamellae" are thin (about 0.2 μm) aragonitic crystals, and are organized into the higher structural units, second and first order lamellae. Such a hierarchy of the structural units in the *cl* is nearly the same as that in the *pcl*, but irregularity in both the shape and arrangement of the higher structural units has disappeared in the *cl*. Highly ordered arrangements of each structural unit in the *cl* imply that the structure results from the mineralogy of aragonitic crystals. On the other hand, the almost uniform width of the largest unit suggests that physiological periodicity of the animal, such as rhythmic secretion or change of secretion rate by the epithelial cells, is reflected in the structure.

The shape and arrangement of each structural unit in the *cl* remain unchanged during the ontogeny of this species.

Formation of Calcitic Microstructure

Blocky structure (bl). — Calcite is probably deposited by about 28 days after spawning, about 5 days before metamorphosis. The calcitic *bl* layer constructs the outer layer of the pc-II.

Early form of the *bl* is typically observable near the shell edge (Fig. 5); scattered calcite depositions of less than 5 μm in diameter are formed at the distal front of mineralization. The depositions grow to about 20 μm or more in diameter to cover almost the entire area of the periostracum near the shell edge. The *bl* layer grows by adding newly formed calcite crystals and is overlain by aragonitic layers in more proximal parts (Fig. 6). SEM observations reveal in detail the growth of calcite crystals; thin (0.3 to 0.5 μm) crystals with plate-like (or occasionally prismatic) appearance oriented parallel with respect to their corresponding predominant faces grow into a larger crystal aggregation by means of an intergrowth or parallel growth within organic membrane (Fig.7). "Striations" with two main orientations are sometimes recognized on the vertical surfaces of this layer; the striae intersect at an angle of about 102.5° to 105°, being indicative of cleavage surface of calcite.

The growing habit of the calcite crystals mentioned above is similar in morphology to that of some carbonate crystals, for example, of non-biogenic calcite or rhodochrosite ($MgCO_3$). It is likely from such growth morphology that the *bl* is formed by the precipitation of calcite crystals under conditions less

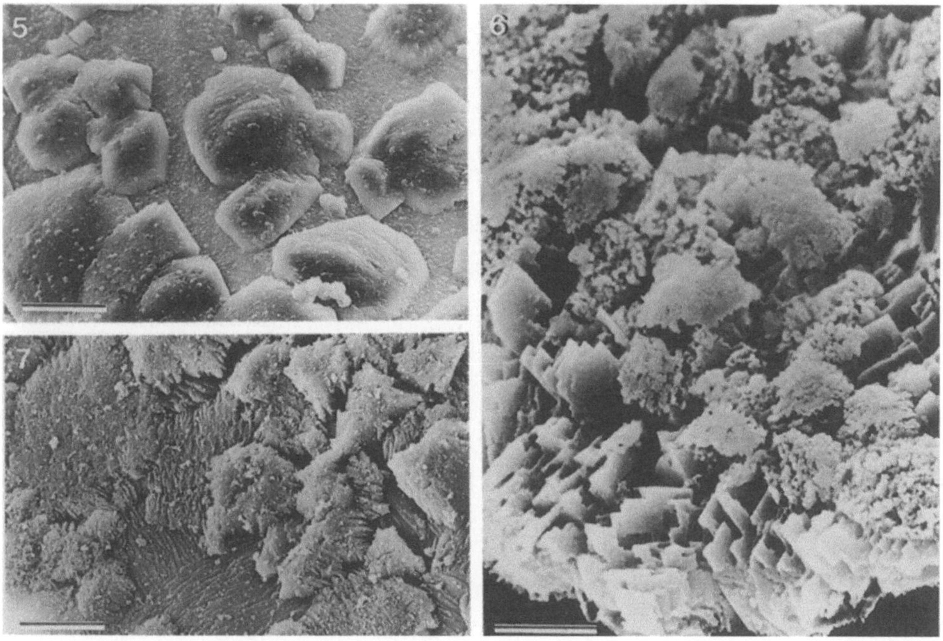

Figs. 5-7: calcite depositions near the shell-edge; 5, 31-day-old larva (air dried, bar = 10 μm); 6, early aragonitic deposition on calcitic layer with rhombohedral appearance, 32-day-larva (organic materials removed: 2.5% NaClO, 20 hours, bar = 4 μm); 7, development by intergrowth or parallel growth of smaller plate-like crystals, 33-day-old larva (organic materials removed: 2.5% NaClO, 20 hours, air dried, bar = 10 μm).

influenced by organic materials, or is a consequence of the mineralogy of calcite. The ability of calcite formation in this species may be derived from some physico-chemical changes of secretion, such as the amounts and kinds of cations, anions and organic materials [15] secreted from the epithelium, and mortion of the extra-pallial fluid.

REFERENCES

[1] Carter JG (1980) In: Rhoads DC, Lutz A (eds) Skeletal Growth of Aquatic Organisms. Plenum, New York and London, pp. 69-113.
[2] Togo Y, Suzuki S (1988) In: Omori M, Suga S, Goto M (eds) Biomineralization and Phylogeny of Marine Organisms. Tokai University Press, Tokyo, pp.113-134.
[3] Erben HK, Krampitz G (1972) Biomineralization 6: 12-31.
[4] Batten RL (1975) Am Mus Novitates no. 2567: 1-29.
[5] Arnaud PM, Bandel K (1976) Tethys 8: 213-230.
[6] Togo Y (1984) J Geol Soc Japan 90: 565-576
[7] Eyster LS (1986) Biol Bull 170: 211-231.
[8] Ivester MS (1972) Amer Zool 12: 717.
[9] Uozumi S, Ohta S (1977) J Geol Soc Japan 83: 425-432.
[10] Blackwelder PL, Watabe N (1977) Biomineralization 9: 1-10.
[11] Togo Y (1981) J Geol Soc Japan 87: 519-526.
[12] Waller TR (1981) Smithsonian Contributions to Zoology no. 328: 1-70.
[13] Suzuki, S (1983) J Geol Soc Japan 89: 433-442.
[14] Grigor'ev (1965) Ontogeny of Minerals. Israel Prog. Sci. Transl. Jerusalem, pp.1-250.
[15] Kitano Y (1986) J Oceanogr Soc Japan 42: 402-420.

CHAPTER 2.6

Using Plasma-Etching and Proteolytic Enzymes in Studies of Molluscan Shell Ultrastructure

H. Mutvei
Swedish Museum of Natural History, S 10405 Stockholm, Sweden

Key words: Shell ultrastructure, Plasma-etching, Enzyme-etching.

INTRODUCTION

As in all mineralized tissues, the mineral components of the molluscan shell are associated with complex organic matrices. In order to study shell ultrastructure with the SEM the shell must first be etched. However, etching solutions used hitherto produced artefacts by partially dissolving both the organic and the mineral components. Using new plasma-etching technique the organic components have been removed without causing any serious alteration to the mineral components (1). Results from the plasma-etching technique are compared with those from enzyme-etching.

MATERIALS AND METHODS

Materials used in this study were: (1) the nacreous layer of the shell of the cephalopod Nautilus pompilius from the Solomon islands; and (2) the outer aragonitic prismatic shell layer from the freshwater bivalve Margaritifera margaritifera from Hälsingland, Sweden.
Plasma-etching was carried out with a Cemex (Durham, U.K.) machine for 6 hours, at a temperature of about 40°C and in an atmosphere of oxygen and tetrafluormethane (R14) at low pressure.
For enzyme-etching the following enzymes were used in cacodylate buffer at pH 8.0: Neutral Proteinase, Trypsin, and Proteinase-K. Neutral Proteinase and Proteinase-K gave the best results because they removed the organic matrix more gently and efficiently than Trypsin. The duration of enzyme-etching was between 24 and 48 hours.
The etched preparations were examined with a Philips SEM 505 in the Swedish Museum of Natural History, Stockholm.

RESULT

Nautilus Septal Nacreous Layer.

Thin glycoprotein sheets always cover the inner, growing shell surface where precipitation of mineral matter takes place. These have hindered direct observation of crystal morphology. But using plasma-etching, these sheets can be gently removed without affecting the morphology of crystalline elements.
In the shell septa of Nautilus, inter-lamellar organic sheets were removed from surfaces of nacreous tablets by plasma-etching. After etching, the tablets show parallel rows of numerous low, elongate tubercles, less than 0.1 microns wide. In a central circular area of each tablet, the tubercles are much more numerous and are smaller (ca, Fig. 1).
In the preparations etched with Neutral Proteinase, not only were the interlamellar sheets dissolved but also the surfaces of the nacreous tablets, and parallel mineral laths became visible /2/ (Fig. 3). These laths correspond to the parallel rows of tubercles on the tablet surface, seen after plasma-etching (compare

Fig. 1 with Fig. 3).
The arrangement of rows of tubercles on the nacreous tablets also corresponds to the pattern of inter-lamellar organic sheets which cover the tablets and which, during subsequent growth of the nacreous layer, become embedded between the nacreous lamellae. In transmission electron microscope preparations the inter-lamellar sheets reveal a composition of parallel thick trabeculae (about 0.1 microns thick), separated by elongate, extremely thin, inter-trabecular areas /3, 4/ (Fig. 2). The inter-trabecular areas are comparatively large in the central portion of each sheet, forming a transparent area (tr, Fig. 2). A comparison of the structure of the inter-lamellar organic sheets and the morphology of the nacreous tablets reveals that the inter-trabecular areas correspond to the tubercles on the tablet surface, and that the transparent area of the sheet corresponds to the central area of the tablet. The latter area is the nucleation site of a new tablet, and contains an accumulation of organic matrix, probably sulphated acid polysaccharide /5/. As previously stressed /2/, the consecutive nacreous tablets are crystallographically continuous with each other through the inter-trabecular areas of the organic sheets.

Margaritifera Outer Aragonitic Prismatic Shell Layer.

After plasma-etching, the periostracum lost its compact structure and appeared as a mesh of coarse organic fibres, about 1 micron in diameter (pr; Fig. 5), associated with somewhat larger, thin platelets. The organic sheaths which encase the prisms are much thickened adjacent to the periostracum where the initial portions of the prisms have a small diameter (sh, pr, Fig. 4). Also these sheaths lost their compact structure after plasma-etching, and appeared to be composed of much thinner fibres than those in the periostracum, being about 0.1 microns in diameter (sh, Figs 5,6).
The initial portions of the aragonite prisms often form thin tablets with hexagonal outlines, 2-4 microns in diameter and resembling nacreous tablets (Fig. 7). The prisms, and their initial portions, are composed of elongated mineral granules, about 0.5 microns in diameter (Fig. 8). These granules are arranged in parallel, more or less distinct rows. Some of the initial hexagonal tablets are subdivided by indistinct furrows into a varying numbers of sectors. The latter feature also strongly resembles the subdivision of the sectors of nacreous tablets.
Enzyme-etching for several days did not reveal the fibrous structure of the periostracum and prismatic sheaths, as did plasma-etching.

REFERENCES

/1/ Arnold JM, Mutvei H, Landman NH, Kuzirian AM (1990) Am Mus Novitates 2974: 1-9
/2/ Mutvei H (1979) Scanning Electron Microscopy 1979/II:457-462
/3/ Mutvei H (1969) Stockholm Contr Geol 22:1-17
/4/ Gregoire Ch (1972) Haliotis 2:51-79
/5/ Crenshaw MA, Ristedt H (1975) Biomineralization 8:1-5

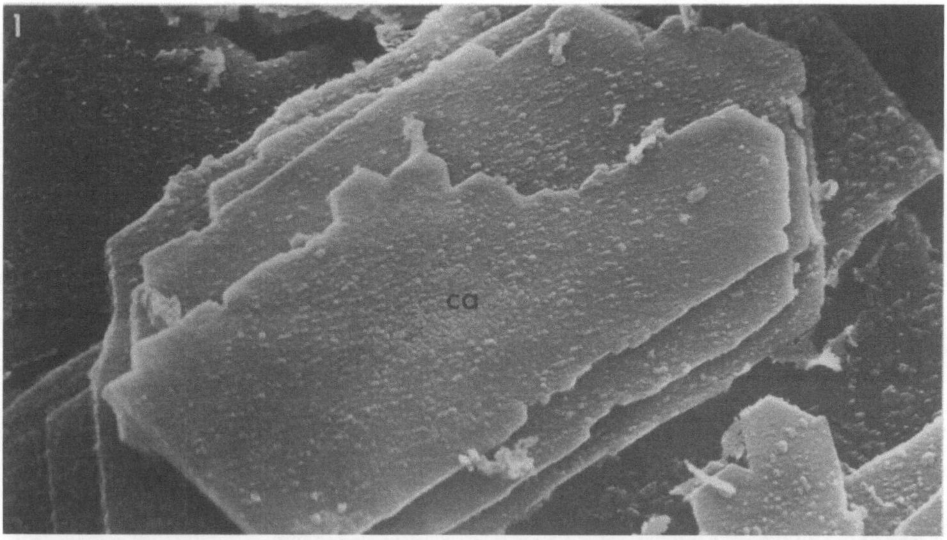

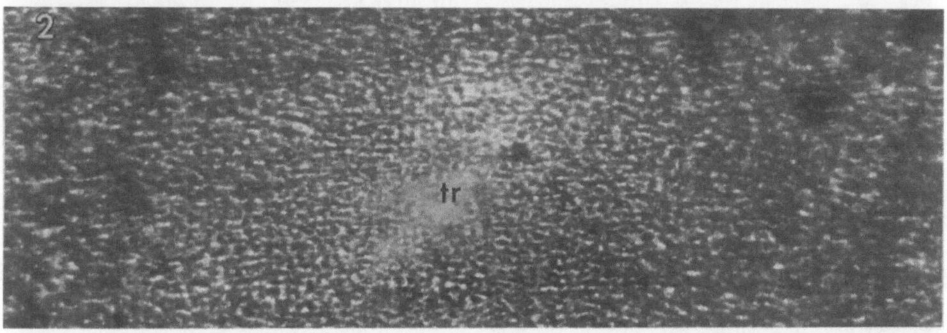

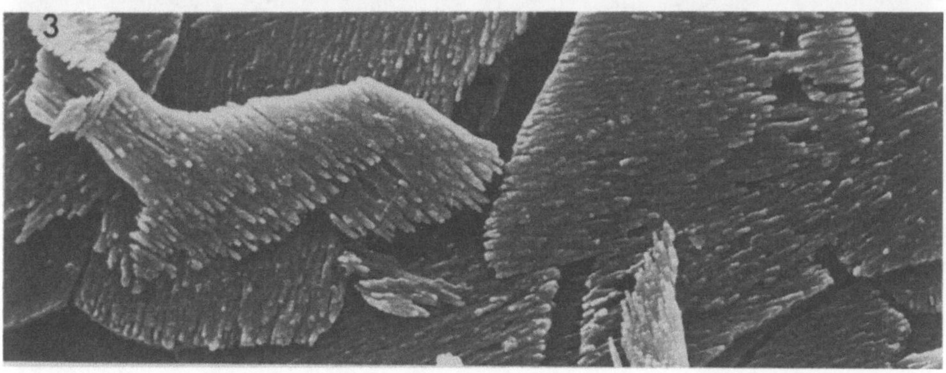

<u>Nautilus</u>. Fig. 1. Stack of nacreous tablets on the septal surface, treated for 6 hours in plasma-etcher; note rows of elongated tubercles on surfaces of tablets, and a central area (ca) where tubercles are smaller and more numerous. Fig. 2. Transmission electron microscope preparation of an inter-lamellar organic sheet, demineralized in EDTA-formaldehyde, and stained with uranyl acetate; note parallel trabeculae (dark), separated by elongated inter-trabecular areas (light), and central transparent area (tr). Fig. 3 Septal nacreous layer after etching with Neutral Proteinase for 36 hours; note that each tablet is composed of parallel mineral laths. Bar = one micron.

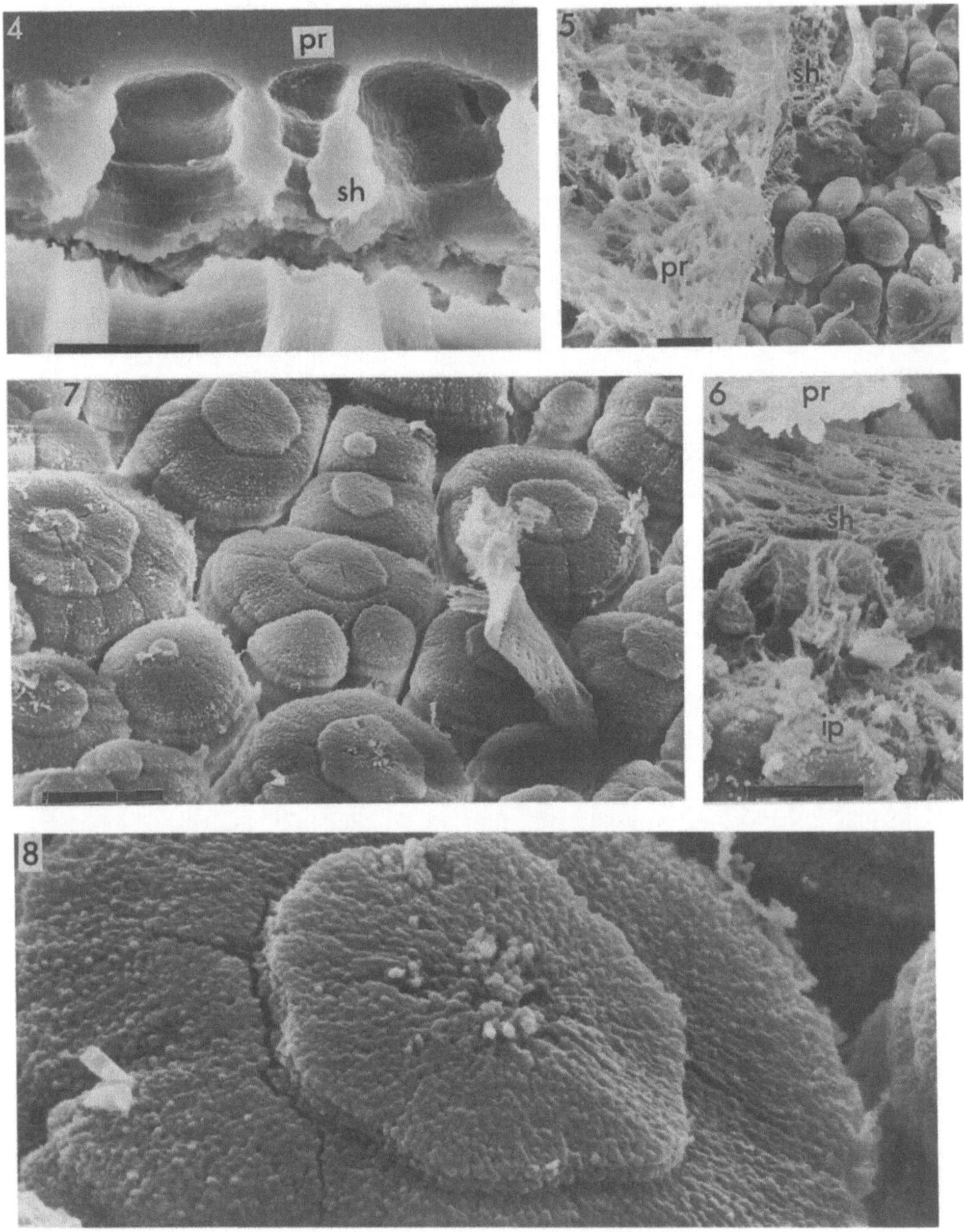

Margaritifera. Fig. 4. Vertical section of outer prismatic layer and periostracum, demineralized in 25% glutaraldehyde; note that periostracum (pr) and prisma sheaths (sh) do not show structural details. Figs.5,6. Shell surface in oblique view, plasma-etched for 6 hours; note coarse fibrous periostracum (pr), fine fibrous prisma sheaths (sh), and initial portions of aragonitic prisms (ip). Fig. 7. Initial portions of aragonitic prisms after removal of periostracum by plasma-etching; note that several initial portions form hexagonal tablets. Fig. 8. Detail of Fig. 7, to show initial portion of a prism forming an hexagonal tablet; note that prism is composed of elongated granules. Bar = 10 microns: Figs. 4, 6, 7, 8; 1 mm: Fig. 5.

CHAPTER 2.7

Hormonal Control of Confirmed Shell Growth in the Snail *Helisoma duryi* (Mollusca: Gastropoda)

A.S.M. SALEUDDIN, H.R. KHAN, M. SEVALA, and V. SEVALA
Department of Biology, York University, Toronto, Canada M3J 1P3

Key words: Calcification, Periostracum, Insulin-like Material, Neurosecretion.

INTRODUCTION

Many molluscs grow by a regular or irregular incremental growth of the shell margin. The shell growth is influenced by photoperiod, lunar cycle, temperature, salinity and other environmental conditions. For example, Helisoma duryi, a basommatophoran pulmonate, grow significantly faster in 00L:24D photoperiod than in 12L:12D photoperiod. Furthermore, within the laboratory stock population fast and slow growth rates, based on mean daily shell deposition rate, are observed. It has been shown that the neurosecretory (NS) mediodorsal cells (MDC) in the cerebral ganglia stimulate periostracum synthesis and calcium transport by the dorsal mantle epithelium and that the lateral lobes of the cerebral ganglia exert an inhibitory influence on MDC (reviewed in [1,2]). Homologous cells in another snail, Lymnaea stagnalis, known as light green cells (LGC) are also involved in body and shell growth [3], and a molluscan insulin-like peptide (MIP) has been isolated and sequenced from these cells [4]. MIP has 40% homology with vertebrate insulin A chain and 20% homology with B chain [5]. Since the MDC of Helisoma are morphologically and functionally homologous to the LGC of Lymnaea, and since LGC of Lymnaea produce MIP, we have studied immunoreactivity of vertebrate anti-insulin in the MDC of Helisoma. The effects of photoperiod and shell repair on the rate of exocytosis of NS granules (NSG) by the MDC in the neurohaemal area has been studied using the tannic acid ringer incubation (TARI) technique. Vertebrate insulin and anti-insulin have been tested in growth and in vitro periostracum synthesis bioassay [2].

MATERIALS AND METHODS

Snails were reared in 40 l tanks containing dechlorinated tap water at 22°C, 16L:8D photoperiod and fed boiled lettuce ad libitum. For immunocytochemistry, brains were fixed in Bouin's fixative, dehydrated and embedded in paraffin wax. Hydrated sections were incubated in anti-porcine insulin (1:50)(Sigma; raised in guinea pig) at 37°C for 45 min, rinsed and similarly incubated in fluorochrome conjugated anti-guinea pig IgG, mounted in glycerol containing anti-fade. For electron microscopy and TARI technique, see [6]. To induce shell repair, about 5mm x 5mm shell was removed from the growing edge of the shell with a fine saw (Dremel Tools, Wisconsin). In 24-48 h, the shell was completely repaired. Shell growth was measured as the distance from a painted reference line parallel to the growing edge of the shell using a calibrated ocular micrometer. Water-proof reference lines and numbers were painted on snails with water-soluble polymer acrylic paint using a fine brush, and then coated with clear lacquer (nail polish). Levels of insulin-like material in the blood and the homogenate of brain tissue were measured using a commercially available anti-human insulin radioimmunoassay kit (Intermedico, Canada). Insulin-like material was detected in immunoblots of blood and brain homogenate using anti-bovine insulin as primary antibody (raised in guinea pig) and alkaline phosphatase conjugated anti-guinea pig IgG (raised in goat) as secondary antibody. Periostracum synthesis was measured as described in [2].

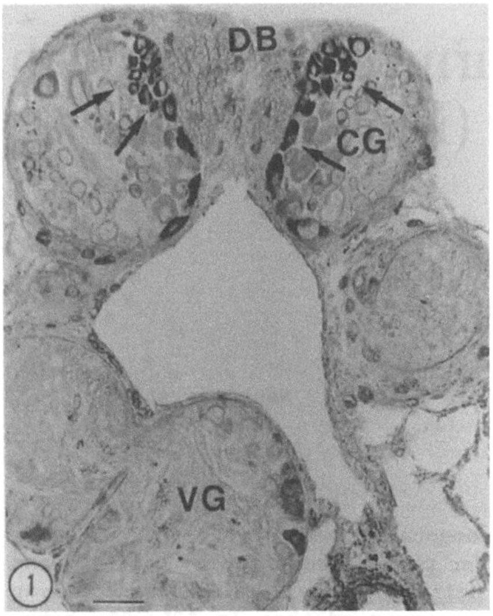

Fig. 1. Showing MDC (arrows) in cerebral (CG) and NSC in visceral (V) ganglia, PAF stain. Bar=100μm.

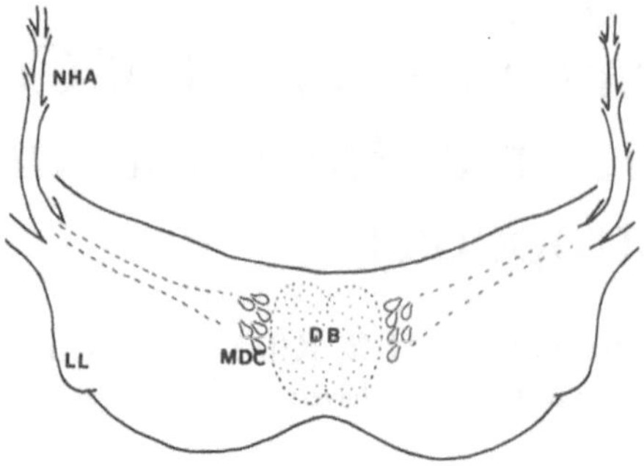

Fig. 2. Diagram showing MDC and its neurohaemal area (NHA) in labial nerve, DB, and lateral lobe (LL). Not to scale.

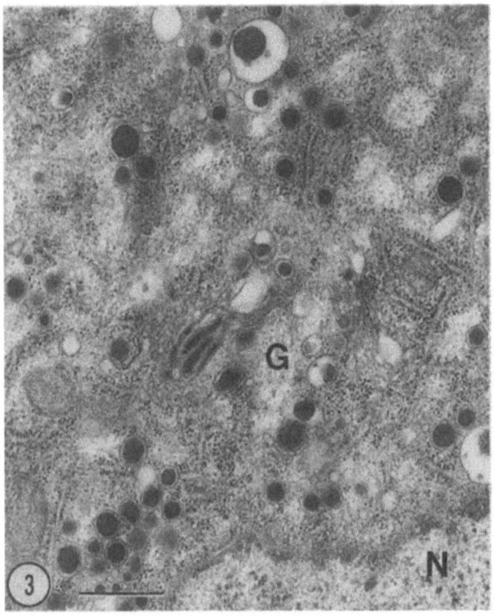

Fig. 3. Electron micrograph of MDC showing Golgi complex (G), nucleus (N) and granules. Bar=1μm.

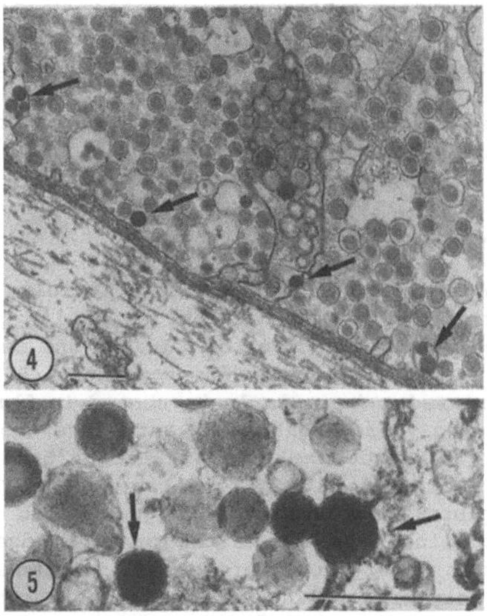

Figs. 4,5. MDC terminals in labial nerve in a shell repairing snail showing release of NSG (arrows) by TARI technique. Bars=1μm.

RESULTS

The peptide secreting NS cells (NSC)in the brain of Helisoma, which encircles the oesophagus, are identified using histochemical stains specific for proteins enriched with disulphide links, since most neurohormone and neurohormone-binding proteins contain many cysteine residues. Fig. 1 is a section of paraldehyde fuchsin stained brain showing the cerebral NSC, the MDC, and some visceral NSC. There are 25-40 MDC near the endocrine dorsal bodies (DB) in each cerebral ganglion. The axons of MDC have been traced in the labial nerve from light and electron microscope serial sections and by micro-injecting stains in the MDC (Fig. 2). The caudodorsal (CDC) NSC which produce egg-laying hormone [6], and lateral lobes, which affect both MDC and CDC, are also present in the cerebral ganglia.

The MDC contain typical dense-cored membrane bound 150-250 nm NSG, scattered rough ER, lysosomes and multivesicular bodies (Fig. 3). Most of the Golgi bodies of the MDC are observed with electron lucent cisternae, very few are seen to contain dense cisternae as in Fig. 3. Synthetic activity of the MDC could not be predicted by the features of Golgi bodies or ER. Some MDC contain electron dense and lucent granules. The NSG of the MDC are variable in diameter, spherical and have characteristic enclosing membrane, thus are easy to distinguish from NSG of other NSC. The perikarya of MDC display innervation from some unknown centre and close association with DB cell processes. The labial nerve, which is the neurohaemal area of the MDC, contain about five morphologically different classes of NSG including NSG of MDC. However, release of NSG, as revealed by TARI was seen mainly in the MDC axon terminals (Figs. 4,5). Quantitative studies on the release of NSG by the MDC terminals with TARI technique showed reduced rate of NSG release during day time compared to that during the night time. Maximum rate of release was seen between 10-11 pm. The difference between rates is statistically significant. The rate of NSG release is significantly higher in shell repairing snails than in control snails both in day and night time.

Insulin-like immunoreactivity in the MDC is shown in Fig. 6, using immunofluorescent staining. The MDC stain positively with both anti-bovine and anti-porcine insulin. The labial nerves also stain positively for insulin-like material. Besides the MDC, anti-insulin immunoreactive cells were seen in the lateral lobes, visceral and left parietal ganglia. For control, in which anti-insulin serum was preadsorbed with excess insulin no staining was seen. Other known NSC of cerebral and visceral ganglia did not stain for insulin immunoreactivity. The insulin-immunoreactive NSC are fuchsinophilic and contain about 200 nm sized NSG. Insulin-like material has been detected on nitrocellulose transfer blots from blood and brain extracts ran in SDS poly-acrylamide gel by immunostaining (Fig. 7). However, insulin-like staining is seen in a protein band of molecular weight of about 14K. The levels of insulin-like material in the blood have been measured using anti-human insulin radioimmunoassay. Higher levels of insulin-like material were detected in the evening than in the morning; also shell repairing snails had higher levels than the control snails.

After injections of porcine insulin (2 µg/snail), and dialysed porcine insulin antiserum (2 µl/snail), physiological saline (2 µl/snail) as a control in groups of 5 snails, shell growth was monitored for two weeks; the results are shown in Fig. 8. Porcine insulin and anti-insulin significantly stimulated and inhibited growth rates respectively compared to rates in the control snails.

Studies on the effects of insulin on the rate of periostracum synthesis in vitro, are under way, in which isolated mantle is maintained in culture medium containing a mixture of radio-labelled amino acids for 48-72 hours. While in culture, the mantle produces the periostracum (Fig. 9). After removal of the periostracum from the mantle and the culture medium, the rate of incorporated labelled amino acids in the periostracum is counted in a scintillation counter. When porcine insulin (10 ng/ml) was added in the culture medium the rate of amino acid incorporation into the periostracum increased.

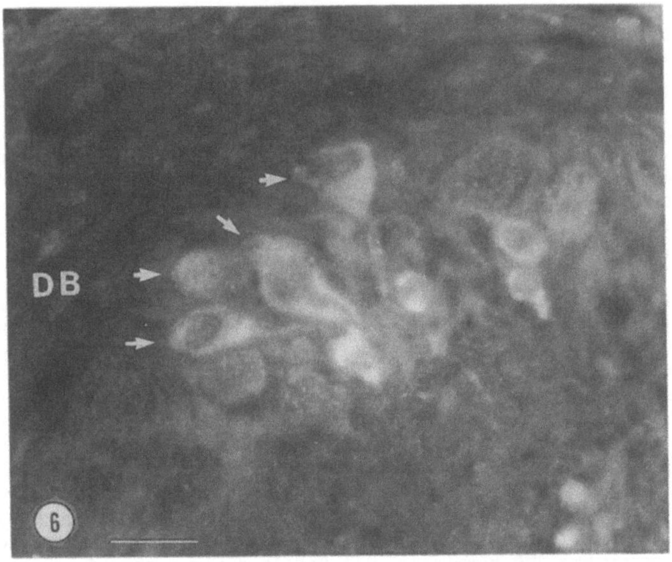

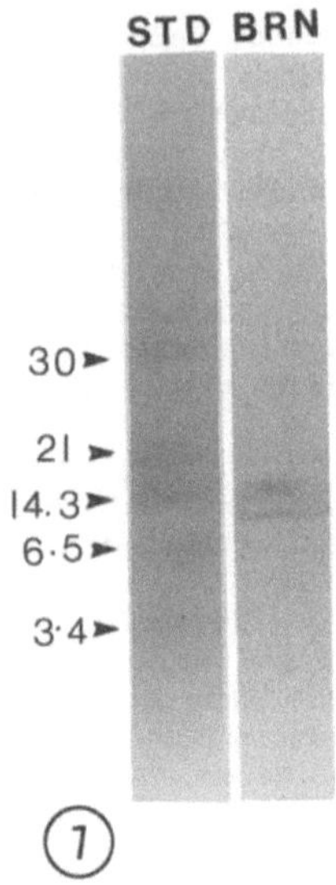

Figs. 6,7. Showing localization of insulin-like material. Fig. 6. MDC (arrows) are stained with anti-insulin, indirect FITC staining. Bar=100µm. Fig. 7. Immunoblot of brain homogenate (BRN) showing presence of insulin-like material at 14.3 kDa. STD, molecular weight markers in kDa.

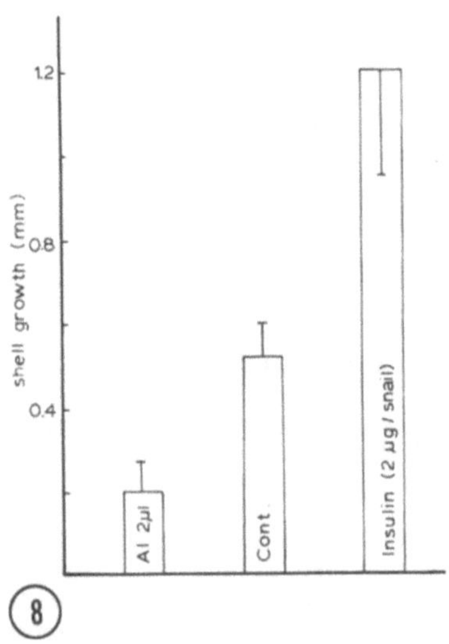

Fig. 8. Histogram showing effects of anti-insulin (AI), saline (cont) and insulin injection on growth.

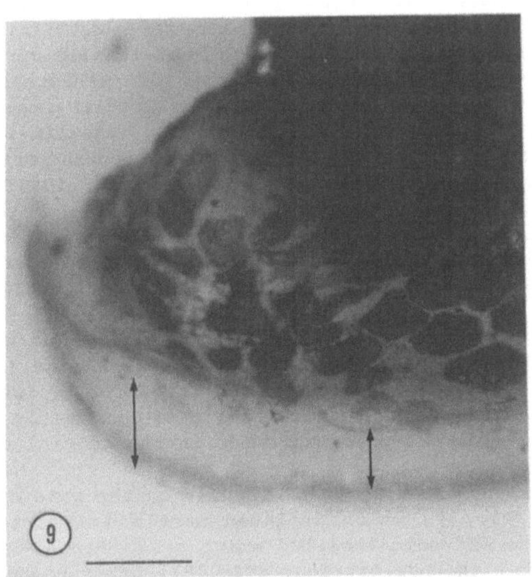

Fig. 9. Mantle edge in culture medium showing periostracum (wide band shown by arrows) produced in vitro. Bar=1.0mm.

DISCUSSION

Previous studies on Helisoma have shown that brain factor(s) and photoperiod regulate shell growth. A putative neurohormone from the cerebral ganglia is involved in shell growth, and a putative neurohormone in the visceral and parietal ganglia is involved in repair of damaged shell (reviewed in [1,2]). The present study suggests that for the MDC axon terminals, the rate of release at their neurohaemal area in the labial nerve is influenced by photoperiod and shell repair. In Lymnaea the increased rate of release by the homologous LGC during evening is also known [3]. The fine structural features of MDC and LGC are similar. However, the MDC show a close association with the DB processes which is not known in LGC. Growth and reproduction in Helisoma are inversely related and it has been suggested that the activity of the MDC, CDC, DB and the lateral lobes are integrated [6].

The LGC of the snail Lymnaea produces MIP which has similarity to mammalian insulin. The present studies provide immunological and physiological evidence for the presence of insulin-like material in Helisoma. Insulin-immunoreactive material is present in the MDC and blood, and in latter, the titre of insulin-like material appears to fluctuate with growth and photoperiod. The data on injection of anti-insulin and insulin into snails, which inhibited and stimulated shell growth respectively, suggest a physiological function of insulin-like material. The possible source of this insulin-like material is the MDC, since the MDC contain insulin-like material, and since the rate of release by MDC terminals is maximum when the level of insulin-like material in blood is also maximum. Injections of extracts made from cerebral and from visceral ganglia increased shell growth; the former was three fold more effective than the latter [1]. The present study shows that there are many more insulin immunoreactive cells in the cerebral ganglia than in the visceral ganglion. Recent studies have shown that in many invertebrates insulin-like peptide is a hormone, and is produced by the nervous system [5]. In insects insulin-related peptide bombyxin controls secretion of moulting hormone ecdysone. In the snail Lymnaea stagnalis, three insulin-like peptides are known. One is produced by the brain and is involved in body and shell growth, and the other two are produced by the gut and are involved in glucose metabolism [5]. The present study also shows the presence of insulin-like material in another basommatophoran snail. Synthesis of periostracum by the mantle edge in vitro is stimulated by mammalian insulin. This assay system will be used for further purification and characterization of insulin-like growth hormone of Helisoma.

REFERENCES

Saleuddin ASM, Kunigelis SC, Khan HR, Jones GM (1980) In: Omori M, Watabe N (eds) The Mechanisms of Biomineralization in Plants and Animals. Tokai University Press, Tokyo, pp 121-129.

Saleuddin ASM, Kunigelis SC (1984) Am Zool 24:911-916

Joosse J (1988) In: Laufer H, Downer RGH (eds) Endocrinology of Selected Invertebrate Types. Alan R Liss, New York, pp 89-140.

Smit AB, Vreugdenhil E, Ebberink RHM, Geraerts WPM, Klootwijk, Joosse J (1988) Nature 331:535-538

Ebberink RHM, Smit AB, Van Minnen J (1989) Biol Bull 177:176-182

Khan HR, Ashton ML, Saleuddin ASM (1990) Can J Zool 68:1233-1240

CHAPTER 2.8

Epizootic Disease in the "Black lip" Pearl Oyster *Pinctada margaritifera* in French Polynesia: Ultrastructural Alterations of the Nacreous Layer

Y. DAUPHIN and J.P. CUIF

URA 723 du CNRS, Pétrologie sédimentaire et Paléontologie, bât. 504, Université Paris 11 Orsay, F 91405 Orsay cedex

Key words : Nacre, diseased oyster, ultrastructure, Polynesia

INTRODUCTION

The development of pearl culture in French Polynesia is based on the "black lip oyster", a medium sized shell, well characterized by the persistent secretion of black pigmentation (usually restricted to the prismatic external layer) in the external zone of the nacreous layer. This is the source of the black pearls, a high value product in the present situation of the pearl industry [1].

Since few years, epizootic disease has ben observed in some atolls of Tuamotu or Gambier archipelagos, where pearl culture is especially important. Under supervision of local institutions, a research project has been organized by ministerial authorities, in order to define better sanitary growth conditions for the artificial oyster populations, and to improve the management of aquatic farms.

This paper summarizes the results of analytical observations carried out on various populations of cultivated oysters, in order to establish a typology of the degradation of nacreous crystals under the influence of epizootic diseases.

MATERIALS AND METHODS

In each locality, the studied specimens have been selected on the basis of macroscopic observations, to obtain complete series of increasing alterations of the shells. The main part of the shells provided by the E.V.A.A.M office in Papeete, is originated from the Takapoto atoll (northern Tuamotu). These specimens have been obtained during the collect of pearls in year 1989. Other black lip oysters have been studied from Manihi [2]. In the field, the selected shells were immediately rinced in distilled water and gently air dryed.

In laboratory, studies have been carried out by continuous observations on transect zones of the nacreous layer internal surfaces. Special attention was paid to the transition zones, between the normal appearance nacreous layers and the obviously deficient material. Most of the specimens have been observed with a PHILIPS 505 SEM, without special preparation. In some cases, a plasma etching preparation has been done, to get better informations on the morphology of nacreous units, under the amorphous organic layer covering the internal side of the epizooty affected shells. Data have been obtained by hydrochloric acid hydrolysis under standard conditions, followed by a PITC derivation. UV detection of amino-acids was achieved at 254 nm with a Beckman system [3].

RESULTS

Macroscopic aspects of the shell alteration

Recognition of the biomineralization disease is easy at the macroscopic level in the affected shells, because it results in a very apparent growth perturbation.

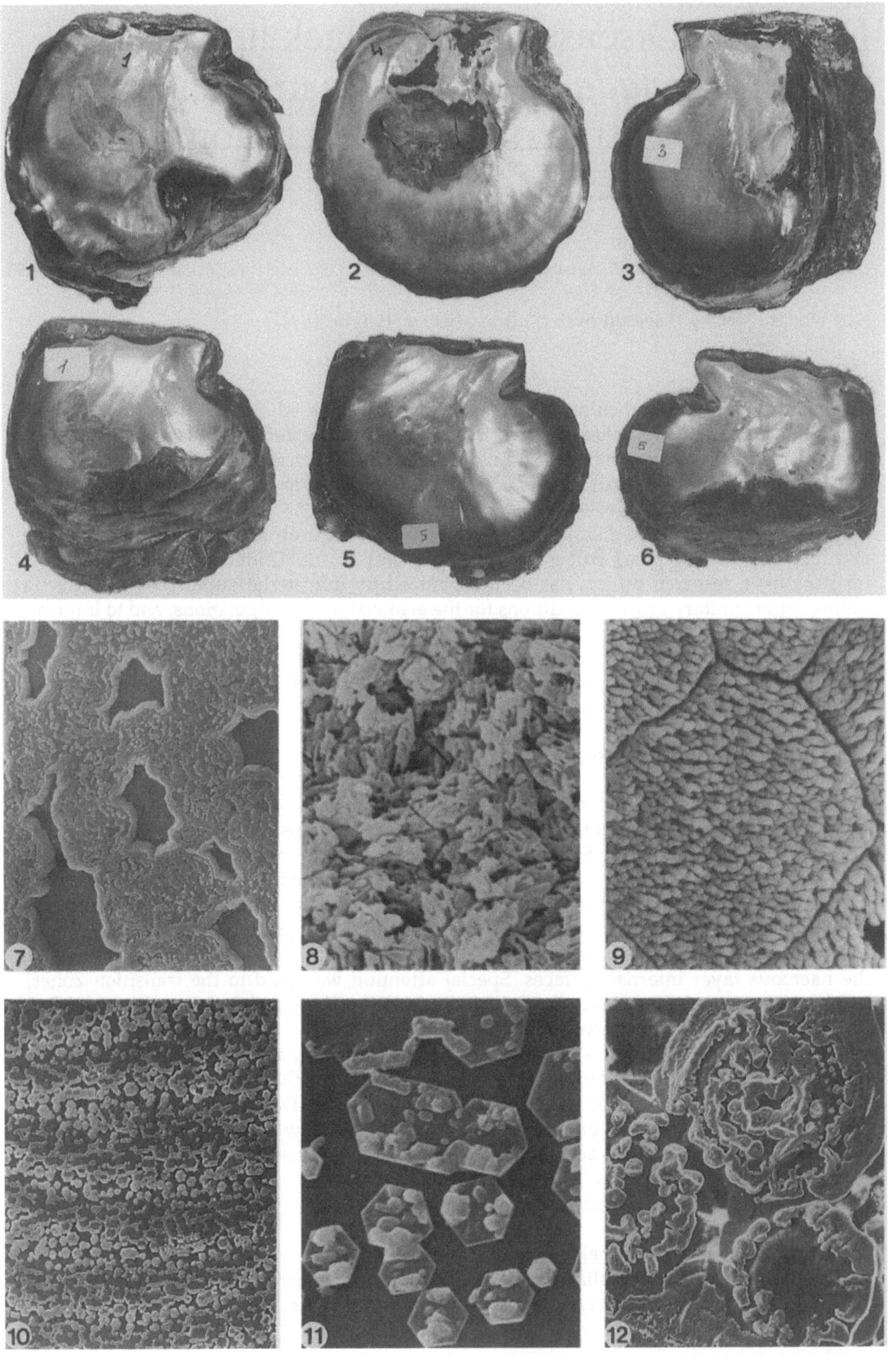

Fig. 1 - 2 : Inner views of the Manihi shells with the diseased brown zones. x 0,5.
Fig. 3 - 6 : Inner view of various Takapoto shells, with their thick organic lamellae. x 0,5 - x 0,55 - x 0,5.
Fig. 7. The spiral pattern and regular hexagonal tablets are absent. Manihi. x 680.
Fig. 8. Diseased nacre with flexuous crystallites and edges poorly delimited. Manihi. x 3120.
Fig. 9. Mature tablets composed of parallel acicular crystallites. Envelopes and inner organic matrix seem disappeared. Manihi. x 7840.
Fig. 10. Spiral pattern with growth zonation are sometimes preserved at Takapoto. x 375.
Fig. 11. New tablets on the edges of 'old' tablets, as in healthy layer. Takapoto. x 2960.
Fig. 12. Abnormal nacre : hexagonal or rounded tablets are absent. Takapoto. x 100.

In the Manihi atoll, unhealthy shells are devoid of outer growth lamellae but show abnormal local thickness of the shell. The brownish degradation of the shell layers seems to be limited to inner parts of the shells (Fig. 1), more precisely the muscular insertion surfaces (Fig. 2).
The external parts of the Takapoto shell layers are reduced to a superposition of brown organic laminae (Fig. 3 - 6), strongly overgrown by epibiontic fauna. In the most diseased specimens, this alteration is spreading on very important surfaces of the shells, up to nearly the third in diameter. In this case, the animal is already dead when the shell is collected. The two valves of a shell are inequally diseased.

Results of the SEM observations

In Manihi shells, the ultrastructure of the diseased nacreous tablets is clearly visible [2], because of the apparent destruction of the surrounding organic envelope and intra-crystalline organic matrix (Fig. 7 - 9). Each tablet seems to be composed of numerous acicular vertical crystallites of about 0,14 µm diameter. These acicular crystallites seem to be composed by rows of granular components. In a given tablet, crystallites are parallel (Fig. 9), but their orientation differs in contiguous crystals. In other perturbed tablets, the crystallites have broken off and lacunae are visible. The shape of the growing tablets is indistinct, their surface is granular. The structure of the diseased nacreous tablets is similar to that produced by chemical etching.
The Takapoto shells show a large diversity of alterations (Fig. 10 - 21). However, SEM observations are made uneasy because of the abundance of the organic matrix. A thin organic layer covers the inner surface and the tablet shape is unclear. Some spiral patterns are composed of regular zonation of tablets (Fig. 10 - 11). In some places, new tablets are developed on the edges of the "old" tablets, as in healthy nacre [4, 5] (Fig. 11). However, the abundance of the underlaying organic matrix layer is abnormal (Fig. 14 - 16 - 17 - 20) and new growing tablets seem to be absent. The ultrastructure of nacre is hidden by organic matrix. The main part of the nacreous tablets is devoid of the regular hexagonal or rounded shape (Fig. 13 - 15 - 16 - 17). The classical structure of nacreous layer is disappeared. A plasma etched nacreous surface shows the abundance of the organic matrix, and the inner structure of tablets (Fig. 21).

Results of the amino-acid analyses

The amino-acid contents of the brownish membranes, the insoluble part and the soluble one of the nacreous layer have been compared to the contents of healthy nacre. The salient features are summarized in Fig. 22. The alanine contents of the brownish membranes are highly variable (Fig. 22 Left). The tyrosine contents of the insoluble matrix are also irregular (Fig. 22 Middle). There is a general agreement that the aspartic acid plays a main role in the calcification processes. It must be noticed that the aspartic acid contents of the diseased shells are lower than in the healthy one in the two studied atolls (Fig. 23 Right). However, it is too soon to establish a connection between the structural aspects and the amino-acid contents of the nacreous layer.

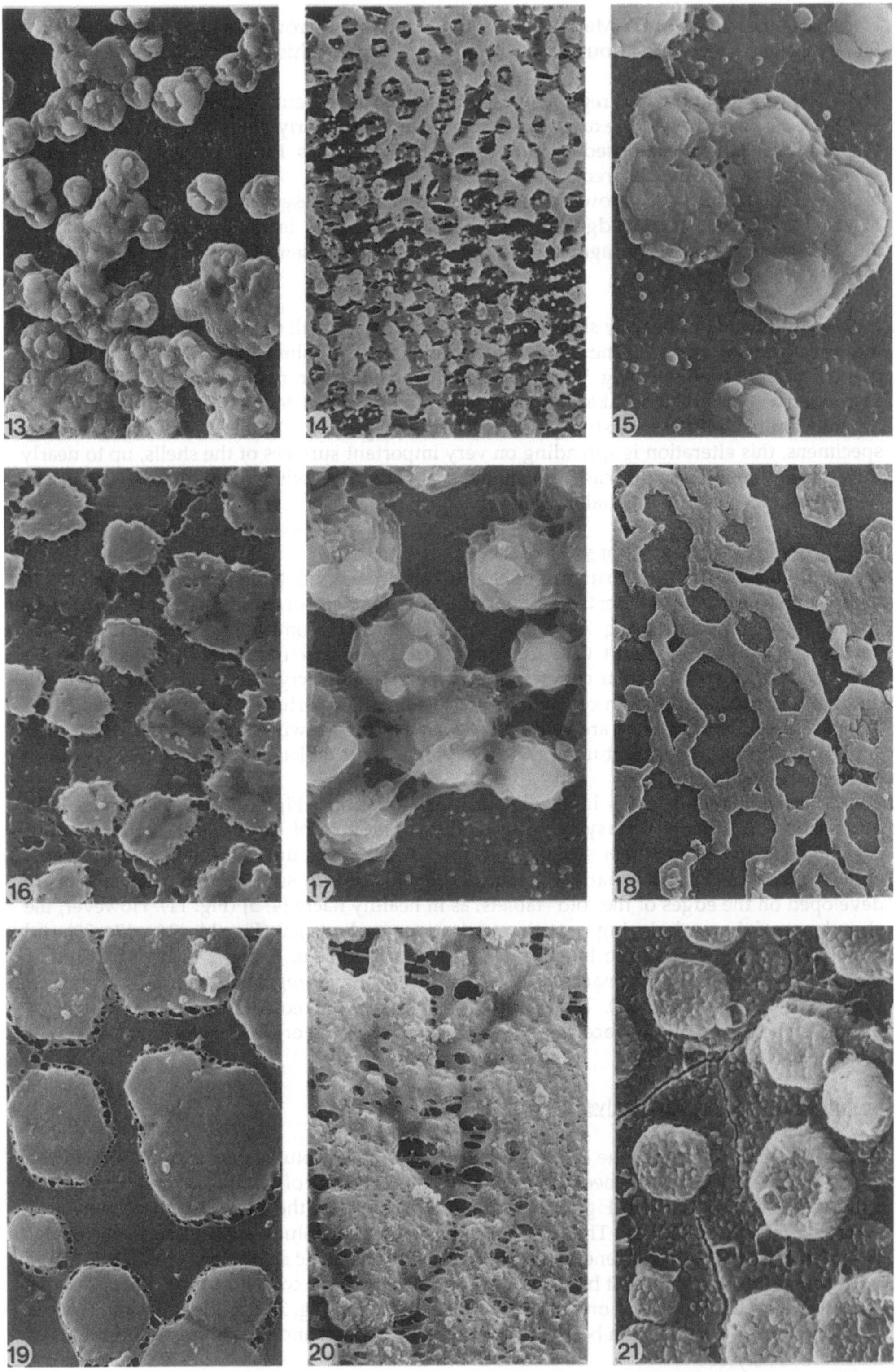
13
14
15
16
17
18
19
20
21

Fig. 13 - 20 : Various aspects of the inner surface of nacreous layer showing the abundance of the organic matrix and the abnormal shapes of the tablets - Takapoto. 13 : x 2035 - 14 : x 1485 - 15 : x 3115 - 16 : x 2980 - 17 : x 3135 - 18 : 2285 - 19 : x 4575 - 20 : x 7135.

Fig. 21. Plasma etched nacre (by courtesy of Dr. H. Mutvei, Stockholm), showing the ultrastructure of hexagonal tablets. The organic matrix is perforated by etching. x 6110.

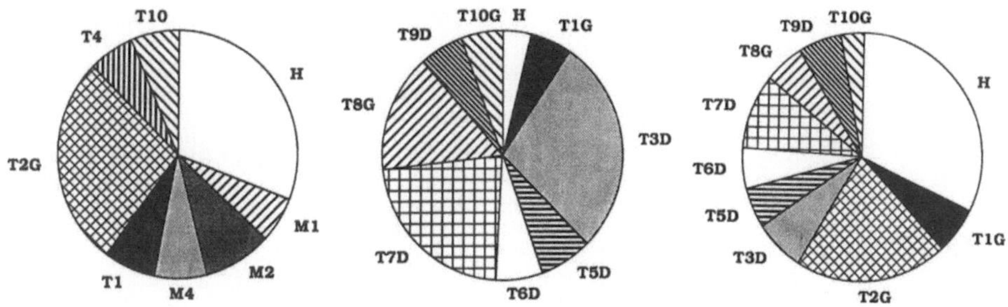

Fig. 22. Left : Alanine contents of the brownish membranes from Takapoto (T) and Manihi (M) shells, compared to the insoluble organic matrix of a healthy nacre (H).

Middle : Tyrosine contents of the insoluble matrices of the Takapoto shells. H : healthy nacreous layer.

Right : Aspartic acid contents of a healthy nacre (H) and diseased nacreous layers from Takapoto.

DISCUSSION

The epizooty of the black lip pearl oyster affects three characteristics of the animal : the morphology with the brown organic membranes, the structure and ultrastructure of the nacreous layers, and the amino-acid composition this layer. However, it is too soon to establish a connection between the morphological, structural and ultrastructural aspects and the amino-acid contents of the nacreous layer. Moreoever its appears some variations in different atolls. But the cause of these malformations is not clear. The use of antifouling paints containing tributyltin is too rare in French Polynesia to have caused toxic damages.

[1] Coeroli M, de Gaillande D, Landret KP, Coatanea D (1984) Aquaculture 39 : 45-67

[2] Dauphin Y (1987) C R Acad Sci Paris, 305, sér. III : 649-654

[3] Marin F & Dauphin Y (in press) N. Jb. Geol. Pal. Mh.

[4] Wada K (1957) Bull Natl Pearl Res Lab, 2 : 74-85

[5] Wada K (1972) In : Biomineralization, Erben HK (ed) International Symposium on Problems in biomineralization, Mainz 1970, 6 : 141-159

CHAPTER 2.9

Inorganic Composition of Cystoliths Isolated from Leaves of Higher Plants

M. OKAZAKI, H. SETOGUCHI, and E. HISANAGA
Department of Biology, Tokyo Gakugei University, Tokyo, 184 Japan

Key words: Calcification, Cystolith, Electron probe microanalysis, Silicification

INTRODUCTION

Cystoliths are well known as intracellular mineralized inclusions which are formed in specialized cells called lithocysts in leaves. They occur in the following five families; Urticaceae, Moraceae, Ulmaceae, Cucurbitaceae and Acanthaceae [1-3]. In these plants, a remarkable number of cystoliths are contained in their leaves and reaches several thousands per square centimeter in the leaves [4,5]. Thus, cystolith formation is the most prominent calcium carbonate deposition in higher plants [3]. Hiltz and Pobeguin [6] studied the inorganic composition of cystoliths of Ficus elastica and found amorphous calcium carbonate as a main component of cystolith together with small amounts of silicon and magnesium with traces of titanium, aluminum and iron. At the same time, they showed a restricted distribution of silicon in the stalk of cystolith by microincineration technique and suggested the silicification of the stalk. Arnott [3] and Setoguchi et al.[5] reported small amounts of vaterite in the amorphous calcium carbonate in Morus alba and M. bombycis. In the present work, we isolated cystoliths from the leaves of F. retusa and Celtis sinensis and studied their major inorganic components with their distribution in cystolith.

MATERIALS AND METHODS

The cystoliths were isolated from the mature leaves of F. retusa (Moraceae) and C. sinensis (Ulmaceae) by the method of Setoguchi et al.[5]. Isolated cystoliths were coated with palladium-gold by a vacuum evaporation and observed by an electron scanning microscope, JSE F-15 at an accelerating voltage of 15kV.

Calcium and magnesium in the cystoliths were analyzed by EDTA-chelate titration after combustion of cystoliths with perchloric acid. Carbonate content was obtained by a manometric method using sodium bicarbonate as a standard.

Distribution of calcium, magnesium and silicon in the cystoliths was determined by electron probe microanalysis, e.g., by surface and line scan analysis. The isolated cystoliths of F. retusa and C. sinensis were mounted on an aluminum block with carbon paint and coated with carbon by a vacuum evaporation. The cystoliths on the block were scanned with an electron beam of 1 μm in diameter and characteristic Kα (X-ray) for above each element was simultaneously detected by wavelength dispersive spectrometory. The analysis was performed with an accelerating voltage of 20kV and specimen current of 0.05 μA by using Shimazu-ARL, type EMX-SM, electron microprobe X-ray analyzer.

RESULTS AND DISCUSSION

Major Inorganic Components in Cystoliths

Figure 1 shows scanning electron micrographs of isolated cystoliths of C. sinensis (A) and F. retusa (B). These cystoliths contained rarely thickened cell walls and leaf fragments. The typical cystoliths (arrowhead in the figure) consisting of main body and a stalk are seen in the figure. Table I lists analyzed major inorganic components in cystoliths of F. retusa and C. sinensis. Cystoliths from two species contain large amounts of calcium and carbonate (8.5 and 7.9 mmoles calcium and 7.4 and 7.7 mmoles carbonate/g cystoliths). Magnesium is also determined in small amounts (0.9 and 0.2 mmoles/g cystoliths). Calcium is contained in approximately equimolar amounts to carbonate, indicating the presence of calcium carbonate. The excess calcium may be bound to some organic matrices such as acid polysaccharides because the latter was histochemically detected in the cystolith body [7]. Thus, calcium carbonate content in cystoliths is estimated from carbonate content and it reaches about 75% in a dry weight basis in two species. Silicon was not determined directly, but small amounts of undecomposed material was retained even after the combustion of cystoliths with perchloric acid. It stained blue with ammonium molybdate and sodium hydrogensulfite in perchloric acid, suggesting the presence of silicon oxide.

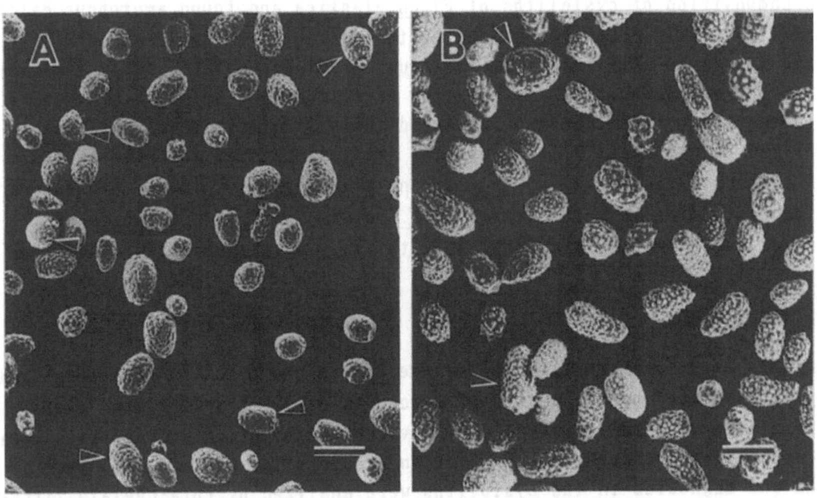

Fig. 1. Scanning electron micrographs of isolated cystoliths from the leaves of Celtis sinensis (A) and Ficus retusa (B).
Note the typical cystoliths (arrowheads) consisting of main body and a stalk. Scale = 100 μm (A,B).

Table I. Contents of major inorganic components of cystoliths from F. retusa and C. sinensis.

Components	Contents	
	Ficus retusa	Celtis sinensis
Ca^{2+}+ Mg^{2+}	9.4 ± 0.1* m moles/g cystolith	8.1 ± 0.1 m moles/g cystolith
Ca^{2+}	8.5 ± 0.0	7.9 ± 0.0
Mg^{2+} **	0.9	0.2
CO_3^{2-}	7.4 ± 0.4	7.7 ± 0.4
$CaCO_3$ (+ $MgCO_3$?)***	0.74 g/g cystolith (74 %)	0.77 g/g cystolith (77 %)

* Mean values ± SD (n= 3-6).
** Mg^{2+}=(Ca^{2+}+ Mg^{2+})- Ca^{2+}.
*** Calculated from CO_3^{2-} contents.

Distribution of Calcium, Magnesium and Silicon in Cystolith

Electron probe analysis was applied to determine the distribution of calcium, magnesium and silicon in the cystoliths of C. sinensis and F. retusa. The results of surface scan analysis were obtained as a secondary electron image (SEI) and X-ray distribution image pairs. Figures 2 and 3 show a result from C. sinensis cystoliths, but a similar result was obtained from cystoliths of F. retusa (data is not shown). Calcium is densely and almost evenly distributed through the cystolith body except for the stalk and the basal part of the body to which a stalk is attached (Fig.2). Magnesium is also contained in the body, but it is much less dense than calcium. On the other hand, silicon is restricted in the stalk and mainly in basal part of the body, suggesting a silicification in those parts (Fig. 2). This characteristic distribution of silicon in cystoliths was supported by a line scan analysis from the top of body to the stalk (Fig.3). As shown in Fig. 3, silicon gradually increases its concentration from the basal part of cystolith body toward the stalk. Its maximal peak is found on the stalk. It seems most probable that at least the stalk is mineralized by silicon. Calcium and magnesium are widely distributed in the body, but they are never detected in the stalk, showing a sharp contrast with silicon.

The calcifing mechanisms of cystoliths and their physiological roles are obscure. Raven and Smith [8] pointed out the relation between cystolith formation and nitrate reduction. When nitrate is assimilated in the plant leaves as the N-source, one hydroxyl ion per one nitrate molecule would be expected, and these excess hydroxyl ion would induce calcium carbonate deposition. They proposed that cystolith formation was one of biological pH stat. On the other hand, cystoliths may be considered as the product of a detoxification system [9]. However, further study is necessary to elucidate the mechanisms of cystoliths formation and its physiological functions.

Fig. 2. (next page, upper) Electron probe micrographs showing the distribution of calcium, magnesium and silicon in the cystoliths of C. sinensis.
The X-ray distribution images of Ca, Mg and Si are presented with a secondary electron image (SEI) of cystoliths. Scale = 50 μm.

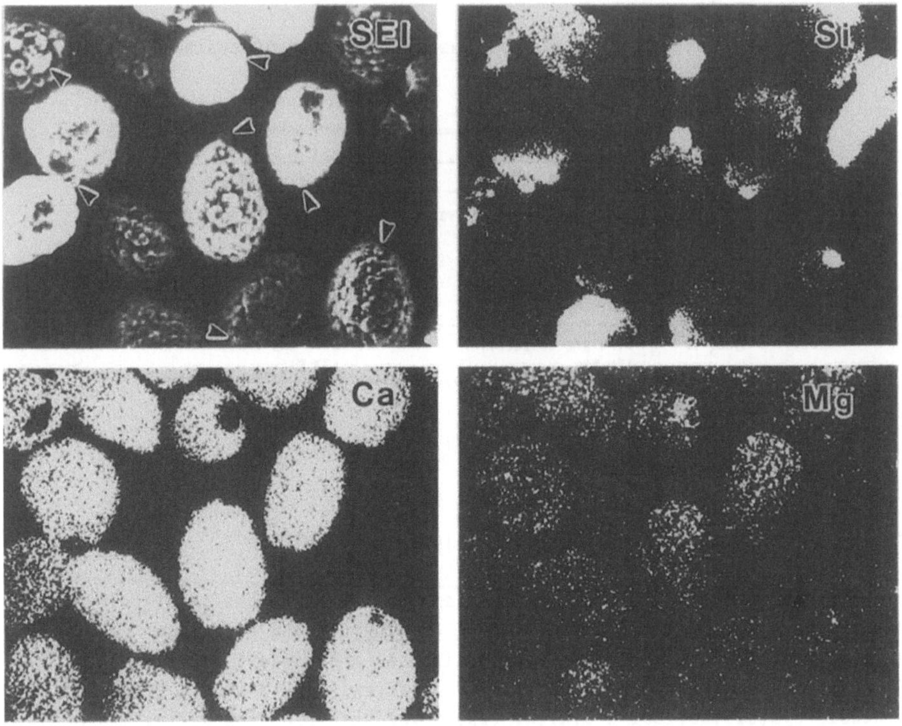

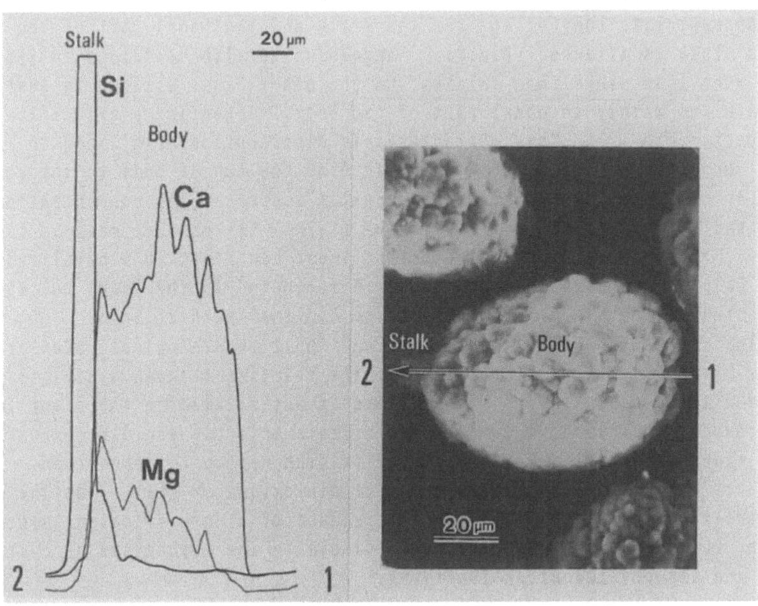

Fig. 3. Line scan profile through the top of body toward the stalk on a cystolith of <u>C. sinensis</u>.
The cystolith was analyzed from symbol 1 (top of body) toward symbol 2 (stalk). Note Si exclusively associated with the stalk of cystolith.

ACKNOWLEDGEMENTS

We wish to thank Prof. S. Suga, Department of Pathology, Nippon Dental University, for his assistance in electron probe microanalysis of cystoliths. Thanks are due to Prof. H. J. Arnott, Department of Biology, University of Texas at Arlington, for reviewing the manuscript.

REFERENCES

[1] Ajello L (1941) Cytology and cellular interrelations of cystolith formation in Ficus elastica. Amer J Bot 28: 589-594
[2] Scott FM (1946) Cystoliths and plasmodesmata in Beloperone, Ficus and Boehmeria. Bot Gaz 107: 372-378
[3] Arnott HJ (1980) In: Omori M, Watabe N (eds) Mechanisms of Biomineralization in Animals and Plants. Tokai Univ Press, pp. 211-218
[4] Okazaki M, Setoguchi H, Aoki H, Suga S (1986) Application of soft X-ray microradiography to observation of cystoliths in the leaves of various higher plants. Bot Mag Tokyo 99: 281-287
[5] Setoguchi H, Okazaki M, Suga S (1989) In: Crick RE (ed) Evolution, and Modern Aspects of Biomineralization in Plants and Animals. Plenum, New York, pp. 409-418
[6] Hiltz P, Pobeguin T (1949) Sur la constitution des cystolithes de Ficus elastica. C R H S Acad Sci 228: 1049-1051
[7] Okazaki M, Hisanaga E, Ishida Y (1986) Histochemical study on organic matrix of cystoliths. Bull Tokyo Gakugei Univ Sec 4 38: 47-54
[8] Raven JA, Smith FA (1976) Nitrogen assimilation and transport in vascular land plants in relation to intracellular pH regulation. New Phytol 76: 415-431
[9] Simkiss K, Mason AZ (1983) In: Hochachka PW (ed) The Mollusca. vol 2, Academic Press, London, pp. 101-164

CHAPTER 2.10

Formation and Solubility of Carbonated Tooth Minerals

E.C. MORENO and T. AOBA
Forsyth Dental Center, 140 Fenway, Boston, MA 02115, USA

Key words: Amelogenesis, Enamel Mineral, Carbonate, Stoichiometry, Solubility

INTRODUCTION

The carbonate ion is consistently found in the crystalline tooth minerals in quantities that vary across species, from higher than 6 %wt in secretory porcine enamel [1] and mature human dentin [2] to about 0.6 %wt for mature shark enameloid [3]. Although part of the carbonate may not be occupying sites in the apatitic crystal lattice of the minerals, there is little doubt that a significant amount [4] resides in lattice sites. Thus, the dental apatites should be considered as carbonatoapatites similar to the mineralogical group denominated dahllites [5], i.e., carbonated apatites having a fluoride content of less than 1 %wt. The chemical and biological importance of these compounds, recognized for a long time, is reflected by the numerous articles [6-9] on the structure and chemical properties of these compounds. However, investigators [10-14] studying the chemical properties of dental minerals, particularly their solubility and stability, invariably consider them as hydroxyapatite, $Ca_5OH(PO_4)_3$, HA. Thus, the solubility product constant for dental enamel has been calculated [12-14] as the ionic activity product for stoichiometric HA in solutions equilibrated with enamel powder; the enamel is then assimilated to an HA displaying a higher solubility than synthetic preparations. Such a simplification facilitates experimental designs and interpretation of results, but it introduces undesirable inconsistences, particularly when comparing the chemical properties of different tissues, e.g., enamel and dentin demineralization in organic buffers. In such a case, it is highly unlikely that a buffer having a given degree of undersaturation with respect to HA will exhibit the same driving force for the dissolution of both tooth minerals. However, without knowing the actual composition of the minerals and their respective solubilities, it is impossible to define the proper driving forces for de- or remineralization.

Incorporation of carbonate into the apatitic lattice can occur in at least two sites [6,15,16]: substituting for OH^- ions (type A substitution) or substituting for PO_4^{3-} ions (type B substitution); both substitutions have been reported in mature human [4] and porcine [1,9] enamels. These substitutions result in stoichiometries different from that of HA but they are the real stoichiometries that should be used to define solubility product constants. The present work examines a) the formation of carbonated apatite during porcine amelogenesis, b) the stoichiometry of porcine and human enamels, and c) the solubilities of enamel minerals.

MATERIALS AND METHODS

<u>Biomaterials:</u> The enamel samples used in the present investigation comprised a) mature human enamel, b) secretory porcine enamel, c) maturing porcine enamel, and d) mature porcine enamel. The dimensions of the enamel organ in a piglet (about 6 month old) allow to take representative samples of very early enamel tissue (S1) in the secretory state (near the

ameloblast layer) and, within the same stage, older enamel (S2), near the dentin-enamel junction of the unerupted incisors. Displacing the sampling zone from the cervical toward the apical region, it is possible to sample "early" maturing enamel from the transition zone [17] and slightly beyond (apically) (M1), and frank mature enamel (M2) from the incisal edge of half-erupted incisors. Details of the dissection procedure are published [1]. The human dental enamel (H.E.) was obtained by pooling enamel from caries-free extracted teeth [2]. Each pooled sample in a frozen state was pulverized with an agate mortar and pestle (to minimize possible conversion of acid phosphate to pyrophosphate). The resulting powder was passed through a 270 mesh sieve. In order to remove most of the organic matter, the samples were ashed at low temperature (about 60°C) by exposing them to oxygen activated by a radiofrequency. This low-temperature ashing was continued until a constant weight was obtained, which usually took 48 hr.

Analytical: Major constituents of each enamel sample were determined by conventional methods: Ca, atomic absorption; total P, spectrophotometry; and carbonate, Conway's micro-diffusion technique [18]. Acid phosphate was analyzed by the Gee and Deitz [19] procedure. Enamel, 30 mg, was heated at 500°C for 24 hr, then dissolved in 10 mL of 0.2 M perchloric acid and diluted to 100 mL with distilled, deionized water. An aliquot of this solution was made 1 N in $HClO_4$ and heated at 100°C for 1 hr, to hydrolyze any pyrophosphate formed during the heating of the solid. Total phosphate was then determined colorimetrically in this solution and in aliquots of the original solution, not subjected to the hydrolytic treatment. The amount of acid phosphate was calculated from the difference in the phosphate analysis of the two aliquots. The recovery of acid phosphate, by this procedure, from a sample of $CaHPO_4.2H_2O$ was 100% and the X-ray diffraction pattern indicated pure calcium pyrophosphate. After heating enamel samples in the manner described, no beta-$Ca_3(PO_4)_2$ was detected by X-ray diffraction.
An additional problem in studying the solubility of carbonate-containing apatites is that it is not known a priori whether a significant fraction of the carbonate in the solid is at sites different from those in the crystalline lattice, e.g., adsorbed on the surfaces of the crystallites. Only constituents in the lattice determine the stoichiometry to be used in defining the solubility of the solid (see below). The assessment of "surface carbonate" was conducted by a dissolution technique. Enamel samples, 20 mg/mL, were treated with HCl solutions of increasing concentrations (10^{-5} to 10^{-2}) for 0.5 hr to dissolve progressively larger amounts of the solid. The residual enamel was then analyzed for Ca, P, and carbonate. If a significant fraction of the carbonate is adsorbed onto the crystal surfaces, then as the enamel sample dissolves, there should be a sensible reduction in the carbonate content of the residual solid and, more important, significant changes in the CO_3/Ca (or CO_3/P) ratio. Similarly, after determining the content of acid phosphate in the mineral, it was necessary to ascertain whether it was a lattice constituent or it was adsorbed on the crystal surfaces. To this end, enamel, 5 mg, was equilibrated with 1 mL of a 3 mM solution of potassium phosphate titrated with KOH to a pH value of 11.0. At the high pH of the equilibration it is anticipated that any acid phosphate on the crystal surfaces will be deprotonated. The suspension was vortexed for 10-60 min, centrifuged, a sample of the sedimented solid taken for HPO_4^{2-} determination, fresh solution added and dispersed again; the process was repeated five times. The acid phosphate was determined on the centrifuged solid [19]. The labile pool of HPO_4^{2-} was assessed by the difference between the HPO_4^{2-} contents of the original and treated samples.

Equilibrations: The solubility of any ionic crystal is defined in terms of the effective concentrations, i.e., activities, of its constituent ions in solution. One of the experimental difficulties in studying the solubility of enamel mineral is that the carbonate ion in solution is a contributor to the solid-solution equilibrium but, at the same time, has to be in equilibrium with the partial pressure of CO_2 in the system; the latter quantity is determined, not only by the thermodynamic properties of the solid phase but also by the composition of the atmosphere in contact with the system. Consequently, unless the activity of the carbonate ion is stabilized, the results may be affected by losses of carbonate (to the atmosphere in the form of CO_2) accompanied by shifts in the solution composition and a great difficulty in obtaining reproducibility. For this reason, we studied the solubility of dental enamel using an experimental design [2] in which ground enamel is equilibrated with dilute solutions of phosphoric acid and mixtures of N_2-CO_2 in adequate proportions to attain the desired partial pressure CO_2. Porcine and human enamel samples, 100-200 mg, were equilibrated with each of six phosphoric acid solutions, 100 mL, covering the range from 0.06 mM to 1.5 mM in wide-mouth bottles partially immersed in water thermostated at 25±0.1°C. A N_2-CO_2 gas mixture (1.86% in CO_2) was bubbled through two saturation towers (in series) containing a 0.1 M solution of KNO_3 (similar ionic strength to the experimental systems) which was 5% (v/v) in chloroform (to minimize bacterial growth). The gas mixture was passed through each experimental system. At intervals of 3-4 days the pH value of the solution was determined and 1 mL of the suspension was withdrawn and filtered through a cellulose membrane (Millipore, 0.45 um); the filtrate was chemically analyzed. Equilibrium conditions were assumed when the solution composition did not change significantly for a period of 3-4 days which usually took from 20 to 25 days.

Stoichiometry of Enamel and Solubility Model. A complete elemental analysis of the enamel mineral, including constituents in locations other than the crystal lattice, would yield the stoichiometry to be considered in solubility studies. However, the variety of isomorphic substitutions that the apatitic lattice accommodates and the heterogeneity of biological specimens make the elemental analysis an impractical approach. Furthermore, not all of the constituents in enamel affect its solubility to the same extent. For example, when the incorporation of one mol of carbonate into the apatite lattice of HA is accompanied by the replacement of one calcium by one sodium and the loss of one mol of phosphate, 96% of the change in the standard free energy of solution (directly related to the solubility product constant) is contributed by the carbonate ion [2]. Therefore, it seems appropriate to assume a stoichiometry on the basis of the major constituents of enamel, particularly those anticipated to affect its thermodynamic properties to a significant degree. In the present report, the stoichiometry adopted for the mature enamels corresponds to the formula $Ca_{5-x}(HPO_4)_v(CO_3)_w(PO_4)_{3-x}(OH)_{1-x}$ with the condition (imposed by electrical neutrality) x=v+w. The three variables in the stoichiometric coefficients can be calculated from the molar ratios Ca/P = m, HPO_4/P_{total} = n, and CO_3/Ca = c. The values of the three molar ratios are obtained from the chemical analysis of the enamel mineral. Making t = m(1+c)+n, the variables in the coefficients are given by w = 5mc/t, v = 5n/t, and x = 5(n+cm)/t. According to the adopted stoichiometry, any solution saturated with respect to enamel should fulfill the condition

$$(Ca^{2+})^{5-x}(HPO_4^{2-})^v(CO_3^{2-})^w(PO_4^{3-})^{3-x}(OH^-)^{1-x} = K_{EN} \qquad (1)$$

in which parentheses stand for the activities of the species enclosed and K_{EN} represents the solubility product constant of the enamel mineral. The procedure to calculate ionic activities in the equilibrium solution was published [20]. Through simple algebraic manipulations, equation (1) can be made to read

$$\log[(Ca^{2+})(OH^-)^{2}] = 1/(5-x)\log[(K_{EN})(K_3)^v(K_w)^{9-x}(Kg)^{-w}] - w/(5-x)\log[Pco_2]$$
$$+ (3-w)/(5-x)\log[(H^+)^3(PO_4^{3-})] \qquad (2)$$

in which $K_3 = 4.52\times10^{-13}$ [21] is the third ionization constant of phosphoric acid, $K_w = 1.008\times10^{-14}$ is the ionization constant of water [21], and $K_g = 7.07\times10^{-19}$ [22] is the equilibrium constant for the reaction $H_2O + CO_2 = 2H^+ + CO_3^{2-}$. The experimental results indicated that, during equilibration, an apatitic phase precipitated. It is reasonable to assume that such a phase is a carbonated apatite similar to the one formed [2] when equilibrating HA under comparable experimental conditions. This carbonated apatite is assumed to have the composition $(Ca)_{5-}$

$_{p/2}(HPO_4)_p(CO_3)_{p/2}(PO_4)_{3-p}(OH)_{1-p}$. There is no way to determine the coefficient p independently of the solubility determinations [2]. The condition of equilibrium with respect to the precipitating phase is given by the definition of its solubility product constant K_{CA}

$$(Ca^{2+})^{5-p/2}(HPO_4^{2-})^p(CO_3^{2-})^{p/2}(PO_4^{3-})^{3-p}(OH^-)^{1-p} = K_{CA} \quad (3)$$

Equations (2) and (3) can be combined to yield the condition of equilibrium of the solution with the two apatitic minerals,

$$\log[(Ca^{2+})(OH^-)^2] = 1/B \log[(K_{EN}/K_{CA})] + I/B + (p/2B) \log[Pco_2]$$

$$- (w/B) \{[\log[Pco_2] - \log[(H^+)^3(PO_4^{3-})]\} \quad (4)$$

in which $B = p/2 - x$ and $I = \log[(K_3)^{v-p}(Kg)^{p/2-w}(Kw)^{-x}]$. According to equation (4), a plot of the left-hand term versus the right-hand term enclosed by curly bracket should yield a straight line having the slope defined by $w/(p/2 - x)$. This procedure makes it easier to visualize differences in the solubility of the enamel minerals. The solubility product constant for each enamel mineral can be calculated according to equation (1) using the composition of the equilibrium solutions. In the present publication, the solubility constants are reported in terms of mean activities $a^s\pm$, i.e., $a^s\pm = K_{EN}^{1/f}$ in which s denotes the saturation condition and f is the sum of the exponents used in the definition of K_{EN} (see equation (1)).

RESULTS AND DISCUSSION

Formation of Carbonated Apatite in Porcine Enamel. It was reported [23] that the carbonate concentration in the enamel fluid (in which mineralization commences) is considerably higher than those of Ca and phosphate which are the major components of the mineral. Consequently, it is not surprising that the mature enamel contain a significant amount of carbonate (see below). There is no information, however, on the fraction of the carbonate occupying lattice positions or on possible changes in the stoichiometry of the mineral precipitating at the various stages of amelogenesis.

The analyses of pooled samples from the four regions already described, after removal of organic matter by low-temperature ashing, are shown in Table 1. If the reasonable assumption is made that most of the carbonate, the Ca and the P is associated with the mineral, it must be concluded that the mineral phase precipitating at the very early stages of mineralization has a carbonate content about double that of the mineral precipitating in the maturation stage. The CO_3/Ca molar ratio decreases from 0.10 in the secretory outermost (younger) enamel to

Table 1. Chemical composition of enamel minerals

Sample	% wt			molar ratios			
	Ca	P	CO_3	Ca/P	CO_3/Ca	HPO_4/P	HPO_4/P^*
S1	31.1	15.7	4.7	1.53	0.10	0.15	0.06
S2	32.2	15.9	4.5	1.57	0.09	0.11	0.06
M1	35.8	17.6	3.9	1.58	0.07	0.08	0.05
M2	36.5	17.6	3.2	1.61	0.06	0.04	0.03
H.E.	35.6	17.2	3.6	1.60	0.07	0.03	0.03

0.06 in the mature specimen; the corresponding CO_3/P molar ratios are 0.15 and 0.08, respectively, consistent with the assumption made. The carbonate enrichment of the early precipitated mineral may be a generalized phenomenon since it has been reported [24] also in the development of rat enamel.

Besides the changes in the carbonate content of the mineral, there are also differences in the actual location of the carbonate in the crystal lattice. In Fig. 1 are shown spectral regions for the S1 (youngest) and M2 (mature) enamel minerals (after low-temperature ashing) in which characteristic carbonate absorption bands appear. In both spectra there are absorption maxima at 1460, 1415, and 873 cm^{-1} which are wave numbers assigned [6,15,16] to the carbonate when it occupies sites usually occupied by phosphate ions. The maxima at 1545 and 879 cm^{-1}, corresponding to carbonate substituting for hydroxyl, are not seen in the youngest sample but only seen in the older enamel mineral. It is not clear why the A-type substitution does not occur in the earliest precipitated mineral, although there are various hypotheses on this matter. For example, if it is assumed that the very first mineral formed is octacalcium phosphate, OCP, (or an OCP-like acid phosphate) it has been shown [7] that the hydrolysis of this phosphate, in the presence of carbonate, results in an apatite having only type B substitutions. An alternative explanation may be that, although the initial mineral is apatitic, the very rapid growth of the mineral ribbons observed [25,26] in the early secretory stage results in a highly strained lattice (numerous defects) which can not accommodate further expansion by the A-type substitution. Such a substitution takes place only in more advanced stages when the rates of crystal growth are lower and the lattice defects fewer. The different characteristics of the initially formed mineral gives rise to the "dark" line often observed in cross sections of mammal enamel [3]. The enamel crystallites grow from the central dark line outward but much more slowly than the rate at which the initial ribbons (the cross section of which correspond to the dark line) grow. The third possible explanation for the absence of A-type carbonate substitution in the early secretory mineral is a change in the composition of the fluid from which the mineral precipitates. There is no information on this matter but it is conceivable that changes in fluid composition (e.g., lower carbonate concentrations in the enamel fluid) may enhance the stability of the mineral having carbonate in positions along the hexad axis. The X-ray diffraction patterns of the four samples listed in Table 1 corresponded to those of an apatite. However, the unit cell parameter a expanded from 9.433±0.003 A in the early secretory stage to 9.443±0.002 A in the mature tissue. Although several lattice substitutions occur during enamel development, it seems reasonable to ascribe the increase in the a parameter to the carbonate occupying the positions that, in HA, are occupied by hydroxyl ions [4,6].

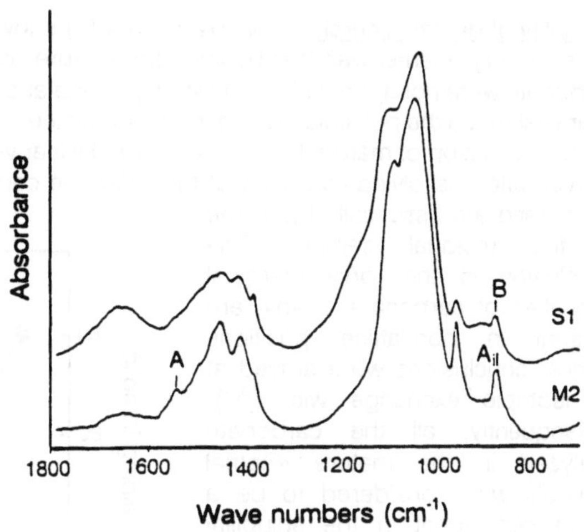

Fig.1 FTIR absorption spectra of the outer secretory (S1) and mature (M2) porcine enamel samples. Note the absence of the A-type CO_3^{2-} bands at 1545 and 879 cm^{-1} in secretory enamel (S1).

Table 2. Changes of the carbonate content of enamel after partial-dissolution with HCl

Acid conc. mol/L	Enamel dissolved %	CO_3 % wt	Ca % wt	CO_3/Ca molar ratio
0*	< 0.02	3.3	35.9	0.06
10^{-5}	0.25	3.3	35.5	0.06
10^{-4}	2.5	3.2	33.9	0.06
10^{-3}	18.9	3.0	32.3	0.06
10^{-2}	32.5	3.0	32.2	0.06

* Enamel sample treated with deionized water.

Solubility of dental minerals. Previous studies [1] showed that about 30 % of the carbonate in the secretory enamel was lost by low temperature ashing, whereas such labile pools of the carbonate were not appreciable in mature porcine and human enamels. In the present studies, we investigated the possible existence of "non-lattice" carbonate in mature enamel. An example of the dissolution of mature human enamel in HCl at various concentrations is shown in Table 2. Even after dissolving over 30% of the solid, the carbonate content, as well as the CO_3/Ca molar ratio are essentially the same in the residual sample. The conclusion is that only marginal quantities of carbonate, if any, are present in non-lattice positions (similar conclusions were arrived at by isotopic exchange with ^{13}C); consequently, all the carbonate analyzed in the mature enamel minerals was considered to be a lattice constituent in the solubility calculations. Analysis of acid phosphate in the enamel samples disclosed that very significant amounts were present in early secretory mineral. It was also ascertained that a substantial part of the HPO_4^{2-} was lost after successive equilibrations in the alkaline media. In Fig. 2 is plotted the percentage of acid phosphate remaining in the solid (S1) as a function of repeated extractions. After the third extraction, the acid phosphate content was close to constant, amounting to about 40 % of the acid phosphate originally present in the secretory (S1) enamel mineral. In the mature enamels, less than 10 % of the original acid phosphate content was in non-lattice positions, i.e., 90 % of

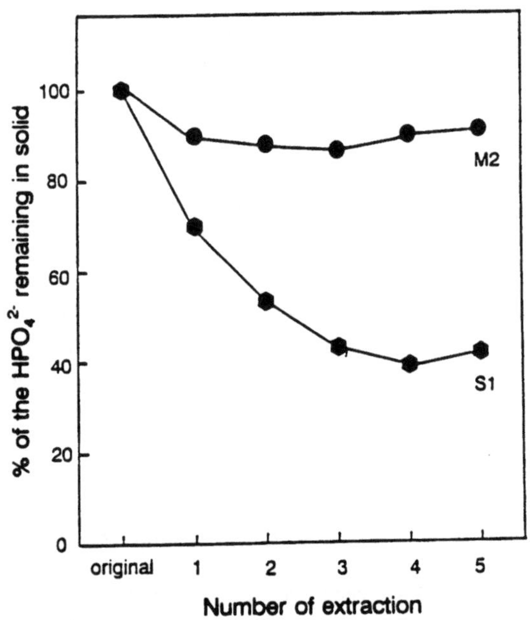

Fig.2 Changes in the acid phosphate content of porcine enamel samples (S1 and M2) as a function of extractions in 3 mM phosphate solution at pH 11. Note a significant decrease of the HPO_4^{2-} content in early secretory enamel mineral in contrast to the mature enamel mineral.

the ions were in the bulk crystal. The value of the acid phosphate in the lattice of each enamel sample is given in the last column of Table 1; these values were used in the assessment of the stoichiometry of the enamel minerals.

In Table 3 are given the stoichiometries of the various enamel mineral samples, which were determined on the basis of the adopted model and the analytical results. There are clear differences in Ca, carbonate and phosphate contents of the various enamel minerals. It may also be significant, and related to the formation process, that the early mineral is low in OH^- content by comparison with the mature specimens. However, the hydroxyl was not determined directly, although a value of about 0.6 has been reported for human dental enamel [27]. Until confirmed experimentally, however, the differences in OH^- should be considered tentative. The decrease in acid phosphate as maturation proceeds may also reflect the initial formation of more acidic phases than the apatitic phase precipitating during maturation.

Table 3. Stoichiometry and solubility product constant of enamel minerals

Sample	Stoichiometry	$a^s_\pm$ x10^6
S1	$(Ca)_{4.38}(HPO_4)_{0.17}(CO_3)_{0.45}(PO_4)_{2.38}(OH)_{0.38}$	2.22
S2	$(Ca)_{4.43}(HPO_4)_{0.16}(CO_3)_{0.41}(PO_4)_{2.43}(OH)_{0.43}$	1.74
M1	$(Ca)_{4.53}(HPO_4)_{0.14}(CO_3)_{0.33}(PO_4)_{2.53}(OH)_{0.53}$	1.45
M2	$(Ca)_{4.65}(HPO_4)_{0.10}(CO_3)_{0.26}(PO_4)_{2.65}(OH)_{0.65}$	1.17
H.E.	$(Ca)_{4.62}(HPO_4)_{0.07}(CO_3)_{0.31}(PO_4)_{2.62}(OH)_{0.62}$	1.36

In Figure 3 are plotted the experimental results for the solubility of porcine enamel according to equation (4). The points obtained for secretory porcine and mature porcine and human enamels describe straight lines (correlation coefficients from 0.97 to 0.99). Interestingly, the points for the mature enamels defined the same line, suggesting that the solubility of the mineral in mature porcine and human enamel are similar but different from the solubility of enamel minerals at earlier developmental stages. The solubility product constants were not determined from the plots because two parameters (p and K_{CA}) remain to be defined unequivocally. We calculated the solubility product constants for the enamel samples

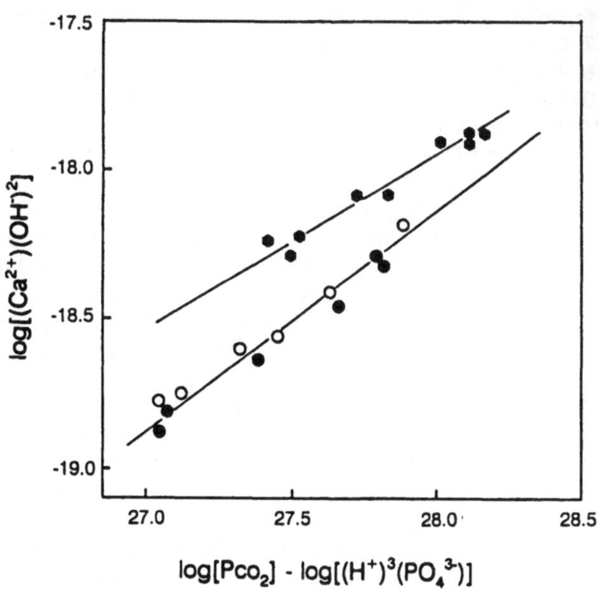

Fig.3 Results of enamel equilibrations in dilute H_3PO_4 solutions (Pco_2 = 1.86 %) plotted according to equation (4). Hexagons, outer secretory porcine enamel; filled circles, mature porcine enamel; open circles, mature human enamel. Regression lines for the secretory and mature enamels have slopes of 0.53 and 0.67, respectively.

based on the solution composition at equilibrium; the values in terms of $a^s\underline{+}$ are shown in the last column of Table 3. It is seen that the solubility product constant decreases steadily with tissue age, again reflecting the higher stability of the phase precipitated much more slowly than in the early stages of amelogenesis.

Although the approach and model adopted seem reasonable to interpret the solubility results, some uncertainty remains requiring further investigation. Thus, the model ignores some of the ions present in lattice positions, e.g., Mg^{2+}, Na^+, and F^-. Although the inclusion of these ions complicate substantially the model, efforts are being made to include them in future work. Also, additional work is being conducted in our laboratories to ascertain the stability of the enamel mineral in solutions simulating the composition [23] of the enamel fluid; it is conceivable that the presence of moieties such as Mg^{2+} may prevent the precipitation of the carbonated apatite postulated in the present model. Finally, work has started to ascertain the actual composition of the carbonated phase precipitating during the equilibrations reported here.

ACKNOWLEDGEMENTS: This work was supported by grants DE08670, DE03187 and DE07009 from the National Institute of Dental Research and a grant from the Colgate-Pulmolive Company.

REFERENCES

1. Aoba T and Moreno EC (1990) Calcif Tissue Int (in press).
2. Moreno EC and Aoba T (1990) Calcif Tissue Int (in press).
3. Aoba T, Moreno EC, Shimoda S, Miake Y, Prostak K, Suga S (1990) (this volume)
4. Elliott JC, Holcomb DW and Young RA (l985) Calcif Tissue Int 37:372-375.
5. McConnell D (l960) Arch Oral Biol 3:28-34.
6. LeGeros RZ, Trautz OR, Klein E and LeGeros JP (l969) Experientia 24:5-7.
7. Chickerur NS, Tung MS and Brown WE (l980) Calcif Tissue Int 32:55-62.
8. Vignoles M, Bonel G, Holcomb DW, and Young RA (l988) Calcif Tissue Int 43:33-40.
9. Rey C, Shimizu M, Collins B, and Glimcher MJ (1990) Calcif Tissue Int 46:384-394.
10. Biltz R., Pellegrino E.D., Miller S.T. and Moffitt A. (1970) Clin Orthopaedics 71, 219-228.
11. Moreno EC, Kresak M, Zahradnik RT (1974) Nature 247:64-65.
12. Moreno EC and Zahradnik RT (1974) J Dent Res 53:226-235.
13. Patel PR and Brown WE (1975) J Dent Res 54:728-736.
14. Larsen MJ and Jensen SJ (1989) Arch Oral Biol 34:957-961.
15. Bonel G and Montel G (l964) CR Acad Sc Paris 258:923-926.
16. Elliott JC (l964) Ph.D. Thesis, University of London.
17. Robinson C, Fuchs P, Deutsch D, Weatherell JA (1978) Caries Res 12:1-11.
18. Conway EJ (l962): Microdiffusion Analysis and Volumetric Error (5th. Ed.), Crosby Lockwood & Son LTD, London, pp. 201-214.
19. Gee A and Deitz VR (1953) Anal Chem 25: 1320-1324.
20. Moreno EC and Margolis HC (l988) J Dent Res 67:1181-1189.
21. Gregory TM, Moreno EC and Brown WE (l970) J Res Natl Bur Stand 78A:667-674.
22. Harned HS and Owen BB (1957) Physical chemistry of electrolytic solutions. New York; Reinhold Pub Co., pp. 691-693 and 755.
23. Aoba T and Moreno EC (1987) Calcif Tissue Int 41:86-94.
24. Hiller CR, Robinson C and Weatherell JA (l975) Calcif Tissue Res 18:1-12.
25. Nylen MU, Eanes ED, and Omnell KA (l963) J Cell Biol 18:109-123.
26. Kerebel B, Daculsi G, and Kerebel LM (l979) J Dent Res 58(special issue B):844-850.
27. Young RA and Spooner S (1969) Arch Oral Biol 15:47-63.

CHAPTER 2.11

Effects of Calcium and Phosphate Deficiencies on Bone Metabolism in a Teleost, Tilapia *(Oreochromis niloticus)*: A Histomorphometric Study

Y. TAKAGI and J. YAMADA

Faculty of Fisheries, Hokkaido University, Hakodate, 041 Japan

Key words: Bone metabolism, Histomorphometry, Calcium-deficiency, Phosphate-deficiency, Teleost fish.

INTRODUCTION

Bone evolved first in fish as a specialized form of the connective tissue. Therefore, fish bone is considered to be most primitive in morphological as well as metabolic aspects. On the other hand, fish have evolved a characteristic type of bone, the acellular bone, which does not contain any osteocytes in its matrix [1,2]. In order to investigate the metabolism of acellular bone, we examined the effects of calcium and phosphate deficiencies on the activities of formation and resorption of the pharyngeal bone in a teleost, tilapia (*Oreochromis niloticus*).

MATERIALS AND METHODS

Young tilapia, weighing 4-10g, were obtained from a commercial dealer (Niigata, Japan). They had been masculinized by a methyltestosterone treatment for about one month after hatching. Before the experiment, they were acclimatized to tap water at 25°C while feeding on an artificial calcium and phosphate-containing (normal) diet for one week. The feeding rate was about 3% body weight per day.

At the start of the experiment, fish were divided into 3 groups; control, calcium-deprived, and phosphate-deprived. Control fish were kept in tap water and fed the normal diet. Calcium-deprived fish were transferred to artificial calcium-deficient water and fed on an artificial calcium-deficient diet. Phosphate-deprived fish were kept in tap water and fed on an artificial phosphate-deficient diet. In all groups, fish were fed on the diets at a ration of 3% body weight per day. Ambient water was changed twice a day. The artificial diets were made largely following Takagi *et al.* [3], with modifications of calcium and phosphate contents. The measured calcium contents of normal, calcium- and phosphate-deficient diets were 1.06, 0.08 and 1.23%, respectively. The measured phosphate contents of these diets were 1.44, 1.37 and 0.06%, respectively. For the artificial calcium-deficient water, deionized water containing 0.5mM NaCl, 0.06mM KCl, and 0.08mM $MgCl_2$ was used. The calcium concentrations of the tap water and the calcium-deficient

water were 0.2mM and <0.02mM, respectively.

After 20 days of rearing, fish were anesthetized in a solution of 0.04% tricaine methansulfonate with 0.04% $NaHCO_3$. Blood was collected in non-heparinized hematocrit tubes from caudal vessels and plasma was separated by centrifugation (3000 rpm, 15min.). The plasma total calcium concentration was determined by atomic absorption spectrophotometry (Hitachi 180-30) and the inorganic phosphate by spectrophotometry (Hitachi UV 2000), following Concustell *et al.* [4]. The pharyngeal bone was dissected out for histomorphometric analyses as described below.

Pharyngeal bones were fixed in 2% paraformaldehyde 2.5% glutaraldehyde in a 0.05M cacodylate buffer at pH 7.4, and then post-fixed with 1% osmium tetraoxide in the same buffer containing 5% sucrose. Fixed samples were dehydrated through graded ethanol and embedded in a methacrylate resin (methyl methacrylate : n-butyl methacrylate=1:4). One micron transverse sections were cut with an ultramicrotome (Ivan Sorvall, MT-1) using glass knives. Sections were stained with silver nitrate-toluidine blue and examined with a light microscope. From one section of each bone, the bone outline was sketched using a drawing device attached to the microscope to measure the following perimeters. 1) Osteoid surface perimeter (OP): bone perimeter covered by the osteoid seam with and without osteoblasts; 2) Osteoblast surface perimeter (ObP): osteoid surface perimeter lined with osteoblasts; 3) Eroded surface perimeter (EP): bone perimeter without the osteoid seam showing an irregular or scalloped outline (Howship's lacunae), with or without resorptive cells; 4) Resorptive cell surface perimeter (RcP): eroded surface perimeter lined with resorptive cells; and 5) Total bone perimeter (BP): total outline of bone comprised of OP, EP and resting surface perimeter as the remainder. The measurements were performed by use of a digitizer (Kanto Denshi K-500) connected to a personal computer (NEC 9801). From the measurements of the perimeters, the following parameters were calculated as percentages to BP. 1) Osteoid surface (OP/BP); 2) Osteoblast surface: (ObP/BP); 3) Eroded surface (EP/BP); 4) Resorptive cell surface (RcP/BP).

Data were represented as mean $\pm$ SE. The Student's t-test was used to compare differences among the groups. Significance was accepted at the 95% confidence level.

RESULTS

Histomorphometric results obtained from measurements of the pharyngeal bone are shown in Table 1. Compared with the control fish, calcium deprived fish developed osteoporotic features, showing a significantly larger eroded surface (EP/BP) and resorptive cell surface (RcP/BP), and an unchanged osteoid surface (OP/BP) and osteoblast surface (ObP/BP). On the contrary, phosphate deprivation caused a decrease in bone resorption and an increase in bone formation; EP/BP decreased significantly, while both OP/BP and ObP/BP increased significantly compared with those observed in the control fish.
The most evident histological feature of the pharyngeal bone in phosphate deprived fish was extensive osteoid characterized by a very thick osteoid

layer (Fig. 1C) in comparison with poorly developed osteoid in control (Fig. 1A) and calcium deprived fish (Fig. 1B).

Table 2 shows plasma calcium and phosphate concentrations in the three fish groups. Calcium deprived fish maintained plasma calcium and phosphate levels. On the other hand phosphate deprived fish showed significant hypercalcemia and hypophosphatemia.

Table 1. Effects of calcium- and phosphate-deficiencies on the histomorphometric features of the pharyngeal bone in tilapia.

	Parameter			
	Formation		Resorption	
Group	OP/BP	ObP/BP	EP/BP	RcP/BP
Control	23.4 + 1.5	14.9 + 1.5	52.8 + 2.0	21.0 + 1.7
CaD	20.4 + 2.7	10.5 + 1.9	68.1 + 3.8*	34.0 + 4.1*
PD	53.6 + 3.4*	28.9 + 2.5*	38.4 + 3.3*	19.6 + 2.5

Mean ± SE, n=5.
*: $P<0.05$ compared with control.
Control: reared in tap water with normal diet.
CaD: reared in Ca-deficient water with Ca-deficient diet.
PD: reared in tap water with Pi-deficient diet.
OP/BP: osteoid surface; ObP/BP: osteoblast surface; EP/BP: eroded surface; RcP/BP: resorptive cell surface.

Table 2. Effects of calcium- and phosphate-deficiencies on the plasma concentrations of calcium and phosphate.

Group	Ca	Pi
Control	3.42 ± 0.11	3.43 ± 0.28
CaD	3.33 ± 0.11	3.09 ± 0.26
PD	4.04 ± 0.10*	2.26 ± 0.24*

Mean ± SE, n=5.
*: $P<0.05$ compared with control.
Control: reared in tap water with normal diet.
CaD: reared in Ca-deficient water with Ca-deficient diet.
PD: reared in tap water with Pi-deficient diet.

DISCUSSION

Calcium is one of the principal minerals contained in bone. In calcium deprived fish, both parameters which were related to bone resorption, eroded surface (EP/BP), and resorptive cell surface (RcP/BP), increased significantly in the pharyngeal bone. These indicate that the size and/or number of resorptive cells increased and, as a result, resorption occurred on broader area of the bone surface compared with that observed in the control fish. It is

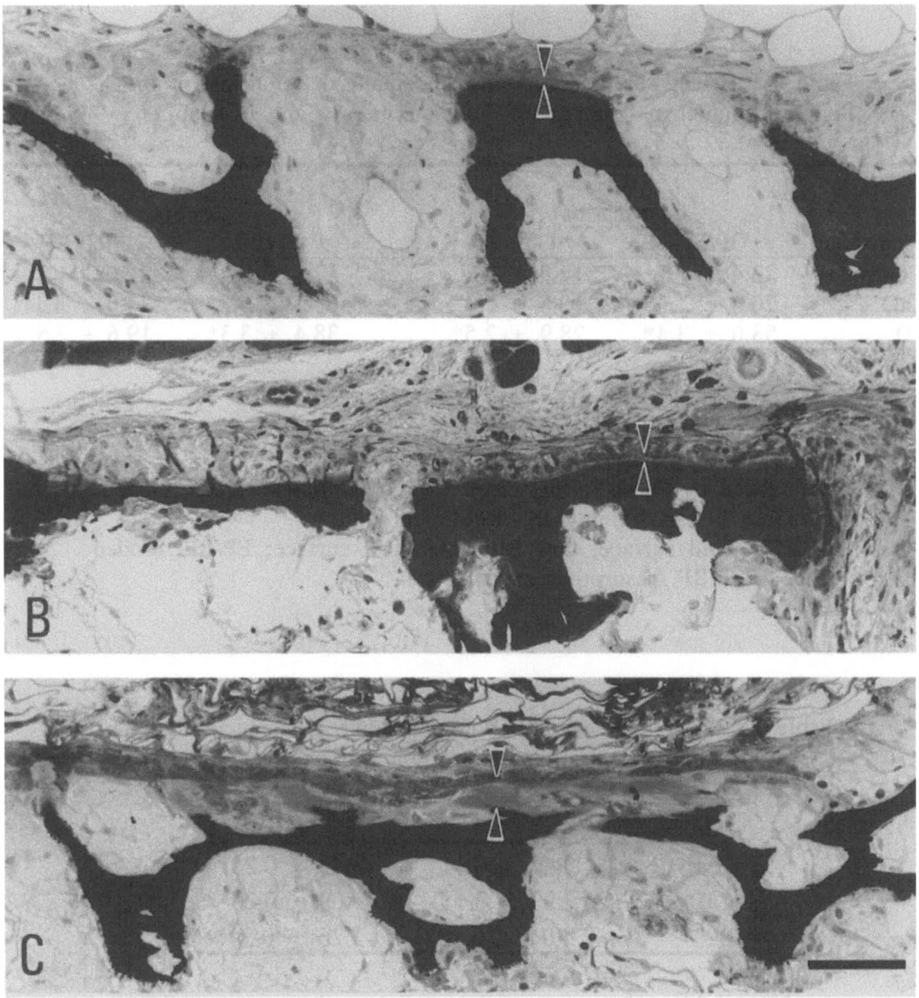

Fig. 1. Photomicrographs of sections of the pharyngeal bones from control fish (A), from calcium deprived fish (B) and from phosphate deprived fish (C). A thick osteoid layer (arrow heads) is noted at actively forming sites in phosphate deprived fish. The bar represents 50um in all figures.

suggested that bone calcium, mobilized into plasma by bone resorption, maintained the plasma calcium level in these fish. Enhanced bone resorption was observed in acellular-boned sunfish (*Lepomis macrochirus*) and in cellular-boned golden shiners (*Notemigonus crysoreucus*) following the removal of scales from one side of the body [5]. Brehe and Fleming [6] showed that calcium of acellular bone in *Fundulus kansae* moved easily into the exchangeable calcium pool. Present results and those previous reports clearly indicate that the acellular bone is metabolically active and is closely related to the calcium homeostasis in fishes.

Phosphate is another main constituent of the bone mineral. In the phosphate deprived tilapia, EP/BP decreased significantly with no observed change in RcP/BP. This indicates that resorptive cavities decreased in size, remaining the size and/or number of resorptive cells unchanged. Therefore, the activity of each resorptive cell is considered to have reduced in phosphate deprived fish. This is in contrast to similar studies on mammalian bone. In rats, phosphate deprivation rapidly stimulates bone resorption [7-9].

Osteoid surface (OP/BP) and osteoblast surface (ObP/BP) in phosphate deprived fish increased significantly, indicating that size and/or number of osteoblasts increased and as a result, bone formation occurred in a broader area of the bone. However, a very thick osteoid layer in these fish suggested that mineralization of newly formed matrix was inhibited by phosphate-deficiency. In mammalian bone, osteoid is suggested to undergo a number of chemical modification steps or to "mature" before mineralization [10]. Baylink *et al.* [7] suggested that hypophosphatemia inhibits the "maturation" process of osteoid and results in a mineralization defect. Similar inhibition of osteoid maturation could have occurred in the bone of tilapia because hypophosphatemia was observed in the phosphate deprived fish.

REFERENCES

1. Moss ML (1961) Acta Anat 46: 343-462.
2. Moss ML (1963) Ann N Y Acad Sci 109: 337-350.
3. Takagi Y, Hirano T, Yamada J (1989) Comp Biochem Physiol 92A: 605-608.
4. Concustell E, Cortes M, Ferragut A, Gener J (1977) Clin Chem Acta 81: 267-272.
5. Weiss RE, Watabe N (1978) Comp Biochem Physiol 60A: 207-211.
6. Brehe JE, Fleming WR (1976) J Comp Physiol 110: 159-169.
7. Baylink D, Wergedal J, Stauffer M (1971) J Clin Invest 50: 2519-2530.
8. Bruin WJ, Baylink DJ, Wergedal JE (1975) Endocrinology 96: 394-399.
9. Cuisinier-Gleizes P, Thomasset M, Sainteny-Debove F, Mathieu H (1976) Calcif Tissue Res 20: 235-249.
10. Baylink D, Stauffer M, Wergedal J, Rich C (1970) J Clin Invest 49: 1122-1134.

CHAPTER 2.12

Nanospace Theory for Biomineralization

N. KATSURA
Department of Oral Biochemistry, School of Dentistry, Nagasaki University, Nagasaki, 852 Japan

Keywords:Nanospace Theory,Critical space,Nanospace,Mineralization,Collagen

INTRODUCTION

The cell and the extracellular matrix have some protective mechanisms against mineralization and body fluids contain widely varying macromolecules that inhibit mineralization. The precipitation of inorganic crystals from body fluid requires: (A)the elimination of the macromolecules inhibitory for crystal formation, (B)sufficient supersaturation of the relevant ions and (C)local restriction of molecular movement. All of these requirements must be performed in a several nano-meter space. In order to accomodate these requirements, the author proposes the "Nanospace Theory" for biomineralization in which extremely thin crystallites are formed in the ultramicrospace at the very beginning of mineralization[1,2]. Only hydroxyapatite mineralization will be discussed here.

CRITICAL SPACE FOR HYDROXYAPATITE

The first crystallite is extremely thin with a thickness of 3~4nm (nano-crystal) which may be the minimum size recognizable as a crystal. If 3nm is a critical size for apatite crystallite, a space which is equal to a crystallite and its hydration shell may be specified as a "Critical Space". The size of the critical space should be 5~6nm.

Adsorbed water molecules on hydrophilic surface are oriented on the surface and this orientation becomes looser with increasing distance from the surface. These oriented water layers are termed the hydration shell and consist of pure water from which solutes are excluded. The thickness of the shell is about 1.5nm and the water has a considerably restricted molecular movement and a stable structure.

The formation of critical space in an ionic solution causes a concentration of ions in the center of the space. The retention time of a monolayer water molecule adsorbed on the hydrophilic surface is 10^{-7}s and that of free water is 10^{-12}s . That of the intermediate layer is 10^{-9}s, so that the outer layer of the hydration shell restricts the molecular movement of water by as much as 10^3 times less than that of free water(Fig.1)[3,4,5]. This means that the structure temperature of vicinal water is low, and so the solubilities of Ca ion and phosphate ion are decreased. Stable ion clusters are formed easier in the nanospace than in the bulk water(Fig.2). Particularly, the restriction of molecular movement in the critical space may induce effective apatite nucreation because the extracellular fluids are supersaturated and are ultrafiltrated spontaneously.

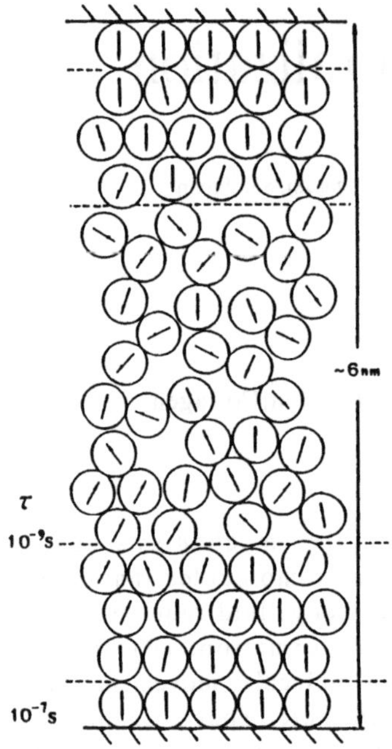

Fig.1
Hydration model of polar surface (H.Uedaira,modified).

τ : retention time of water molecule.
↑ : direction of water molecule.
Molecular movement of the central water between both surfaces is restricted as low as 10^{-9} s in the "Critical Space"(~6nm).

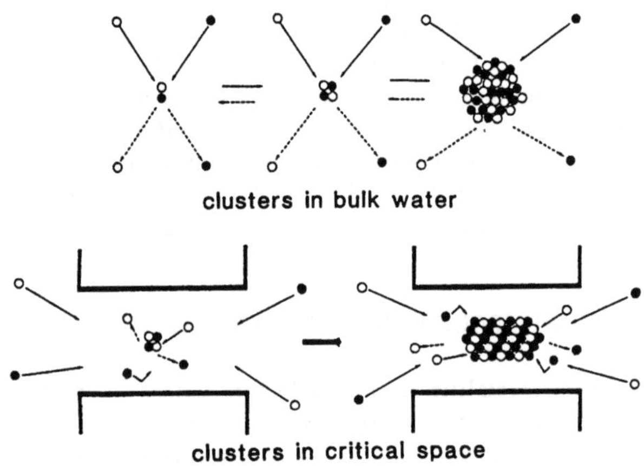

Fig.2 Cluster formation in a critical space should be irreversible.

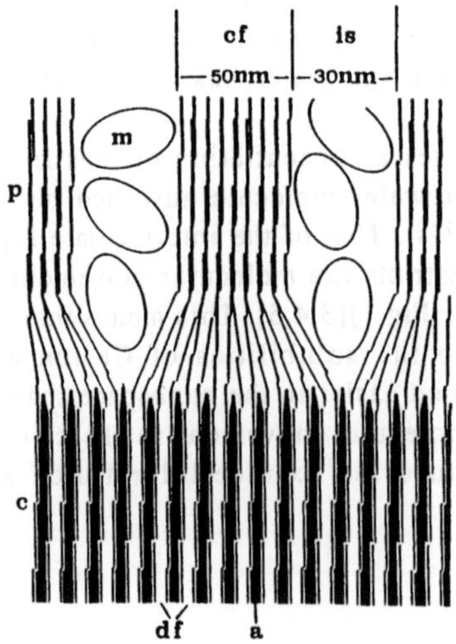

Fig.3 Collagen fibril at the unmineralized-mineralized interface.

p: periodontal ligament,
c: cementum,
m: macromolecule in interfibrillar matrix,
cf; compact fibril,
is: interfibrillar space,
df: deployed fibril,
a: apatite crystallite.

NANOSPACE IN BIOLOGICAL MATRICES

Collagen fibril has about a 9nm unit cross section composed of 5 × 5 tropocollagen molecules. This is described as the "filament" of collagen.
There are at least two modes for the formation of nanospace with biofibers. The first, an interfibrillar triangular space is formed between cylindrical fibrils getting very close. The range of fibers with larger diameters or fibers with thinner diameters getting separated from each other are well over the critical space and allow the existence of anti-mineralization macromolecules.The second critical space, an intrafibrillar space, would be formed within a microfibril composed of filament aggregation.
A typical example of calcification in both the interfibrillar space and intrafibrillar space has been observed in goldfish scale fibrillary plate (Fig.8). Curved thin crystallites appear in inter-fibrillar spaces at first and then intrude into intra-fibrillar spaces. Finally, proliferated crystallites show pin-wheel features.
In amelogenesis, ameloblasts excrete ultramicrotubules and fine needle-like apatite crystallites extend into the tubules immediately from dentin.
Lingula unguis, a Brachiopod, has a shell which consists of alternately distinct calcified and uncalcified layers, but no pre-matrix material. Lamellar cracks may be caused by hardening of the matrix provide inter-lamellar spaces in which apatite crystallites grow and proliferate. A fine, thread-like mineral deposit in cytoplasmic vacuoles of human gingival fibroblast cell line (HGF-7) was surrounded by the vesicle membranes[6]. There is an interesting report that the interlamellar repeat distance for hydrated dimyristoyl lecithin at 37℃ increases with increasing hydration to its limiting value of 6.22nm at 40% water, remaining constant thereafter[7]. This value coincides with the critical space.

DEPLOYED STRUCTURE OF COLLAGEN FIBRILS

The highly magnified electron micrographs of the periodontal ligament and cementum interface are shown in the text book of Ten Cate's oral histology (Fig.13-4, p.221 and Fig. 14-10, p.242)[8]. The periodontal collagen fibrils are thin and are clearly separated from each other, but the individual fibrils in cementum are undifferentiated and appear swollen. This observation may be interpreted in terms of the sudden dissociation of the compact microfibrils into filaments. A schematic model is shown in Fig.3, in which a compact microfibril of about 50nm in diameter swells up to 80nm with deployed filaments and microcrystallites of hydroxyapatite grow up in the intercalating nanospace between filaments.
The transformation from compact Type I collagen fibril (hooped fibril) to swollen fibril (dishooped fibril) should be a common phenomenon in the process of collagenous calcification. We need, therefore, a precise three dimensional structural model of the collagen fibril to understand intra-fibrillar mineralization.
Some mild treatments induce many declined stripes (15~18° to the fiber axis) on the surface of fibril followed by obscuring or disappearing of the characteristic banding. This indicates that the fibril is a multi-helix composed of many filaments which have no bands.
Figure4-A is freeze-fractured dog meniscus without fixation. Typical bands are clear. After glutaraldehyde fixation, on the other hand, the bands are disappeared and declined many stripes appeared as shown in Fig.4-B. Cross-sections of Fig.4-B are shown in Fig.4-C. They show some swirl features.
The molecular hierarchy of our proposed model[9] is: (1) a pentamer base unit comprising an aggregation of five quarter staggered tropocollagen molecules (Fig.5-A); (2)filamentous rods of unlimited length comprised of overlapping base units with a constant cross-section

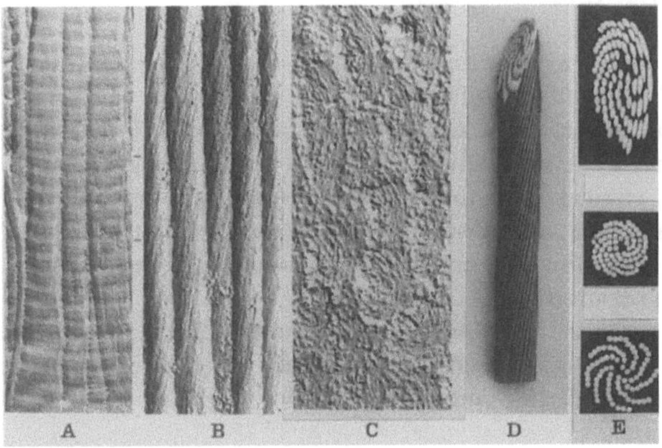

A: unfixed,
B: gultaraldehyde, 30%glycerol,
C: cross section of B,
D: wire model,
E: cross sections of D at 30, 90 and 90(loosed) degrees.

Fig.4 Freeze-fractured replica of dog knee joint meniscus collagen

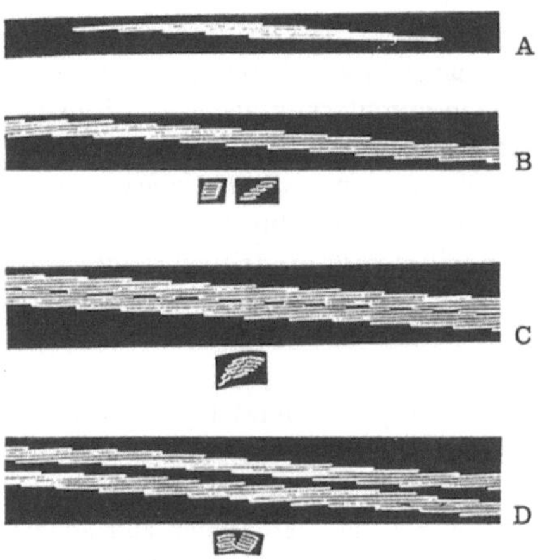

A: pentamer base unit,
B: filamentous rod (5x5 molecules in cross section),
C: banding formation (lateral association of filamentous rods)
D: striation (lateral dissociation of filamentous rods).

Fig.5 Assembly of tropocollagen molecules..

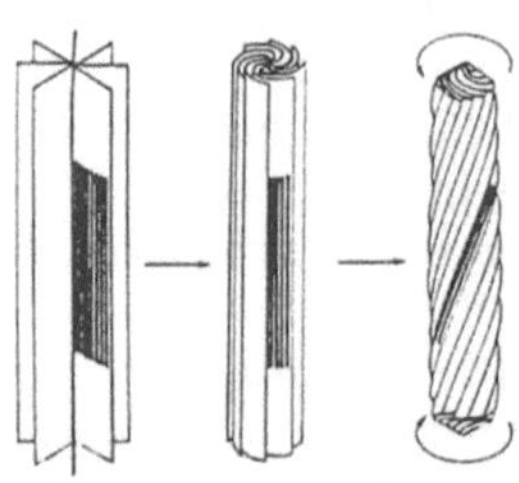

Fig.6 Making process of the wire model.

Fig.7 Banding fibril with hole zones and striated fibril with cracks..

of five base units which are capable of lateral shifting (Fig.5-B); (3) fibrils comprised of 8 to 16 sheet-like elements of parallel filaments the center of which all intersect at a common axis to which and about which the sheet-like elements are rotated so that each filament forms a helix about the central axis, and which display a pin-wheel structure in perpendicular cross-section (Figs.6,7).Tight lateral association of the filamentous rods or sheets in the fibril produces periodic bands (Fig.5-C). Loosening of the association produces helical cracks between sheets (Figs.5-D,7). The outer portion of the cracks in the fibril provide "critical space" of 5~6nm in thickness in which hydroxyapatite microcrystals are induced(Fig.7).

A wire model was made according to the procedures as shown in Fig. 6. Its whole feature is shown in Fig.4-D and cross-sections are shown in Fig.4-E. A deployed (or swollen) fibril should show a pin-wheel feature in cross section as shown in Fig.4-E. This model explains successfully an early mineral deposition in goldfish scale collagenous plate as shown in Fig. 8.

At the onset of the normal calcification, some surface substances of the fibril may be degraded by a matrix vesicle proteinase which would be released by rupture of the membrane accompanied by proliferation of the inner vesicular crystallites[10].

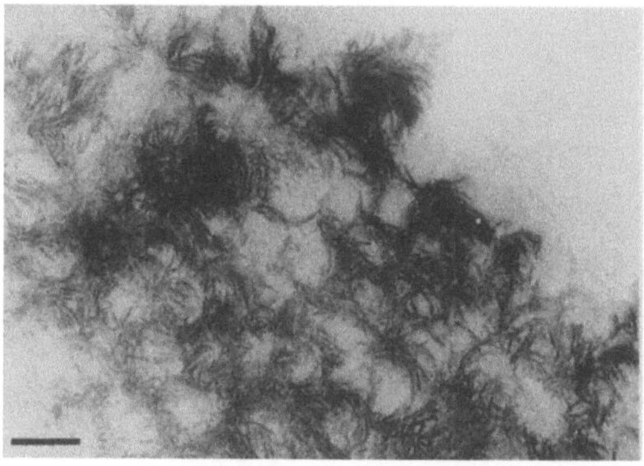

Fig.8 Cross-section of early mineral deposition of gold fish scale.Diameter of collagen fibril is about 90nm. Double stained with uranyl acetate and lead citrate. bar=100nm

REFERENCES

[1]Katsura N(1988)Jap J Oral Biol 30(suppl.):225

[2]Katsura N(1990):Dentistry in Japan.26 :(in press)

[3]Drost-Hansen W (1971)In:Brown HD(ed) "Chemistry of the Cell Interface" part B, Academic Press,NY,Chap.6,pp.1-184

[4]Uedaira H, Ohsaka A(1989) Water in Biological Systems,Kodansha Scientific LTD. Tokyo,1989.(in Japanese)

[5]Belfort G, Sinai N(1980)In:Rowland SP(ed)"Water in Polymers".ACS Symposium Series No.127,ACS Washington DC,pp.323

[6]Yajima T ,Kumegawa M , Hiramatsu M(1984) Arch Histol Jap 47:43-55

[7]Janiak MJ ,Small DM, Shipley GG(1979) J Biol Chem 254,6068-6078

[8]Ten Cate AR(1985) Oral Histology(2nd ed.) The CV Mosby Co.,St.Louis

[9]Katsura N,Tanaka O, Yokoyama M(1991) Connective Tissue 23:(in press)

[10] Katsura N, Yamada K(1986) Bone;7:137-143

CHAPTER 2.13

The Initial Mineralization During Tooth Development in Sharks

I. SASAGAWA
Department of Anatomy, School of Dentistry at Niigata, The Nippon Dental University, Niigata, 951 Japan

Key words: Enameloid, Mineralization, Sharks, Tooth, Tubular vesicle.

INTRODUCTION

In the early stage of enameloid formation in sharks and skates, a great number of "tubular vesicles" appear in the enameloid matrix [1-6]. Initial enameloid crystallites are observed in the tubular vesicles so that the tubular vesicles are probably the sites of initial mineralization in shark enameloid [1,6]. Kemp and Park [2], Goto [3] and Kemp [4] suggested that the tubular vesicles originated from epithelial cells. To the contrary, their odontoblastic origin was proposed by Sasagawa [6] because the tubular vesicles were limited by the unit membrane which sometimes seemed to continue into the cell membrane of odontoblasts. Thus, the aim of this study was to investigate the fine structure of enameloid in the early stage of tooth development, and to reveal further details on the relationship between the tubular vesicles and the surrounding cells, and between the tubular vesicles and the initial enameloid crystallites.

MATERIALS AND METHODS

Two smooth dogfishes, <u>Triakis</u> <u>scyllia</u>, Triakidae (male, total length (TL) 38 cm; female, TL 60 cm) and a Japanese bullhead shark, <u>Heterodontus</u> <u>japonicus</u>, Heterodontidae (male, TL 41 cm) were used in this study.
The tooth-bearing jaws were removed after decapitation, and placed in Karnovsky's fixative (0.1N cacodylate buffer added to 1.7% NaCl, pH 7.4) for 10-24 hours at room temperature followed by a 1% osmium tetroxide solution buffered with cacodylate. Most of the specimens were demineralized with 2.5% EDTA-2Na solution after initial fixation. After dehydration, these specimens were embedded in Araldite-Epon 812 resin. Ultrathin sections were cut by glass or diamond knives, mainly stained with uranyl acetate and lead citrate or lead citrate only, and examined in a transmission electron microscope.

RESULTS

<u>Heterodontus</u> <u>japonicus</u>

In the stage before the appearance of the enameloid matrix,odontoblasts occupied a great portion of the dental papilla, that is, the space surrounded by the dental epithelial cells. The odontoblasts extended their bodies to the apical side of the dental papilla. There were many small projections from the odontoblasts. Longitudinally long mitochondria, rough-surfaced endoplasmic reticulum (rER), tubules and filaments were found in the elongated bodies of odontoblasts. Widely distributed Golgi apparatus were situated in the cytoplasm near the nuclei. There were many vesicles containing electron-dense substances, multivesicular bodies, vacuoles and ribosomes in the odontoblasts (Fig. 1). Tubular vesicles were rarely found in the extracellular space among the odontoblasts, although short collagen fibrils which showed undefined stripes, nearly 60 nm or shorter in interval, were already visible at this stage (Fig. 2). Several large membrane-limited bodies having many small vesicles, which resemble multivesicular body in cytoplasm, were observed around the odontoblasts. Some of these bodies seemed to make contact with the cytoplasm of the odontoblasts (Fig.3a, b). Collapse of the limited membrane of the bodies, which caused the release of the contents, was occasionally seen in this study (Fig. 3c). The diameter of the small vesicles and/or tubules contained in the bodies was 20-40 nm.

Inner dental epithelial (IDE) cells were short columnar cells and the nuclei occupied the approximate centre of the cells. Many mitochondria were concentrated at the distal and proximal portion. Moderately developed Golgi apparatus were situated in the proximal cytoplasm near the nuclei. There were vesicles containing electron-dense substances, lysosomes, ribosomes, rough-surfaced and smooth-surfaced endoplasmic reticulum in both the distal and proximal cytoplasm. Basal laminae were obviously seen at the distal end (Fig. 4).

In the stage of enameloid matrix formation, abundant electron-dense tubular vesicles, 12-20 nm in diameter, which were limited by the unit membrane, and the odontoblast processes containing filaments, tubules, vesicles and vacuoles, constituted most of the enameloid matrix (Fig. 5). Two kinds of collagen fibrils, approximately 60 nm or 15 nm in the interval of stripes, were often found among them. Many tubular vesicles approached the odontoblast processes, and some of them seemed to continue into the processes. In that case, the unit membrane limiting the tubular vesicles seemed to link that of the odontoblast processes (Fig. 6). Organelles in the IDE cells generally became scattered as compared to those in the former stage except for well-developed interdigitation between the IDE cells. Defined straight basal laminae were still found in this stage. The layer beneath the lamina densa probably corresponding to lamina fibroreticularis, which consisted of chiefly fine filaments and some tubular vesicles and was 500 nm in thickness, was clearly seen in this stage.

Triakis scyllia

In the stage before the appearance of the enameloid matrix, a number of odontoblast processes, which contained some mitochondria, vesicles, vacuoles, longitudinal filaments and tubules, occupied the apical area of the dental papilla. The tubular vesicles appeared among the odontoblast processes in the portion near the apex of the dental papilla. The tubular vesicles often came into contact with the odontoblast processes (Fig. 7).

In the middle stage of enameloid matrix formation, a great number of tubular vesicles appeared as the main element of the enameloid matrix. A number of odontoblast processes were still seen in the enameloid matrix in this stage. Septums consisting of flocculent matrix and collagen fibrils, but few tubular vesicles, occasionally separated tubular-vesicle-rich enameloid matrix into several palisades [1] (Fig. 8). Two types of collagen fibrils, approximately 60 nm or 15 nm in the interval of stripes, were found in the septum areas and among tubular vesicles. In the undemineralized sections stained with only lead citrate, a number of electron-dense particles were present at the tubular vesicles. Initial enameloid crystallites appeared at the tubular vesicles. These crystallites were surrounded by the unit membrane which continued into that of the tubular vesicles (Fig. 9).

In the stage of dentine matrix formation, many extremely slender

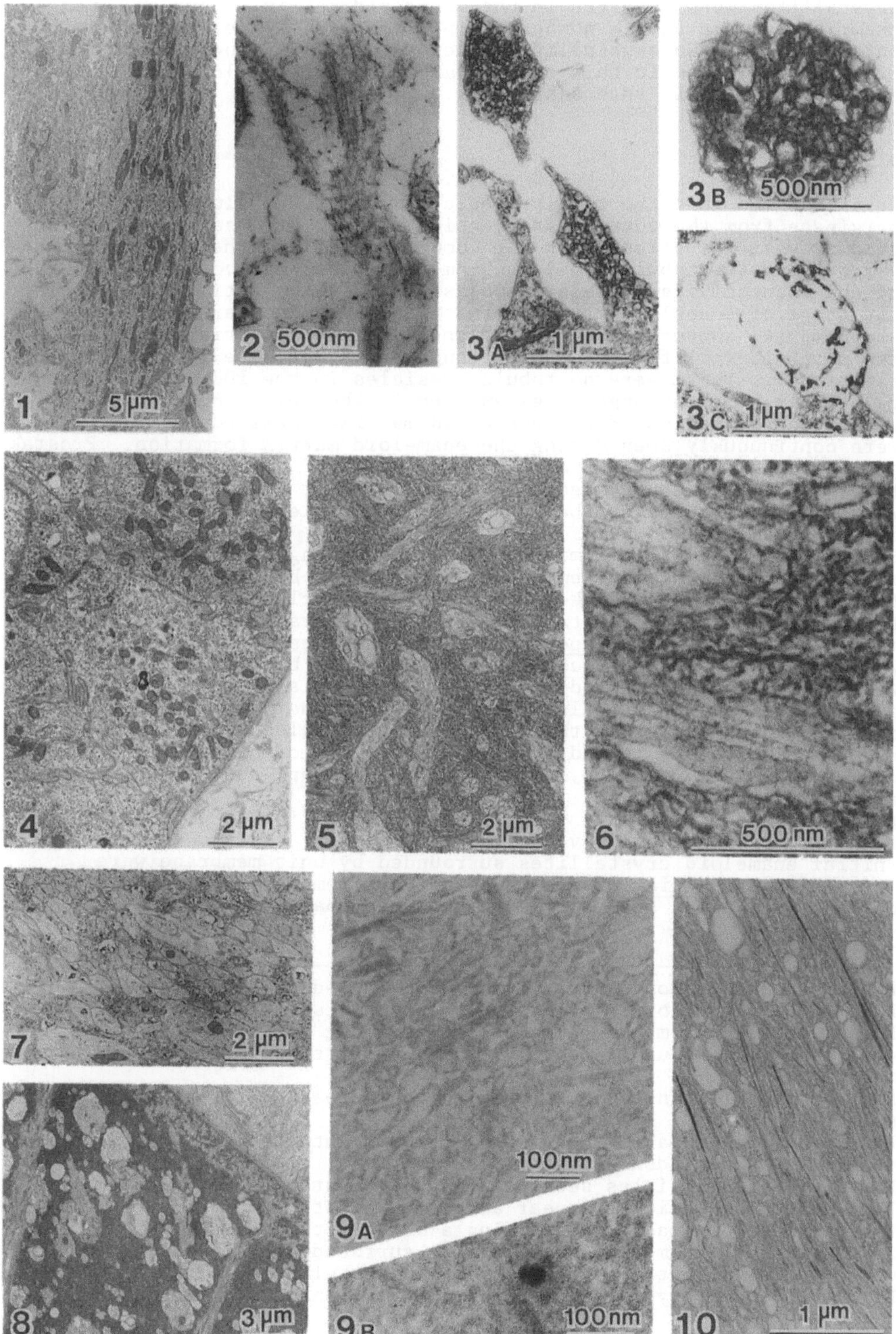
1
5 µm
2
500nm
3A
1 µm
3B
500nm
3C
1 µm
4
2 µm
5
2 µm
6
500 nm
7
2 µm
8
3 µm
9A
100nm
9B
100nm
10
1 µm

crystallites were observed in the enameloid matrix. However, among them, there were also a number of initial crystallites surrounded by unit membrane. No crystallites concentrated along the collagen fibrils were found in this study. Many vacuoles developed in the enameloid matrix in this stage (Fig. 10).

DISCUSSION

It is assumed that the tubular vesicles in the enameloid matrix originate from the odontoblasts, chiefly their processes, because the tubular vesicles often touch the odontoblasts, and the unit membrane surrounding the tubular vesicles sometimes seems to continue into that of the odontoblasts. The odontoblasts seem to be active cells during the stage of enameloid matrix formation. Kemp and Park [2], Goto [3] and Kemp [4] conversely suggested their epithelial origin. However, few morphological features supporting their idea were investigated in this study. There were no tubular vesicles in the IDE cells, although relatively numerous organelles existed in the cells in the stage before the formation of the enameloid matrix. Defined basal laminae were continuously seen during the enameloid matrix formation. Prostak and Skobe [5] suggested that "the giant fibers" which were odontoblastic in origin disaggregated to form the tubular vesicles. I agree with them when they say that the tubular vesicles originate from the odontoblasts.

Several large membrane-limited bodies containing many small vesicles and/or tubules, which resembled multivesicular bodies in cytoplasm and the contents of which looked like tubular vesicles, were found among the odontoblasts in the stage before the appearance of enameloid matrix in Heterodontus japonicus. Some of these bodies made contact with part of the odontoblasts. Multivesicular bodies and lysosomes were often seen in such odontoblasts. It is also possible that the odontoblasts secrete tubular vesicles by secondary lysosomal bodies, in addition to the idea that the tubular vesicles are formed by budding from the cells and degeneration of the odontoblast processes. In Heterodontus japonicus, collagen fibrils among the odontoblasts appeared earlier than did the tubular vesicles. The odontoblasts in the early stage may begin to produce collagen fibrils prior to the formation of the tubular vesicles.

Initial enameloid crystallites surrounded by unit membrane were investigated in Triakis scyllia. The unit membrane surrounding the crystallites often seemed to continue into that of the tubular

---

Fig.1. Extended odontoblasts containing well-developed organelles.
Fig.2. Collagen fibrils among the odontoblasts.
Fig.3. A, large membrane-limited bodies among the odontoblasts. B, enlarged view of such a body. C, collapse of the limited membrane of the body.
Fig.4. IDE cells in the stage before the appearance of the enameloid matrix.
Fig.5. Enameloid matrix consisting of abundant electron-dense tubular vesicles and the odontoblast processes.
Fig.6. Tubular vesicles seem to link the odontoblast processes.
Fig.7. Tubular vesicles appear among the odontoblast processes.
Fig.8. Enameloid matrix containing a few crystallites.
Fig.9. Initial enameloid crystallites surrounded by the unit membrane which seems to continue into that of the tubular vesicles.
Fig.10. Slender crystallites develop in the enameloid matrix.

vesicles. It is therefore likely that the initial enameloid crystallites are formed in the peculiar closed spaces, that is, the tubular vesicles.
The size of enameloid crystallites may rapidly increase after the initiation because thick and slender crystallites were already seen in the enameloid matrix after the beginning of dentine matrix formation. On the other hand, initial enameloid crystallites were still found among the large crystallites. The initiation of enameloid crystallites probably occurs over a long period, from the middle stage of enameloid matrix formation through to the stage of dentine matrix formation.
Two kinds of collagen fibrils in enameloid matrix were also observed in this study, although the quantity of these fibrils was small. It is credible that the collagen fibrils are not connected with the mineralization process in shark enameloid because there are no crystallites accumulating along the collagen fibrils.
The finding mentioned above corresponds with data from gummy sharks, Mustelus manazo [6]. It is therefore probable that these results on enameloid matrix formation and mineralization, such as the appearance of the tubular vesicles constituting most of the enameloid matrix, and the initiation of enameloid crystallites in the tubular vesicles, are common features in sharks. If so, there are many differences found between the initial mineralization stages of sharks and teleosts. The differences are summarized as follows: the site of initial mineralization is the boundary between the enameloid and dentine in teleosts, but that is in the enameloid matrix near the apex in sharks; a number of matrix vesicles are found at the initial mineralization site in teleosts, while in sharks, no matrix vesicles but many tubular vesicles are found; crystallites are small and slender in the early stage in teleosts, but large crystallites can be observed even in the early stage in sharks; additional mineralization occurs along the collagen fibrils in teleosts, but not in sharks; the enameloid matrix consists of mainly collagen fibrils in teleosts, but the tubular vesicles occupy most of the matrix in sharks [6, 7]. I believe that the mechanism for initial mineralization is different in sharks and teleosts. Shark enameloid and teleost enameloid, although similar in final form, each has its own peculiar evolutionary development.

I wish to thank Professor A. Chiba, Department of Biology, Nippon Dental University, Niigata, and Mr. Y. Tsuchiya and Mr. S. Tasaka, Shimoda Marine Aquarium for their valuable advice and kindness in offering their facilities for the collection of specimens. This study was partially supported by grants from the Scientific Research Fund of the Ministry of Education of Japan (No.60771487 and No.62771442).

REFERENCES

[1] Garant PR (1970) J Ultrastruct Res 30: 441-449
[2] Kemp NE, Park JE (1974) Arch Oral Biol 19: 633-644
[3] Goto M (1978) J Stomatol Soc Jpn 45: 527-584
[4] Kemp NE (1985) J Morphol 184: 215-230
[5] Prostak K, Skobe Z (1988) Am J Anat 182: 59-72
[6] Sasagawa I (1989) J Anat 164: 175-187
[7] Sasagawa I (1988) Arch Oral Biol 33: 75-86

CHAPTER 2.14

Effects of 1-Hydroxyethylidene-1, 1-Bisphosphonate (HEBP) on the Maturation Enamel in Rat Incisor

K. OHYA, S. MATAKI[1], H. WAKAMATSU, and H. OGURA

Department of Pharmacology Faculty of Dentistry, Tokyo Medical and Dental University, Tokyo, 113 and Department of Pharmacology, School of Dentistry, Nagasaki University[1], Nagasaki, 852 Japan

Key words: Enamel, Enamel Crystallites, Mineralization, Bisphosphonate, Rat

INTRODUCTION

Bisphosphonates are compound which contain a central P-C-P bond and are thus structurally related to the inorganic pyrophosphates, but are resistant to metabolic destruction (Fig.1). They bind strongly onto the hydroxyapatite crystal and reduce crystal nucleation, aggregation, growth and dissolution *in vitro*. Further, they decrease the mineralization of bone and dentin and inhibit bone resorption *in vivo* [1].

Previously we have reported the inhibitory effect of 1-hydroxyethylidene-1,1-bisphosphonate (HEBP) on the formation process of secretory stage of enamel in rat incisor [2]. HEBP exerted their effect not only to affect the mineralization of enamel crystallites, but also to inhibit the synthesis of enamel matrix proteins, resulting the disordered formation of enamel in secretory stage. In this study, we investigate the effect of successive injections of HEBP on the maturation stage of enamel in order to clarify the pharmacological action of the drug on enamel formation process.

MATERIALS AND METHODS

Male Wistar rats, each with a body weight of 150-160 g, were used. The rats in experimental group were injected subcutaneously with HEBP at a dose of 4 or 8 mgP/kg/day for 7 days. The rats in control group were injected with an equal volume of physiological saline. At 24 hours after the last injection of the drug, each rat was anesthetized by sodium pentobarbital and perfused for 15 minutes through the ascending aorta with a 2.5 % glutaraldehyde-2 % paraformaldehyde solution (pH 7.4). The maxillae were extracted and processed either for the scanning electron microscopy (SEM) or light microscopy (LM) or contact microradiography (CMR). For the SEM, mid-sagittal ground surface of incisor were treated with 5% NaOCl for 30 min and etched with 0.05 M HCl solution for 10 sec. For the LM, the isolated maxillae were decalcified with 10% EDTA solution and embedded in Epon 812. CMR was made in mid-sagittal plane of incisor. In a separate experiment, calcein (15 mg/kg) was injected intravenously 1 hr after the last injection of HEBP. Then, after 1 hr, mandibular incisor was isolated and labeling pattern of enamel surface was observed under a fluorescent microscope. A paired half incisor was stained with glyoxal bis (2-hydroxyanil) (GBHA) solution.

RESULTS

The SEM observations were performed in the three regions, early, middle and late

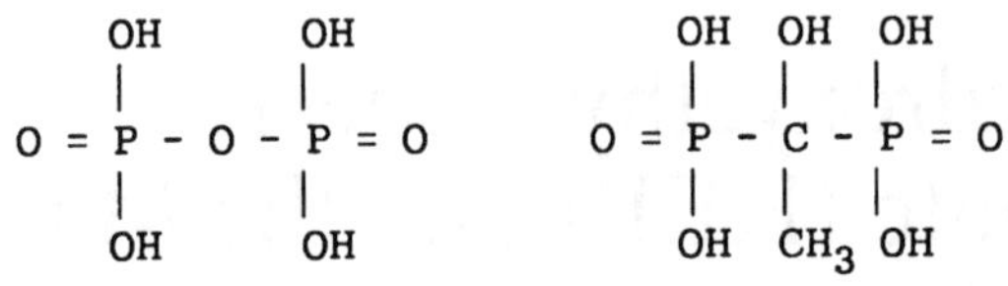

Inorganic Pyrophosphate

1-Hydroxyethylidene-1, 1-bisphosphonate (HEBP)

Fig. 1 Chemical structures of pyrophosphate and HEBP.

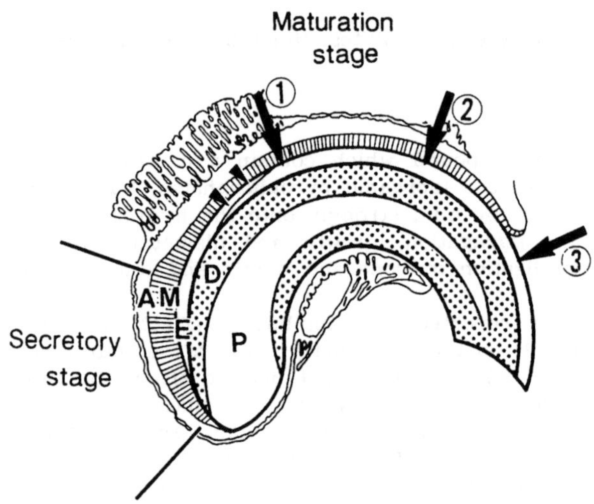

Fig.2 Schematic illustration of mid-sagittal plane of the maxillary incisor. SEM observations were performed at the stage of early (1), middle (2), and late (3) maturation stage. AM: ameloblasts, E: enamel, D: dentin, P: pulp, arrow head: hypomineralized line.

enamel maturation stage as shown in Fig.2. The hypominearlized line, which might be caused at the interface between ameloblasts and enamel surface by the daily injections of HEBP, was refer to the boundary line between secretory and maturation enamel. The region incisal to this line was examined as a maturation enamel at the start of HEBP injection.

In a ground surface of enamel in the early maturation stage, control rats showed a well ordered prismatic enamel in the middle layer of enamel (Fig.3a,b). The enamel rods were composed of many needle-like crystallites and crossed each other. Inter-rod enamel separated each enamel rod. In the innermost layer of the enamel, there was a thin layer of aprismatic enamel. The enamel rods changed their course in the outer enamel and inclined to the incisal direction.

In the ground surface of enamel in experimental rats injected 8 mgP/kg of HEBP, the cross pattern of enamel rods were disarranged in the middle layer of enamel (Fig.3c,d). Enamel rods were composed of fine and slender crystallites and the number of crystallites of each rod seemed to be decreased. Crystallites were bent like as fiber and were not well packed as an enamel rod. The formation of inter-rod

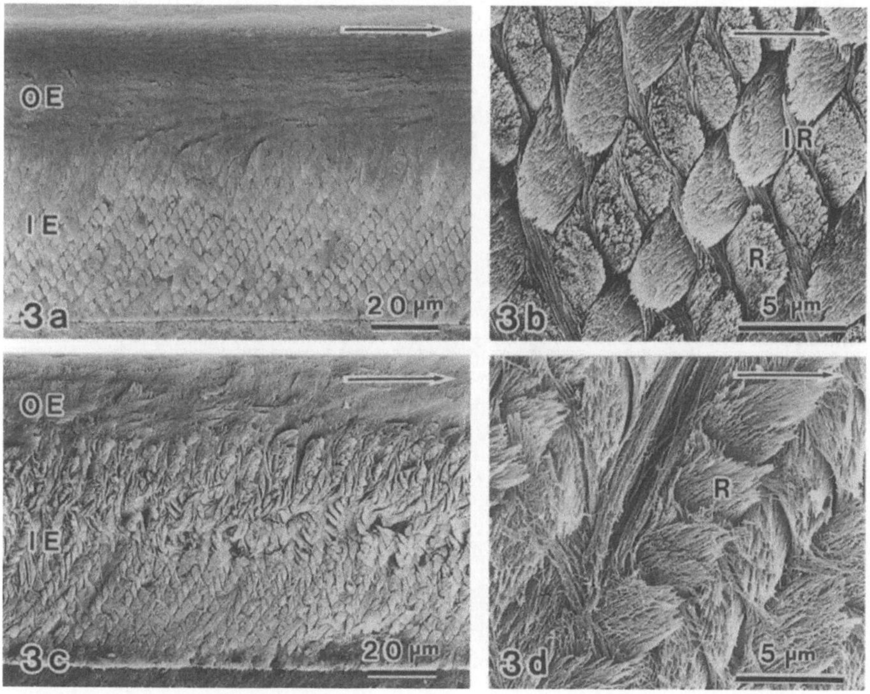

Fig.3 The SEM photomicrographs of a longitudinal surface of early maturation stage of enamel. (a),(b) control rats. (c),(d) experimental rats injected 8 mgP/kg HEBP for 7 days. OE: outer enamel, IE: inner enamel, R: enamel rod, IR: inter-rod enamel, arrow: incisal direction.

enamel was reduced and enamel rods were not clearly separated each other. The width of outer enamel seemed to be decreased and the inclined enamel rods were lost their oriented pattern. These changes were less pronounced in rats injected 4 mgP/kg than 8 mgP/kg and thus the effects of HEBP were dose-related.

In the middle maturation stage, the ground surface of enamel in control rats showed a similar enamel view to that in the early maturation stage (Fig.4a,b). In the experimental rats injected with 8 mgP/kg, the surface of enamel was observed to be smooth compared with the control rat, indicating that enamel crystallites were resistant to the treatment of acid-etching (Fig.4c,d). Crystallites were finer and more slender than control crystallite. In each enamel rod, crystallites were decreased in number and the pattern of decussating enamel rods was obscured.

In the late stage of enamel, the SEM view of enamel both in control and experimental rats was similar.

Histological observations of maturation ameloblasts showed no changes in their appearance between both groups (Fig.5). By CMR, the degree of mineralization in the experimental rats seemed to be similar to the control rats. The calcein labeling pattern and the GBHA staining of enamel surface were not different in both groups.

DISCUSSION

The SEM observation in this study demonstrated that enamel crystallites as well as enamel rods formed under the influence of HEBP were disarranged and reduced in

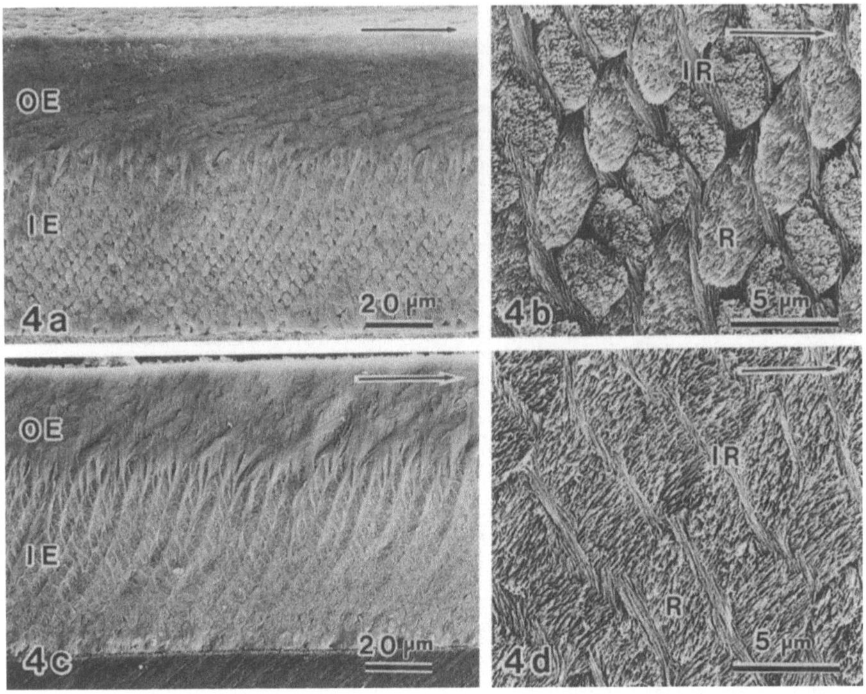

Fig. 4 The SEM photomicrographs of a longitudinal surface of middle maturation stage of enamel. (a),(b) control rats. (c),(d) experimental rats injected 8 mgP/kg HEBP. OE: outer enamel, IE: inner enamel, R: enamel rod, IR: inter-rod enamel, arrow: incisal direction.

their appearance. These actions of drug seemed to mainly be due to its physico-chemical interaction with calcium phosphate crystal formation. The binding of HEBP to the growth sites of the mineral phase interrupted the orderly process of further mineral deposition onto these sites.

The effects of HEBP on maturation stage of enamel are different from that on secretory stage of enamel. As previously reported, HEBP inhibits both the mineral deposition and the formation of enamel matrix in the secretory stage of enamel [2]. These actions of the drug are at least in part due to its effect on cellular functions of secretory ameloblasts. On the other hand, ameloblasts in maturation stage seem to be not affected. This suggests that the action of HEBP on the formation process of maturation enamel is not mediated through the cellular effect. Direct action of the drug on the growth sites of enamel crystallites may induce the disarrangement of maturation enamel. This is further supported by the observation of the labeling pattern or the GBHA staining of enamel surface. The enamel formation of maturation stage is related to the cyclical modulation of ameloblasts as revealed by these methods [4], but HEBP did not influence the modulation pattern.

Disarrangement of enamel rods and crystallites was more pronounced in the early maturation stage than the other stages. It is known that mineralization of enamel progresses rapidly in the transitional stage between secretory and maturation enamel where the enamel matrix absorption and the formation of enamel crystallites take place in the same time [5]. HEBP seems to decrease formation of enamel crystallites more severely in this stage than other stages.

In the late stage of maturation enamel, the effect was not apparent, because the

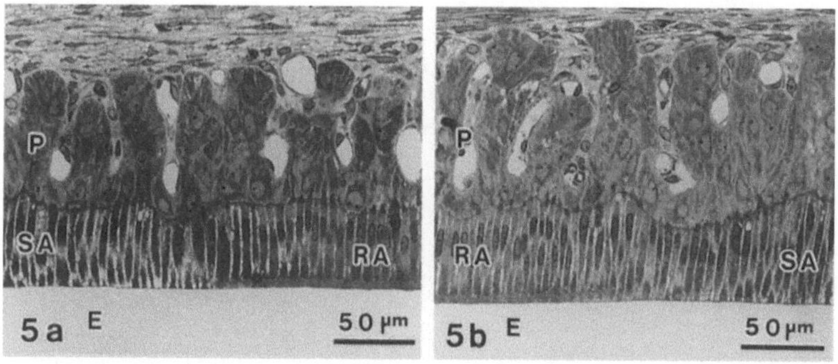

Fig. 5 The LM photomicrographs of decalcified longitudinal sections of maturation ameloblasts. (a) control rats and (b) experimental rats injected 8 mgP/kg HEBP. P: papillary cell, E: enamel space, SA: smooth-ended ameloblasts, RA: ruffle-ended ameloblasts.

mineralization of enamel has already completed at the start of experiment and thus HEBP could not exert its action.

From the results of the present and previous studies, it is suggested that successive injections of HEBP interfere the formation of enamel through the different action on the secretory and maturation enamel. During the secretory stage, HEBP affects both the enamel matrix formation and the deposition of enamel crystallites and during the maturation stage, it only reduces the mineral deposition and the enamel crystallites growth.

REFERENCES

1. Fleisch H (1983) In: Peck WA (ed) Bone and Mineral research annual I. Excerpta Medica, Amsterdam, pp. 319-357
2. Ohya K, Mataki S, Wakamatsu H, Ogura H (1989) In: Fearnhead RW (ed) Tooth Enamel V, Florence Publishers, Yokohama, pp. 227-231
3. Ishige N (1987) Jpn. J. Oral Biol. 29: 107-116
4. Takano Y, Crenshaw MA, Bawden JW, Hammarstrom L, Lindskog S (1982) J. Dent. Res. 61: 1580-1586
5. Hiller CR, Robinson C, Weatherell JA (1975) Calcif. Tiss. Int. 18: 1-12

CHAPTER 2.15

Ultrastructure of Rat Ameloblasts During Enamel Maturation Observed by Scanning Electron Microscopy

M. KISHINO, T. OGAWA, T. HAMADA, and T. INOUE[1]

Department of Oral and Maxillofacial Surgery and [1]Department of Anatomy, Tottori University School of Medicine, Yonago, 683 Japan

Key words: Ameloblasts, Enamel maturation, SEM, Rat.

INTRODUCTION

The rat incisor is an excellent model system for studying amelogenesis [1, 2]. Its distal surface has particularly received great attention from the viewpoint of enamel formation. The ameloblasts at the stage of enamel maturation are divided into two different groups from their morphological findings: ruffle- and smooth-ended ameloblasts. Since these two kinds of ameloblasts show different activity in calcium transport and endocytosis [1], their ultrastructural study is important to clarify the mechanism of mineralization. Although transmission electron microscopic (TEM) studies have clarified their ultrastructures [3-6], their three-dimensional architecture has not yet been sufficiently understood. Several scanning electron microscopic (SEM) studies on the ameloblasts have been reported [7-11], but their distal surface at the stage of enamel maturation has not been clearly demonstrated because of the presence of a dense lamina lining the distal surface. In this study, we prepared specimens by the aldehyde-osmium-dimethyl sulfoxide-osmium (A-O-D-O) method in combination with the decalcification technique used for preparing bone tissues [12]. As a result, the dense lamina having been completely removed, the surface structures of the ameloblasts were clearly demonstrated. Furthermore, their intracellular structures were simultaneously demonstrated together with the surface morphology. We have previously reported the distal surface of the ruffle-ended ameloblasts obtained by this specimen preparation procedure [13]. In this paper, we describe the three-dimensional morphology of the distal surface obtained by high-resolution SEM with a particular reference on the difference between ruffle- and smooth-ended ameloblasts.

MATERIALS AND METHODS

Rats of the Wistar strain from 3 weeks old were used as materials. Specimens were fundamentally prepared by the A-O-D-O method [14]. After anesthesia by intraperitoneal injection of Nembutal, a mixture of 0.5% glutaraldehyde and 0.5% paraformaldehyde in 1/15 M phosphate buffer (pH 7.4) was perfused from the left ventricle. Lower incisors and their adherent alveolar bone were dissected from surrounding tissues. They were washed in the buffer, and treated with 0.3 M ethylenediamine tetraacetic acid for 5 days for decalcification. They were then postfixed with 1% osmium tetroxide in the same buffer. After washing in the buffer, they were immersed in 25% and 50% dimethyl sulfoxide (DMSO) for 60 min each. The specimen was then frozen on a metal plate chilled with liquid nitrogen, and split into two pieces with a precooled razor blade and a hammer using a freeze-cracking apparatus (TF-2, Eiko Engineering Co. Ltd., Japan) [15]. The two split specimens were placed and thawed in 50% DMSO at room temperature. They were then rinsed in the buffer and placed in 0.1% osmium tetroxide for 96 hr to remove excess cytoplasmic matrices (the osmic maceration procedure in the A-O-D-O method) [14]. After conductive staining with 2% tannic acid and 1% osmium tetroxide [16], they were dehydrated through a graded series of ethanol, transferred to isoamyl acetate and dried by critical-point drying using dry ice [17] or by the t-butyl alcohol freeze-drying method

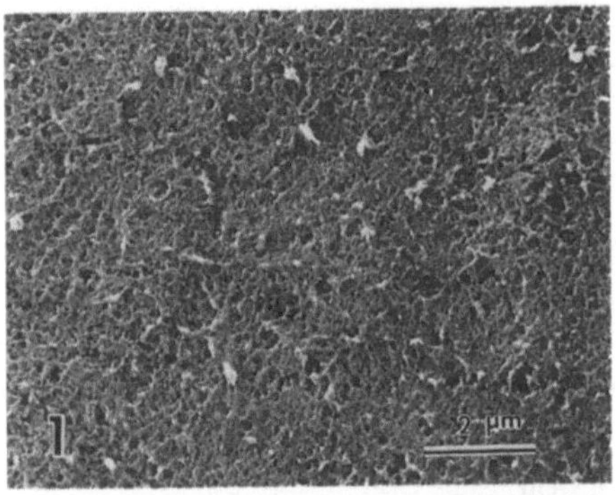

Fig. 1. High-magnification micrograph of the surface view of the dense lamina lining the distal surface of ameloblasts. The dense lamina is composed of a dense meshwork of microfibrils.

Fig. 2. Low-magnification micrograph showing fractured smooth-ended ameloblasts and their distal surface (asterisk). Arrow: Capillary.

Fig. 3. Low-magnification micrograph showing fractured ruffle-ended ameloblasts and their distal surface (asterisk). Arrow: Capillary.

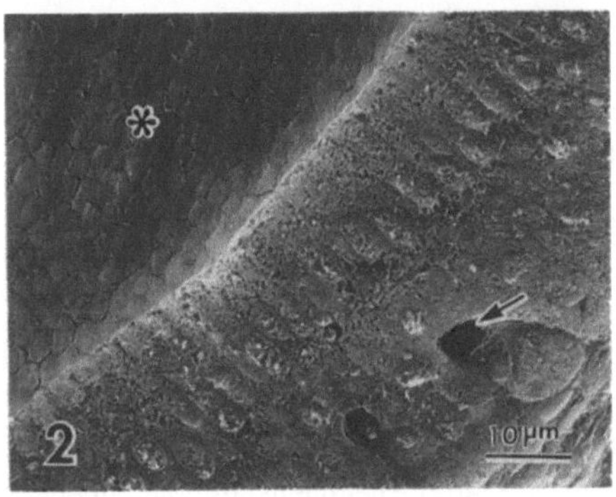

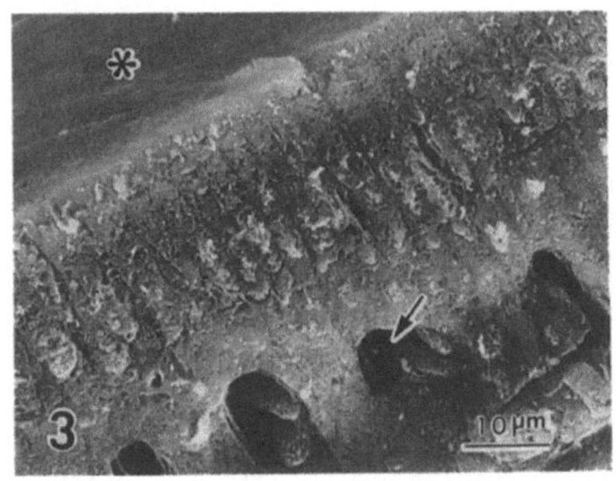

[18]. The dried specimens were coated with platinum of approximately 3 nm by an ion-sputter coater (VX-10R, Eiko Engineering Co. Ltd., Japan), observed with a field emission scanning electron microscope (HFS-2ST, Hitachi Ltd., Japan) at 25 kV, and photographed in combination with stereophotography.

RESULTS

Since the maceration procedure using a dilute osmic solution is a delicate technique, the dense lamina lining the distal surface sometimes remained undissolved prohibiting the demonstration of the true surface of ameloblasts. Figure 1 shows a higher magnification of the dense lamina. The lamina was composed of a dense meshwork of microfibrils. When the osmic maceration was properly carried out, the dense lamina was completely removed and the distal surface was clearly observed. Intracellular membranous structures were also three-dimensionally shown since the excess cytoplasmic matrix was properly removed. Two kinds of ameloblasts (ruffle- and smooth-ended ameloblasts) were easily identified from their distal surface morphology at high magnification. They were arranged in alternating bands at low magnification (not shown here).

Figures 2 and 3 are low-magnification micrographs of the smooth- and ruffle-ended ameloblasts, respectively. Since an SEM image has a deep focal length, both the distal surface and intracellular structures were simultaneously observed. At low magnification of the distal surface of the smooth-ended ameloblasts, a hexagonal pattern of the cell boundaries was clearly visible. On the other hand, the distal surface of the ruffle-ended ameloblasts appeared flat in the low-magnification micrograph under 2,000 X, but highly invaginated at high magnification over 10,000 X.

The smooth-ended ameloblast was characterized by the smooth cell surface without invaginations. The cell boundaries bulged slightly into the enamel side and were clearly distinguished. A large accumulation of the mitochondria was evident in the vicinity of the distal surface (Fig. 4). The mitochondria were round in shape associating with the tubular or lamellar endoplasmic reticulum. Small vesicular structures were

often observed under the surface cell membrane measuring 100-200 nm in diameter (Fig. 4). Round or oval depressions measuring 50-150 nm in diameter were scattered on the distal surface. Microvillous protrusions were often observed on its surface (Fig. 5).
The ruffle-ended ameloblasts were characterized by a ruffle border with channel invaginations. On the cracked surface of the ameloblasts, the invagination of the plasma membrane was clearly seen (Fig. 6). An accumulation of the mitochondria was evident in the deeper layer of the ruffled border (Fig. 6). The mitochondria were slightly elongated compared with those of the smooth-ended ameloblasts. A tubular smooth endoplasmic reticulum formed a loose network around the mitochondria. At low magnification, the cell boundary of the ameloblasts was difficult to discern (Fig. 3), but a higher-magnification micrograph shows that the boundary was highly interdigitated, forming a complex labyrinth formed by the protrusion and invagination of the plasma membrane (Fig. 8). Protrusions measuring 0.1 - 0.2 um in width formed interdigitated structures with each other. At higher magnification of the distal cell surface, two kinds of minute granules were visible: small and large granules measured as 10-20 nm and 70 nm in diameter, respectively (Fig. 9). The former were more numerous than the latter. These granules were partially absent showing a round denuded area which bulged slightly. These microfibrils appeared to connect the protrusions of the plasma membrane with each other. Another characteristic of the distal surface of the ruffle-ended ameloblast was the presence of microfibrils connecting the protrusions of the plasma membrane. The microfibrils measured 7 nm in width when observed on the specimens coated with 3 nm of platinum.

DISCUSSIONS

According to the previous TEM findings, the distal cell membrane is in close apposition to a dense lamina which is easily discerned in demineralized specimens [4,5]. As was pointed out by Skobe [7], it has been difficult to observe the distal surface by SEM due to the presence of the dense lamina. In the present specimen preparation, however, the dense lamina having been dissolved during the osmic maceration procedure, the distal cell surface was clearly demonstrated, as shown in Figs. 2 through 9. Several authors reported that the dense lamina is similar to the basal lamina recognized between epithelial cells and underlying connective tissues [4,5]. However, the basal lamina is intact against the osmic maceration procedure [16]. Furthermore, the fine structures of the basal lamina differ from those of the dense lamina in having finer microfibrils.
A further advantage of the present SEM was the simultaneous demonstration of both

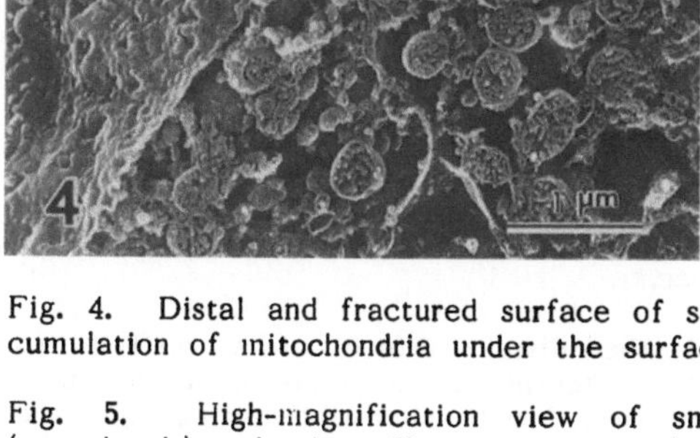

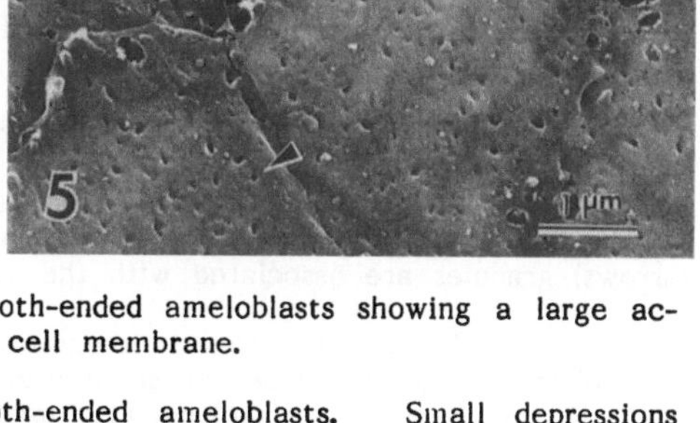

Fig. 4. Distal and fractured surface of smooth-ended ameloblasts showing a large accumulation of mitochondria under the surface cell membrane.

Fig. 5. High-magnification view of smooth-ended ameloblasts. Small depressions (arrowheads) and microvillous projections (arrows) are visible on their surface.

surface and intracellular structures, which was useful in understanding the relationship between both structures.

Cyclic changes of ameloblasts are well known during amelogenesis: ruffle-ended ameloblasts altered their shape to become smooth-ended ones and subsequently into ruffle-ended ones again. The most significant changes of these ameloblasts are recognized in the distal cell surface which alters in a short time [19]. The distal cell membrane forms deep invaginations in ruffle-ended ameloblasts and becomes flat in smooth-ended ameloblasts. The present SEM study demonstrated the complex labyrinth on the distal surface in ruffle-ended ameloblasts. The labyrinth of the cell membrane fused with each other and became flat as the ruffle-ended ameloblasts became smooth-ended ones. The distal surface of the smooth-ended ameloblasts, however, had many surface depressions on its surface.

On the distal surface of ruffle-ended ameloblasts, two kinds of minute granules were

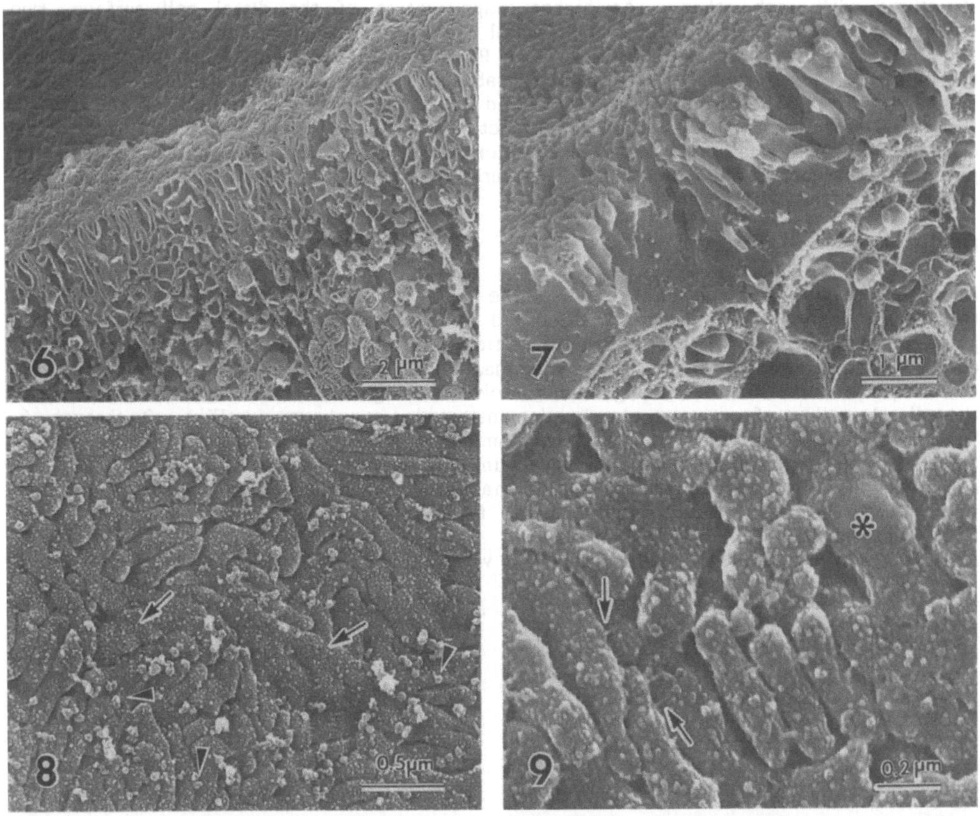

Fig. 6. Fractured surface and distal surface of ruffle-ended ameloblasts showing the invaginations of the surface cell membrane referred to as a ruffle border.

Fig. 7. Lateral view of the ruffled border of ruffle-ended ameloblasts.

Fig. 8. Distal surface of a ruffle-ended ameloblast. A labyrinth formed by protrusion and invagination of the plasma membrane is clearly seen. Small (arrowheads) and large (arrows) granules are associated with the surface.

Fig. 9. High-magnification micrograph of the distal surface of a ruffle-ended ameloblast. A great number of small granules is discerned on its surface, but are partially absent, showing a round denuded area (asterisk). Note microfibrils connecting the protrusions of the plasma membrane (arrows).

visible in this study. The smaller granules resembled the granular substances described on the basal cell membrane facing the basal lamina of mesothelial cells [20]. Hence, these granules may be an inherent structure which anchors the dense lamina with the distal surface of the ameloblast. Josephsen and Fejerskov [5] reported membrane-bound granules about 40-60 nm in size between the dense lamina and distal cell membrane in the areas of smooth- to ruffle-ended ameloblasts, and in the incisal part of the preceding smooth-ended areas. Judging from the size, their granule is similar to the large granules found in this study, but differs in the location. The larger granules are likely to engage in enamel formation, but their precise significance is unknown. In smooth-ended ameloblasts, no granular structures were seen on the distal surface. This finding indicates that there are no direct connections between the dense lamina and smooth-ended ameloblasts. The round denuded area recognized in the ruffle-ended ameloblasts was free of granular substances. In our preliminary study, such a denuded area was commonly observed on the transitional ameloblasts that fall in a category between ruffle-ended ameloblasts and smooth-ended ameloblasts. The denuded area is likely to spread, forming a flat surface as the ruffle-ended ameloblasts altered into smooth-ended ones.

The microfibrils connecting the protrusions of the labyrinth were observed only in ruffle-ended ameloblasts. These microfibrils may be derived from the dense lamina and may engage in packing the ruffle border.

REFERENCES

1. Takano Y, Ozawa H (1980) Arch Histol Jpn 43: 385-399
2. Reith EJ (1960) Arch Oral Biol 2: 253-262
3. Reith EJ (1970) J Ultrastruct Res 30: 111-151
4. Warshawsky H (1971) Anat Rec 169: 559-584
5. Josephsen K, Fejerskov O (1977) J Anat 124: 45-70
6. Warshawsky H, Josephsen K, Thylstrup A, Fejerskov O (1981) Anat Rec 200: 371-399
7. Skobe Z (1976) Calcif Tiss Res 21: 83-103
8. Boyde A, Reith EJ (1977) Cell Tiss Res 178: 221-228
9. Skobe Z (1980) Arch Oral Biol 25: 395-401
10. Sasaki T, Debari K, Higashi S (1984) J Electron Microsc 33: 168-171
11. Ohmi S (1987) Jpn J Oral Biol 29: 332-362 (in Japanese with English abstract)
12. Hayashi K (1987) J Yonago Med Assoc 38: 387-399 (in Japanese with English abstract)
13. Kishino M, Ogawa T, Inoué T, Hamada T (1989) J Electron Microsc 38: 394-398
14. Tanaka K, Mitsushima A (1984) J Microsc 133: 213-222
15. Tokunaga J, Edanaga M, Fujita T, Adachi K (1974) Arch Histol Jpn 37: 165-182
16. Murakami T (1974) Arch Histol Jpn 36: 189-193
17. Tanaka K, Iino A (1974) Stain Technol 49: 203-206
18. Inoué T, Osatake H (1988) Arch Histol Cytol 51: 53-59
19. Smith CE, McKee MD, Nanci A (1987) Adv Dent Res 1: 162-175
20. Inoué T, Osatake H (1989) J Submicrosc Cytol Pathol 21: 215-227

CHAPTER 2.16

Changes of the Ratio of Calcium and Phosphate in Mineralizing Tissues

T. KAWAMOTO and M. SHIMIZU
Department of Biochemistry, School of Dental Medicine, Tsurumi University, Yokohama, Japan

Key words: Ca/P ratio, enamel, dentine, bone.

INTRODUCTION

Our previous autoradiographic study using freeze-dried sections has shown that in the formation of rat incisal enamel the secretory ameloblasts extrude calcium to the capillary side to maintain the enamel at a low level of calcification, whereas in the maturation stage the ameloblasts control the entry of calcium and phosphate into the enamel by changing the tightness of intercellular junctions and/or their activities for ion transport.[1]

In the present study an attempt was made to examine whether some similar processes were operating during the mineralization of dentine and bone.

MATERIALS AND METHODS

<u>Autoradiography</u> Autoradiographic procedures suitable for the study of water-soluble materials in whole body sections, described previously by the authors were used.[1,2]

Fifty minutes after an intraperitoneal(i.p.) injection of 1% calcein solution (0.1ml/100g body weight), 9 day old Wistar strain rats were injected i.p. with $^{45}CaCl_2$(18x10^4Bq/g body weights) or $H_3{}^{32}PO_4$ solution(3.7x10^4Bq/g body weight). Five minutes after the injection of the radioisotope the animals were frozen in hexane (-90°C). The frozen animal was embedded in 5% carboxymethyl cellulose solution(CMC), and was completely frozen in the coolant. Freeze-dried sagittal sections(5μm thick) of the lower incisor and the femur from the whole body section were placed in close contact with X-ray film, and autoradiographed.

<u>Measurement of radioactivity in the different mineralized tissues</u> Nine day old rats were injected i.p. with 1% calcein solution 30 min prior to an i.v. injection of a mixture of $^{45}CaCl_2$(3.7x10^{6}Bq) and $H_3{}^{32}PO_4$(3.7x10^{6}Bq) solution. The animals were frozen in the hexane 5 min after the injection of the radioisotopes. Serial freeze-dried sections of 20μm thick were made by the method described above. Samples of enamel, dentine and bone in various developing stages were dissected with a razor blade from serial sections, and the samples collected were placed in a vial containing a liquid scintillator(ACS-2, Amersham Co. Ltd, USA). The radioactivity was measured for 5 min every day during a period of 4 weeks. The radioactivity of ^{45}Ca and $^{32}PO_4$ was estimated from their decay curves respectively.

<u>Energy dispersive X-ray analysis</u> A 5μm thick freeze-dried section was mounted on an aluminum stub and coated with carbon by evaporation. The element analysis of the mineralizing areas of interest in the section was recorded with an energy dispersive X-ray analysis system attached to a scanning electron microscope. The accelerating voltage was adjusted to 15Kv and the counts of Ca and P Kα1 X-ray peaks were collected from an area of 1x1μm^2 for 2 min using a beam current of 2x10^{-9}A. The ratios of Ca/P in the analyzed areas were calculated using values obtained from the analysis of the standard samples of known Ca/P ratios in the same conditions.

<u>Glyoxal bis(2-hydroxyanil)(GBHA) staining</u> The freeze-dried sections were stained with a GBHA solution for 30 sec. The GBHA solution consisted of 20mg GBHA and 2ml of 75% ethanol containing 0.105 N NaOH.

RESULTS AND DISCUSSION

Since calcein does not become incorporated into cells[1], the incorporation of calcein in the enamel in the secretory stage, in the enamel of the zone of smooth-ended ameloblasts in the maturation stage and in the mineralizing front of the developing dentine and the developing bone, this indicates an intercellular pathway for ions and molecules from the blood to these mineralizing areas.

As shown in Fig-1, the developing enamel of the lower incisor incorporated ^{45}Ca periodically. The most intensively labelled zone was the apical side of the ruffle-ended ameloblast zone. In the developing dentine, although all the surface was labelled, the labelling intensity was not uniform but rather periodic. In the bone, almost all of the surfaces were labelled, but the labelling intensities were not uniform. The highest intensity was located on the calcification zone of the metaphysis of long bones. On the other hand, the labelling patterns in ^{32}P autoradiography were slightly different from those in the ^{45}Ca autoradiography. The variation in the labelling intensity in the ^{32}P autoradiography was much smaller than that in the ^{45}Ca autoradiography. Table 1 shows the radioactivities and the ratios of ^{45}Ca to ^{32}P in the samples dissected from various mineralizing zones of the enamel, dentine and bone as designated in Fig. 2. Although the ratio in each region was different, the high ratio in the zone of calcification of the metaphysis and in the enamel in the region of the ruffle-ended ameloblasts was noticeable. In the dentine, although it was difficult to dissect the highly labelled area selectively, the ratio in the Dn-3 area was higher than that in the Dn-1 area. These results indicate that in these regions in addition to the transport of ions by an intercellular pathway the calcium ions are transported from the blood to these regions preferentially by the activity of cells. The molar ratios of Ca to P in various regions of the enamel, dentine and bone, which were calculated from the results of EDXA, showed that except for the surface layer of the trabeculae in the zone of calcification of the metaphysis, the Ca/P molar ratios in the regions of the early mineralizing stage were smaller than those in the regions of the more advanced stage of mineralization. Furthermore, the regions having a Ca/P molar ratio in the range of 1.0-1.4 were stained with GBHA staining, whereas the regions having a higher Ca/P molar ratio than 1.5 were not stainable. The examination on the stainability with GBHA staining for the standard samples, such as synthesized amorphous calcium phosphate(ACP), dicalcium phosphate(DCP), octacalcium phosphate(OCP), tricalcium phosphate(TCP) and hydroxyapatite(HAP) showed that ACP and DCP were stained. These results would indicate that in the mineralizing processes of enamel, dentine and bone the calcium phosphate is deposited as a precipitate with a lower Ca/P ratio similar to ACP and DCP than the ratio of hydroxyapatite. This precipitate may then transform to hydroxyaptite by a preferential supply of calcium ions by the cells. In the trabeculae in the zone of calcification of the metaphysis, the outer surface layer of the trabeculae where the Ca/P molar ratio was high(approximately 1.6) was not stained with GBHA staining, whereas the inner part of the trabeculae where the Ca/P molar ratio was low(approximately 1.4) was stainable. Since the $^{45}Ca/^{32}P$ ratio in the region Bn-2(Table 1) was high(approximately 1.9), these findings could indicate that in this region a rapid transformation of calcium phosphate deposits is necessary.

Acknowledgment. We are indebted to Professor R.W. Fearnhead, Department of Anatomy, Tsurumi University, for his valuable comments and advice.

Reference

1) Kawamoto T, Shimizu M (1990) Calcif Tiss Int 46:406-414
2) Kawamoto T, Shimizu M (1986) Stain Technol 61:169-183

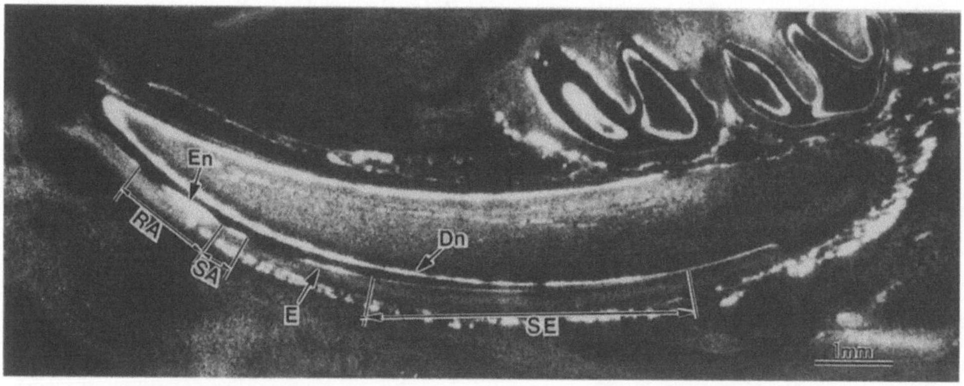

Fig. 1 Contact autoradiography of the lower incisor of a 9-day-old rat 5 min after an i.p. injection of $^{45}CaCl_2$. En= enamel; Dn= dentine; RA= zone of ruffle-ended ameloblasts; SA= zone of smooth-ended ameloblasts; SE; secretory stage.

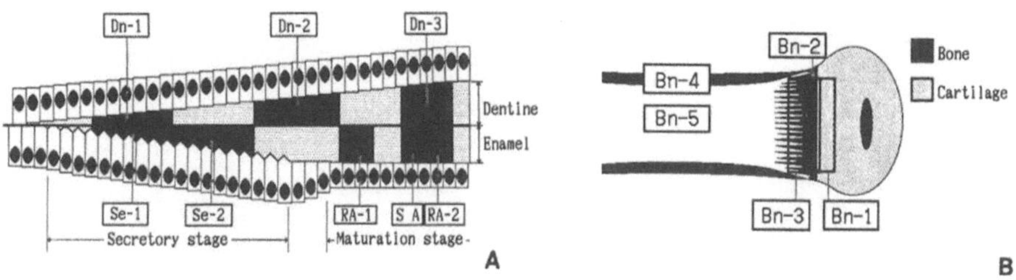

Fig. 2 Schematic drawings showing the position of sample dissected from the freeze-dried section of lower incisor(A) and femur(B) of rat.

	Se-1	Se-2	RA-1	S A	RA-2	Dn-1	Dn-2	Dn-3
^{45}Ca (cpm)*	0.88±0.02	1.59±0.02	6.32±0.06	2.26±0.05	3.72±0.05	2.69±0.03	4.28±0.03	4.45±0.04
$^{32}PO_4$(cpm)*	1.10±0.02	1.28±0.02	2.57±0.04	1.53±0.04	1.64±0.03	1.92±0.03	2.96±0.03	2.79±0.03
Ratio(^{45}Ca/^{32}P)	0.81±0.02	1.24±0.02	2.46±0.04	1.48±0.05	2.27±0.05	1.40±0.03	1.45±0.02	1.59±0.02

	Blood	Bn-1	Bn-2	Bn-3	Bn-4	Bn-5	Bn-6
^{45}Ca (cpm)	835±13	2140±103	17402±295	9707±220	13300±258	765±62	3207±127
$^{32}PO_4$(cpm)	784±13	419±46	9047±213	7242±190	8849±210	1394±83	2142±103
Ratio(^{45}Ca/^{32}P)	1.07±0.02	5.10±0.61	1.92±0.06	1.34±0.05	1.50±0.05	0.55±0.06	1.50±0.09

Table-1 The radioactivities and ratios of ^{45}Ca and $^{32}PO_4$ in different mineralized tissues. The positions of the sample dissected from the lower incisor and the femur are indicated in Figure 2, and Bn-6 is dissected from calvaria. $^{45}Ca^*$ and $^{32}PO_4^*$ indicate the amount of radioactive calcium and phosphate transported through a unit area(500 μm^2) of cell layer into the enamel or the dentine.

CHAPTER 2.17

Two Patterns of Calcification in Rat and Rabbit Incisor Dentin

H. Mishima, Y. Kozawa, and T. Sakae

Department of Anatomy, Nihon University School of Dentistry at Matsudo, Chiba, Japan

Key words: Incisor, Dentin, Rat, Rabbit, Calcification.

INTRODUCTION

The authors have reported previously that there are some differences of crystal orientation and calcification pattern between the enamel-covered labial and cementum-covered lingual dentin in rat incisors [1-2]. As in the rat, the incisors of the rabbit are continuously growing, and some reports have indicated that the dentin of incisors in both animals shows significant differences between the labial and lingual areas [3-6]. The purpose of this study was to investigate the structural differences between labial and lingual dentin of incisors from both rats and rabbits.

MATERIALS AND METHODS

Bilateral incisors extracted from 90 male Wistar-strain rats (weighing 150 to 450 g each) and 20 male Japanese white rabbits (weighing 2.5 to 3.5 kg each) were used in the present study. The animals were decapitated following ether anesthesia, and their incisors were extracted and fixed immediately in 10% neutral formalin. Decalcified and undecalcified specimens were then prepared. These specimens were studied using light microscopy, polarizing light microscopy, UV fluorescence microscopy, microradiography, scanning electron microscopy and micro-Laue diffraction.

RESULTS

<u>Labial dentin</u>: In the labial dentin of both rat and rabbit, the calcospherites in the predentin were globular in form, and fusion of the calcospherites could be observed in the mineralization front. Interglobular dentin and calcospherites were clearly seen in the superficial (s) and middle (m) layers of labial dentin from rat (Fig. 1). Incremental lines were not observed in the labial dentin. The interglobular dentin and calcospherites were evident in the superficial layer (s) of labial dentin from rabbit (Fig. 2). Unlike rat, incremental lines were prominent in the middle (m) and deep (d) layers of rabbit labial dentin. In both rat and rabbit, the labial dentin was more highly calcified than the

lingual dentin. Crystals were randomly oriented in the labial dentin of rat, and slight orientation was noted in the deep layer. On the other hand, in the middle (m) and deep (d) layers of rabbit dentin, the crystals were oriented almost parallel to the long axis of the incisor (Fig. 3). The fluorescent labelling lines obtained with tetracycline (TC) in labial dentin were wavy from the middle layer (m) to the superficial layer (s) but were linear in the deep layer (d) (Fig. 4). Collagen fibers were arranged randomly in the labial dentin of rat (Fig. 5), whereas in the rabbit, collagen fibers became more oriented near to the deep layer.
Lingual dentin: In the lingual dentin of both rat and rabbit, calcospherites in the predentin were small and flat in form. The mineralization front in the lingual dentin appeared to be a combination of a linear and globular form. Incremental lines stained densely with hematoxylin were evident in the rat lingual dentin (Fig. 6). Though incremental lines were observed as in rat dentin, the vasodentin was observed in the middle layer (m) of rabbit lingual dentin (Fig. 7). Dentinal tubules of the lingual dentin were sharply bent before and after the formation of vasodentin. Crystals were regularly oriented almost parallel to the long axis of the incisor in the lingual dentin of both rat and rabbit (Fig. 8). The fluorescent labelling lines produced by TC in lingual dentin were almost linear (Fig. 9). Collagen fibers in lingual dentin were arranged regularly almost perpendicular to the dentinal tubules and parallel to the long axis of the incisor (Fig. 10).

DISCUSSION AND CONCLUSION

The results of the present study have shown that labial dentin and lingual dentin of both rat and rabbit incisors exhibit distinct structural differences. Steinfort *et al.* (1989) [7] found that considerable differences existed between the two dentins (enamel-related dentin and cementum-related dentin) with respect to the quantity of the various phosphoproteins. Steinfort [1990] (8) reported that faster-growing crystals (cementum-related dentin) were less pure and incorporated more Mg^{2+} which, in turn, slowed down their growth rate. From these findings and the present observations, the following conclusions can be made;

1. The structural differences between labial and lingual dentin are due to their different patterns of calcification.
2. Globular calcification is predominant in labial dentin, whereas linear calcification is predominant in lingual dentin.
3. When incremental lines are clearly observed, crystals and collagen fibers are regularly oriented.
4. When calcospherites are globular, crystals and collagen fibers are randomly oriented.

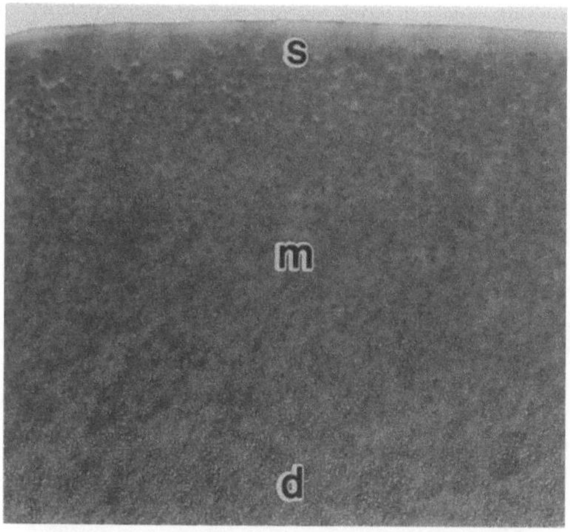

Fig. 1 Longitudinal section of labial dentin from rat (H.E. staining). Interglobular dentin and calcospherites can be clearly seen. s: superficial layer, m: middle layer, d: deep layer. x50.

Fig. 2 Longitudinal section of labial dentin from rabbit (H.E. staining). Interglobular dentin and calcospherites can be seen in the superficial layer (s). x50.

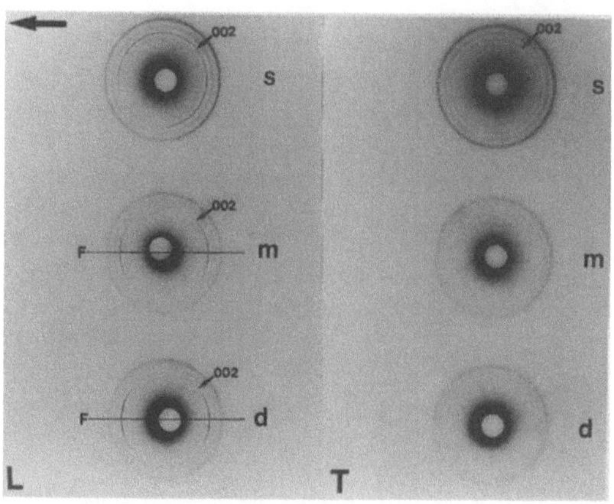

Fig. 3 X-ray microdiffraction patterns of labial dentin from rabbit. In the middle (m) and deep (d) layers, crystals are oriented almost parallel to the long axis of the incisor (arrow). Crystals are randomly oriented in the superficial layer (s). L: longitudinal section, T: transverse section, F: fiber axis, 002: diffraction from plane (002).

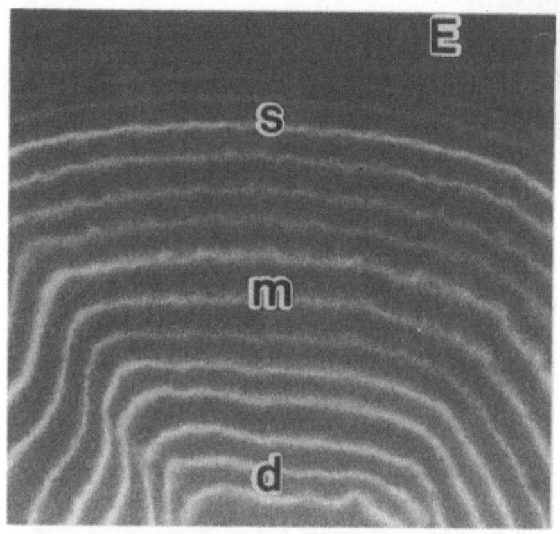

Fig. 4 Transverse section of labial dentin from rat (tetracycline labelling). Tetracycline was injected subcutaneously every other day. Note undulating fluorescent labelling lines. E: enamel. x25.

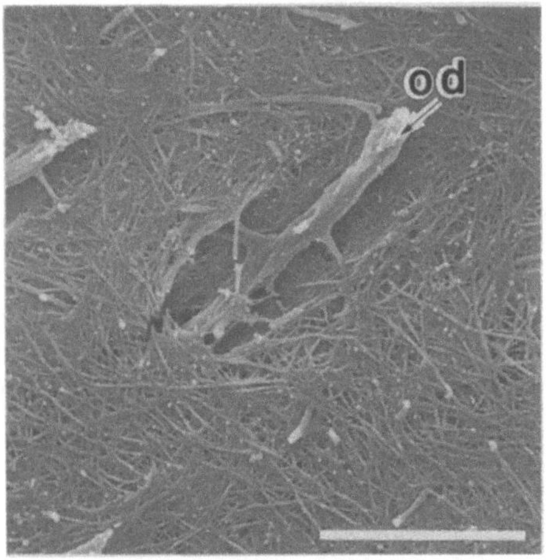

Fig. 5 Longitudinal section of labial dentin from rat (SEM, 1 N HCl etched and treated with collagenase Type I for 24 h). Collagen fibers are randomly arranged. od: odontoblast process. Bar= 5 μm.

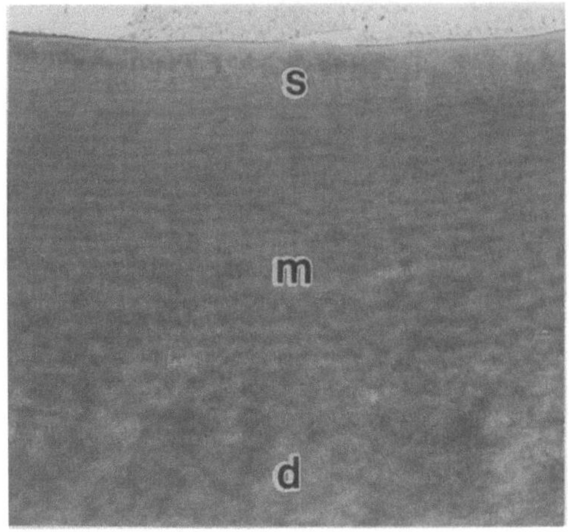

Fig. 6 Longitudinal section of lingual dentin from rat (H.E. staining). Incremental lines are evident. x50.

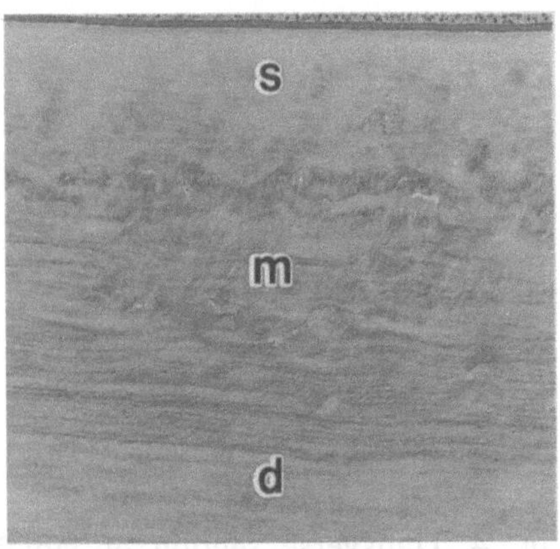

Fig. 7 Longitudinal section of lingual dentin from rabbit (H.E. staining). incremental lines are observed as in the case of rat dentin. x50.

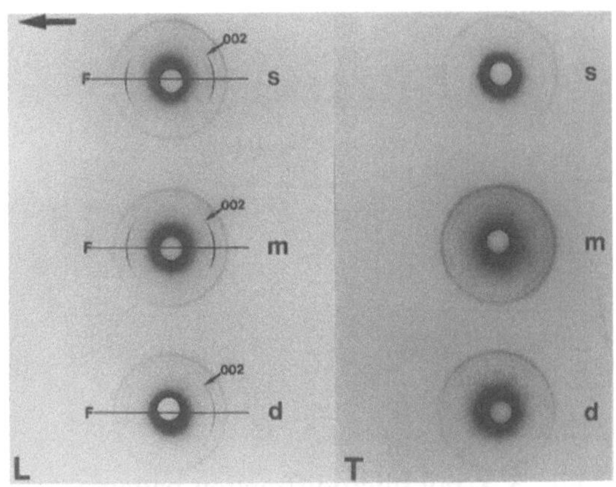

Fig. 8 X-ray microdiffraction patterns of lingual dentin from rabbit. Crystal are oriented almost parallel to the long axis of the incisor (arrow).

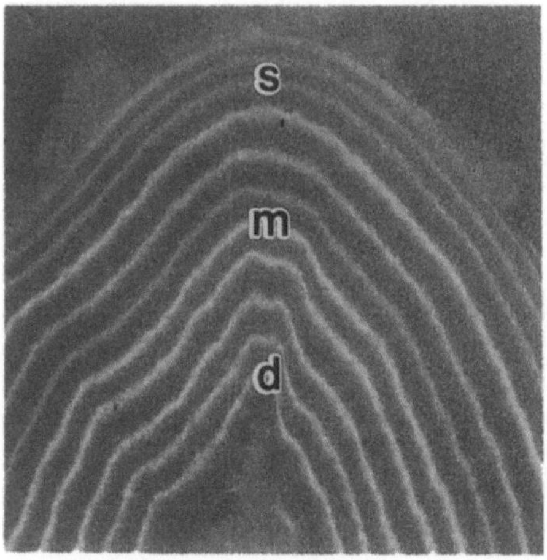

Fig. 9 Transverse section of lingual dentin from rat (tetracycline labelling). Note linear fluoresent labelling lines. x25.

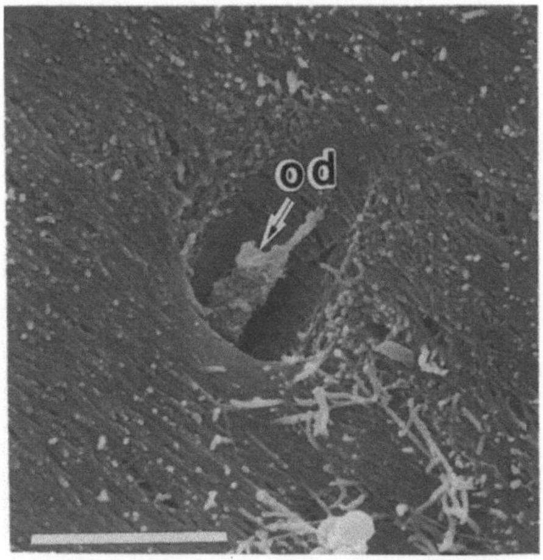

Fig. 10 Longitudinal section of lingual dentin from rat (SEM, 1 N HCl etched and treated with collagenase Type I for 24 h). Collagen fibers are regularly arranged. od: odontoblast process. Bar= 5 μm.

REFERENCES

1 Mishima H, Sakae T (1986) Demonstration of structural variation in rat incisor dentin as determined by the x-ray Laue method. J Dent Res 65: 932-934

2 Mishima H, Sakae T, Kozawa Y, Hirai G (1988) Structural variation in labial dentin and lingual dentin in the rat incisor. J Nihon Univ Sch Dent 30: 1-10

3 Kozawa Y, Terajima T, Mishima H, Sakae T (1989) The intranuclear rodlet in the odontoblast. Nihon Univ J oral Sci 15:295-299

4 Rasmussen P (1972) Effect of extreme calcium deprivation on degree of mineralization of alveolar bone, dentin and enamel in rat. Scand J Dent Res 80: 327-333

5 Rosenberg GD, Simmons DJ (1980) Rhythmic dentinogenesis in the rabbit incisor: allometric aspects. Calcif Tissue Int 32:45-53

6 Yonaga T (1978) Action of parathyroid hormone, with special reference to its anabolic effect on different kinds of tissues in rats (III). Bull Tokyo Med Dent Univ 25: 259-268

7 Steinfort J, van den Bos T, Beertsen W (1989) Differences between enamel-related and cementum-related dentin in the rat incisor with special emphasis on the phosphoproteins. J Biol Chem 264: 2840-2845

8 Steinfort J (1990) The possible role of noncollagenous matrix components in dentin mineralization. The rat incisor as a model. Kaal Boek. Amsterdam, pp102-106

CHAPTER 2.18
Spontaneous Mineral Aggregations in the Circumpulpal Dentine Formation

H. SAWAMURA and I. MASHIMOTO
Tsurumi University, School of Dental Medicine, Yokohama, 230 Japan

Key words: Dentine formation, Mineralization, Root dentine.

INTRODUCTION

One of the characteristic phenomena in dentinogenesis is the formation of globular mineralized masses called calcospherites. Mineralizing points appearing in outer layer of the predentine grow and are penetrated by usually over ten dentinal tubules. Calcospherites are found in the circumpulpal dentine.
It has been reported that in each calcospherite collagen associate crystals and radial crystals exist [1, 2]. However, the mechanism of formation of the radially orientated crystals and the characterisics of dentine mineralization that form the spheritic mineral aggregates is not clear.
In the dentine of the floor of pulp chamber of mammallian molar teeth we observed a mineralization pattern which has not been described before as far as we are aware [3]. In this report we present a description of the pattern using light and scanning electron microscopy and X-ray diffraction.

MATERIALS AND METHODS

Mature and developing molar teeth of dog, pig and crab-eating macaque monkey were examined. After extaction teeth were partially cut to expose pulp tissues and fixed in 10 per cent formaldehyde solution.
Light Microscopy: Sections decalcified in 5 per cent nitric acid were stained with Haematoxilin-Eosin. Undecalcified ground sections were prepared thinner than 50µm under the wet condition for contact microradiography (CMR). For CMR, EMS type(Softex company, Tokyo) was used with a copper anode operated at 7kV and 3mA tube current and a 4cm distance between the X-ray source and the specimen.
Scanning Electron Microscopy: Forming surfaces were examined using a Jeol T300. For observation of the predentine surface the pulp tissues were swept away mechanically and cell fragments and non-fibrous matrices were removed in 1 percent Trypsin solution, phosphate-

buffered at pH 7.2 and incubated at 37°C for 6 hours. To expose the mineralization surface the predentine was removed in 7 percent NaOCl solution.
The orientation of the crystal c-axis(long axis) was examined by X-ray diffraction using a flat plate camera with a collimator 100 μm in diameter, under similar conditions to those used by Mishima and Sakae [4].

RESULTS

In longitudinal sections through the centre of the floor of the pulp chamber and the root apex, the circumpulpal dentine shows perpendicular stripes in the dentine layer. In decalcified HE stained sections, the parallel lines unstained by hematoxilin have about 10-20 μm intervals (Fig.1). In developping materials the predentine surface is straight and the mineralizing surface shows the serrations. The unstained lines intersect the mineralization front at the points of the serrations (Fig.2). In the CMR of ground sections the lines corresponding to the unstained lines in HE stained sections are shown to be highly mineralized (Fig.3). The dentinal tubules cross the dentine layer and the lines obliquely (Fig.1).
Scanning electron micrographs of the predentine surface show the matrix collagen fibres lay in the plane of the predentine surface and have a preferred orientation. In perspective the fibers are arranged radially from the central area of the floor of the pulp chamber. The dentinal tubules have elliptical openings along the fibres on the predentine surface.
On the mineralization surface, ridges also have a preferred orientation, 90° to the fibre direction. They were approximately 10-20 μm in width, 10 μm in height and 50-100 μm in length. This length differs between animals, being longer in dogs and shorter in monkeys (Fig.6-12).
By X-ray micro-diffraction two different crystal orientations were detected. In longitudinal section two pairs of 002 arcs were found, one meridional and the other equatorial. The former indicates that the long axes of the crystals are perpendicular to the predentine surface or parallel to the highly mineralized lines, whereas the latter are parallel to the predentine surface or parallel to the matrix fibres (Fig.4). In the section parallel to the predentine surface the long axes of the crystals were found to be parallel to the matrix fibre orientation (Fig.5). Therefore, in this dentine two different orientated crystals exist, one is perpendicular to the collagen fibres and to the predentine surface, the other is collagen associated.

DISCUSSION

The highly mineralized lines in the section are lamellae-like structures in three dimension. The lamella-like structure is situated

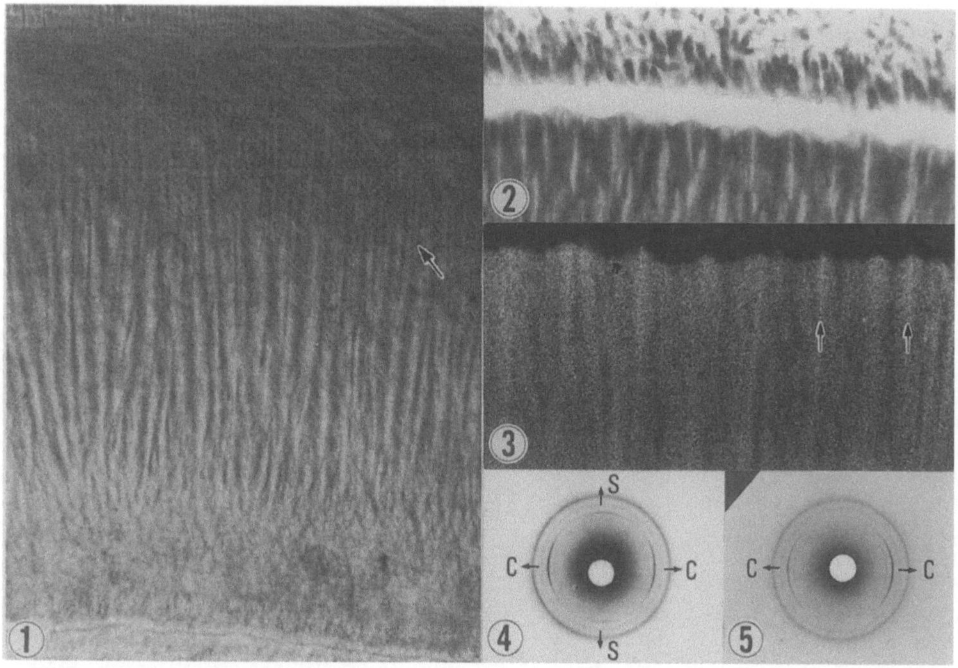

Fig.1 Longitudinal decalcified HE stained section of dentine of the floor of the pulp chamber of pig molar dentine. Arrow: dentinal tubules.

Fig.2 Predentine and mineralization surfaces of pig molar. Note unstained lines crossing at the points of serrations of mineralization front.

Fig.3 CMR corresponding to Fig.2. Arrows: highly mineralized lines.

Fig.4 X-ray micro-diffraction pattern of longitudinal section of pig molar dentine. C-C: direction of collagen fibre, S-S: perpendicular to predentine surface.

Fig.5 X-ray micro-diffraction pattern of section parallel to the predentine surface (in pig). C-C: direction of collagen fibres.

in the centre of the ridge which results from the higher mineralization at the ridge-crest during the dentine formation.

In the calcospherite of circumpulpal dentine, the highly mineralized centre and spoke structure in the radial spread are demonstrated by back scatterd electron microscopy and it is presumed that the radial crystallites are located in the interfibrillar ground substance [5]. The random arrangement of matrix fibres leaves free interfibrillar spaces in which radial crystals grow. In the sphere formed by the radial crystals, collagen associated crystals are formed.

On the other hand, in the dentine of the floor of the pulp chamber the matrix fibres and also the interfibrillar spaces have a preferred orientation. The radial crystal growth may therefore be inhibited. By

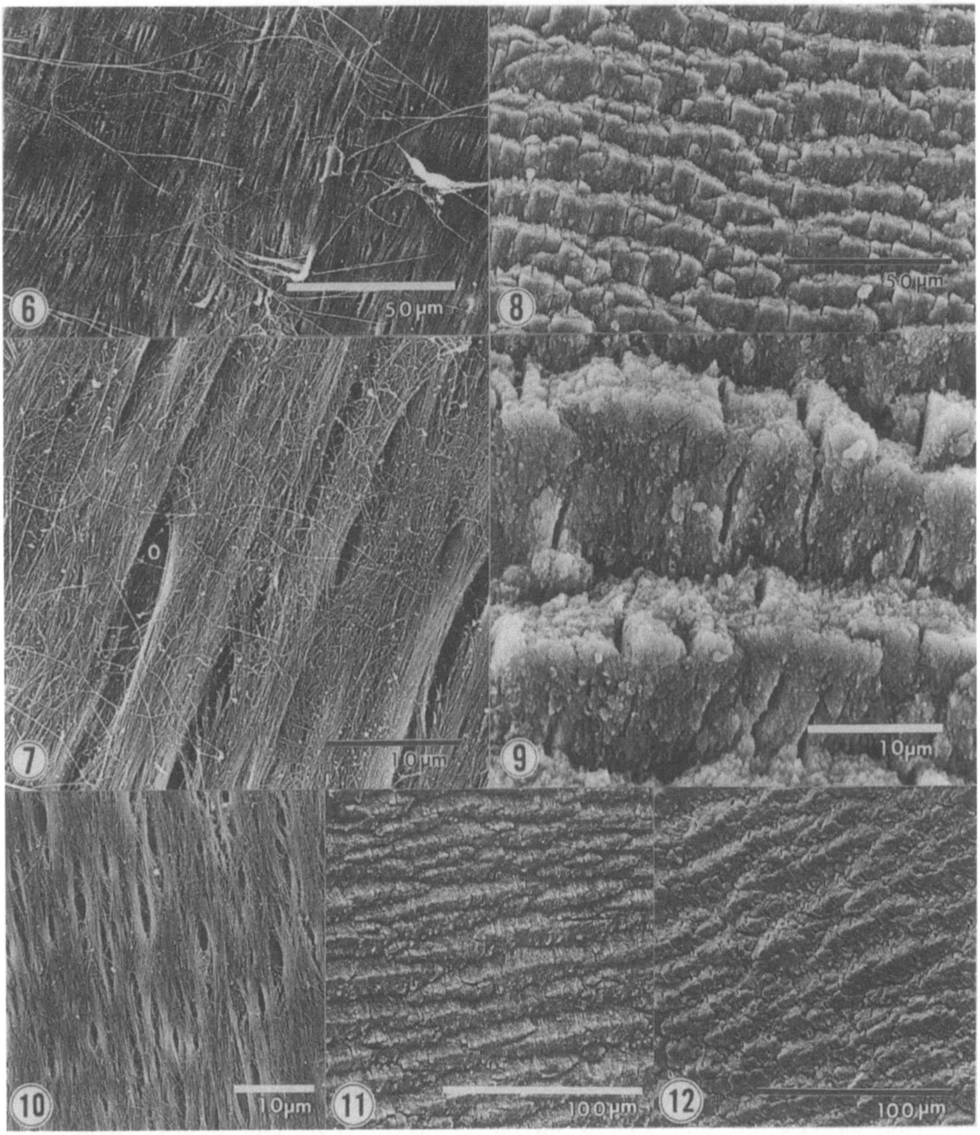

Fig.6 SEM of predentine surface of the floor of the pulp chamber of pig molar.

Fig.7 High magnification of Fig.6. O: opening of dentinal tubule.

Fig.8 SEM of mineralization surface of the dentine corresponding to Fig.6.

Fig.9 High magnification of Fig.8.

Fig.10 Predentine surface in dog.

Fig.11 Mineralization surface in dog.

Fig.12 Mineralization surface in monkey.

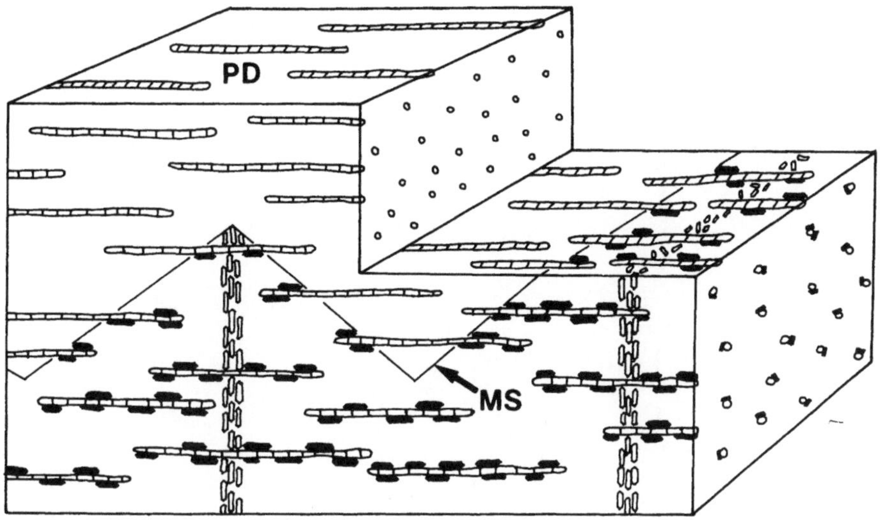

Fig.13 Hypothetical scheme of the structure of the ridge pattern mineralization. Highly mineralized lamellae are composed of the collagen independent crystals.
: collagen fibre, : collagen independent crystals,
: collagen associated crystals,
PD: predentine surface, MS: mineralization surface.

presumption, collagen independent crystals must be formed with an orientation not influenced by the matrix collagen fibres. Namely perpendicular to the fibres, and the predentine surface, these crystals gather together the highly mineralized central lamellae in the ridges.

In both calcospheritic and ridge patterns, collagen independent crystals form the structures which lead to the shape of the unit patterns. The formation of collagen associated crystals are restricted within the area determined by the advancing collagen independent structures.

REFERENCES

1 Schmidt WJ, Keil A transl Poole DFG, Darling AI (1971) Polarizing Microscopy of Dental Tissues. Pergamon Press.Oxford.
2 Shellis RP (1983) Archs oral Biol 28: 85-95
3 Sawamura H (1989) Tsurumi Univ Dent J 15: 487-513
4 Mishima H, Sakae T (1986) J Dent Res 65: 932-934
5 Jones SJ, Boyde A (1984) Dentin and Dentinegenesis. CRC Press, Boca Raton.

CHAPTER 2.19

The Role of the Organic Matrix in Dentine (Re)mineralization

J.M. TEN CATE, B. KLONT and R.A.M. EXTERKATE
Department of Cariology and Endodontology, Academic Centre for Dentistry, Amsterdam, the Netherlands

Keywords: Teeth, dentine, mineralization, collagen

INTRODUCTION

In the past decades much attention has been given to studying the process of enamel caries. More recently this has focused on the role of the saliva mediated repair of incipient lesions, the remineralization process. Caries essentially comprises of enamel being dissolved during periods of low pH resulting from the fermentation of carbohydrates by bacteria in dental plaque. Additionally, calcium phosphates from saliva may reprecipitate into enamel lesions during periods of near neutral pH. This process is driven by the degree of supersaturation of saliva with respect to hydroxyapatite. A second tissue of interest in the oral cavity is dentine, which is exposed to the oral environment when teeth are subject to or have been treated for periodontal infections. Dentine is quite vulnerable [1] in the oral environment, presumably due to a higher solubility [2] and small crystallite diameters. Moreover the treatment of caries in the tooth root is difficult, this making a search to prevent the formation of dentinal lesions relevant.
The research presented here is concerned with remineralization of dentinal (root) caries lesions, in particular regarding the role of the organic matrix. Three separate questions were studied. Firstly, do dentinal lesions remineralize when subjected to a saliva like solution? Secondly, does the demineralized collagen matrix pose a restriction on ionic diffusion, thereby limiting dentine remineralization. Lastly, can the collagen matrix in dentine act as a site of precipitation.

MATERIALS AND METHODS

Dentin specimens were prepared from bovine incisors, which were stored at -20C immediately after slaughter. Remnants of periodontal tissue were removed by a short HClO treatment followed by thorough rinsing in distilled water. The pulp tissue was extirpated mechanically. For remineralization studies 4 mm thick slices were cut using a water cooled diamond coated wheel. The cut surfaces were protected with acid resistant nail varnish. Next the specimens were subjected to a demineralization challenge resulting in caries like defects observed clinically. This comprised of either a 0.1 M acetic acid buffer at pH=4.0 (to create etched or eroded defects) or a 0.1 M lactic acid buffer adjusted at pH=5.0 with 0.2 mM methylenehydroxy diphosphonate to enhance surface layer formation (subsurface lesions). Remineralization of specimens was done in a 20 mM cacodylate buffer at pH=7.0 with 130 mM KCl, and varying amounts of calcium and phosphate, always having a molar calcium to phosphorous ratio of 1.6. Remineralization was assessed by measuring the calcium levels in solution using Atomic Absorption Spectroscopy. Measurement of the amounts of accessible collagen or its removal from the lesion was accomplished by subjecting the specimens to bacterial collagenase for one day [3]. This procedure solubilized the collagen matrix of completely demineralized dentine specimens. Solubilized collagen was measured by determining the hydroxyproline content of the supernatant.
Dentine matrix mineralization studies were performed with 400 µm thick sections obtained

from the substrate described above. The sections were demineralized during 3 weeks in twice weekly refreshed aliquots of one of the following three solutions : 10% EDTA at pH=7.4, 0.6 M acetic acid at pH=4.0 or 0.6 M HCl. These three demineralization schemes remove different non-collagenous proteins from the dentinal matrix during demineralization. In the case of acetic acid buffer treatment phosphoproteins are not extracted from the specimens, unlike in the other two solutions. Complete demineralization was checked by hydrolysing samples of batches of sections and assessing the concentrated hydrolysate for calcium. Next the specimens were subjected to a mineralizing solution comprising 3.75 mM $CaCl_2$, 1.7 mM KH_2PO_4, 22mM $KHCO_3$ at pH=7.2. After 1 week the specimens were centrifuged, washed in distilled water and the mineral dissolved in 0.5 M HCl overnight. The calcium concentration in the acid solution was determined with Atomic Absorption Spectroscopy. For all calcium determinations phosphate interference being suppressed by adding 1.56% $La(NO_3)_3.H_2O$ in 50 mM HCl.

RESULTS AND DISCUSSION

Mineral Deposition during Remineralization of Root Lesions

Table I shows the rate of calcium uptake for etched (surface eroded) dentine, and for dentine lesions. In these experiments the remineralization solution used had a Ca*PO4 concentration product of 1.35 mM^2. In the case of *etched specimens* the rate of remineralization was found to depend linearly on the amount of mineral removed during demineralization. This can be explained from the increasing number of sites becoming available for deposition of mineral with increasing demineralization times.

Table I: Rate of remineralization ($\mu mol/cm^2/day$)

removed during demineralization ($\mu mol/cm^2$)	etched	lesion
50	0.59	0.98
100	1.11	1.05
150	1.57	1.14

For *dentinal lesions* the rate of remineralization was found to be higher for the early lesions (low values of mineral loss) but constant with increasing severity. Since lesions are characterized by a subsurface loss of mineral it is conceivable that diffusion of ions to the site of precipitation become rate limiting for the more advanced lesions. The higher rate of remineralization for the 'early' lesions indicates a favourable condition for remineralization inside the lesion, presumably resulting from a higher local pH, as suggested for similar enamel lesions [4].
The effect of supersaturation of the remineralization solution on the initial rates of calcium deposition is shown in Figure 1. Extrapolation of the dashed lines shows that the presumed zero rate of deposition occurs at about pI=54. Obviously no definitive conclusions can be drawn about the nature of the deposited material from the little data available. Nevertheless the results indicate that the solubility of the deposit is much less than that of the original dentinal tissue for which the pK_{sp} was determined at 52.7[2]. The experimental methods and results of these remineralization experiments will be been described in more detail in a separate publication [3].

The Role of Collagen in Root Lesions: Diffusion Inhibition?

A comparison of groups that were either treated or non treated with collagenase prior to remineralization showed that removal of the accessible collagen did not affect the rate of remineralization. This was observed for both etched and lesion specimens with differing severities. We conclude that the presence of non apatite covered collagen

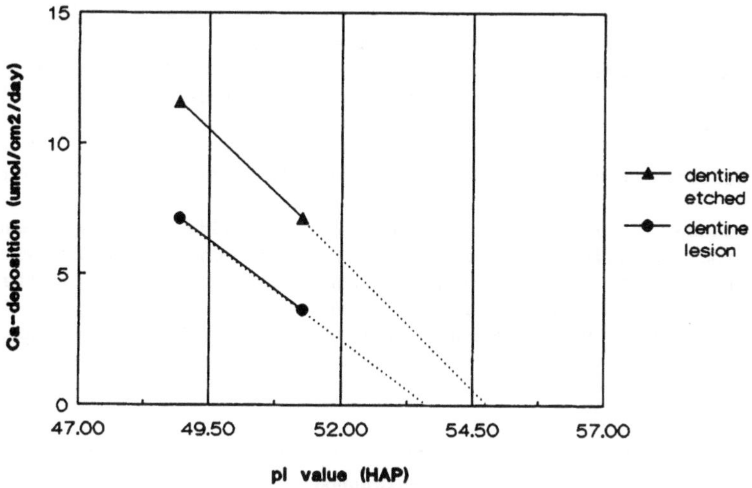

Figure 1: Initial rate of calcium deposition in etched dentine and dentine lesions from solutions with different degrees of supersaturation.

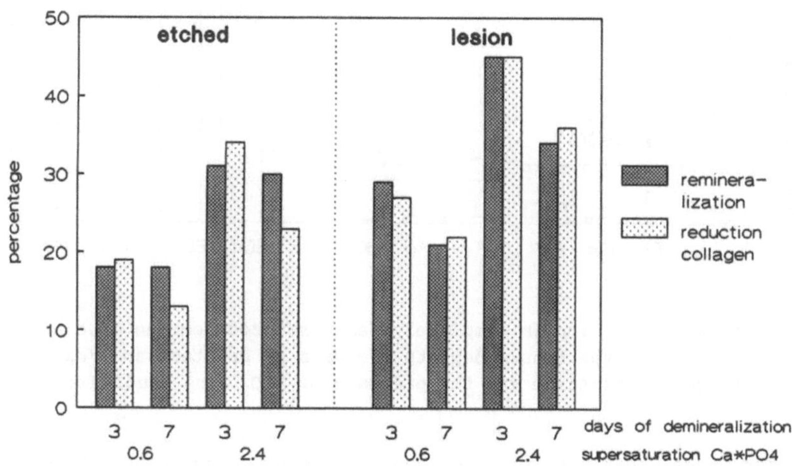

Figure 2: Comparison of the percentages of *remineralization* of etched and lesion specimens with the *reduction in collagen accessibility*. Remineralization was assessed from the depletion of calcium from solution. The amounts of accessible collagen before and after remineralization were determined by collagenase treatment (of different specimens). The duration and method of demineralization of the specimens and the degree of supersaturation of the remineralizing solution were studied as parameters.

inside the lesion did not inhibit diffusion of ionic species to the sites of precipitation. A comparison of the values before and after remineralization, however, showed that the amounts of accessible collagen had decreased (Figure 2). This could suggest that collagen had acted as a nucleus for precipitation. We calculated that this decrease was of similar magnitude as the degree of repair of the lesions. We therefore hypothesize that the decrease in collagen accessibility could equally well be attributed to an entrapment of the matrix by outgrowing crystals, which withstood lesions formation. Increase of crystallite dimensions in natural lesions to values exceeding those of sound dentine has been observed by using TEM by Daculsi et al [5]. They suggested that remineralization of root lesions occurred by growth of crystallites rather than by de novo precipitation.

The Role of Collagen in Root Lesions: Matrix for Calcium Phosphate Precipitation?

With the abundance of crystallites in the partly demineralized root specimens (either etched or lesion) it is difficult to determine definitively whether also the collagen matrix acts as a matrix for de novo precipitating crystallites inside the dentinal lesions. This was studied in completely demineralized thin dentin sections. The results of the mineral deposited during a one week immersion in a highly supersaturated solution are depicted in Figure 3.

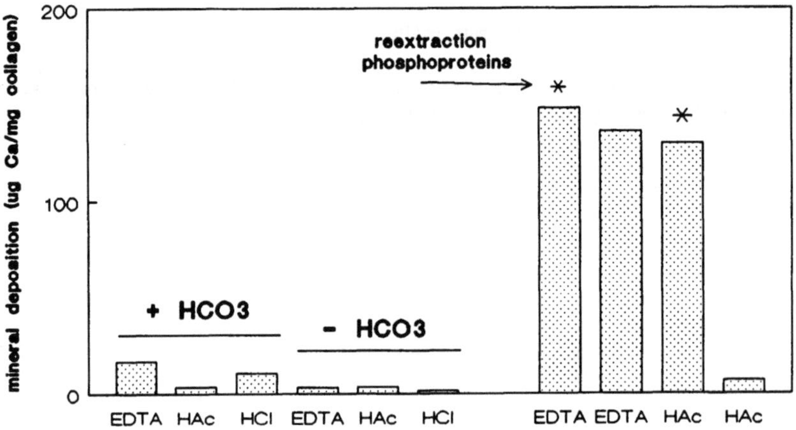

Figure 3: Mineral deposition in completely demineralized dentine specimens.
The right hand side of the figure depicts the data by De Steno and Feagin [6].
The method of demineralization and the presence of HCO_3 were studied as parameters.

Our findings reveal that precipitation does not occur in any appreciable amount under these conditions, which are favourable to mineral deposition when compared with those usually prevailing in the oral environment, in particular in plaque fluid.
For comparison the data by De Steno and Feagin [6] using a similar experimental design have also been included in the figure. These authors treated specimens with a similar nucleating solution as used in this study, and observed significant calcium phosphate precipitation onto the dentin matrices unless demineralization conditions had not extracted soluble phosphoproteins.
In the literature there is controversy whether collagen as such can promote mineral deposition or only in conjunction with non collagenous proteins. Koutsoukos and Nancollas [7] observed apatite formation on Type I collagen from solutions with a pI=48.0. (1.5 mM Ca, 0.9 mM PO_4). Moreover they observed that the induction time for initial precipitation in these solutions was reduced from 10 hours to 5 minutes when adjustment in the phosphate level had been made to correct for the large phosphate adsorption onto collagen [8]. Hunter et al [9], on the other hand, concluded that much higher Ca and PO_4 levels are required (calculated at pI=42.3-44.3) to achieve precipitation in collagen gels. Their experimental model, using a collagen gel

compartment containing phosphate and a connecting liquid compartment with calcium ions, is developed to determine diffusion processes in the gel but is difficult to characterize in physico chemical terms. They concluded that fibrillar collagen is not per se a promoter of apatite deposition and suggest that extracellular matrix macromolecules may influence calcification by restricting the diffusion through the connective tissues. Recently, Beertsen and van den Bos [10] also reported that calcium phosphate deposition in demineralized dentine did not occur in a systematic pattern (i.e. different from random precipitation) from supersaturated solutions containing calcium and inorganic phosphate in the concentration range studied in this paper. In contrast the specimens could be readily mineralized if organic phosphate and a tissue producing alkaline phosphatase replaced the inorganic phosphate. The exact mechanism behind this precipitation reaction has not been cleared.

The state of the art on the role of the many non collagenous matrix proteins involved in calcification in bone, cartilage and dentin has recently been reviewed by Boskey [11]. She concluded that the removal of these proteins from native collagen may cause it to loose much of its mineralizing capacities. Apparently the discrepancy in the results of dentine mineralization is caused by the difference in 'purity' of the dentinal matrix.

Regarding the initial question whether a collagen matrix inside a developing dentin lesion can be remineralised two final considerations may be given. Irrespective whether proteins make collagen a nucleator for calcium phosphate precipitation the degree of supersaturation required for initiating mineralization are not likely to be found in plaque, where, in spite of elevated calcium and phosphate levels compared to saliva, major fractions are bound to organic or inorganic species (bicarbonate). Secondly, up to date little data is available regarding the extraction of possible promoters of calcium phosphate precipitation during the caries process. Although phosphoproteins and proteoglycans are solubilized during complete demineralization of dentine, this was not observed during partial demineralization[12]. We presume that these non-collagenous components are redeposited onto apatite crystals rather than on the collagen matrix. In view of the observation that the physical status of a protein (solubilized versus adsorbed) transforms a precipitation inhibitor into a promoter it is impossible to predict in what way the matrix proteins released or redeposited during root caries would interfere with possible mineralization phenomena.

IN CONCLUSION

Root surface lesions (etched or subsurface type) are remineralized when treated with supersaturated calcium phosphate solution. The presence of demineralized collagen does not impose a diffusion barrier for precipitation. Experiments with completely demineralized dentine show that collagen (under these conditions) does neither act as a nucleator for precipitation.

REFERENCES

1. Nyvad B, ten Cate JM, Fejerskov O (1989) Caries Res 23:218-224
2. Hoppenbrouwers PMM, Driessens FCM, Borggreven JMPM (1987) 32:319-322
3. Klont B, ten Cate JM (1991) Caries Res In press
4. Ten Cate JM, Arends J (1978) Caries Res 12:213-222
5. Daculsi G, Kerebel B, le Cabellec M-T, Kerebel L-M (1979) Caries Res 13:190-202
6. De Steno CV, Feagin FF (1975) Calcif Tiss Res 17:151-159
7. Koutsoukos PG, Nancollas GH (1987) Coll Surf 28:95-108
8. Koutsoukos PG, Nancollas GH (1986) Coll Surf 17:81-90
9. Hunter GK, Nyburg SC, Pritzker PH (1986) Coll Rel Res 6:229-238
10. Beertsen W, van den Bos T (1989) Matrix 9:159-171
11. Boskey AL (1989) Bone and Mineral 6:111-123.
12. Klont B, ten Cate JM (1990) J Dent Res 69:896-900.

CHAPTER 2.20
Pathologic Mineralization in Humans

P.-T. CHENG

Departments of Pathology and Clinical Biochemistry, Mount Sinai Hospital, University of Toronto, Toronto, Canada, M5G 1X5

Key words: Pathologic mineralization, Chondrocalcinosis, Urolithiasis, Thermodynamics, Kinetics.

INTRODUCTION

Physiologic mineralization in humans involves principally the formation of carbonated hydroxyapatite [$Ca_{10}(PO_4,CO_3)_6(OH)_2$, HAP] crystals in bone and teeth [1]. In contrast, pathologic mineralization in humans is more complicated, involving HAP and several other inorganic and organic biominerals. HAP, the most ubiquitous pathologic biomineral, is frequently observed in atherosclerosis, nephrocalcinosis, chondrocalcinosis and other ectopic calcifications, and in renal and dental calculi. Besides HAP, there are other biologic calcium phosphates including brushite or calcium hydrogen phosphate dihydrate [$CaHPO_4.2H_2O$, Bru], whitlockite or magnesium-substituted tricalcium phosphate [$(Ca,Mg)_3(PO_4)_2$, Whi] and octacalcium phosphate [$Ca_8H_2(PO_4)_6.5H_2O$, OCP]. These biominerals are less frequently observed and are mainly found in renal and dental calculi [2 - 6]. Other important calcium containing biominerals are triclinic (T) and monoclinic (M) calcium pyrophosphate dihydrates [$Ca_2P_2O_7.2H_2O$, CPPD], calcium oxalate monohydrate [$CaC_2O_4.H_2O$, COM] and dihydrate [$CaC_2O_4.2H_2O$, COD], and three forms of calcium carbonate [$CaCO_3$]. CPPD(T) and CPPD(M) are commonly associated with chondrocalcinosis or pseudogout in the aged [7]. COM and COD are commonly observed in renal calculi [3], but will also deposit in other tissues in patients suffering primary or secondary oxalosis. Although calcium carbonate has been reported as a rare renal stone component, it is mainly observed in gallstones, together with cholesterol and calcium bilirubinate [8]. A non-calcium phosphate frequently observed in renal stones formed during infections is struvite or magnesium ammonium phosphate hexahydrate [$MgNH_4PO_4.6H_2O$] [3]. Important organic biominerals include monosodium urate or sodium acid urate monohydrate [$C_5H_3N_4NaO_3.H_2O$, MSU], ammonium acid urate [$C_5H_3N_4NH_4O_3$, AAU], uric acid [$C_5H_4N_4O_3$, UA] and uric acid dihydrate [$C_5H_4N_4O_3.2H_2O$, UAD]. AAU, UA and UAD are commonly observed in renal calculi [3]. While MSU has been infrequently observed as a renal stone component, it is usually associated with gout [9].

Biominerals are crystalline solids. Therefore, not only can they be analyzed chemically, but can also be identified by their unique physical properties. Morphologically, some biominerals have characteristic crystal habits and sizes recognizable under light or electron microscopy. The ability of some of them to rotate polarized light also helps in their identification [10, 11]. However, in most cases, definitive identification requires confirmation by an analytical method that examines them crystallographically or spectroscopically. The common methods employed are x-ray and electron diffractions, and x-ray and infrared spectroscopies [10 - 12]. Such physicochemical data are very important as they serve to provide information beyond routine identification; i.e. they provide information on the crystals themselves, such as size, crystallinity, lattice defects and impurities, which will provide insight into the ambient physicochemical environment in which they were formed. For example, very fine, poorly crystalline biominerals with lattice or inclusion impurities would suggest very rapid formation in an environment with high concentrations of lattice and non-lattice constituents.

It is probably true that each pathologic mineralization process is unique. However, if considered in a very broad sense, many of these processes do have similarities, especially in the areas of crystal nucleation and growth. A five-step generalized theory is applicable to many pathologic mineralization processes: 1. Supersaturation - the thermodynamic prerequisite; 2. Nucleation - either by homogeneous nucleation (Homo Nu) or heterogeneous nucleation (Hetero Nu) [13]. 3. Transformation - some initial precipitates are unstable [14]; 4. Growth; (steps 2 to 4 are kinetically controlled [15] and can be modified by promoters and inhibitors); 5. Accumulation - crystals are retained (as opposed to excreted) in the tissues. Since each of the five steps of the theory is an immense topic by itself, and this paper is concerned only with pathologic mineralization in humans, the discussion on each step will be necessarily limited and will be illustrated only by examples frequently addressed in the literature. In particular, two very different mineralization processes will be employed as major examples: 1. Chondrocalcinosis -a local enzymatic defect in the joint cartilage leading to the accumulation of inorganic pyrophosphate ions (PPi) followed by a slow accumulation of well-formed CPPD(T) and CPPD(M) crystals in the cartilage and in the synovial fluid in the adjacent joint space [9, 16]; 2. Calcium phosphate urolithiasis - a urinary disease arising from more systemic conditions such as hypercalciuria or high [Ca]X[Pi] in urine (Pi = inorganic phosphate) resulting in the formation of pure or mixed calculi containing various calcium phosphates [17].

SUPERSATURATION

A body fluid is saturated with respect to a solute when it is in equilibrium with an excess amount of the solute. The equilibrium concentration is variable; its value depends on the ambient conditions. Important ambient factors that will affect the solubility include temperature, pH, ionic strength or osmolality and concentrations of general or specific inhibitors that may be present. Most body fluids can be considered as aqueous solutions maintained at 37°C. But, since different body fluids have different functions, their organic and inorganic compositions, pH, osmolality and viscosity, etc. assume different values [18]. Extracellular fluids such as synovial fluid and serum maintain an ionic strength close to 0.14 mol/L or 280 mosM (isotonic) and a slightly alkaline pH near 7.4. Saliva is hypotonic (<100 mosM) and acidic (pH 5 - 7). Depending on the fluid intake, urine osmolalities can have values ranging from hypotonic (<100 mosM) to highly hypertonic (>1000 mosM); normal urine is hypertonic. Depending on the diet, normal urine can range from acidic to slightly alkaline (pH 5 - 7.5); however, during urinary tract infections, urine can become alkaline owing to bacterial conversion of urea to ammonia [18]. In general, fluids with higher osmolalities have higher equilibrium concentrations; this explains why solute concentrations in urine are usually higher than those in sera and saliva. Also, acidic fluids can dissolve more of a basic solute than alkaline fluids, and vice versa; that uric acid is less soluble in acidic urine is a good example.

As mentioned above, usually a body fluid will not produce crystals unless it is highly supersaturated. This is so because work has to be done to produce crystals and the energy required comes from the chemical potential of the solution achieved by virtue of the supersaturation. This potential to precipitate can be measured by the Gibb's free energy change (ΔG) for the reaction solute → solid in an aqueous solution [19].

$$\Delta G = -RT\ln(a/a_{eq})$$

where a is the activity of the solute in the supersaturated solution at T°K and a_{eq} is the activity of the solute at equilibrium at the same temperature. For electrolytes,

$$\Delta G = -RT\ln(IP/SP)$$

where IP is the product of ionic activities of the lattice ions in the supersaturated solution and SP is the solubility product at equilibrium [19]. The more supersaturated is the solution, the more negative the ΔG and the higher the potential to precipitate. If a solution is supersaturated with respect to more than one solute, then there will be more than one ΔG to be considered. In fact, it is not uncommon for a body fluid to be supersaturated with respect to several calcium phosphates at the same time, and since each calcium phosphate has a different IP/SP, depending on its molecular formula, the fluid will have a different ΔG for each calcium phosphate. Since, for a given temperature, the value for SP, but not IP, of each calcium phosphate is a constant independent of pH, the corresponding ΔG value in a fluid with given [Ca] and [Pi] will vary with pH. For HAP which contains PO_4^{3-} and OH^- ions, ΔG increases with increasing pH; for Bru which contains HPO_4^{2-} ions, ΔG decreases with increasing pH.

NUCLEATION

Crystals can be nucleated either homogeneously or heterogeneously [13]. Homo Nu requires a higher ΔG value for the solute-solid transformation than Hetero Nu. In the latter case, the activation energy for the transformation is lowered by the presence of a catalytic surface. Therefore, for a given solubility, the threshold solute concentration for Hetero Nu is considerably lower than that for Homo Nu. Consequently, Hetero Nu should be the preferred path for the precipitation of biominerals from fluids which are not highly supersaturated due to either limited supply of lattice constituents as in the case of CPPD deposition in cartilage, or to relatively high solute solubility as in the case of Bru. But, there is not a fixed threshold ΔG value for the Homo Nu of all solutes; nor is there one for Hetero Nu. A good example is given by the different ΔG values obtained for different biologic calcium phosphates at 37°C in a certain set of synthetic body fluids, e.g. $\Delta G_{Homo,HAP}$ = -7.7 KJ/mole, $\Delta G_{Homo,Bru}$ = -2.2 KJ/mole and $\Delta G_{Hetero,HAP}$ = -7.0 KJ/mole, $\Delta G_{Hetero,Bru}$ = -0.9 KJ/mole [20]. However, it does not follow from these values that HAP is more difficult to form than Bru. Quite on the contrary, most body fluids except the highly acidic ones become readily supersaturated with respect to HAP due to its low solubility; only those fluids with substantially higher [Ca]X[Pi] will form Bru as it is considerably more soluble. Besides energetics, other factors such as the presence of impurities can also contribute to the selection of the nucleation pathway for a biomineral, e.g. Whi forms only by Homo Nu and only in the presence of Mg^{++} ions [12, 20].

Crystals formed by Homo Nu are microscopic and poorly crystalline while those formed by Hetero Nu are usually larger and better formed crystals. Although it is inconclusive to determine the mode of nucleation retrospectively based on the observed crystal sizes and shapes of a mineral deposit, most pathologic biomineralization processes are believed to involve some form of Hetero Nu. An exception to this generalization is perhaps given by pathologic HAP deposition. Like its physiologic counterpart observed in bone, pathologic HAP deposits consist mainly of microscopic spherulites of very fine crystallites which can only be examined individually under a transmission electron microscope. In contrast, other biominerals such as Bru, CPPD, MSU, cholesterol, COD, and COM are always observed as large well-formed crystals, suggesting that Hetero Nu has been involved in their formation. Surfaces which promote Hetero Nu can do so in a specific or non-specific manner. Non-specific Hetero Nu is well demonstrated by non-specific organic tissues such as cellular debris acting as nidi for various kinds of kidney stones [21]. Even foreign bodies are occasionally responsible for the nucleation of urinary and non-urinary calculi [22]. On the other hand, the existence of specific Hetero Nu has been suggested only by in vitro experiments, e.g. only CPPD(T) crystals grown in native collagen gels or in gelatin gels seeded with MSU, but not in unseeded gelatin gels, develop the characteristic prismatic morphology as observed in chondrocalcinosis [23].

TRANSFORMATION

There are several in vitro models for Homo Nu of biominerals including calcium phosphates and calcium pyrophosphates. In vitro experiments have shown that microscopic HAP crystallites are in fact a transformation product derived from an unstable amorphous calcium phosphate (ACP) which is the original precipitate formed rapidly by Homo Nu in highly supersaturated solutions or gels [14].

Similarly, amorphous calcium pyrophosphate (ACPP) is the initial precipitation product of solutions and gels highly supersaturated with respect to CPPD [24, 25]. In both cases, it is believed that the crystalline phase grows at the expense of the amorphous material which redissolves. In the presence of Mg^{++}, the conversion of ACP to a crystalline phase is delayed [26], and the Mg doped ACP can transform to HAP or Bru or Whi depending on the ambient Mg/Ca and [Pi] values [12]. The ACP → HAP transformation can also be inhibited by impurities such as PPi ions [27]. Similarly, Pi and Mg^{++} ions can inhibit the ACPP → CPPD transformation, and Mg^{++} favours CPPD(M) as the transformation product [24, 28]. Other calcium binding compounds such as glycosaminoglycans can also inhibit the nucleation and transformation of ACP or ACPP [25, 29]. All these in vitro findings are important to the understanding of pathologic biomineralization if ACP and ACPP can be confirmed to exist in vivo. However, while ACP has been observed intracellularly, ACPP has not yet been observed in human tissues. This may stem from the fact that in cells ACP is readily formed and there is abundant PPi and Mg^{++} around to stabilize it, but as already mentioned, in cartilage, the ambient [Ca]X[PPi] is not expected to be high enough to support the formation of ACPP. Even so, the existence of ACPP in vivo can not be ruled out totally as its detection may have simply eluded us so far because either the search has not been exhaustive or the right methodology has not been employed. Although ACP, ACPP and other amorphous materials do not have any long range structural order - they do not have discernable electron diffraction patterns - they do have short range order and can be analyzed by suitable methods such as radial distribution function (RDF) analysis [30]. Individual amorphous deposits can be readily identified spectroscopically, e.g. ACP and ACPP have different infrared spectra [31].

GROWTH

Crystals formed either by Homo Nu or Hetero Nu will grow as long as the body fluid remains supersaturated. In cases where the source of solutes does not constitute a large reservoir, crystal growth is limited and the crystal deposits so formed are focal in nature. However, if a constant ambient solute level is maintained either by an accelerated local production or by a continuous supply from flowing body fluids, the crystal deposits should become larger with time as more crystals are formed by secondary nucleation and more crystals mature in size. Occasionally, cyclical growth patterns develop as the depletion of solutes is not replenished at once, but only after sometime when a new wave of solutes arrives. Similarly, calculi with two or more components can display cyclical alternating growth patterns with alternating layers of different components which are formed sequentially as the ambient concentration of each component becomes elevated and then depleted in a cyclical manner [32].

ACCUMULATION

Discrete crystals formed pathologically do not necessarily become a major health hazard as long as they are effectively removed from tissues, either by cellular action such as phagocytosis or by excretion from the body. Physiologic bone formation and resorption demonstrate exactly how biomineralization can be efficiently regulated by the body. If crystals are allowed to accumulate in a tissue, e.g. cartilage, then a pathologic mineralization disease state, e.g. chondrocalcinosis, will result. In human kidneys, crystalluria is quite common and does not always develop into urolithiasis. However, if discrete flowing crystals become aggregates, either by mutual adhesion or with the help of a cementing material [33], they may become attached to tubular walls and develop into calculi with time. Sometimes, the gradual accumulation of mineral deposits may eventually lead to local tissue degeneration or even tubular stasis, hence further promoting mineralization as cellular debris are gathered and fluids are retained [21, 34].

SUMMARY

Pathologic mineralization in humans involves various inorganic and organic biominerals. The common ones are calcium containing phosphates, pyrophosphates, oxalates and carbonates. Most pathologic mineralization processes can be categorized by a five-step generalized theory. Specific or non-specific heterogeneous nucleation, continuous or intermittent crystal growth support, and means for crystal retention are necessary factors for pathologic mineralization.

ACKNOWLEDGEMENT

Financial support from the Medical Research Council of Canada is gratefully acknowledged.

REFERENCES

1. LeGeros RZ (1981) Prog Crystal Growth Charat 4: 1-45.
2. Cheng P-T (1988) Rheum Dis Clin N Amer 14: 341-351.
3. Sutor DJ (1968) Br J Urol 40: 20-28.
4. Schroeder HE (1969) Formation and inhibition of dental calculus. Huber, Berne.
5. Legeros RZ (1974) J Dent Res 53: 45-50.
6. Rowles SL (1967) Bull Soc Chim Fr 1797-1802.
7. McCarty DJ (1976) Arthritis Rheum 19: 275-285.
8. Bogren H (1964) Acta Radiologica Suppl 226.
9. McCarty DJ, Hollander JL (1961) Ann Intern Med 54: 452-460.
10. Cheng P-T, Pritzker KPH, Kandel RA, Reid A (1983) Scanning Electron Microsc I: 369-377.
11. Pritzker KPH, Cheng P-T, Grynpas MD, Holmyard DP (1988) Scanning Microsc 2: 1471-1478.
12. Cheng P-T, Grabher JJ, LeGeros RZ (1988) Magnesium 7: 123-132.
13. Walton AG (1979) The formation and properties of precipitates. Krieger, New York.
14. Eanes ED,Gillessen IH, Posner AS (1965) Nature 208: 365-367.
15. Nancollas GH, Tomson MB (1976) Faraday Disc Chem Soc 61: 175-183.
16. Howell DS, Martel-P J, Pelletier JP, Morales S, Muniz O (1984) Arthritis Rheum 27: 193-199.
17. Pak CYC, Eanes ED, Ruskin B (1971) Proc Nat Acad Sci USA 68; 1456-1460.
18. Diem K, Lentner C (1970) Scientific tables, 7th ed. Ciba Geigy, Basle.
19. Moore WJ (1957) Physical chemistry, 3rd ed.
20. Cheng P-T (1989) In: Walker VR, Sutton RAL, Cameron ECB, Pak CYC, Robertson WG (eds) Urolithiasis. Plenum, New York, pp. 225-226.
21. Cheng P-T, DiGregorio RC, Li F, Pritzker KPH (1988) Proc IVth Asia-Pacific Conference and Workshop on Electron Microscopy, Bangkok, pp. 741-746.
22. Cheng P-T, Pritzker KPH, Richards J, Holmyard D (1987) Scanning Microsc 1: 2025-2032.
23. Mandel N, Mandel G (1988) Rheum Dis Clin N Amer 14: 321-340.
24. Cheng P-T, Pritzker KPH (1983) J Rheumatol 10: 770-777.
25. Hunter GK, Grynpas MD, Cheng P-T, Pritzker KPH (1987) Calcif Tissue Int 41: 164-170.
26. Boskey AL, Posner AS (1974) Mat Res Bull 9: 907-916.
27. Fleisch H, Russell RGG, Bisaz S, Termine JD, Posner AS (1968) Calcif Tiss Res 2: 49-59.
28. Cheng P-T, Pritzker KPH (1981) J Rheumatol 8: 772-782.
29. Hunter GK, Allen BL, Grynpas MD, Cheng P-T (1985) Biochem J 228: 463-469.
30. Grynpas MD, Bonar LC, Glimcher MJ (1984) Calcif Tissue Int 36: 291-301.
31. Cheng P-T (unpublished).
32. Cheng P-T, Pritzker KPH (1985) In: Schwille PO, Smith LH, Robertson WG, Vahlensieck W (eds) Urolithiasis and Related Clinical Research. Plenum, New York, pp. 919-922.
33. Cheng P-T, Reid AD, Pritzker KPH (1985) Scanning Electron Microsc I: 201-207.
34. Cheng P-T, Pritzker KPH, Oreopoulos DG (1984) In: Ryall R, Brockis JG, Marshall V, Finlayson B (eds) Urinary Stone. Churchill Livingstone, Melbourne, pp.318-323.

CHAPTER 2.21
Organization of Crystals in Bone

S. Weiner[1] and W. Traub[2]

Departments of Isotope Research[1] and Structural Chemistry[2], The Weizmann Institute of Science, Rehovot 76100, Israel

Key words: Bone, Collagen, Apatite, Tendon, Mineralization

INTRODUCTION

The structure of bone is not fully understood, despite the fact that the first studies were initiated some 300 years ago! [1]. The major gap in our knowledge lies in the structural organization intermediate between the resolving capability of a good light microscope (few microns) and the molecular scale organization determined primarily by the collagen framework itself. The underlying reason why bone structure has been so difficult to resolve, as compared to the structures of most other mineralized tissues, is that the crystals are exceedingly small - some few hundred Å long and wide and only 20 to 30 Å thick [reviewed in 2].

The major milestones along the road to resolving bone structure, at the light microscope level, are well described by Martin and Burr [3]. At the "molecular" level, the identification of the apatitic nature of the mineral phase [4], and the observed alignment of the mineral c crystallographic axes with the collagen fibril axes [5,6], represent the first major steps in the elucidation of bone fine structure. The earliest transmission electron microscope (TEM) studies of bone showed that the crystals all have the form of thin tablets or plates [7] and are intimately associated with the collagen framework, as they reveal the characteristic banding pattern of unmineralized stained collagen [8]. The significance of this observation was understood by Hodge and Petruska [9] when they proposed the 2-dimensional staggered array model for collagen structure. In their model there is a gap or hole between the end of one triple helical molecule and the beginning of the next. They noted that this would be the most likely site for the location of the mineral phase as well.

Exactly how the crystals are arranged in the fibril, and whether or not all the crystals in bone are associated with the collagen fibrils, are subjects which have been discussed at length since the pioneering work of Hodge and Petruska [9]. Here we present answers to both these questions based on our own studies, and in addition propose a model for the manner in which the mineralized fibrils are ordered up to the level of individual lamellae. In so doing we reduce the gap in our knowledge between the known structure of bone at levels of a few microns and larger, and the disparate information available at the sub-micron to nanometer structural level.

INDIVIDUAL CRYSTALS

The key to understanding the basic structure of bone is an appreciation of the structure, shapes and sizes of the individual crystals. Lattice imaging [10,11] and electron diffraction [12] of individual crystals show that each particle is a relatively well ordered single crystal of carbonate apatite or dahllite. There is no evidence for a stable amorphous or poorly ordered mineral phase in addition to the crystals. A series of electron microscope studies have confirmed Robinson's [7] original description of the crystals as being irregular shaped thin plates [reviewed in 2]. Their sizes as measured directly by imaging in the TEM vary from about a hundred up to more than a thousand Ångstroms in length and width, with average dimensions of roughly 500x250 Å [7,13,14]. Their thicknesses are estimated directly by TEM [15] or indirectly by X-ray diffraction [16] as being between 20 and 30 Å.

THE STRUCTURAL DILEMMA

The fact that in poorly mineralized bone the crystals reveal the collagen banding structure, implies that they are preferentially located in the gap regions between the ends of triple helical molecules. Individual gaps are smaller by at least one order of magnitude than individual crystals [13], implying that either the crystals are not located at all within the collagen fibrils, or if they are, that they have a very special organization that somehow allows them to be accommodated within the collagen structure.

THE ORGANIZATION OF CRYSTALS ASSOCIATED WITH COLLAGEN FIBRILS

A study of individual mineralized collagen fibrils from turkey leg tendons using the TEM [17] showed that the plate-shaped crystals are indeed located within the fibrils, and that they are preferentially arranged in parallel layers across the fibril (Fig. 1). Layers are separated by about 40 Å [unpublished observation]. This corresponds to the 40-50Å measurements reported by Hohling et al., [18] for center-to-center distances between parallel needlelike apatite crystallites in bone. This organizational motif was confirmed by the TEM examination of mineralized fibrils embedded in a thin layer of vitreous ice, thus excluding the possibility that it was due to a drying artefact during freeze-drying of the original preparations [19].

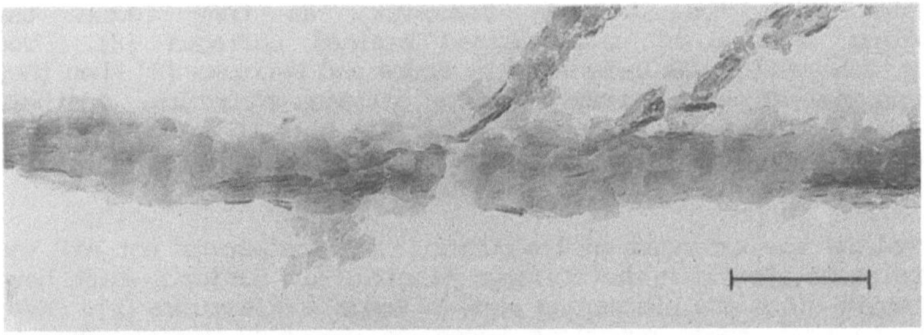

Fig.1 Transmission electron micrograph of mineralized collagen fibril from turkey leg tendon, showing layers of platey crystals. The banding pattern is due to more mineral being present in the gap regions of the collagen than in the overlap regions. The fibril is unstained and embedded in a thin layer of vitreous ice. (Bar=0.2μm).

Furthermore, electron diffraction patterns demonstrated that the crystal layers can extend for distances of almost a micron, which include several individual fibrils. Thus the basic organizational motif of crystals within collagen fibrils in turkey tendon is a layered structure, illustrated schematically in Fig. 2A. Indirect evidence based on an SEM study of the fracture surfaces of a variety of bones suggested that this organizational motif may well be applicable to bone as well [14].

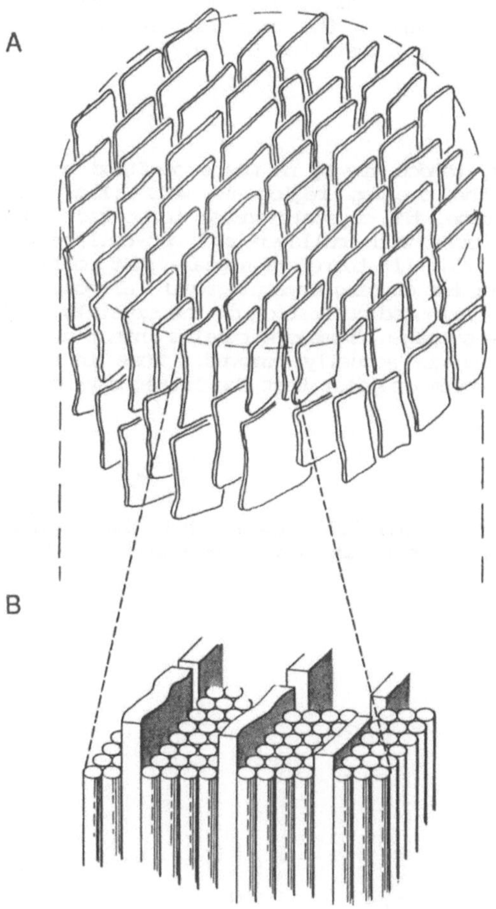

Fig.2 Schematic illustration of distribution of apatite crystals within a collagen fibril of mineralized turkey tendon. In (a) plate-shaped crystals are represented by rectangles, all similarly oriented with c axes along the fibril axis. The crystals are arranged in parallel coplanar arrays forming grooves through the fibrils. In (b) these grooves are shown to be separated by four layers of triple-helical collagen molecules. This figure is not drawn to scale.

THE STRUCTURE OF THE COLLAGEN FIBRIL AS IMPLIED BY THE LAYERED CRYSTAL MOTIF

The crystals are organized in a 3-dimensional array within the fibril, and hence the fibril itself must have 3-dimensional order. The 3-dimensional structural model for collagen proposed by Katz and Li [20] predicts that the gaps are contiguous and arranged in parallel layers or grooves; properties consistent with the observed crystal organization [17]. Independent evidence supporting this structure has been obtained from TEM image studies of transverse sections of rat tail tendon [21] and more recently from careful stoichiometric analyses of the collagen cross-links in bone [22]. The implications of this latter study with regard to the mineralization of collagen are discussed by Katz et al., [23], and are consistent with the observed layered structure of the crystals. The structure of mineralized collagen fibrils is schematically illustrated in Fig. 2B.

THE ONGOING MINERALIZATION OF COLLAGEN FIBRILS

The crystals in collagen fibrils of turkey tendon are by no means confined to the gap regions [24,14], but extend into the overlap region as well (Fig. 1). Furthermore, in more heavily mineralized bone, the crystals appear to contact each other to form large 3-dimensional structures with collagen and non-collagenous proteins trapped within the mineral phase [25]. It therefore appears that collagen does not act as a rigid confining framework for ongoing crystal growth, although it is somehow instrumental in determining the oriented nucleation of the crystals. An interesting question is to what extent the collagen framework remains intact during mineralization and whether or not collagen is actually removed. Some indirect measurements of collagen content per unit volume suggest that the latter may indeed occur [26].

THE ORGANIZATION OF THE CRYSTALS IN BONE

TEM studies of small crushed rat bone particles [27] reveal that the layered structure observed in mineralized tendon fibrils is also predominant in bone (Fig. 3).

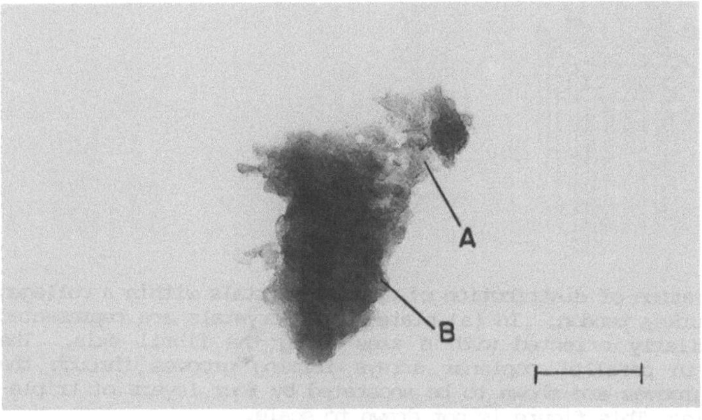

Fig.3 Transmission electron micrograph of particle of crushed rat tibia. Outer portions (A) show layers of flat crystals, but thicker inner portions (B) show parallel striations from crystals seen edge-on. The A and B regions show two distinct electron diffraction orientations and presumably correspond to portions of thin and thick lamellae respectively. (Bar=0.2μm).

In fact no other organizational motif for the crystals in bone has, as yet, been observed, implying that all crystals in bone form in close association with the collagen framework. Bone particles in which the c crystallographic axes of the crystals show two distinct orientations, as determined by electron diffractrion, are assumed to be derived from two or more adjacent lamellae. It is well known that collagen fibril orientations (and hence crystal c axis orientations) are offset in adjacent lamellae [28]. Examination of such particles shows that crystal layers in individual lamellae are parallel, but those in adjacent lamellae are not parallel but offset at some angle. Fractured transverse sections of tibiae examined in the scanning electron microscope (SEM) were consistent with this observation, provided that the parallel fracture planes observed (Fig. 4) are assumed to be produced as a result of cleavage along the crystal layers [14].

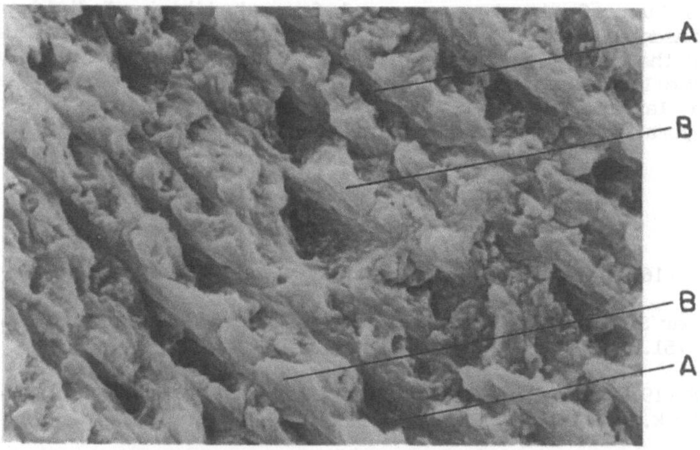

Fig.4 Scanning electron micrograph of surface of 10 month old rat tibia fractured roughly transverse to bone long axis. The bone consists of alternating thin and thick lamellae and the fractures show cleavage mainly along parallel planes of intergrown crystals. In the thin lamellae (A) these are seen edge-on approximately parallel to the interlamellar boundaries. In the thick lamellae (B) these are seen face-on making a large angle with the boundaries. (Bar=10μm).

Thus the crystal layers are offset relative to the lamellar boundaries in a highly ordered 3-dimensional structure, which we call "rotated plywood" [27]. It is schematically illustrated, not to scale, in Fig. 5.

It is not yet known whether the structure observed in rat bone tibiae is representative of other bones in the rat, or bones from other animals. The structure does fill the gap in our knowledge, at least for these tibiae, between the higher order structures of bone involving different arrangements of the lamellae themselves, and the "molecular" scale of crystals associated with a collagen framework.

Fig.5 Schematic illustration of structure proposed for rat tibia. Individual platey crystals are represented by rectangles which are arranged in parallel layers within each lamella. In the central (thin) lamella the crystal planes are roughly parallel to the interlamellar boundaries, but in the outer (thick) lamellae the crystal planes make a large angle with the boundaries. The collagen fiber directions are roughly paralled to the long axes of the crystals. The figure is not drawn to scale.

REFERENCES

[1] Van Leeuwenhoek, A (1674) Phil trans R Soc Lond 106: 121-128

[2] Lowenstam HA, Weiner S (1989) On Biomineralization. Oxford University Press. New York, pp. 149-151.

[3] Martin BR, Burr DB (1989) Structure, function and adaptation of compact bone. Raven Press, New York, pp. 1-17.

[4] De Jong WF (1926) Recl Tran Chim Pays-Bas Belg 45: 445-448

[5] Schmidt WJ (1936) Naturwissenschaften 24:361

[6] Stuhler R (1937) Fortscht Geb Röntgenstrahlen 57: 231-264

[7] Robinson RA (1952) J Bone Joint Surg 34A: 389-434

[8] Robinson RA, Watson ML (1952) Anat Rec 114: 383-410

[9] Hodge AH, Petruska JA (1963) In: Ramachandran GN (ed) Aspects of Protein Structure. Academic Press, New York, pp. 289-300.

[10] Selvig KA (1970) Calcif Tissue Res 6: 227-238

[11] Cuisinier F, Bres EF, Hemmerle J, Voegel JC, Frank RM (1987) Calcif Tissue Int 40: 332-338

[12] Moradian - Oldak J, Weiner S, Addadi L, Landis WJ, Traub W (1990) Conn Tissue Res (in press)

[13] Weiner S, Price PA (1986) Calcif Tissue Int 39: 365-375

[14] Weiner S, Traub W (1989) Conn Tissue Res 21: 259-265

[15] Boothroyd, B (1975) Clin Orthop Rel Res 106: 290-310

[16] Fratzl P, Fratzl-Zelman N, Klaushofer K, Hoffman O, Vogl G, Keller K (1989) Calcif Tissue Int 144: 591

[17] Weiner S, Traub W (1986) FEBS Lett 206: 262-266

[18] Höhling HJ, Ashton BA, Köster HD (1974) Cell Tissue Res 148: 11-26

[19] Traub W, Arad T, Weiner S (1989) Proc Natl Acad Sci USA 86: 9822-9826

[20] Katz EP, Li S (1973) J Mol Biol 80: 1-15

[21] Hulmes DJS, Holmes DF, Cummings C (1985) J Mol Biol 184: 473-477

[22] Yamauchi M, Katz EP, Kazunori O, Teraoka K, Mechanic GL (1989) Conn Tissue Res 21: 159-169

[23] Katz EP, Wachtel E, Yamauchi M, Mechanic GL (1989) Conn Tissue Res 21: 149-158

[24] Arsenault AL (1988) Calcif Tissue Int 43: 202-212

[25] De Niro MJ, Weiner S (1988) Geochim Cosmochim Acta 52: 2415-2423

[26] Lees S (1987) Conn Tissue Res 16: 305-322

[27] Weiner S, Arad T, Traub W (1990) Submitted for publication

[28] Ruth EB (1947) Am J Anat 80: 35-53

CHAPTER 2.22

Histological Studies on the Otolith of *Huso huso*, Chondrostei

M. Takahashi

The Nippon Dental University, Niigata, Japan

Key words: Otolith, Sturgeon, Calcification, Granule, Scanning electron microscopy.

INTRODUCTION

In Agnatha, Chondrichthyes, Dipnoi, Amphibia, Reptilia, Aves and Mammalia, the otolith is composed of many granules [1-8]. But in Teleostei of Actinopterygii, it is done of the crystals and organic matters which deposit periodically in concentric circles around one or a few spherical granules [3,4,7-9]. In the present study, the histological structure of the otolith in Chondrostei which is considered to be the most primitive fish in Actinopterygii was investigated to clarify the origin of the calcification mode of the otolith in Teleostei which is peculiar in Vertebrata.

MATERIALS AND METHODS

The otolith of the two larvae and three adults of sturgeon *Huso huso* of Chondrostei were used to perform the following procedures. Surfaces and fractured surfaces of them were etched with 10 % Wako NaOCl liquid for an hour, washed, dried with critical point dryer, coated with Pt ion and observed with HITACHI S-800 scanning electron microscope. Sagittal ground sections of them were etched with 0.05 *N* Wako HCl liquid for 45 seconds and observed with the same microscope after the same preparation.

RESULTS

Shapes of the sagitta otoliths of three adults were uniform but those of two larvae various. Surfaces of the sagitta and asteriscus otoliths in adults and larvae were uneven (Fig.1). Many large granules concreted together, whose diameters were 10 to 40 μm, were observed on the convex surfaces. Many free small granules, whose diameters were 1 to 10 μm, existed on the concave surfaces (Fig.2). Shapes of them were various, such as spherical, hexagonal plate and red blood corpuscle-shaped. Moss-shaped solid substances were attached on some large granules (Fig.3). Some parts of the surface in the sagitta otolith were stairs-shaped (Fig.4). Some small granules were fused on the sagitta otolith (Fig.5). Many small granules were arranged in a certain direction on some parts of the surface in the asteriscus otolith (Fig.6). The granules were concreted by the solid substances on the fractured surface at the outer layer of the asteriscus otolith. Outlines of the granules were clear at the outer layer (Fig.7), while they weren't recognized and the uniform structure was observed at the middle and inner layers. Outlines of the concreted granules were recognizable but

obscure on the fractured surface at the outer layer of the sagitta otolith. They weren't recognized and the rods-shaped structure, whose diameters were 1 to 3 µm, was observed at the middle and inner layers (Fig.8).
Many granules, whose diameters were 1 to 10 µm, were concreted together on the ground surface at the outer layer of the sagitta otolith (Fig.9). Outlines of them weren't recognized and many thin growth bands, about 1 µm wide, were clear at the middle and inner layers (Fig. 10).

DISCUSSIONS

In Agnatha, the otolith is composed of many spherical granules which are concreted by the solid substances; in Holocephali of Chondrichthyes, many spherical which are done by the solid substances; in Rajiformes of Chondrichthyes, many spherical and many football-shaped which are done by the solid substances; in Galeiformes of Chondrichthyes, many spherical, many football-shaped, a few plate-shaped and a few pyramid-shaped which are done by the solid substances; in Dipnoi of Choanichthyes, many football-shaped among which the liquid substances fill; in Amphibia, Reptilia, Aves and Mammalia, many crystal-shaped among which the liquid substances fill [1-8]. In Teleostei of Actinopterygii, it is composed of the crystals and organic matters which deposit periodically in concentric circles around one or a few spherical granules, and many thin growth bands are clear on the ground section of it [3,4,7-9]. In Crossopterygii of Choanichthyes, it forms a solid mass but its histological structure hasn't yet been investigated [4]. My present study indicates that many spherical, a few hexagonal plate

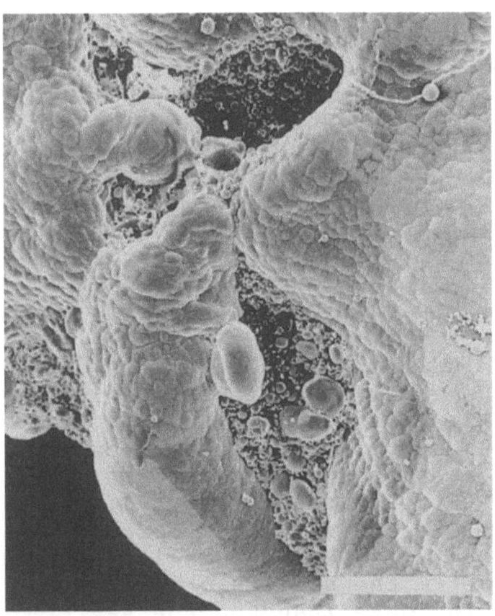

Fig.1. Scanning electron micrograph of the surface of the sagitta otolith in adult sturgeon *Husu huso* etched with NaOCl. The surface is uneven. Scale bar, 200 µm.

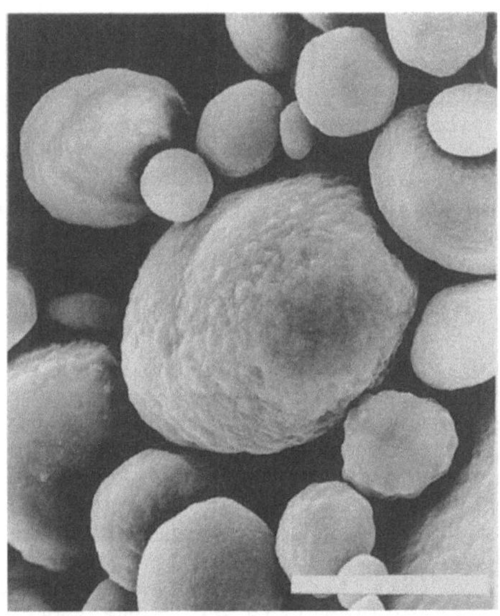

Fig.2. High power view of a concave surface in Fig.1. Many small granules exist on the concave surface. Scale bar, 6 µm.

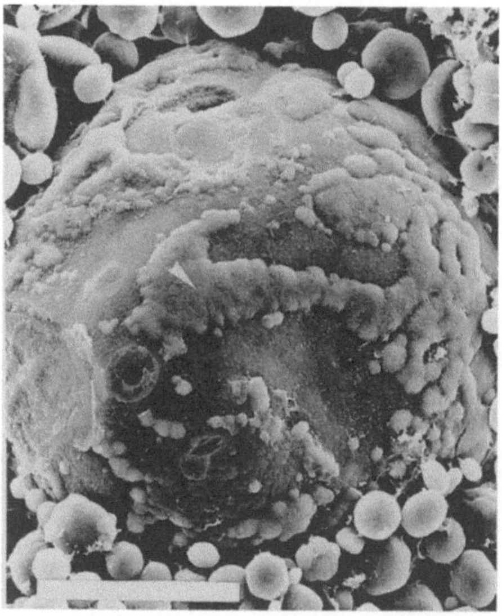

Fig.3. The same technical photograph of another concave surface as Fig.2. Arrow indicates the moss-shaped solid substance on a large granule. Scale bar, 20 μm.

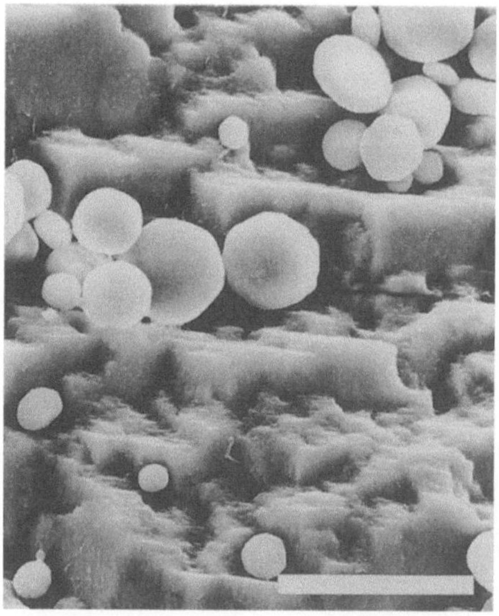

Fig.4. The same technical photograph of the stairs-shaped surface as Fig.1. Scale bar, 10 μm.

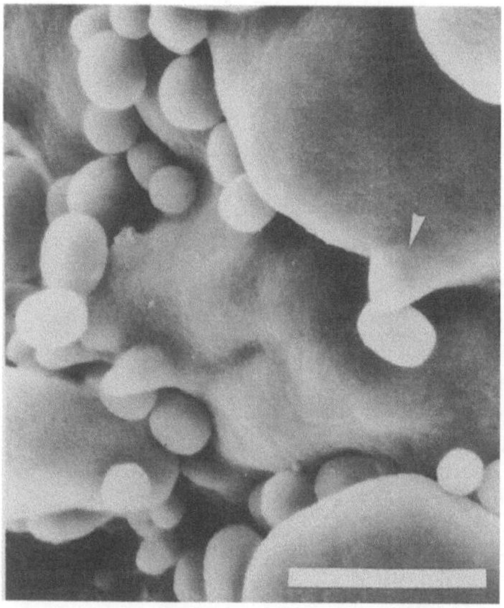

Fig.5. The same technical photograph in larval specimen as Fig. 1. Arrow indicates a small granule fused on the sagitta otolith. Scale bar, 10 μm.

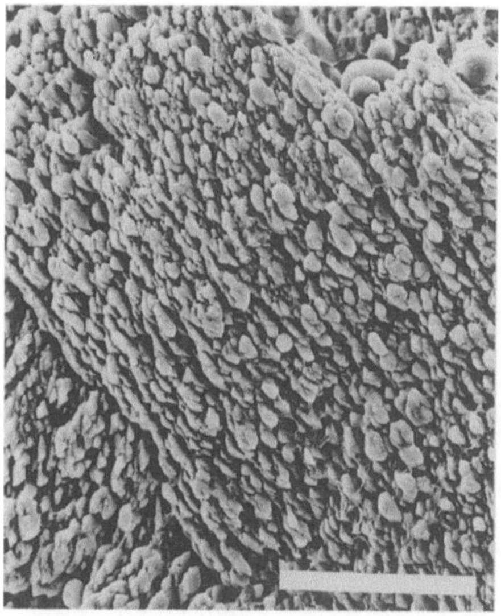

Fig.6. The same technical photograph of the asteriscus otolith as Fig.1. Many small granules are arranged in a certain direction. Scale bar, 30 μm.

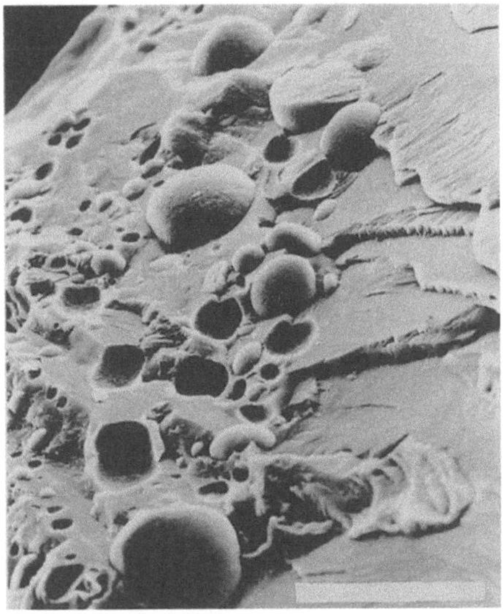

Fig.7. The same technical photograph of the fractured surface at the outer layer of the asteriscus otolith as Fig.1. Outlines of the granules are clear. Scale bar, 30 μm.

Fig.8. The same technical photograph at the inner layer of the sagitta otolith as Fig.7. Rods-shaped structure is observed. Scale bar, 8 μm.

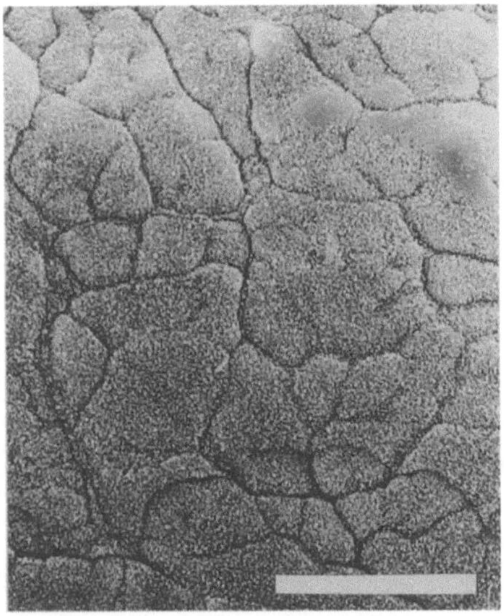

Fig.9. The same technical photograph of the ground surface at the outer layer etched with HCl as Fig.1. Many granules are concreted together. Scale bar, 10 μm.

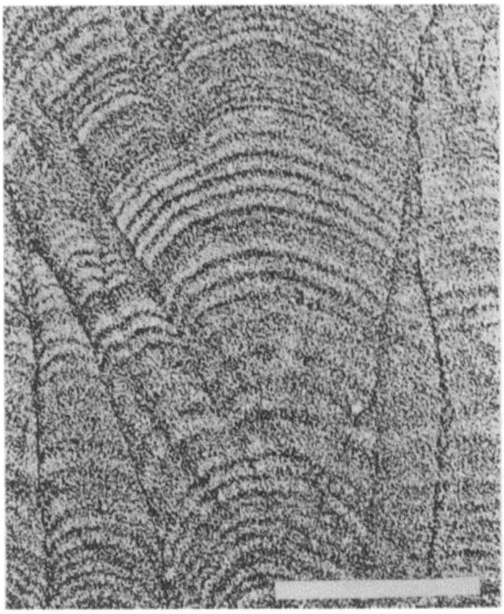

Fig.10. The same technical photograph at the middle layer as Fig. 9. Many thin growth bands are clear. Scale bar, 10 μm.

and a few red blood corpuscle-shaped granules are first formed and concreted by the solid substances, and then the crystals and organic matters of them are rearranged, and the outlines of them disappear, and the thin growth bands become clear in the calcification process of the otoliths in Chondrostei of Actinopterygii.
It is presumed that the liquid substances fill among the granules in the otoliths of the early developmental stage in Chondrostei because the granules are steeped in the liquid substances in the larval otoliths and then they are concreted by the solid substances in Galeiformes. It is considered that the shapes of the sagitta otoliths of adults are uniform but those of larvae various in Chondrostei because the shapes of the sagitta otoliths of sdults reflect those of their statocysts but the sagitta otoliths of larvae are too small to reflect the shapes of their statocysts.
The character of the otolith in Chondrostei that it forms a solid mass agree with them in Agnatha, Holocephali, Rajiformes, Galeiformes, Teleostei and Crossopterygii but differs from them in Dipnoi, Amphibia, Reptilia, Aves and Mammalia. Spherical granules in the otolith of Chondrostei exist in them of Agnatha, Holocephali, Rajiformes, Galeiformes and Teleostei (only at its center) but don't in them of Amphibia, Reptilia, Aves and Mammalia. Hexagonal plate and red blood corpuscle-shaped granules in the otolith of Chondrostei are considered to be characteristic of this fish like the football-shaped of Rajiformes and the same shaped, plate-shaped and pyramid-shaped of Galeiformes. Many clear thin growth bands at the middle and inner layers of the otolith in Chondrostei exist in it of Teleostei but don't in them of Agnatha, Holocephali, Rajiformes, Galeiformes, Dipnoi, Amphibia, Reptilia, Aves and Mammalia. From above facts, it is considered that the calcification mode of the otolith in Chondrostei presents the intermediate stage between the primitive Agnatha and peculiar Teleostei.

CONCLUSIONS

My present study indicates that many spherical, a few hexagonal plate and a few red blood corpuscle-shaped granules are first formed and concreted by the solid substances, and then the crystals and organic matters of them are rearranged, and the outlines of them disappear, and the thin growth bands become clear in the calcification process of the otoliths in Chondrostei. It is considered that the calcification mode of the otolith in Chondrostei presents the intermediate stage between the primitive Agnatha and peculiar Teleostei.

REFERENCES

Hinoura H (1938) J Hokuetsu Med 7: 835-875
Ou T (1948) J Niigata Med Anat 22: 1-74
Aiba S (1950) J Niigata Med Anat 24: 1-27
Carlström D (1963) Biol Bull 125: 441-463
Hoshino T (1976) Igakunoayumi 99: 661-662
Nakahara H, Bevelander G (1979) Anat Rec 193: 233-242
Shichiri K (1985) J Crystal Growth Jap 12: 42-56
Takahashi M (1985) In: Goto M, Takahashi M, Kimura M, Horikawa H (eds) Evolution and Adaptation of Marine Vertebrates. Kokusai-insatsu, Tokyo, pp.5-17.
Takahashi M (1983) Earth Science 37: 1-7

CHAPTER 2.23

Effect of Physiology and Endolymph Composition on the Strontium Content of Bearded Rock Cod *(Pseudophycis barbatus)* Otoliths

J.M. KALISH

Fisheries Research Centre, Ministry of Agriculture and Fisherires, Wellington, New Zealand

Key words: Fish otoliths, Endolymph, Strontium, Plasma, Albumin.

INTRODUCTION

The feasibility of applying data on the trace element chemistry of aragonitic fish otoliths to studies of fish biology has received considerable attention in recent years, however, the utility of these data has not been assessed conclusively [1]. Greatest interest lies in the use of otolith strontium content as an indicator or environmental temperature[2,3]. This interest is derived from laboratory-based empirical evidence which supports the theory that the quantity of Sr present in inorganically precipitated aragonite is inversely correlated with temperature [4]. Subsequent research has shown that temperature related variations in otolith Sr may be related more to physiological factors than to physical properties [1,3].

A recent study showed that otolith Sr content in several fish species is highly correlated with endolymph Sr content [1] and highlighted the need for further investigation into the relationship between otolith and endolymph composition before determining the basis for changes in otolith trace element composition. Previously, only a handful of studies had examined the composition of fish endolymph and they did not consider the relationship between endolymph and otolith trace elements [5,6,7]. Seasonal changes in levels of magnesium and calcium in the endolymph and blood plasma of two fish species have been investigated, but these measurements were not compared with the composition of the otolith [6]. Thus, only two studies have related the trace element composition of fish endolymph to otolith composition [1,8]. This study seeks to provide further information on the relationship between endolymph and otolith Sr content and to determine those physiological and environmental factors that effect, or are related to the composition of the endolymph.

MATERIALS AND METHODS

Bearded rock cod (*Pseudophycis barbatus*) were collected from March 1987 to February 1988 by handline fishing in Variety Bay, Tasmania. All sample collection was done at sea immediately after capture of individual fish. Within 30 sec of being hooked fish were brought to the surface where they were killed by a blow to the head. Individuals were weighed to the nearest 10 g with a spring balance and the standard length measured. The tail was cut off in the region of the caudal peduncle for collection of blood from the dorsal aorta. Blood was collected by placing a capillary tube containing about 2 µl lithium heparin (0.25 mg/ml) against the dorsal aorta and collecting blood from the capillary tube in 1.5 ml plastic centrifuge tubes. Whole blood was placed on ice and centrifuged within 2 h of collection. Endolymph was collected immediately after blood collection. The brain was exposed and removed with forceps and any fluid remaining in the brain cavity was absorbed with lint-free paper making the sacculi containing the sagittal otoliths visible. The membrane between the brain cavity and the sacculus was punctured and the endolymph was removed with an autopipette with a standard low volume (10 to 250 µl) tip. Immediately after collection the endolymph was placed on ice and subsequently stored at 4° C prior to analysis. Standard biomedical procedures were used for analysis of metabolites and ions in blood plasma and endolymph [8]. Sagittal otoliths were extracted from the sacculus and the adhering otolith capsule was removed. Sagittae were rinsed in freshwater and then stored in individual glass vials. Preparation of otolith samples for microprobe analysis is discussed elsewhere [1].

Multiple regression models were investigated in an attempt to explain the variability in otolith and endolymph composition. Data on blood plasma, endolymph and otolith composition, fish length, weight, condition, reproductive state, age and water temperature were used to derive multiple regression models of the form:

$$Y = a + b_1X_1 + b_2X_2 + \cdots + b_nX_n$$

Multiple regression models were constructed using an interactive stepwise method with a significant F-statistic ($p \leq 0.05$) as the major statistical criterion for inclusion in the model. However, only those independent variables that were deemed biologically relevant and that would ultimately aid in the prediction of the dependent variable were considered. It was not believed appropriate, for example, to utilize a constituent of the endolymph to aid in the prediction of another endolymph variable. In such instances, it would be more appropriate to measure the dependent variable directly. Of course, the measurement of one endolymph variable might help to predict one or more dependent endolymph variables. Also, before consideration for inclusion in the model independent variables were plotted against dependent variables, along with the resultant standardized residuals and these plots were inspected to confirm that the assumptions of the regression analysis were not violated. Where skewness was evident, a log transformation was applied to the independent variable data.

RESULTS

Numerous significant correlations were found among endolymph and otolith Sr data and the environmental and physiological variables (Table 1). Although the most significant correlations with endolymph Sr were other endolymph variables, these data were not considered in the models for reasons discussed earlier. Furthermore, endolymph variables such as protein, glucose, calcium, phosphate and triglycerides were all highly correlated with gonadosomatic index (GSI), a variable considered for inclusion in the endolymph Sr models.

Multiple linear regression models were clearly more effective at explaining the variation in endolymph Sr measured in female rather than male fish. Details of the variables included in the models are shown in Table 2. A model containing 3 variables (temperature, GSI and plasma Sr) was able to explain 84% of the variance in the endolymph Sr data from female cod, whereas a multiple regression model, also containing 3 variables (temperature, plasma Sr and plasma Ca) could only explain 49% of the variance in male endolymph Sr data. When both sexes were combined the multiple regression model included 5 variables (temperature, gonad weight, plasma Sr, fish weight and plasma K) and explained 69% of the variance in the data (Table 2). All three models involved ambient temperature and plasma Sr.

Like the endolymph Sr models, the otolith Sr multiple linear regression models were more effective at explaining the variance in the female cod (Table 3). A model containing 6 variables (fish length, GSI, plasma Sr, plasma Ca, condition factor and age) explained 69% of the variance in the female cod otolith Sr data (Table 3). The best model for male otolith Sr data explained only 44% of the variance and contained 3 variables (fish weight, endolymph Sr and endolymph protein). A model to predict otolith Sr in both males and females explained 52% of the variance with 6 variables (fish weight, gonad weight, endolymph Sr, plasma Sr, endolymph protein and condition factor) (Table 3).

Cellulose acetate electrophoresis showed that there were large differences in the relative protein composition of cod plasma over the year and somewhat smaller differences within samples from the same collecting trip. The potential variation in the calcium-binding capacity of fish plasma from various individuals was evident based on the differences in the quantities of the albumin and globulin fractions in the plasma and, the albumin/globulin ratio (A/G ratio) derived from elution diagrams. Spectrophotometric measurement of albumin in the plasma samples also highlighted the variability of the plasma protein complement [8]. Incorporation of the variable A/G ratio into the multiple regression model to predict the endolymph Sr made it possible to explain 98.5% of the total variance in female endolymph Sr levels (Table 4). The model included the variables A/G ratio, plasma protein, fish length, fish weight, gonad index and plasma Sr.

Table 1. Correlation coefficients for physiological and environmental variables significantly correlated <0.05) with endolymph and otolith strontium variables (Total N=115; Females N=54; Males N=61).

	Endolymph Sr			Otolith Sr/Ca		
	Females	Males	Both sexes	Females	Males	Both sexes
Temperature	-.768	-.635	-.685	-.38		-.295
Length				-.281		-.265
Weight				-.267	-.295	-.275
Condition factor					-.261	-.234
Age			.234			
Gonad weight	.623	.337	.538	.263		.19
GSI	.745	.31	.589	.467		.329
Plasma protein	.533		.312			
Plasma Sr	.549	.474	.512	.547	.363	.468
Plasma Ca	.296	.393	.329			
Plasma Na				.285		
Plasma K						
Plasma glucose						
Plasma triglyceride	.574		.377	.36		.215
Plasma phosphate	.338		.231	.371		.272
Endolymph protein	.823	.673	.754	.48		.327
Endolymph Sr	1.00	1.00	1.00	.583	.534	.562
Endolymph Ca	.388		.263			
Endolymph Na		.284				
Endolymph K						
Endolymph glucose	.734	.377	.616	.456		.319
Endolymph triglyceride	.80	.42	.692	.478		.347
Endolymph phosphate	.793	.547	.718	.52		.391

Table 2. Multiple linear regression model coefficients and associated statistics for estimation of endolymph Sr (mM) in cod.

	Variable	Coefficient	SE	P
Females (N=51)				
	Temperature	-0.00045	8.43×10^{-5}	<0.0001
	GSI	0.001	9.55×10^{-5}	<0.0001
	Plasma Sr	0.142	0.034	<0.0001
	Constant (a)	0.008		
	Multiple correlation, R=0.917			
Males (N=64)				
	Temperature	-0.00036	8.01×10^{-5}	<0.0001
	Plasma Sr	0.117	0.039	<0.005
	Plasma Ca	0.001	0.000224	<0.025
	Constant (a)	0.007		
	Multiple correlation, R=0.703			
Both sexes (N=115)				
	Temperature	-0.00038	6.52×10^{-5}	<0.0001
	Fish weight	-2.08×10^{-6}	4.81×10^{-7}	<0.0001
	Gonad weight	9.71×10^{-5}	1.22×10^{-5}	<0.0001
	Plasma Sr	0.152	0.028	<0.0001
	Plasma K	-0.00016	5.74×10^{-5}	<0.01
	Constant (a)	0.009		
	Multiple correlation, R=0.832			

Table 3. Multiple linear regression model coefficients and associated statistics for estimation of otolith Sr/Ca ratios in cod.

	Variable	Coefficient	SE	P
Females (N=51)				
	Fish length	-6.96×10^{-5}	1.55×10^{-5}	<0.0001
	GSI	0.000219	3.48×10^{-5}	<0.0001
	Plasma Sr	0.073	0.011	<0.0001
	Plasma Ca	-0.000218	9.23×10^{-5}	<0.025
	Condition	-0.087	0.034	<0.020
	Age	7.58×10^{-5}	3.00×10^{-5}	<0.020
	Constant (a)	0.005		
	Multiple correlation, R=0.829			
Males (N=64)				
	Fish weight	-6.97×10^{-7}	2.32×10^{-7}	<0.005
	Endolymph Sr	0.25	0.04	<0.0001
	Endolymph protein	-0.000129	3.60×10^{-5}	<0.001
	Constant (a)	0.003		
	Multiple correlation, R=0.662			
Both sexes (N=115)				
	Fish weight	-8.28×10^{-7}	1.78×10^{-7}	<0.0001
	Gonad weight	1.49×10^{-5}	5.12×10^{-6}	<0.005
	Endolymph Sr	0.128	0.036	<0.001
	Plasma Sr	0.04	0.011	<0.001
	Endolymph protein	-7.05×10^{-5}	2.94×10^{-5}	<0.025
	Condition	-0.07	0.027	<0.020
	Constant (a)	0.003366		
	Multiple correlation, R=0.720			

Table 4. Multiple linear regression model coefficients and associated statistics for estimation of endolymph Sr/Ca ratios in female cod. Unlike the model presented in Table 2, blood protein data are included in the model.

	Variable	Coefficient	SE	P
Females (N=12)				
	Albumin/globulin ratio	0.001	9.44×10^{-5}	<0.0001
	Total plasma protein	4.08×10^{-5}	8.79×10^{-6}	<0.0001
	Fish length	-0.00027	4.99×10^{-5}	<0.0001
	Fish weight	1.35×10^{-6}	5.78×10^{-7}	<0.01
	GSI	0.001	3.52×10^{-5}	<0.0001
	Plasma Sr	0.08	0.022	<0.0001
	Constant (a)	0.01		
	Multiple correlation, R=0.992			

DISCUSSION

Attempts to explain the variations in otolith Sr/Ca ratios in terms of temperature have not given detailed consideration to the mechanism that results in variations in otolith Sr/Ca ratios [2,3,9]. Results presented here show a negative correlation between temperature and both plasma Sr and endolymph Sr in cod (Table 1). On this basis, it seems that, in this species, physiological changes are resulting in changes in the level of Sr in both the plasma, the endolymph and, ultimately the otolith and this supports the findings of an earlier study [1]. The temperature related changes in plasma Sr, in the case of adult cod, were probably associated with gonadal development and, thus, can be related to changes in plasma Ca [1,8]. The relationship between endolymph Sr and temperature was probably a further manifestation of the physiological changes associated with reproduction and the related changes in the plasma protein complement [8].

Although there is clearly a relationship among several of the measured variables and otolith Sr/Ca ratios, the multiple correlations are not as good as had been anticipated (Table 3). It is important to indicate the difficulties that arise when using measurements of the outer otolith margin composition as an indicator of the composition of the most recently formed portions of the otolith. Because of sample destruction during electron microprobe analysis of readily vaporized minerals such as carbonates, it is necessary to use a scanning raster or a defocussed electron beam. Ideally, the minimum size for a square window that would also minimize sample destruction is 10 x 10 µm. In the slow growing adult cod, 10 µm of otolith material may encompass aragonite deposited in the last several months, particularly in the winter periods when the growth rate would be lowest. The result is the comparison of a temporally discrete endolymph sample with an integrated sample of otolith composition, not an ideal situation.

The results of the linear multiple regression models, while of little practical use in actual prediction, provide a valuable insight into those factors that ultimately influence the Sr content of both fish endolymph and otoliths. Most importantly, in view of the present interest in fish otoliths as predictors of environmental temperature, is the fact that the variable ambient temperature was the first variable selected in models to predict the Sr content of fish endolymph (Table 2). Where plasma albumin and globulin ratios were considered the temperature variable was not included (Table 4). Furthermore, temperature does not appear as a variable in any of the multiple regression models to predict otolith Sr (Table 3). This illustrates that, in the otoliths of adult cod at least, temperature does not appear to directly influence the Sr content of fish otoliths. Alternatively, the effect of temperature or, perhaps more generally, the effect of seasonal changes in physiology which are associated with changes in temperature, are the primary force in altering the composition of fish plasma, which ultimately affects the compositon of the endolymph and otoliths. This conclusion is highlighted by the relationship that also exists between the Sr content of blood plasma and endolymph, evident in both male and female fish (Table 1).

Gonadosomatic index was the second most important variable in the prediction of endolymph and otolith Sr in female cod (Tables 2 and 3), whereas neither GSI nor gonad weight was used to predict the dependent variable in male cod. In view of these results it would be important to investigate variations in endolymph and otolith composition in a species that spawned in the summer to conclusively determine the roll of gonadal development in altering otolith Sr/Ca ratios and to investigate other variables that might have a similar efffect.

REFERENCES

1 Kalish JM (1989) J Exp Mar Biol Ecol 132: 151-178
2 Radtke RL (1989) Comp Biochem Physiol 92A: 189-193
3 Townsend DW, Radtke RL, Morrison MA, Folsom SD (1989) Mar Ecol Prog Ser 55:1-13
4 Kinsman DJJ, Holland HD (1969) Geochim Cosmochim Acta 33: 1-17
5 Mugiya, Y (1964) Bull Jap Soc Sci Fish 30: 955-967
6 Mugiya Y (1966) Bull Jap Soc Sci Fish 32: 549-557
7 Mugiya Y, Takahashi K (1985) Bull Fac Fish Hokkaido Univ 36: 57-63
8 Kalish JM (1989) Fish otolith chemistry an an indicator of physiological, ecological and environmental events. Doctoral dissertation, University of Tasmania, 368 pp.
9 Radtke RL (1984) Flodevigen Rapp 1: 273-298

CHAPTER 2.24

Modelling Biomineralization: Studies on the Morphology of Synthetic Calcite

J.M. Didymus[1], S. Mann[1], N.P. Sanderson[1], P. Oliver[1], B.R. Heywood[1], and E.J. Aso-Samper[2]

[1]School of Chemistry, University of Bath, Bath, BA2 7AY, UK and [2]Departmento de Ciencias de la Tierra, Universidad de Zaragoza, Spain

Key words: Morphology, Calcite, Additive, Stereoselective adsorption

INTRODUCTION

Many organisms possess the remarkable ability to deposit single crystals of calcite ($CaCO_3$) with morphologies not normally observed in the inorganic world [1]. Whilst it is true that single geological crystals of calcite can exhibit an enormous range of different habits, all these forms have common interfacial angles and symmetry as described by the $R\bar{3}c$ space group. By contrast, the external forms of some biological single crystals of calcite have symmetries that are non-crystallographic. The coccolith segments deposited by the unicellular marine alga *Emiliania huxleyi* illustrate this phenomenon particularly well:-

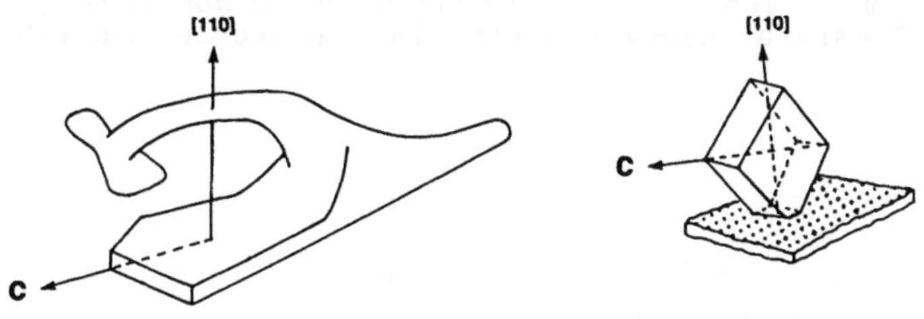

Fig. 1 Single crystal segment of *E. huxleyi* [2] (left) compared to a similarly oriented geological calcite crystal (right)

Current theories of biomineralization suggest that calcifying organisms have adopted strategies for controlling morphology based on the deployment of functional organic molecules. For example, proteins rich in aspartate and glutamate residues and also phosphoserine, are common for molluscs [3] whilst coccoliths of *E. huxleyi* are deposited along with sulphated and carboxylated polysaccharides [4]. Thus, carboxylate groups and, to a lesser extent, sulphates and phosphates play an important role in the biomineralization of calcite.

Surprisingly few studies have been concerned with the mechanisms by which $CaCO_3$ habit is modified. Berman *et al* noticed that sea urchin proteins present in solution could change the habit of single crystals of calcite by expressing $\{1\bar{1}0\}$ faces not observed in the control system [5]. However, most investigations have focused on the inhibitory effects of various macromolecules in solution [6-9]. Little is known about the molecular interactions taking place between polyanionic molecules and the calcite crystal lattice. In this regard, we have recently undertaken studies to model these processes through an investigation of the morphological effects of simple oxyanions present in the crystallizing solution [10]. This has enabled the importance of features such as spatial charge density, stereochemistry and conformational freedom to be assessed. The molecules investigated included dicarboxylates, phosphates, phosphonates, sulphate and nitrate.

RESULTS

$CaCO_3$ crystals were grown from seedless supersaturated bicarbonate solution (S ≈ 10^2; Ca = 9 mM; pH ≈ 6) over a period of several days in crystallization dishes. This preparation was adapted from the method of Kitano [11].

In the absence of any additives, calcite crystals with a rhombohedral habit comprising six smooth {104} faces were deposited (fig. 2a). In contrast, the dicarboxylate additives gave spindle-shaped crystals. Their potency was critically dependent on molecular structure. For example, malonate [$CH_2(COO^-)_2$] at a Ca/malonate molar ratio of 3:1, induced the formation of crystals which were elongated along the c axis, had {104} end faces and a rounded central body of roughened, curved, prismatic faces (fig. 2b). Increasing the chain length between the terminal carboxylates (eg succinate, glutarate, adipate etc), reduced this morphological effect at equivalent concentrations. Introducing an unsaturated carbon bond into the additive molecules to give maleate [*cis* $C_2H_2(COO^-)_2$] and fumarate [*trans* $C_2H_2(COO^-)_2$] had a pronounced effect. For example, maleate was much less effective than malonate but more potent than succinate. Fumarate, however, had minimal influence on morphology.

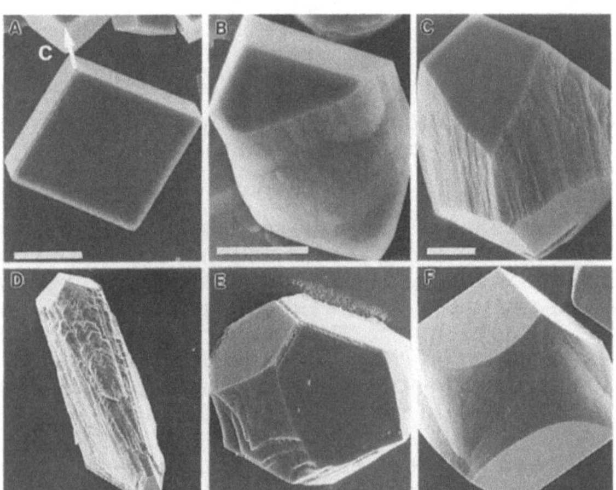

Fig. 2 SEM micrographs of calcite crystals (a) control (b) grown in the presence of malonate (c) γ-carboxyglutamate (d) orthophosphate (e) phenyl phosphonate (f) phenyl phosphate. Scale bar = 10 µm.

Functionalization of the α,ω-dicarboxylates resulted in an increased morphological effect. Substitution of the α-hydrogen of succinate for an amino group to give aspartate $[H_3N^+C_2H_4(COO^-)_2]$, resulted in a dramatic enhancement of potency. Again, prismatic {110} faces were observed but these appeared to be smoother and better defined than those for succinate and malonate indicating a more specific interaction at the crystal/solution interface. An ω-carboxylate functionalization of glutamate to give γ-carboxyglutamate, resulted in further improvement in this specificity (fig. 2c).

The effect of other oxyanions was also investigated. Orthophosphate as $H_2PO_4^-$ was effective at Ca/additive ratios as high as 10,000:1. The morphological interaction, however, appeared similar to that of the dicarboxylate anions (fig. 2d). At 100:1, the crystals were distinctly elongated with tapered side faces slightly inclined to the c axis. These faces were not strictly $\{1\bar{1}0\}$ but a higher index set possibly {401} [12]. Analysis by infrared spectroscopy suggested that the phosphate was bound to the surface in a bidentate manner. Interestingly, oxyanions of stronger acids were less effective in modifying crystal growth. Sulphate produced similar morphologies to maleate and nitrate was ineffective even at high concentrations. Ester derivatives of orthophosphate $[ROPO_3H^-]$ showed a marked reduction in morphological effect as the R group increased in size and hydrophobicity (H > butyl > phenyl > naphthyl). The phenyl phosphoester $[C_6H_5OPO_3H^-]$ was slightly more potent than its phosphonate counterpart $[C_6H_5PO_3H^-]$ but the latter produced a remarkably different morphology. At 10:1, phenyl phosphate gave spindles similar to sulphate but with smoother, rounded faces (fig. 2f). In contrast, phenyl phosphonate gave roughened side faces which were indexed as {018} (fig. 2e). These are of negative rhombohedral form and subtend an angle of 63.8^o to the c axis. A further $-PO_3H^-$ functionalization of the methyl and ethyl phosphonates to give the corresponding diphosphonates resulted in a substantial enhancement of potency comparable to orthophosphate and with a similar mechanism of interaction.

DISCUSSION

A common feature of the calcite faces aligned parallel to the c axis, is that the planar carbonate anions (CO_3^{2-}) protrude perpendicular to the surface. Thus, insertion of HCO_3^- ions from solution into these faces is primarily through bidentate binding.

The dicarboxylates were all fully ionized at pH 6 thus enabling the planar bidentate COO^- groups to mimic HCO_3^- surface incorporation. The distance between carboxylate groups in malonate and maleate is *ca.* 4 Å close to that for adjacent layers of carbonate sites in the prismatic faces. It is, therefore, feasible that these molecules can adsorb on the surface in a bridging configuration whereby the carboxylate groups interact with calcium atoms in different layers. The results with maleate and fumarate suggest that cooperative binding of both carboxylates is necessary for significant change in crystal habit. This is because the C=C double bond confers rigidity on the molecule and, whereas the *cis* isomer (maleate) can still bind with both carboxylates, the *trans* isomer (fumarate) can only interact with one. Data obtained with the longer-chain dicarboxylates indicate that charge separation is a critical factor in determining the extent of this cooperative interaction. It is geometrically possible for aspartate and γ-carboxyglutamate to span three layers of carbonate sites on the $(1\bar{1}0)$ face. It is also stereochemically possible for the cationic amino group to simultaneously occupy a Ca^{2+} site in the surface. These possibilities allow for a greater contact between the adsorbed molecule and the crystal face and so may explain the observed enhanced specificity of aspartate and γ-carboxyglutamate.

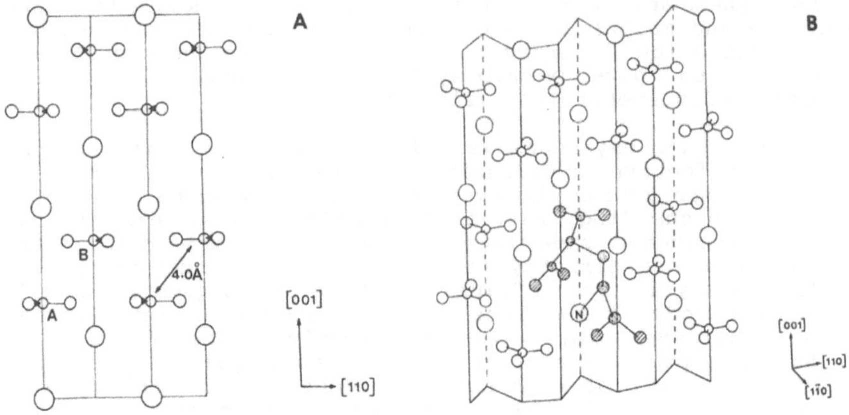

Fig. 3 (a) View of {1$\bar{1}$0} face showing carbonate stereochemistry (b) Perspective drawing of the same face showing possible γ-carboxyglutamate binding site.

Nitrate has the same stereochemistry as bicarbonate but does not compete effectively for lattice sites due to its weaker affinity for Ca^{2+} (log stability constants: ($CaHCO_3^+$) = 1.4; ($CaNO_3^+$) = 0.68).

Phosphate is predominantly singly charged at pH 6 [$H_2PO_4^-$] and tetrahedral (T_d). It can therefore interact with a crystal face in either a bidentate or tridentate fashion. The morphological results here suggest that the bidentate mode of interaction was favoured as the {401} faces are effectively identical in their surface characteristics to {1$\bar{1}$0}. The {401} faces are inclined at 3.6^o to {1$\bar{1}$0} which reflects the inability of phosphate to completely mimic the bidentate stereochemistry of carbonate. The phosphoester series demonstrated the importance of charge density and the hydrophobic nature of the substituent attached to the active functional group. An increased hydrophobicity leads to a reduction in the strength of the additive/crystal interaction due to the repulsive forces between the charged surface and the hydrophobic part of the molecule. This phenomenon can also be used to explain the observed difference in recognition between the phenyl additives and the crystal lattice. In the phosphonate, the bulky phenyl group attached directly to the P atom, is at a maximum distance from the crystal surface when the molecule is bound in a tridentate manner. The {018} faces expressed by phenyl phosphonate have carbonates arranged at an angle of 26.3^o to the plane of the face. Thus, insertion of anions into the surface proceeds essentially in a tridentate manner. The {001} which also has similar stereochemistry is probably not selected because it is dipolar and thus much more unstable than the neutral {018}. By comparison, rotation about the C-O-P linkage in phenyl phosphate enables the hydrophobic group to be at a maximum distance from the surface for a bidentate mode of binding on the prismatic faces.

Sulphate is tetrahedral and doubly charged at pH 6. Despite this, its potency was substantially inferior to the phosphate monoanion. This may be because the surfaces processes are dominated by chemical rather than electrostatic interactions. In particular, the ability of phosphates to undergo ligand (proton) exchange reactions on binding to Ca^{2+} ions in the crystal surface is an important feature contributing to the high affinity of the anions for the calcite lattice.

Relevance to Biomineralization

This work indicates that specific morphological changes can be induced in calcite crystals by anionic species in solution. These effects are critically dependent upon the geometric, stereochemical and electrostatic features of the interfacial processes. For example, simple changes in molecular structure can give rise to profound changes in calcite morphology.

From this study, it can be concluded that phosphoester and carboxyl groups on proteins could be responsible for controlling calcite crystal morphology. It appears that in the case of carboxyl groups, cooperative interactions between neighbouring anions is required in order for the effect to be significant. The interaction of sulphate moieties with calcite may proceed via a similar mechanism since the potency of this oxyanion is comparable to carboxylate. Interestingly, otoconia crystals found in the inner-ears of vertebrates are closely associated with proteins containing high levels of acidic amino acids including γ-carboxyglutamate [13]. Thus, the morphology of otoconia crystals could be attributed to interactions between such carboxyl-functionalized macromolecules and the prismatic faces of calcite.

REFERENCES

1. Watabe N (1974) J Cryst Growth 24/25:116-122
2. Mann S, Sparks NHC (1988) Proc R Soc London Ser B 34: 441-453
3. Wheeler AP, Rusenko KW, Sikes CS (1988) In: Sikes CS, Wheeler AP (eds) Chemical Aspects of Regulation of Mineralization. University of Alabama Publication Services, Mobile, Alabama, pp 9-13
4. Borman AH, de Jong EW, Huizinga M, Kok DJ, Westbroek P, Bosch L (1982) Eur J Biochem 129: 197-183
5. Berman A, Addadi L, Weiner S (1988) Nature 331: 546-548
6. Wheeler AP, George JW, Evans CA (1981) Science 212: 1397-1398
7. Sikes CS, Wheeler AP (1988) In: Sikes CS, Wheeler AP (eds) Chemical Aspects of Regulation of Mineralization. University of Alabama Publication Services, Mobile, Alabama, pp 15-20
8. Addadi L, Weiner S (1986) Mol Cryst Liq Cryst 134: 305-322
9. Addadi L, Moradian J, Shay E, Maroudas NG, Weiner S (1987) Proc Nat Acad Sci USA 84: 2732-2736
10. Mann S, Didymus JM, Sanderson NP, Heywood BR, Aso Samper EJ (1990) J Chem Soc Faraday Trans 86(10): 1873-1880
11. Kitano Y (1962) Bull Chem Soc Japn 35(12) 1973-1980
12. Ford WE (ed) (1932) Dana's textbook of mineralogy. John Wiley & Sons, p 122
13. Ross MD, Pote KG (1984) Philos Trans R Soc London Ser B 34: 445-452

CHAPTER 2.25

Physical Chemical Studies of Mineralization and Demineralization of Apatites

A. EBRAHIMPOUR, L. PASCHALIS, J. ZHANG, and G.H. NANCOLLAS

Departments of Chemistry and Biomaterials, State University of New York at Buffalo, Buffalo, New York 14214, USA

Key words: Constant Composition, Dentin, Hydroxyapatite, Mineralization, Demineralization, Kinetics.

INTRODUCTION

The formation, remineralization, and dissolution of hard tissue such as bones and teeth are complex processes not only because of the presence of multiple components in the solution media but also because of the numerous calcium phosphate phases that may be involved. Moreover, other phases such as amorphous calcium phosphate (ACP) dicalcium phosphate dihydrate (DCPD) and octacalcium phosphate (OCP) as well as defect apatites may participate. Calcium phosphates are also used in many orthopaedic and dental implant applications such as HAP-plasma coated titanium (HPCTI) prosthetic devices. Such materials are of considerable interest because of their osteoconductive and possible osteoinductive properties [1-3].

There is little doubt that many biomineralization processes involve the formation of metastable intermediates which may subsequently transform into thermodynamically more stable phases. Most biominerals such as dentin and HPCTI are mixtures of calcium phosphates as evidenced by X-ray analysis. Physical chemical studies of the de- and re-mineralization reactions of such materials therefore involve the simultaneous growth and/or dissolution of multiple calcium phosphate phases. Since, in the calcium phosphate system, solutions are often supersaturated with respect to more than one phase, simultaneous precipitations, proceeding through DCPD-OCP-HAP phases, are possible. However, in many cases the precipitation of one of the phases is kinetically favored. Kinetic studies of simultaneous crystallizations has been severely limited in conventional kinetics methods because of the decreasing driving force or supersaturation with respect to each of the phases as the reactions proceed. Based on the Constant Composition (CC) method [4-10], the recent development of the Dual Constant Composition method (DCC) [11, 12] has enabled studies to be made of concurrent dissolution and/or mineralization of mixtures of phases.

In the present study the highly reproducible DCC method has been used to study the concurrent calcium phosphate dissolution reactions involved in the demineralization of ground human dentin as a function of undersaturation, ionic strength, and in the presence of added magnesium ions. In addition, the de- and re- mineralization of HPCTI prosthetic devices have been investigated in saline solution.

Materials and Methods

The development of the DCC method has been discussed in detail elsewhere [11]. Briefly, this technique employs two potentiostatic titrators (Metrohm Impulsomat 614, MultiDosimat 645, Dosigraph 625 and Potentiostat 605, Brinkmann Instruments) in conjunction with two sensors to provide two constant composition devices for the control of the simultaneous reactions. For each growth or dissolution process a set of titrant solutions was prepared having concentrations calculated so as to maintain constant driving force in each case. The rate of the individual reactions could be calculated from the recorded titrant consumption as a function of time following the addition of seed crystals. It has been previously confirmed [11-13] that even by using an electrode sensitive to a lattice ion common to each of the seed materials, the rate of the individual reactions could be evaluated with a precision approaching that of a conventional CC experiment. Solutions, prepared in grade A volumetric glassware, using reagent grade chemicals (Fisher Scientific) with deionized, triple-distilled, CO_2-free water, were filtered (0.22 μm, Millipore filters) twice before use. Solutions of calcium chloride were standardized by EDTA titration, by atomic absorption spectrometry (Perkin-Elmer Model 503), or by inductively coupled argon plasma (ICAP) (Thermo Jarrel Ash). Phosphate concentrations were measured spectrophotometrically (Cary 210, Varian Instruments) as the phosphovanadomolybdate complex [14]. Potassium hydroxide solution was prepared in a nitrogen atmosphere from washed pellets and standardized against potassium hydrogen phthalate. The apatitic nature of ground human dentin seed material and HPCTI samples was verified by X-ray powder diffraction analysis (Nicolet/NIC, Siemens).

In the dentin dissolution experiments, the undersaturated solutions were prepared by mixing calcium chloride, potassium phosphate, potassium hydroxide and sodium chloride solutions. The relative supersaturation with respect to HAP, σ_{HAP}, is given by Eq. (1),

$$\sigma_{HAP} = (\{Ca\}^5 \{PO_4\}^3 \{OH\} / K_{SO(HAP)})^{1/9} - 1, \qquad (1)$$

where the braces represent ion activities and $K_{SO(HAP)}$ is the thermodynamic solubility product of HAP. In the dentin dissolution studies, the values of σ_{HAP} were -0.7 and -0.3. The ionic strength in these experiments was maintained constant at 0.15 mol L^{-1} using NaCl as background electrolyte. The reactions were initiated by the addition of known amounts of powdered human dentin. During the experiments, glass (Orion) and calcium (Orion) electrodes, coupled with a thermostatted Bronsted Ag/AgCl reference electrode with intermediate junction, were used to monitor the hydrogen and calcium ion concentrations. The pH electrode was used to control acidified sodium chloride titrant in order to maintain both the hydrogen ion activity and ionic strength constant during the dissolution reactions. The calcium electrode was used to control the addition of sodium chloride titrant in order to maintain the calcium ion activity constant during the experiments. Periodically, aliquots of suspension were removed from the reaction cell, filtered and analyzed for calcium and phosphate in order to verify constancy of concentration. For the dentin dissolution, the experiments were also made at 0.05 and 0.30 mol L^{-1} (NaCl) ionic strengths, as well as in the presence of 5×10^{-5} mol L^{-1} magnesium ion.

In studies of HPCTI dissolution, titanium coupons coated with the apatitic phase were introduced into the undersaturated solutions and the activities of ionic species were controlled using similar titrant pairs with glass and calcium electrodes. In this way, it was possible to monitor the initial ion release from these surfaces prior to regular dissolution. Following this free drift period, DCC studies were made of the dissolution rates.

In the crystal growth experiments, solutions supersaturated with respect to various calcium phosphate phases were prepared by the careful mixing of calcium chloride, potassium phosphate, potassium hydroxide and sodium chloride solutions and the dual specific ion electrodes were again used to control the addition of titrants in order to maintain ionic activities constant during the experiments.

RESULTS AND DISCUSSION

To be able to interpret the kinetics of the growth and dissolution reactions of the various calcium phosphate phases, it is necessary to calculate the free ionic concentrations in the solutions taking into account ion pair and complex formation. These computations may be made as described previously [15] using expressions for mass balance, electoneutrality and the appropriate conditional equilibrium constants, by successive approximations for the ionic strength. Typical solubility isotherms of the calcium phosphate phases at 37.0°C are shown in Fig.1 as logarithmic plots of the product of calcium and phosphate concentrations as a function of hydrogen ion activity.

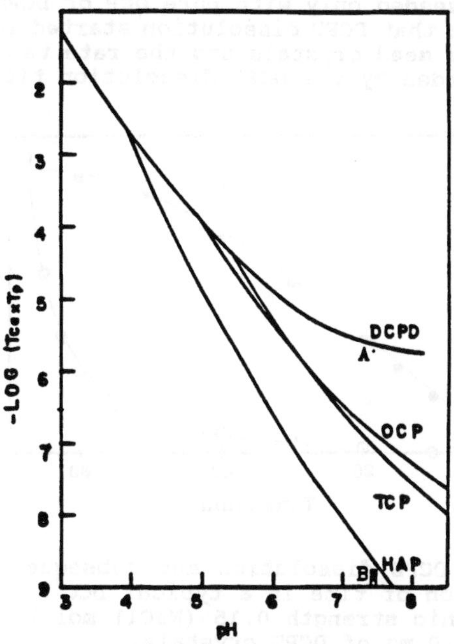

Fig.1: Solubility isotherms for calcium phosphate phases in equimolar total calcium and phosphate solutions in the ternary system $Ca(OH)_2$-H_3PO_4-$NaCl$-H_2O at 0.15 mol L^{-1} ionic strength and 37.0°C.

In order to demonstrate the usefulness of the DCC method for investigating the dissolution and transformation of DCPD to OCP, solutions at point A on the solubility isotherm (Fig.1), undersaturated in DCPD and supersaturated in OCP were seeded with DCPD crystallites. In order to control the DCPD dissolution, a titrant solution having the following composition was used.

$$[CaCl_2]_t = [CaCl_2] - C_{eff} \quad (2)$$
$$[KH_2PO_4]_t = [KH_2PO_4] - C_{eff} \quad (3)$$
$$[KCl]_t = [KCl] + [KOH] + C_{eff} \quad (4)$$
$$[HCl]_t = C_{eff} - [KOH] \quad (5)$$

In these equations the square brackets represent the molar concentrations of the electrolytes in the working solution and the subscript t represents the titrant concentration. C_{eff} is the effective DCPD concentration of the titrants. The addition of this mixed titrant was controlled by a glass electrode. In order to maintain constant ionic activities during OCP growth, two titrant burets were used, one containing calcium chloride and sodium chloride, and the other potassium hydroxide and potassium dihydrogen phosphate, of concentrations given by Eqs.(6-9).

$$[CaCl_2]'_t = 2\ [CaCl_2]' + 4\ C'_{eff} \quad (6)$$
$$[KH_2PO_4]'_t = 2\ [KH_2PO_4]' + 3\ C'_{eff} \quad (7)$$
$$[KCl]'_t = [KCl]' - 8\ C'_{eff} \quad (8)$$
$$[KOH]'_t = 2\ [KOH]' + 5\ C'_{eff} \quad (9)$$

Here, the primed symbols refer to the titrants for OCP growth and C'_{eff} is the effective OCP titrant concentration. The results of a simultaneous DCPD dissolution and OCP growth experiment at pH = 7.40, in solutions having total calcium and phosphate concentrations of 1.80×10^{-3} mol L^{-1} each and ionic strength of 0.15 (NaCl) mol L^{-1}, are illustrated in Fig.2. In this experiment the individual growth rates of OCP and dissolution of DCPD were comparable with their respective single CC experiments seeded only with pure OCP or DCPD crystals [11]. It can be seen in Fig. 2 that DCPD dissolution started immediately after the introduction of the seed crystals and the rate was calculated from the CC profile as recorded by the DCPD dissolution titrant addition.

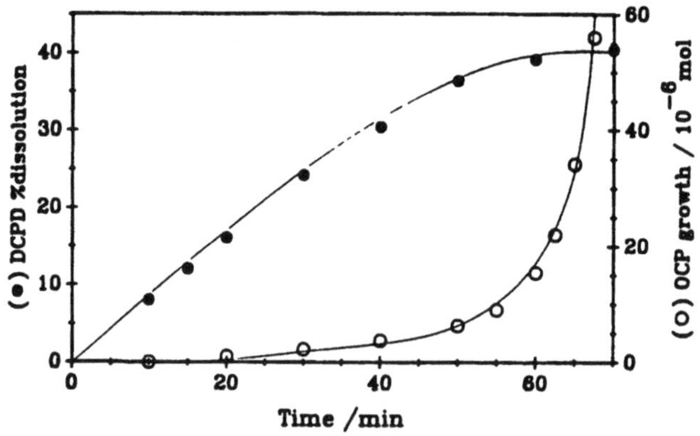

Fig.2: The Concurrent DCPD dissolution and subsequent OCP nucleation and growth as a function of time in a typical DCC experiment at TCa = TPO_4 = 1.80×10^{-3} and ionic strength 0.15 (NaCl) mol L^{-1}, pH = 7.40, and 37.0°C, seeded with 40.0 mg of DCPD crystals.

OCP growth titrant addition commenced after an induction period of about 20 minutes. X-ray powder diffraction and chemical analysis confirmed that the growing phase was OCP and the concentrations of calcium and phosphate could be controlled to within ±2 percent while the pH was maintained constant to within 0.003 units. These results demonstrate the usefulness of the DCC method for investigating simultaneous dissolution and growth reaction rates.

In a typical dissolution experiment of powdered human dentin, solutions of concentration in region B of Fig.1 were prepared and the calcium and hydroxyl ion activities were maintained constant. In the initial experiments, only the phosphate concentration was allowed to vary. Samples of suspension, withdrawn periodically from the reaction medium, were filtered and analyzed for calcium and phosphate in order to determine the actual calcium/phosphate and calcium/hydroxide molar ratios as a function of time as well as extent of dissolution. These ratios were then used to determine the composition of the dissolving phase. The rate of dentin dissolution, determined from the moles of acid added as a function of time, is plotted in Fig.3 for σ_{HAP} values of -0.3 and -0.7. It can be seen that the dentin dissolution rate increased with increasing relative undersaturation. Moreover, the results of the dissolution experiments indicated that one dissolving dentin mineral phase was of OCP stoichiometry. The results of synthetic HAP crystallite dissolution in DCC experiments are included in Fig.3 for comparison. It can be seen that the sensitivity of dentin dissolution rates to changes in undersaturation was considerably less than that for HAP.

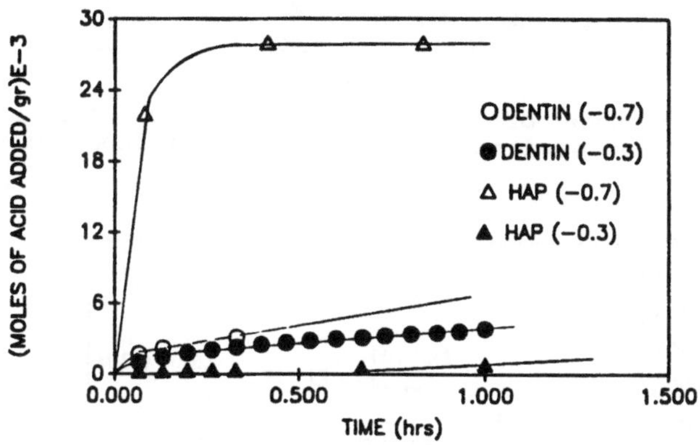

Fig. 3: Dissolution of HAP and powdered human dentin as a function of time at different σ_{HAP}.

The results of experiments in the presence of 5×10^{-5} mol L^{-1} magnesium ion are plotted in Fig.4. The difference between dentin and HAP is quite striking in that the dissolution rates of the latter are markedly inhibited by the presence of magnesium ion while dentin is almost unaffected. Certainly, during the dissolution of dentin the release of macromolecular components may markedly influence the concentration of calcium ion in the solution by complexation. Experiments were also made of the DCC kinetics of dentin dissolution as a function of ionic strength. The experimental data at two ionic strengths 0.05 and 0.3 mol L^{-1}, maintained by the addition of sodium chloride, are presented in Fig.5. It can be seen that the rate of reaction was markedly increased at higher ionic strength probably reflecting an effectively higher solubility of dentin under these conditions.

The DCC method was also used to investigate the dissolution of plasma deposited apatitic phases from titanium implants (HPCTI). When HPCTI disks were introduced into saline solutions, the coatings released significant amounts of calcium ion (Fig.6) as determined by atomic absorption and ICAP. When these disks were equilibrated in the saline solutions the final pH increased to a value of about 10. In these initial stages, however, no phosphate release could be detected in the solutions. Rather, scanning electron micrographs indicated reprecipitation of a calcium phosphate phase during the dissolution experiments and the results suggest a very reactive basic surface, probably containing calcium oxide or oxyapatite, formed by decomposition during the plasma coating.

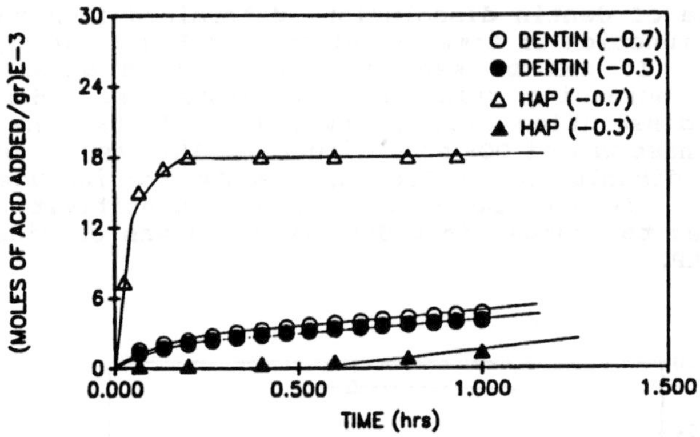

Fig. 4: The influence of magnesium ion (5×10^{-5} mol L^{-1}) on the dissolution of HAP and powdered human dentin as a function of time at different σ_{HAP}.

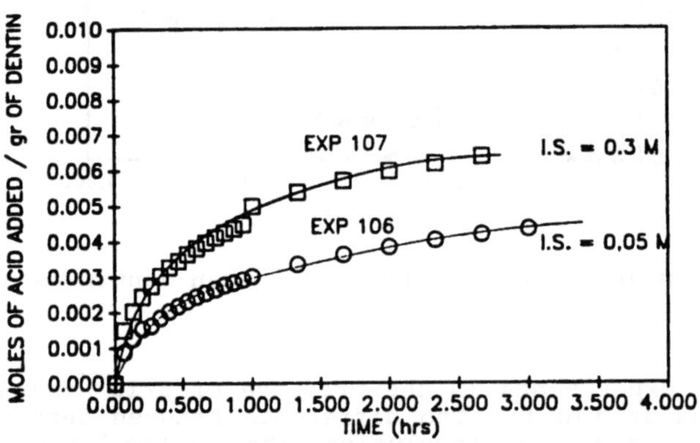

Fig. 5: Dissolution of powdered human dentin as a function of time at σ_{HAP} = -0.50 and different ionic strengths.

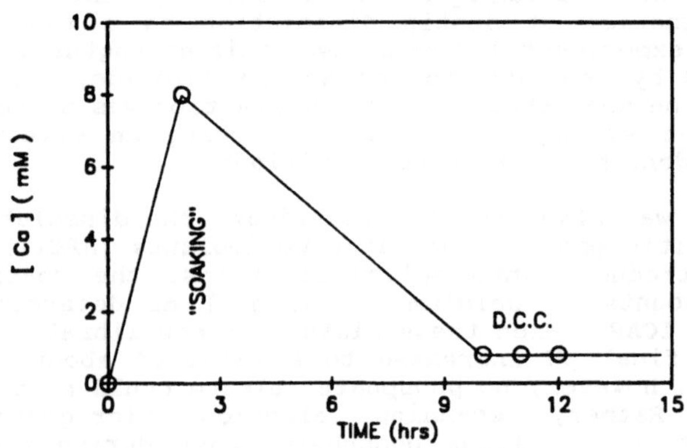

Fig.6: Calcium concentration profile as a function of time for HPCTI at pH = 5.50 in the free drift (soaking) and DCC periods.

The results from the chemical analysis of the free drift experiment and from the subsequent CC dissolution profiles (Fig.7) showed that the calcium/phosphate molar ratios of each HPCTI sample varied as a function of depth. For example, at pH = 5.5, this molar ratio prior to the addition of the HPCTI discs was 1.68. Introducing the discs produced a change in this ratio to 2.02, 1.72 and 1.69 at 20, 60, and 120 minutes, respectively, indicating that the first few layers had a calcium/phosphate molar ratio greater than 2.

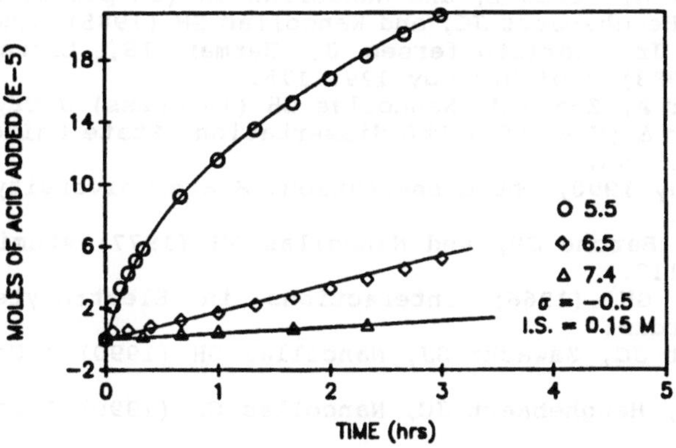

Fig.7: The dissolution of apatite plasma coated HPCTI discs as a function of time at pH 5.5, 6.5, and 7.4.

These surface characteristics may be of great importance in the success or failure of implants _in vivo_ since high surface pH will reciprocate extended osteoclastic activity.
In the DCC mineralization experiments of non-stoichiometric "HAP-like" phases, the growth rates were in general agreement with those determined using the single CC method (of the order 4×10^{-8} mol m^{-2} min^{-1}). However the results from the DCC growth experiments reveal the non-stoichiometry of the calcium deficient apatite crystals grown under physiological conditions. The mineralizing phase had a calcium/phosphate molar ratio of about 1.57, as opposed to a value of 1.67 for stoichiometric HAP. This is in good agreement with previous studies by Zawacki et al [16, 17].

In summary, the CC and now the DCC methods provide powerful tools for investigating both the stoichiometry and kinetics of mineralization and demineralization of calcium phosphate phases. The ability to follow these reactions using mixtures of seed material more closely simulates the reactions that take place with biological apatites.

ACKNOWLEDGMENTS

We thank the National Institute of Dental Research for a grant (DE03223) and the Procter and Gamble Company for grants in support of this work.

REFERENCES

[1] Booustea WD, ten Bosch JJ, Arneds J (1990) J Biol Buccale 17: 43-48.

[2] de Groot K, Geesink R, Kleiu CPAT, Serekiew P (1987) J Biomed Mat Res 21: 1375-1381.
[3] Cook SD (1988) Clin Orth and Rel Res 232: 225-243.
[4] Tomson MB, and Nancollas GH (1978) Science 200 : 1059.
[5] Koutsoukos P, Amjad Z, Tomson MB, and Nancollas GH (1980) J Am Chem Soc 102: 1553.
[6] Perez L, and Nancollas GH (1984) J Crystal Growth 66: 412.
[7] Heughebaert JC, De Rooij JF, and Nancollas GH (1986) J Crystal Growth 77: 192.
[8] Ebrahimpour A, Perez L, and Nancollas GH (in press) Langmuir.
[9] Salimi MH, Heughebaert JC, and Nancollas GH (1985) Langmuir 1: 119.
[10] White DJ Jr, Christoffersen J, Herman TS, Lanzalaco A, and Nancollas GH (1983) J of Urology 129: 175.
[11] Ebrahimpour A, Zhang J, Nancollas GH (in press) J Crystal Growth.
[12] Ebrahimpour A (May 1990) PhD dissertation. State University of New York at Buffalo, USA.
[13] Zhang J (May 1990) PhD dissertation. State University of New York at Buffalo, USA.
[14] Tomson MB, Barone JP, and Nancollas GH (1977) Atomic Absorption Newsletter 16: 117.
[15] Nancollas GH (1966) Interactions in Electrolyte Solutions. Elsevier, Amsterdam.
[16] Heughebaert JC, Zawacki SJ, Nancollas GH (1990) J Coll Interface Sci 135: 20-32.
[17] Zawacki SJ, Heughebaert JC, Nancollas GH (1990) J Coll Interface Sci 135: 33-34.

3 Trace Elements

CHAPTER 3.1

Controlled Deposition and Transformation of Iron Biominerals in Chiton Radula Teeth

J. WEBB, L.A. EVANS, K.-S. KIM, T.G. ST. PIERRE, and D.J. MACEY

Murdoch University, Perth WA 6150, Australia

Key words: magnetite, ferrihydrite, organic matrix, composite materials, *Acanthopleura hirtosa.*

TECHNOLOGICAL PERSPECTIVE

Biosynthesis of solid phases containing iron as a major constituent, that is, iron biominerals, is now known to occur in various unicellular organisms, multicellular tissues and extracellular matrix material, as described in several recent monographs and reviews [1-5]. These phases can be crystalline, amorphous or partly crystalline/partly amorphous. The dimensions of the individual biomineral deposits vary, but are generally less than one micron with the smallest known being the particle within the protein shell of ferritin [6] which is less than 8 nm in diameter.Due to the small scale dimensions of iron biominerals, they are instances of **nanospace materials.** In addition, iron biominerals can be components of **composite materials** where the iron-containing phases are organized with other inorganic components and organic components such as proteins and polysaccharides in the construction of the biological tissue or structure. Both nanospace materials and composite materials are of considerable interest and relevance to the development of new specialist materials and emerging technologies [7] and thus the mineralization of iron continues to attract considerable attention.

Chiton Teeth and Iron Biominerals

In this paper, we present data on the controlled deposition and transformation of iron biominerals in chiton teeth and, in particular, on the role of the organic matrix in these processes. This mineralizing system is a comparatively well-studied instance of a complex biological composite material, containing iron biomineral components that are of nanospace dimensions [8]. Chitons (Phylum Mollusca; Class Polyplacophora] are marine invertebrates whose radula (a tongue-like organ) contains teeth hardened with iron biominerals together with, in many species, calcium biominerals, which are used to scrape their food of encrusting algae and sponges from the intertidal rocks on which they live.

A considerable number of iron biominerals is now known to occur in chiton teeth, following the initial identification by Lowenstam of magnetite (Fe_3O_4) as the black material that forms the prominent capping of the major lateral teeth [9]. The major iron biominerals used as structural components of the teeth are magnetite, lepidocrocite (γ-FeOOH) and, in some species, hydrous iron(III) phosphate. In some species, such as *Acanthopleura hirtosa* considered in detail below, a calcium hydroxyapatite material ($Ca_5(OH,F)(PO_4)_3$ is found in place of this iron phosphate biomineral [2]. Goethite (α-FeOOH) has been identified as a minor deposit [10] while maghemite has been reported as probably an oxidation product of magnetite in the chiton *Liolophura japonica* [11,12]. The ferrihydrite ($5Fe_2O_3.9H_2O$) deposits are almost certainly key components of the process of mineralization since they are the first biomineral observed in the mineralizing teeth. In addition, ferrihydrite is found in the hemolymph (blood) protein ferritin that transports iron from the hemolymph to the superior epithelial cells that surround the teeth at this stage of their development [13].

As can be seen from this variety of mineral forms, biomineralization of iron in chiton teeth is a complex chemical and biochemical process [8]. The *in vitro* synthesis of iron oxide phases can indicate factors that affect the transformations among those phases [14] and that are potentially of importance in the biological mineralization of iron. They will be considered later in this report.

The chiton *Acanthopleura hirtosa*

In our laboratory, we have been engaged for some time in studies designed to elucidate the underlying design principles of the complex microarchitecture of the mineralizing tissue in radula teeth of chitons (and in limpets, which are sympatric with chitons). For this, we have concentrated on the species *Acanthopleura hirtosa* (formerly *Clavarizona hirtosa* [15]), that is common in the intertidal environment near Perth, Western Australia. As in all chitons, the teeth of *A. hirtosa* are arrayed along the radula, shown in the schema of Figure 1, with the oldest teeth being used for feeding. Abrasion during use means that teeth need to be replaced continually, with the radula functioning as a 'conveyer belt' of teeth at various stages of development. Initially, the nonmineralized teeth are composed of a soft organic framework synthesized by odontoblast cells of the radula sac. The teeth stay in contact with superior epithelial cells for much of the time, emerging for use in feeding only at the anterior end of the radula, shown in Figure 1. Through these cells are provided the elements (Fe and Ca) that develop into the biomineral deposits in the teeth. The overall morphology and arrangement of the radula teeth can be seen in the scanning electron micrograph of Figure 2. The smaller teeth are used for sweeping into the mouth the food particles and debris released by the excavating action of the highly mineralized and prominent major lateral teeth.

The biomineral deposits are found in architecturally discrete compartments within the tooth cusp, as can be seen from Figure 3. The clear compartment on the posterior region of the tooth cusp is the region from which the iron has been selectively removed by treatment with dithionite and 2,2'-dipyridyl [16]. The remaining dark region of the tooth cusp has been stained for calcium, with occasional tearing of the section due to its mineralized nature. This compartmentation had been noted previously using chemically untreated material [2, 8, 17].

Characterization of the components present can utilize Mössbauer spectroscopy. This technique is particularly attractive in that it can determine the oxidation states, the phase composition and the magnetic properties of iron present in the sample, including those of poorly crystalline and amorphous materials that are not amenable to diffraction analysis [4]. The room temperature Mössbauer spectrum of whole radulas of *A. hirtosa* is shown in Figure 4. The majority (81%) of the spectral intensity can be assigned to the magnetic sextets of the two sublattices of magnetite, with a central doublet (19% of spectral intensity) due to lepidocrocite and/or ferrihydrite and small particle goethite. A full analysis of the spectra for this biogenic magnetite must take into account the possibility that some maghemite is present, as proposed for *L. japonica* [12]. Low temperature Mössbauer spectra can be used to resolve this complex question, as will be reported elsewhere [18]. Since no sextet component corresponding to the parameters of goethite is observed, the fraction of the total iron present as crystals of goethite larger than about 20nm must be less than about 3%.

The first iron biomineral observed in the teeth is electron dense spherules [10] which are found in close association with the fibrous organic matrix, as seen in Figure 5. The spherules are aggregates of small ferrihydrite particles of diameter 15-50 nm. By the next tooth row, mineralization in the major lateral teeth has changed dramatically from that shown in Figure 5. Thus, on the posterior side of the tooth cusp and towards the top, mineralization in the surface area is complex but apparently poorly ordered. Figure 6 shows the increasing size and density of the mineralized aggregates from the surface to the inner region. Electron diffraction analysis shows that the surface consists of a mixture of magnetite, ferrihydrite and goethite. Immediately below the surface, ferrihydrite is dominant, with the amount of magnetite increasing in this dense, heavily mineralized region at greater depths. These changes occur within micron distances (Figure 6).

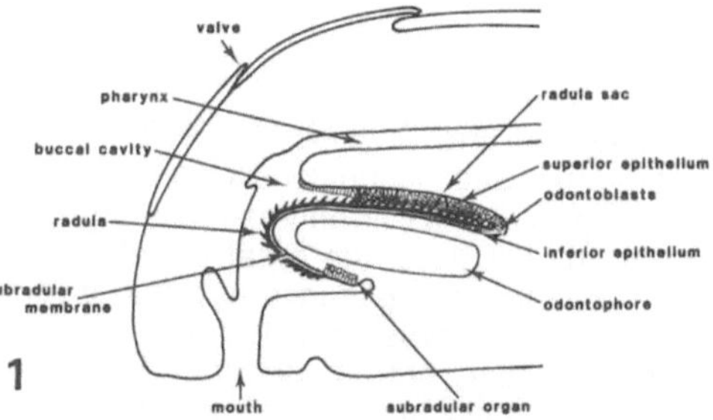

1

2

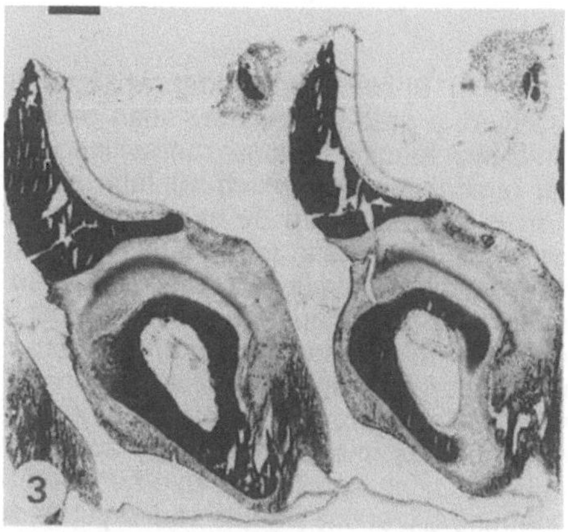

3

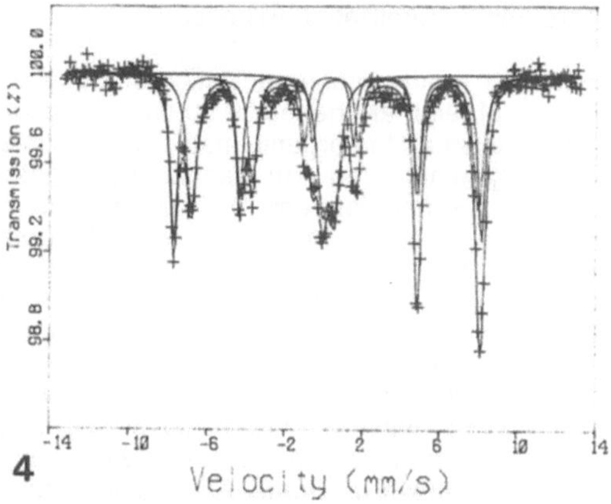

4

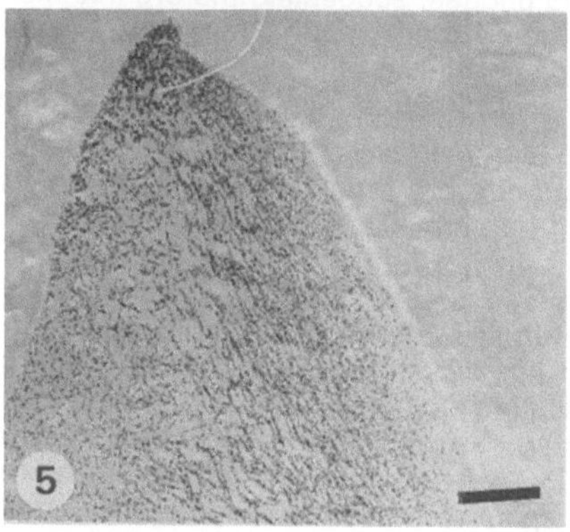

5

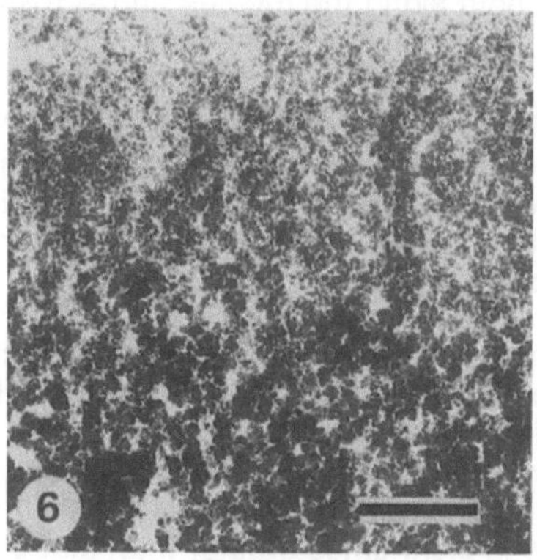

6

Fig 1. Schematic cross-section of anterior end of a chiton. For clarity, radula muscles are omitted.

Fig 2. Scanning electron micrograph of radula of *A. hirtosa*. Major lateral teeth are indicated by arrow. Bar is 500 μm.

Fig 3. Light micrograph stained for calcium of longitudinal section through mature teeth of *A. hirtosa*. Bar is 50 μm.

Fig 4. Mössbauer spectrum of whole radulas of *A. hirtosa* recorded at room temperature.

Fig 5. Unstained section of major lateral tooth cusp of *A. hirtosa* at onset of mineralization. Bar is 1.0 μm.

Fig 6. Unstained section of major lateral tooth cusp of *A. hirtosa* one row later than Fig. 5. Bar is 1.0 μm.

Organic Matrix

The organic matrix in chiton major lateral teeth is largely formed prior to mineralization [19]. The major component of the organic matrix is α-chitin, a polysaccharide consisting of β-(1→4)-linked 2-acetamido-2-deoxy-D-glucose units arranged as an array of anti-parallel hydrogen-bonded chains. Protein makes up approximately 10% by weight of the organic matrix. This protein component is phosphorylated, enriched with aspartic acid, glutamic acid and glycine such that it has marked similarity to proteins occurring in other types of calcium mineralizing systems. Histochemical staining has shown that chitin and protein are found in close association with one another and that protein is located primarily in the anterior region of the cusp (where calcification will occur) with very little protein occurring in the posterior region (where iron mineralization will occur).

Studies of the structural organization of the organic matrix [19] have shown that, while the first few rows of teeth are generally fibrous and poorly formed, a gradual transformation occurs in the tooth cusp as the organic matrix differentiates into two specific regions: that which will be extensively mineralized with magnetite (the posterior region) and that which will later become calcified (the anterior region). The posterior region is characterized by its distinct lack of organic fibres while, in marked contrast, the anterior region contains large amounts of organic material. The matrix also exhibits phenolic hardening processes similar to those observed in tanning of insect cuticle and which may be promoted by zinc-dependent enzymes such as carbonic anhydrase. The changes observed in the organization and distribution of fibres in the early stages of tooth development presumably occur in preparation for the onset of mineralization. Under the electron microscope, the organic matrix in the anterior region can be seen to be highly organized, consisting of parallel tubules interconnected by an organic network of finer fibrils often occurring with fibrous bridges lying at a fixed angle between the tubules, as shown in Figure 7. Following partial demineralization, needle-like crystals are seen aligned both along the tubules and across fibrous bridges, suggesting the organic matrix is intimately involved in mineral deposition. The organic matrix in the magnetite-bearing region, however, is much less organized with the matrix consisting of finer, more sparse fibres (Figure 8). These two regions can be seen also in the micrograph of Figure 9. A fractured fully mineralized tooth reveals a discrete layer of magnetite, approximately 15 µm thick with a large number of densely-packed fibres behind it. These strands are presumably chitin, and they originate in the lower section of the cusp before rising and curving back towards the anterior surface (Figure 10).

The major iron binding protein in the hemolymph is ferritin, with an average iron content of 2000 Fe/molecule [13]. The iron core of this ferritin is poorly crystalline ferrihydrite, comparatively hydrous and low in phosphate. It exhibits a complex temperature-dependent behaviour when studied by Mössbauer spectroscopy, as has been discussed in detail elsewhere [20]. The low crystallinity is consistent with its high reactivity, whereby the initial rate of iron uptake is eight times greater than that of horse spleen ferritin, whose ferrihydrite core is significantly more crystalline [13]. This reactivity, in turn, is related to the function of this ferritin which is to deliver iron to the rapidly mineralizing tissue. Details of the cellular environment around the mineralizing tooth are shown in Figures 11 and 12. The superior epithelial cells accept, by endocytosis, ferritin from the hemolymph that bathes these cells. The ferritin molecules then aggregate into the clusters shown as electron dense circular structures in Figure 12. These, however, do not appear to transfer iron, as ferritin cores, across the membrane into the tooth. Instead, it seems that this ferritn core ferrihydrite is dissolved and transfered, as iron(II), into the tooth where mineralization into ferrihydrite and other more crystalline phases can occur.

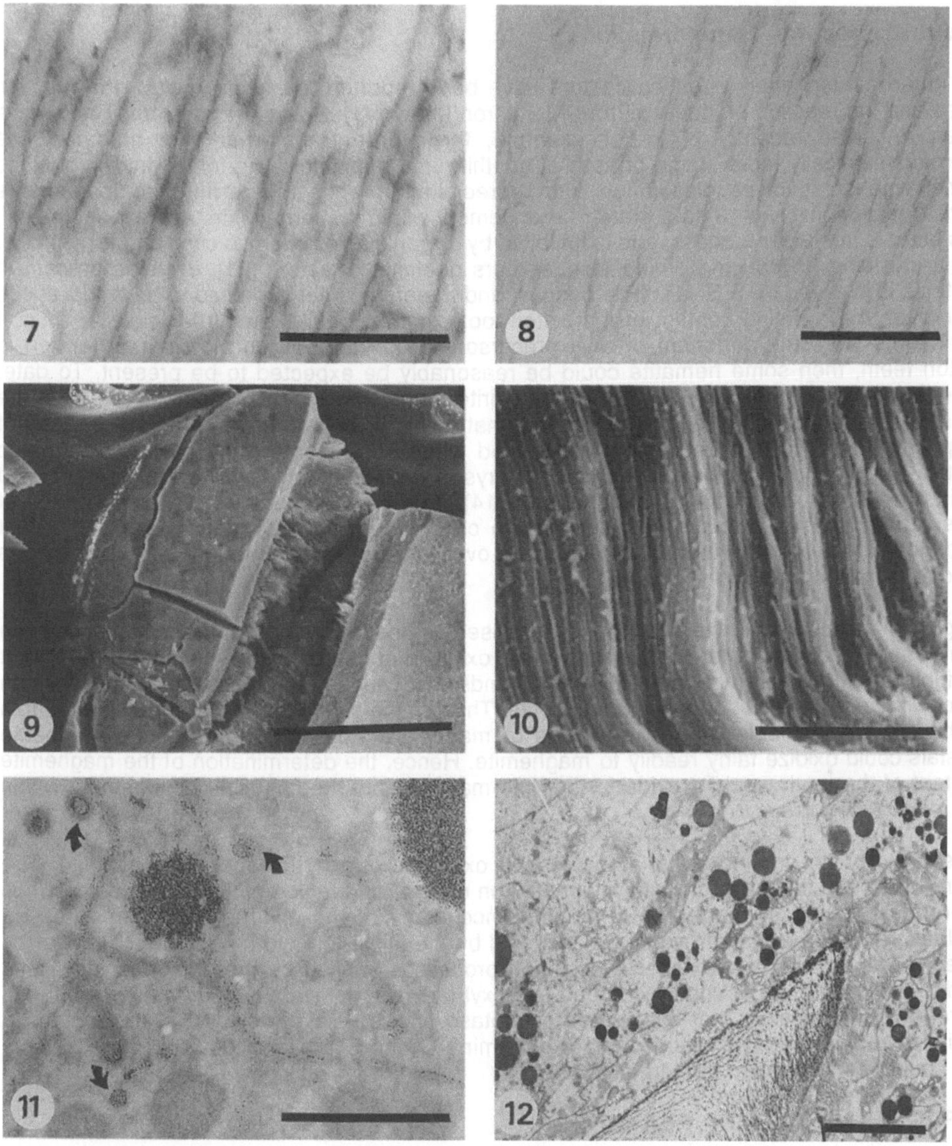

Fig 7. Stained longitudinal section of anterior region of acid-treated mature tooth cusp of *A. hirtosa.* Bar is 0.5 µm.

Fig 8. Stained section of posterior region of acid-treated mature tooth cusp of *A. hirtosa.* Bar is 0.5µm.

Fig 9. Posterior surface of a fractured fully mature tooth of *A. hirtosa.* Bar is 100 µm.

Fig 10. Fibrous structure of anterior region of alkali-cleaned acid-etched tooth cusp of *A. hirtosa.* Bar is 25 µm.

Fig 11. Section of superior epithelial cells of *A. hirtosa* showing hemolymph ferritin capture by endocytosis (arrows). Bar is 1.0 µm.

Fig 12. Stained transverse section of tooth cusp and superior epithelial cells of *A. hirtosa* at onset of mineralization. Bar is 5 µm.

Factors Affecting Transformations

As noted earlier, many chemical factors have been reported to affect, *in vitro*, the reaction rates and the reaction products of iron(II) and iron(III) hydrolysis and precipitation. These have been reviewed recently [14]. For example, ferrihydrite transforms over time into the thermodynamically more stable phases of goethite and hematite. The ferrihydrite to goethite transformation involves dissolution and reprecipitation while the ferrihydrite to hematite transformation involves a solid state rearrangement within the ferrihydrite. At near neutral pH, this latter is the faster process, and is inhibited by organic acids such as citric and lactic acids. While cysteine, a reducing amino acid, favours goethite formation [21], cysteine present in proteins is involved in -S-S- disulfide bridges, and therefore is not expected to have this effect. Moreover, the cysteine content of the chiton tooth matrix is quite low at 0.2 mole%. Thus, if ferrihydrite were the dominant or only precursor of the various iron biominerals formed in chiton teeth, then some hematite could be reasonably be expected to be present. To date, none has ever been found. Generally, ferrihydrite forms where iron is rapidly oxidised in the presence of high concentrations of organic matter, as in certain soils [14]. Hence it is not surprising that it is the first biomineral formed when iron(II) enters the preformed organic matrix. With conditions that do not favour crystal development, ferrihydrite, goethite and lepidocrocite can be formed simultaneously [14]. This may account for the co-existence of these three phases in close proximity in the chiton tooth [10]. Carbonate ion and a fast oxidation rate favour the formation of goethite over lepidocrocite [22] offering another means for differentiation among the phases.

Magnetite (and maghemite) can be synthesised under low temperature conditions when iron(II) solutions of moderate concentration are oxidized at neutral to alkaline conditions [14]. It is significant, however, that under the same conditions, various preformed iron(III) phases can react with iron(II) in solution to yield magnetite. Thus, in the chiton tooth, it is possible that both pathways are available for the synthesis of magnetite. Any small fine-grained magnetite crystals could oxidize fairly readily to maghemite. Hence, the determination of the maghemite content of the tooth would provide further information about the chemical environment within the mineralizing tissue.

When little or no organic material is present, oxidation of iron(II) can yield magnetite. The sparse nature of the organic fibres within the iron mineralizing region of the chiton tooth is thus appropriate to favour magnetite formation since many of the chitin-protein fibre sites for nucleation and precipitation would be occupied by the initial ferrihydrite deposits . It must be noted, however, that the extensive oxidation processes implied by these mechanisms may lead to the generation of superoxide and hydroxyl radicals, particularly in an environment rich in iron(II). It is possible that superoxide dismutase and/or catalase enzymes are present to minimize radical-induced tissue damage. Vitamins C and E are also possible anti-oxidants that could offer protection from free radicals.

STRUCTURAL DESIGN FEATURES

The appropriate design of the chiton radula for feeding has been noted [23] and indeed used as the inspiration for designs of novel dredging equipment [24]. The chiton radula comprises 17 teeth per row, compared to more than 500 in some other mollusc groups. Only 4 of the 17 appear to be involved in feeding and of these, only 2, the major lateral teeth, are used to physically excavate the substratum. The cusp is hardened by magnetite, which has a hardness of 6-6.5 on the Moh scale, to enable feeding on crustose coralline algae. A self-sharpening effect in the tooth as the cusp becomes worn has also been proposed [25]. Tough but flexible support is provided by the calcified region built up predominantly from organized oriented rope-like bundles of chitin fibres [19]. The chitin fibres shown in Figure 10 can be considered as 'shock absorbers' that support the tooth during the impact and stress of feeding.

Calcification occurs in the region where most of the protein is located, suggesting that the protein serves a critical role in nucleation and crystal growth of the calcium biomineral deposits. In the iron mineralized region of the cusp, however, these roles are less evident, and little organization can be observed in the arrangement of the magnetite deposits.

Thus the complex microarchitecture of the tooth is reflected in its overall function. A matrix-poor biomineral-rich region serves as the cutting edge that is complemented by a matrix-rich calcified region that gives it support. Together, they result in the bioconstruction of an effective organ for the excavation of rocky substrates .

ACKNOWLEDGEMENTS

This research has been supported by the Australian Research Council, Murdoch University's Special Research Grant and by the award of Postgraduate Scholarships to KSK and LAE. The assistance and advice of numerous colleagues in Australia and elsewhere over the years are again gratefully acknowledged.

REFERENCES

[1] Mann S, Webb J, Williams RJP (eds) (1989) Biomineralization: Chemical and Biochemical Perspectives. VCH Verlagsgesellshaft, Weinheim.
[2] Lowenstam HA, Weiner S (1989) On Biomineralization. Oxford University Press, Oxford.
[3] Simkiss K, Wilbur KM (1989) Biomineralization. Academic Press, London.
[4] Webb J, St. Pierre TG (1989) In: Long GJ (ed) Mössbauer Spectroscopy applied to inorganic chemistry, Vol 2. Plenum Press, New York, pp 417-444.
[5] Webb J, St. Pierre TG, Macey DJ (1990) In: Frankel RB (ed) Iron biominerals. Plenum Press, New York, in press.
[6] St. Pierre TG, Webb J, Mann S (1989) In: Mann S, Webb J, Williams RJP (eds) Biomineralization: Chemical and Biochemical Perspectives. VCH Verlagsgesellschaft, Weinheim, pp 295-344.
[7] Birchall, JD (1989) In: Mann S, Webb J, Williams RJP (eds) Biomineralization: Chemical and Biochemical Perspectives. VCH Verlagsgessellschaft, Weinheim, pp 491-509.
[8] Webb J, Macey DJ, Mann S (1989) In: Mann S, Webb J, Williams RJP (eds) Biomineralization: Chemical and Biochemical Perspectives. VCH Verlagsgessellschaft, Weinheim, pp 345-388.
[9] Lowenstam HA (1962) Bull. Geol. Soc. Am. 73:435-438.
[10] Kim KS, Macey DJ, Webb J, Mann S (1989) Proc. Roy. Soc. Lond. B 237:335-346.
[11] Mizota M, Maeda Y (1985) J. Phys. Soc. Japan 54: 4103-4106.
[12] Mizota M, Maeda Y (1986) Hyp. Interact. 29:1423-1426.
[13] Kim KS, Webb J, Macey DJ (1986) Biochim. Biophys. Acta 884:387-394.
[14] Schwertmann U, Taylor RM (1989) In: Minerals in Soil Environments, 2nd Ed. Soil Science Society of America, Madison USA pp 379-438.
[15] Ferreira AJ (1986) Veliger 28:221-279
[16] Evans LA, Macey DJ, Webb J, submitted for publication.
[17] Suga S (1984) In: Fearnhead RW, Suga S (eds) Tooth enamel IV. Elsevier, Amsterdam pp 472-477
[18] St. Pierre TG, Evans LA, Webb J, unpublished results.
[19] Evans LA, Macey DJ, Webb J (1990) Phil. Trans. Roy. Soc. Lond. B 329:87-96
[20] St. Pierre TG, Kim KS, Webb J, Mann S, Dickson DPE (1990) Inorg. Chem. 29:1870-1874.
[21] Cornell RM, Schneider W (1989) Polyhedron 8:149-155.

[22] Carlson L, Schwertmann U (1990) Clay Min. 25:65-71.
[23] Steneck RS, Watling L (1982) Mar. Biol. 68:299-319.
[24] Van der Wal P, Videler JJ, Havinga P, Pel R (1989) In: Crick RE (ed) Origin, evolution and modern aspects of biomineralization in plants and animals. Plenum Press, New York pp 153-166.
[25] Bullock, RC (1989) Amer. Malac. Bull. 7:13-19.

CHAPTER 3.2

Mineralization of Iron in Ferritin and Haemosiderin: The Effect of Tissue Environment on Mineral Structure

T.G. St. Pierre[1], K.C. Tran[1], J. Webb[1], D.J. Macey[1], and P. Pootrakul[2]

[1]Murdoch University, Perth WA 6150, Australia and [2]Siriraj Hospital, Mahidol University, Bangkok, Thailand

Key words: ferritin, haemosiderin, thalassaemia, Mössbauer spectroscopy, electron diffraction.

INTRODUCTION

Ferritin and Haemosiderin

Iron is the most abundant trace element in man. It plays an essential role in biochemical processes such as oxygen transport and storage, and electron transfer reactions. However, too much iron in the body can be toxic. Excess iron may precipitate in cells and catalyse adverse reactions such as peroxidation of the lipid membranes of lysosomes [1]. To help maintain the balance between iron-deficiency and iron-overload, nature provides the protein ferritin which is capable of holding between 0 and about 4500 iron atoms in a non-toxic, soluble form. The iron in ferritin can be remobilized when required. Thus ferritin can act as a buffer by temporarily storing iron until it is needed. Ferritin consists of an approximately spherical shell of protein with internal and external diameters of about 8 and 12nm respectively. Channels connect the cavity to the exterior thus allowing ions to pass into and out from the inner space of the protein [2]. Iron is stored in this vesicle in the form of a mineral phase which can be described as a metastable hydrous iron(III) oxide with a variable phosphorus and water content. It has been noted previously that this mineral is an appropriate compound to balance the important but opposing physiological functions of iron storage and iron mobility [3].

Other forms of microcrystalline iron deposits can be found in both normal and iron-overloaded tissues. These deposits are similar in structure to the mineral cores of ferritin with varying degrees of hydration and phosphorus content. Although these deposits are associated with protein material, unlike ferritin the material does not appear to be composed of well-defined protein structures and does not render the iron water soluble. This form of iron is generally known as haemosiderin. While in normal subjects only a small fraction of the total non-haem iron is in this form, in iron-overloaded subjects haemosiderin is the predominant form of non-haem iron [4].

Structure-Function Relationship of the Mineral Core

The inter-relationships among the biological function, reactivity, biological environment, structure, and magnetism of the mineral cores of ferritin and haemosiderin are schematically represented in Figure 1. The scheme is based on the results of many workers and illustrates the variable features of the cores of ferritin and haemosiderin. Samples of ferritin and haemosiderin isolated from various sources, such as mammalian tissue, molluscan haemolymph, and bacteria [5-8], have provided examples of the various types of core structure. The size, degree of crystallinity, and structure of the cores have been probed with techniques such as electron diffraction, transmission electron microscopy (TEM), high resolution transmission electron microscopy, and Mössbauer spectroscopy. Mössbauer spectroscopy has also been used to examine the magnetic properties of the cores. Correlations between the structure and composition of the cores and their magnetic properties allow these physical techniques to give a detailed description of the microcrystalline mineral deposits [9].

Core structures are most often described in terms of variations on the structure of the mineral ferrihydrite ($5Fe_2O_3.9H_2O$). These variations appear to correspond to the different biological functions of ferritin. For example, the well-crystalline ferrihydrite structure of horse spleen ferritin accommodates its function as a storage form of iron whereas the poorly crystalline hydrated ferrihydrite-like structure of ferritin cores from the haemolymph of the marine molluscs *Patella vulgata, Patella laticostata,* and *Acanthopleura hirtosa* (all three of which constantly secrete mineralized teeth with a high iron content) reflect the role of the ferritin as a high capacity iron transport protein. In the latter case the iron has to be deposited and remobilized far more rapidly than in the former. The poor crystallinity of the molluscan ferritins reflects the rapid rate of deposition of the mineral [10] and results in iron that is less strongly bound than in the horse spleen ferritin thus allowing easier remobilization at the site of tooth mineralization.

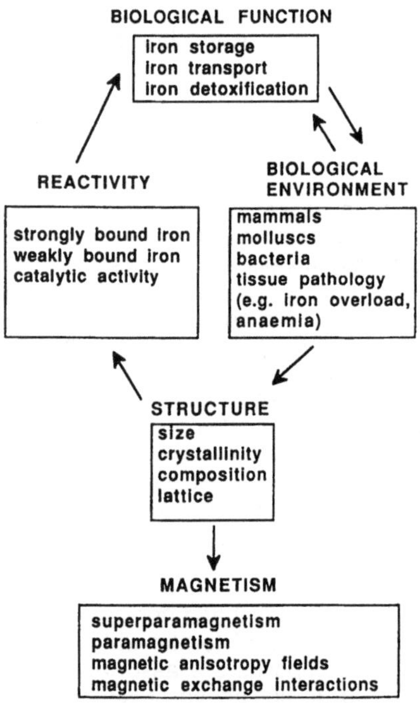

Fig 1. Inter-related features of the mineral cores of ferritin and haemosiderin.

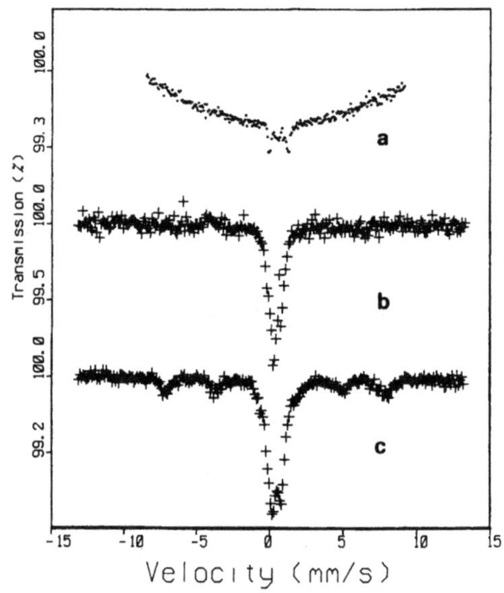

Fig 2. Mössbauer spectra of spleen samples at 78K from (a) normal subject (b) thal. patient and (c) thal. patient who had been treated with regular blood transfusions and iron chelation therapy. (Spectrum (a) reproduced from ref 11.)

THE EFFECT OF TISSUE PATHOLOGY ON IRON DEPOSITS

Just as the different biological environments of different organisms result in different ferritin core structures, so too can the altered environments of organs in various pathological conditions result in several types of iron mineral deposits [8]. Figure 2 is a dramatic example of how the structure and magnetism of iron deposits in the human spleen, as expressed in ^{57}Fe Mössbauer spectra, vary according to the pathology. Mössbauer spectroscopy is well-suited to this type of study since it is sensitive only to the iron in a sample. The figure shows spectra of samples of spleen from (a) a normal subject, (b) a β-thalassaemic patient, and (c) a β-thalassaemic patient who had been treated with regular red blood cell transfusions and iron chelation therapy.

β-thalassaemia is a genetic disease in which there is a reduction in β-globin chain synthesis. This disorder results in chronic haemolytic anaemia. In an attempt to counteract the anaemia, erythropoietic activity increases by up to 10 times normal or more [12] which leads to increased gastrointestinal absorption of iron. The excess iron is deposited in the form of ferritin and haemosiderin in various organs in the body and a condition of iron-overload results.

A comparison of the spectra of normal and thalassaemic spleens in Figure 2 (a) and (b) indeed shows the increased iron content of the thalassaemic spleen. In the normal case there is a significant signal due to haemoglobin while in the thalassaemic case this signal is swamped by that from non-haem iron(III). In fact most of the iron in the thalassaemic spleen is in the form of haemosiderin.

The anaemia of thalassaemia can be relieved by administration of regular transfusions of healthy red blood cells. However, this procedure introduces more iron into the body further contributing to iron-overload. Iron-chelation treatment is usually administered in an attempt to reduce the excess of iron. Both of these treatments will tend to alter the cellular environment in which the iron deposits are found. Figure 2 (c) shows the Mössbauer spectrum of a sample of spleen taken from a β-thalassaemic patient who had received these treatments. The appearance of spectral peaks at the more extreme velocities clearly shows the presence of an iron deposit with a different structure from that in the untreated case.

Thus it can be seen that the use of physical techniques to study of the structure and magnetism of the iron-deposits can give important information regarding the pathological condition and history of the tissues.

FERRITIN CORES FROM β-THALASSAEMIC / HAEMOGLOBIN E ORGANS

As mentioned previously, ferritins are more precisely defined than haemosiderins making ferritin an easier form of iron deposit for comparative studies. We have studied the variations in core structure of ferritins from various organs of β-thalassaemic/ haemoglobin E (β-thal/Hb E) patients, who had received minimal or no blood transfusions, and normal subjects [13]. In β-thal/Hb E disease there is a structural mutation in the β-globin gene as well as a reduction in β-globin chain synthesis.

Mössbauer spectroscopy of the ferritins isolated from spleen, liver, heart and pancreas tissues of β-thal/Hb E subjects and the spleen and liver from a normal subject showed small but significant variations in their mean superparamagnetic blocking temperatures ($<T_B>$). Figure 3 shows a set of spectra of the ferritin from a β-thal/Hb E heart over a range of temperatures. These spectra are typical, the qualitative features being similar to those from the other samples. $<T_B>$ is defined as the temperature at which the sextet and central doublet spectral components are of equal intensity (i.e. equal spectral area). Table 1 shows the value of $<T_B>$ for each sample. The disappearance of the sextet in the wings of the spectra as the temperature is raised indicates a lower limit for the magnetic ordering temperatures of the ferritins of about 55-60K.

Table 1 also shows the number of rings obtained in electron diffraction patterns of the ferritins. The positions of the rings corresponded to those of the mineral ferrihydrite [13,2,14,15]. Ferrihydrite is known to exist with varying degrees of crystallinity giving between 2 (poorly crystalline) and 6 (well crystalline) diffraction lines [14].

Core size distributions of the ferritins were measured by TEM [13]. The distributions were found to approximate to a normal curve (Figure 4). The mean core size and standard error on the mean for each ferritin are given in Table 1.

The data from the electron diffraction and TEM are in agreement with the Mössbauer data in that the ferritins with smaller, less crystalline cores have lower values of $<T_B>$ than the ferritins with larger, more crystalline cores. Details of the correlations between core size and crystallinity and superparamagnetic behaviour have been discussed elsewhere [9].

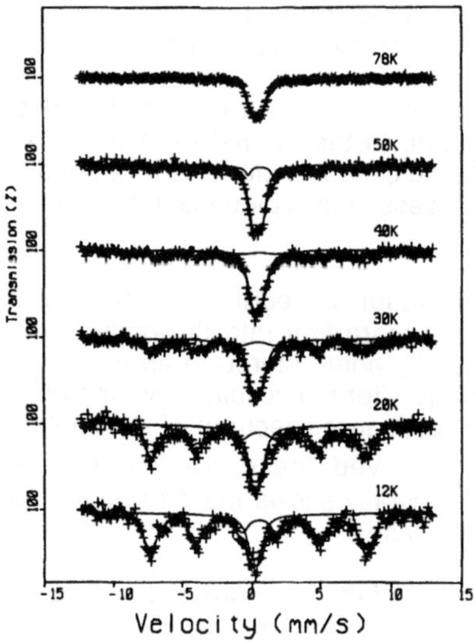

Fig 3. Mössbauer spectra of β-thal/Hb E heart ferritin between 12 and 78K.

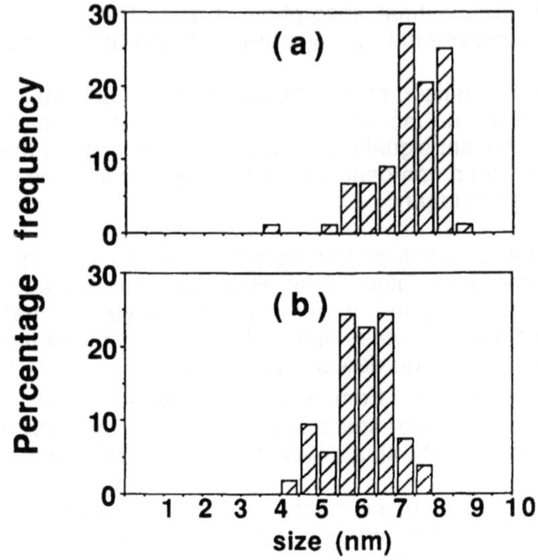

Fig 4. Core size distributions of ferritins from β-thal/Hb E (a) heart and (b) spleen.

Table 1.
Mössbauer spectroscopic, electron diffraction, and transmission electron microscopy data for ferritins isolated from β-thalassaemic/ haemoglobin E and normal organs.

Source organ	$\langle T_B \rangle$ (K)	Mean core size (nm)	Standard error on mean (nm)	№ of diffraction lines
β-thal/Hb E spleen	29	6.19	0.10	3
β-thal/Hb E liver	29	5.80	0.13	3
β-thal/Hb E heart	35	7.37	0.09	5,6
β-thal/Hb E pancreas	32	7.18	0.08	6
normal spleen	31	6.25	0.11	5,6
normal liver	31	7.01	0.13	5

IRON KINETICS IN β-THALASSAEMIA / HAEMOGLOBIN E

The data on ferritin core size and crystallinity almost certainly reflect the kinetics of iron metabolism in the body and the differences between the dynamic equilibrium of iron in normal and in β-thal/Hb E subjects. Iron from old red blood cells is reprocessed by the reticuloendothelial cells of the spleen and liver. The lower degree of crystallinity of the ferritin cores from β-thal/Hb E spleen and liver compared to those from normal organs indicates a more rapid deposition of iron in the diseased state. This is consistent with the shorter mean lifetime of β-thal/Hb E red blood cells and the consequent approximate 10 fold increase in erythroid iron turnover rates [12].

Heart and pancreas iron loadings are increased by about 2.7 and 47 times respectively in β-thal/Hb E subjects [16]. Unlike the spleen and liver, the heart and pancreas are not directly involved in the reprocessing of red blood cell iron. These organs are terminal long term deposition sites for excess iron [17] implying that iron is slowly deposited over a long period without being remobilized. This is consistent with the larger core sizes and higher degree of crystallinity of the ferritins from the heart and pancreas samples compared to those from the liver and spleen samples of β-thal/Hb E subjects.

CONCLUSION

Studies of many and varied ferritins and haemosiderins have built up a detailed picture of structure-function relationships that are summarized in Figure 1. Biological function of these proteins and their environment are intrinsically linked with the environment determining the type of core structure. This in turn determines the chemical properties of the core which are a crucial feature in the way it functions. The ferrihydrite basis for the structures of the various types of cores and the structural variability that this allows seem ideally suited to the balancing of inter-related aspects of this system. Perturbation of this system by pathogenesis in the biological environment results in a feedback via changes in the core structure that can alter the functioning of ferritin and haemosiderin in the cell. In practical terms, a full understanding of the mechanisms involved in the inorganic pathology of iron-overload will be crucial to the evaluation of iron-chelation therapies and blood transfusion regimens.

REFERENCES

[1] Peters TJ, Selden C, Seymour CA (1977) Ciba Symposium NS51: Iron Metabolism. pp317-325.
[2] Ford GC, Harrison PM, Rice DW, Smith JMA, Treffry A, White JL, Yariv J (1984) Phil. Trans. R. Soc. Lond. B 304: 551-565.
[3] Towe KM (1990) In: Crick RE (ed) Origin, Evolution, and Modern Aspects of Biomineralization in Plants and Animals. Plenum Press, New York, pp265-272.
[4] Weir MP, Gibson JF, Peters TJ (1984) Cell Biochemistry and Function 2:186-194.
[5] Mann S, Bannister JV, Williams RJP (1986) J. Mol. Biol. 188: 225-232.
[6] St.Pierre TG, Bell SH, Dickson DPE, Mann S, Webb J, Moore GR, Williams RJP (1986) Biochim. Biophys. Acta 870: 127-134.
[7] St.Pierre TG, Kim KS, Webb J, Mann S, Dickson DPE (1990) Inorg. Chem. 29: 1870-1874.
[8] Dickson DPE, Reid NMK, Mann S, Wade VJ, Ward, RJ, Peters TJ (1988) Biochim. Biophys. Acta 957: 81-90.
[9] St.Pierre TG, Webb J, Mann S (1989) In: Mann S, Webb J, Williams RJP (eds) Biomineralization: chemical and biochemical perspectives. VCH Verlagsgesellschaft, Weinheim, pp295-344.
[10] Kim KS, Webb J, Macey DJ (1986) Biochim. Biophys. Acta 884: 387-394.
[11] Bell SH, Weir MP, Dickson DPE, Gibson JF, Sharp GA, Peters TJ (1984) Biochim. Biophys. Acta 787: 227-236.
[12] Pootrakul P, Kitcharoen K, Yansukon P, Wasi P, Fucharoen S, Charoenlarp P, Brittenham G, Pippard MJ, Finch CA (1988) Blood 71: 1124-1129.
[13] St.Pierre TG, Tran KC, Webb J, Macey DJ, Heywood BR, Sparks NH, Wade VJ, Mann S, Pootrakul P (1990) FEBS lett., submitted for publication.
[14] Murad E, Johnston JH (1987) In: Long GJ (ed) Mössbauer spectroscopy applied to inorganic chemistry, vol 2. Plenum Press, New York, pp507-582.
[15] Towe KM, Bradley WP (1967) J. Colloid Interface Sci. 24: 384-392.
[16] Shuler TR, Pootrakul P, Yarnsukon P, Nielsen FH (1990) J. Trace Elements in Experimental Med. 3: 31-43.
[17] Jacobs A, Worwood M (1981) In: Bronner F, Coburn J (eds) Disorders of Mineral Metabolism, vol 1. Academic Press, New York, pp1-56.

CHAPTER 3.3

Studies on the Transport Mechanism of Iron in Rat Incisor Enamel

S. MATAKI[1], K. OHYA, M. INO, and H. OGURA

Department of Pharmacology, School of Dentistry, Nagasaki University[1], Department of Pharmacology, Faculty of Dentistry, Tokyo Medical and Dental University, Japan

Key words: Enamel, Iron, Ferritin, Transferrin receptor, Hemosiderin

INTRODUCTION

The enamel of the continuous growing incisors of adult rats is characterized by the presence of a orange-yellow pigmentation, which is caused by the content of large amounts of iron. Histochemical studies and electron microprobe analysis have shown that iron-containing pigment is also identified within the rat incisor ameloblasts during the enamel maturation stage [1-4]. Autoradiographic studies using ^{55}Fe have demonstrated that the iron is incorporated from papillary layer to the ameloblasts and accumulated during the progress of maturation stage, and finally secreted onto the completely mineralized enamel surface at the end of maturation[5-7].
It is thought that iron is transported as the form of transferrin in the blood plasma and stored as the form of either ferritin or hemosiderin within the cells.
In the present study we investigated the iron transport process in rat incisor enamel organ by means of immunohistochemical and biochemical studies with respect to ferritin and transferrin receptor.

MATERIALS AND METHODS

Male Wistar rats weighing approximately 180 g were used in this study. In order to avoid the contamination of blood all rats were perfused with 0.05 M phosphate buffered saline containing 0.15 M NaCl (PBS, pH 7.2) for 3 minutes before dissection. Incisor enamel organ was obtained by the methods of Josephsen [8]. Isolated enamel organs were homogenized in PBS at pH 7.2. The homogenate was heated at 90°C for 15 minutes and left until it cooled. After the centrifugation, the supernatant was analyzed by gel filtration chromatography and polyacrylamide gel electrophoresis (PAGE). Gel filtration chromatography was performed on Superose 6 (1 x 30 cm, Pharmacia) eluted with PBS at flow rate of 0.5 ml/min.. Samples were electrophoresed under either native condition or dissociative condition (SDS-PAGE). Purified rat liver ferritin (Sigma) was also analyzed to characterize ferritin isolated from rat incisor enamel organ. Iron content of each fraction of the gel filtration chromatography was determined by the bathophenanthroline method.
For immunological studies, polyclonal antibodies anti-rat liver ferritin were raised in rabbits. Immunological specificity of the antibody was confirmed by ELISA and Western blot analysis. This polyclonal antibody anti-rat ferritin and monoclonal antibody anti-rat transferrin receptor(purchased from Serotec) were also used for immunohistochemistry in the cryosection of rat incisor enamel organ.

RESULTS

Purified rat liver ferritin was eluted at 30 minutes on the gel filtration chromatography(Fig.1A) Similarly, in the elution profile of rat enamel organ extract, a major protein peak corresponding to the rat liver ferritin appeared at the exactly same position (Fig.1B). The most part of iron in the enamel organ extract was detected in this major protein peak(Fig.1B).

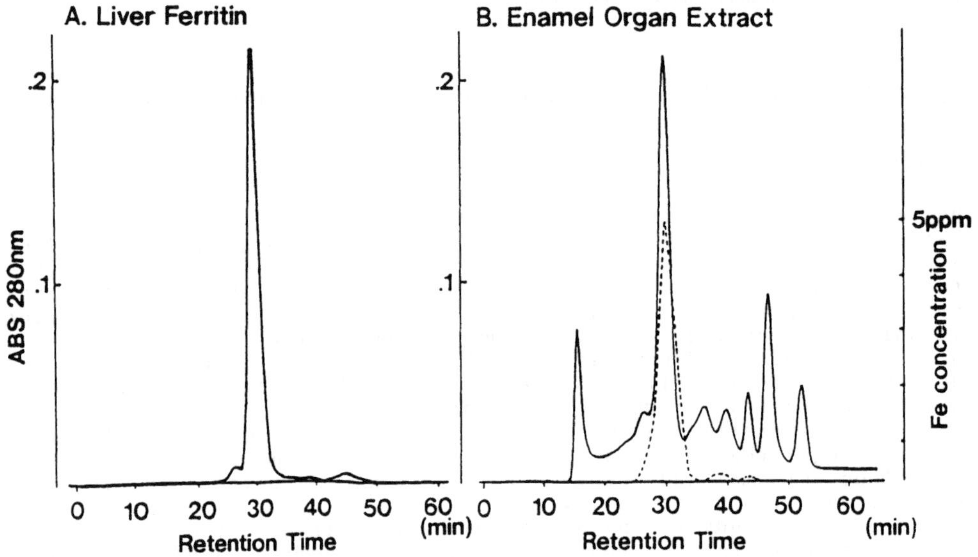

Fig.1 Elution profiles of gel filtration chromatography of rat liver ferritin (A) and rat enamel organ extract (B) gel filtration chromatography on Superose 6. Dotted line in (B) indicates iron concentration.

Ferritin isolated from rat enamel organ has the same mobility as rat liver ferritin in an electrophoresis on 5% gel in native condition(Fig.2A). Analysis of ferritin by SDS gel electrophoresis(5-20% linear gradient gel) revealed subunit structure. Liver ferritin migrated in two bands, heavy (H) subunit and light (L) subunit (Fig 2B lane 1). On the other hand, ferritin in the enamel organ extract has a different subunit ratio, indicating more H subunits than L subunits, as compared with liver ferritin(Fig.2B lane 2). Western blot analysis showed that the antibodies against rat liver ferritin also reacted with ferritin from incisor enamel organ(Fig. 2A). Further the antibodies recognize subunits of ferritin, H subunit and L subunit, respectively(Fig.2B).
Immunohistochemical observations showed that ferritin was detected in the papillary layer and within the cytoplasm of the ameloblasts at an early maturation stage(Fig.3). On the other hand, transferrin receptors were detected mainly in the papillary layer cells (Fig.4). And the immunoreactivity in the papillary layer cells reduced gradually in the incisal direction(Fig.5).

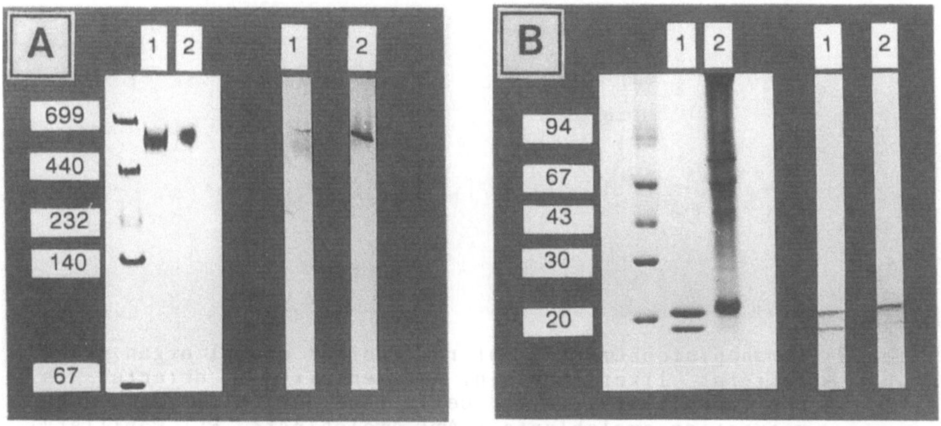

Fig.2 Polyacrylamide gel electrophoresis and the corresponding Western blot of rat liver ferritin(lane 1) and rat incisor enamel organ extract(lane 2). A: 5% native gel B: SDS-PAGE, 5-20% linear gradient gel.

DISCUSSION

In the present study, ferritin was identified biochemically and immunologically in the rat incisor enamel organ and that ferritin is a major iron containing protein in water soluble extract of this tissue. Ferritin in the rat enamel organ has properties similar to rat liver ferritin except for subunit composition. Ferritin is made up of 24 subunits of two types, a slightly heavier H subunit(mol. wt about 21,000)and a lighter L subunit(mol. wt. about 19,000). It is noted that there are tissue specific differences on the ratio of H and L subunits. Further it has been reported that L subunit functionally related to iron storage[9], while the abundance of H subunit mRNA relative to L subunit increases during tissue differentiation[10]. In the present study, the tissue specific difference on ferritin subunit ratio was also observed between rat liver and enamel organ. This difference might reflect the biological function of each tissue that liver store iron, while incisor enamel organ including ameloblasts and papillary layer cells are differentiating continuously in the incisal direction and transport iron rather than store one.

Immunolocalization of transfferin receptor over the papillary layer cells may be closely related to active transport of iron from capillary toward maturation ameloblasts. From another point of view, since transferrin is known as one of the growth factors required during cell differentiation[11], it is possible that transferrin receptor is located abundantly on the papillary cell because of its high proliferation activity.

Positive reaction of Prussian blue stain, detecting Fe^{3+} ion, is usually observed evenly over the cytoplasm of maturation ameloblasts, however, the immunostaining pattern for ferritin did not correspond to that of Prussian blue stain in the present study. These findings might explain that iron within ameloblasts could exist as the forms other than ferritin, presumably as hemosiderin. Hemosiderin is another form of storage iron. Several biochemical studies have shown that hemosiderin originates from ferritin that has undergone intracellular digestion of its protein shell, leaving the iron core, and become water insoluble[12]. Ultrastructural studies also demonstrated that

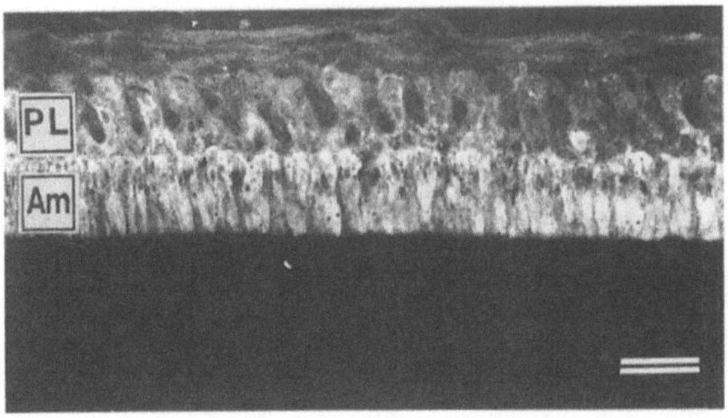

Fig.3 Immunohistochemistry of rat incisor enamel organ using anti-rat liver ferritin. Antigenicity is detected in the papillary layer cells and cytoplasm of the maturation ameloblasts (Am: Ameloblasts, PL: Papillary layer). Bar indicates length of 50 micrometers.

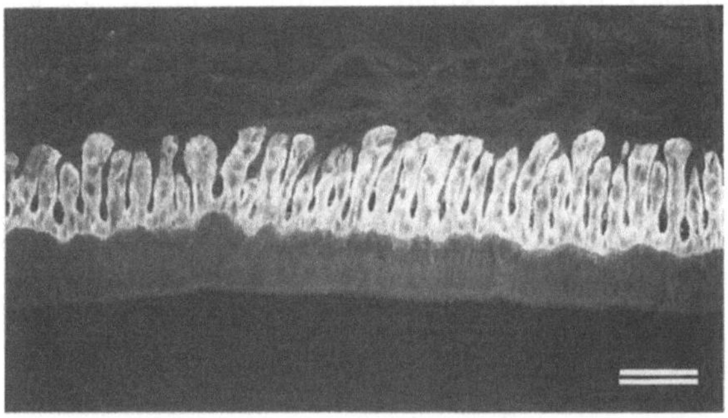

Fig.4 Immunofluorescent localization of rat transferrin receptor in rat incisor enamel organ by using mouse anti-rat transferrin receptor antibody. Intense immunostaining was observed over the papillary layer cells. Bar indicates the length of 50 micrometers.

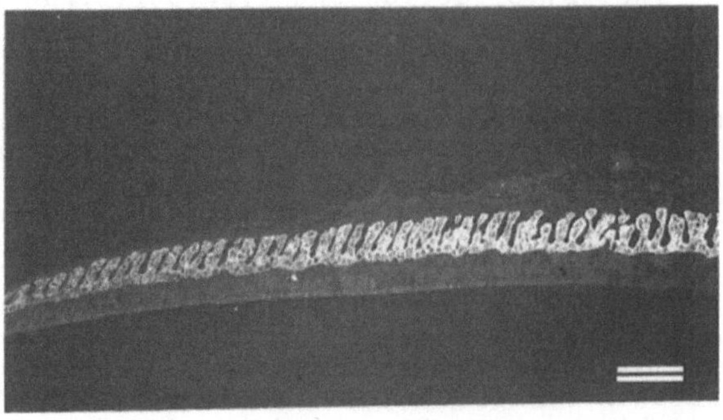

Fig.5 Immunoreaction for rat transferrin receptor over the papillary layer cells decreases gradually in the incisal direction. Bar indicates 100 micrometers.

iron containing ferritin granules are gradually incorporated into lysosomal vesicles and digested by lysosomal enzymes or other lytic bodies, and the iron is eventually transported toward the distal cell membrane of ameloblasts[13]. Further it is reported that the activity of acid phosphatase[14] and H^+-K^+-ATPase[15] are located in the pigment granules in the maturation ameloblasts, suggesting that these enzymes might be involved in the digestion process of ferritin. Although biochemical analysis were performed on only water soluble fraction in the present study, it seems possible that part of iron within the maturation ameloblasts exist in water insoluble fraction as in the form of hemosiderin.
In conclusion, it is speculated that iron is actively incorporated as the form of transferrin by papillary layer cells and transported to maturation ameloblasts. Further it is suggested that iron is stored as the form of ferritin and hemosiderin in maturation ameloblasts and then released onto enamel surface mediated by digestion process of these iron containing proteins.

REFERENCES

1. Pindborg JJ(1947)Odontol Tidskr 55:443-446
2. Stein G, Boyle PE(1959)Arch Oral Biol 1:97-105
3. Reith EJ(1959)Anat Rec 133:75-89
4. Halse A, Selvig KA(1974)Scand J Dent Res 82:47-56
5. Ogura H, Ohya K, Mataki S, Hashimoto K, Kubota M(1984)Tooth Enamel IV, Elsevier, Amsterdam, pp256-260
6. Karim A, Warshawsky H(1984)Am J Anat 169:327-335
7. Kubota M, Ohya K, Ogura H(1987)Adv Dent Res 1:330-338
8. Josephsen K(1974)Scand J Dent Res 82:229-238
9. Bomford AC, Conlon H, Munro HN(1981)J Biol Chem 256:948-955
10. Chou CC, Gatti RA, Fuller ML, Concannon P, Wong A, Chada S, Davis RC, Salser NA(1986)Mol Cell Biol 6:566-573
11. Trowbridge IS, Omary B(1981)Proc Natl Acad Sci USA 78:3039-3043
12. Munro HN, Aziz N, Leibold EA, Murray M, Rogers J, Vass JK, White K (1988)Ann NY Acad Sci 526:113-123
13. Kallenbach E(1970)J Ultrastruct Res 30:38-63
14. Takano Y, Ozawa H(1981)Calcif Tissue Int 33:51-55
15. Sasaki T, Tadokoro K, Yanagisawa T, Higashi S, Garant PR(1988) Anat Rec 221:823-833

CHAPTER 3.4

Elemental Distribution on the Growth Front of Oyster Shell Measured by Synchrotron Monochromatized X-rays

K. OKOSHI[1], M. ISHIKAWA[2] and A. IIDA[3]

[1]Department of Biotechnology, Senshu University of Ishinomaki, Miyagi, 986 Japan; [2]Division of Marine Radioecology, Laboratry of Radioecology, National Institute of Radiological Sciences, Ibaraki, 311-12 Japan; and [3]Photon Factory, National Laboratory for High Energy Physics, Ibaraki, 305 Japan

Key words: Element, Shell, Oyster, Synchrotron, X-rays

ABSTRACT

Elemental analyses on the growth front of the paper shell obtained from a chambered oyster, *Crassostrea gigas*, were carried out using the newly developed synchrotron radiation X-ray fluorescence analysis. This paper discusses the possibility of using the synchrotron monochromatized X-ray scanning microbeam to analyze mineralized samples. The newly formed paper shells were analyzed two dimensionally with an x-z scanner. The number of scanning spots was approximately 3000 to 4000. The distribution maps of Ca, Fe, Zn and Sr were obtained. Ca distribution corresponded well with that of membrane-like shell materials and its pattern was similar to that for Sr. Fe and Zn did not represent any characteristic elemental locality in the sample without jelly-like materials except for a rather high accumulation at some small areas. Particular interest is given to Zn, which was similar to the distribution patterns of Ca and Sr in the sample with jelly-like materials. This analytical system is very promising in the field of biomineralization.

INTRODUCTION

The development of synchrotron radiation (SR) as a research tool was driven largely by the needs of materials scientists and solid-state physicists. However, the availability of SR has also significantly extended the capability of scientists who study biological structure with radiation [1], X-ray spectrochemical analysis, irradiation effect on organs and tissues and so on.

X-ray fluorescence (XRF) analysis has been used in the field of biomineralization. It is expected that application of synchrotron X-rays to the XRF analysis will greatly improve its analytical ability because of the high brightness, polarization, and wavelength tunability of the synchrotron X-rays [2].

Despite these advantages, SR X-ray fluorescence (SRXRF) analysis has been little used in the field of biomineralization. The reason may be due to the fact that the method which requires an SR source does not come into general use. Also it is still unpopular to biologists concerned because it is a very recently developed tool.

This paper discusses SRXRF analysis for calcified biological samples by representing the results of two dimensional imaging of some main and trace elements in the growth front of the paper shell obtained from a chambered Japanese oyster, *Crassostrea gigas*, in order to obtain effective information about the relationship between main elements and trace ones related to their behavior during shell formation, especially shell chamber formation [3].

MATERIALS AND METHODS

Oyster The oyster spat used was collected in August 1985 at Matsushima Bay. In April 1986, they were transplanted to Takenoura in Miyagi Prefecture, Japan. The spat were hung under the floating rafts from April 1986 to June 1989.

Sampling Sampling was carried out in June 1989. The oyster which was observed just in a state of paper shell formation was used in this study. After removal of the soft body, the inner surface of the right valve was carefully washed with ultra-pure water three times.

Sample preparation for the SRXRF analysis The two parts which were observed in a state of the very early to early stage of paper shell growth [4] were cut out by an anatomical knife, then using rubber gloves. The process of paper shell formation occurs on the soft jelly-like materials. One part was cut out without the jelly-like materials (PS1), and the other with the jelly-like materials (PS2). Each part was placed on a Makrofol backing fixed on an aluminium frame [5] and dried in an electric oven at 40 °C. The film-like sample is called a "target".

SRXRF analysis The two dimensional multi-elemental analyses were carried out on beam line (BL)-4A using the SRXRF facilities at the Photon Factory (PF) of the National Laboratory for High Energy Physics, Japan. A schematic beam line layout is shown in Fig.1. The sample was excited by monochromatized X-rays of 19.5 keV using a Si(111) double crystal monochromator. The X-ray fluorescence intensities were measured by a Si(Li) detector with energy dispersive mode. The SR microbeam was obtained by a set of vertical and horizontal slits. The final beam sizes were adjusted to the dimension of 150x150 μm^2. For the two dimensional analysis the sample was mounted on a remote controlled X-Z stage driven by pulse motors (ca. 1 μ / pulse) and the point analyzed was monitored by an optical microscope with a TV camera. The electronic signal from the detector was processed through an amplifier and single channel analyzers. In our system, X-ray fluorescence intensities from 4 elements such as Ca, Fe, Zn and Sr can be recorded on magnetic disks. The PS1 was scanned with 67 points in the horizontal direction, by 47 points in the vertical direction and each point took three seconds. The scanning area is shown in Fig.3a. The PS2 was horizontally 67 points, vertically 60 points, and with a 3 sec. exposure made a total of 4020 points. The scanning area is shown in Fig.3b. The measurements were carried out under the vacuum condition of 10^{-2} torr. The beam currents of the storage ring were 220 mA and 183 mA when the PS1 and PS2 were measured, respectively.

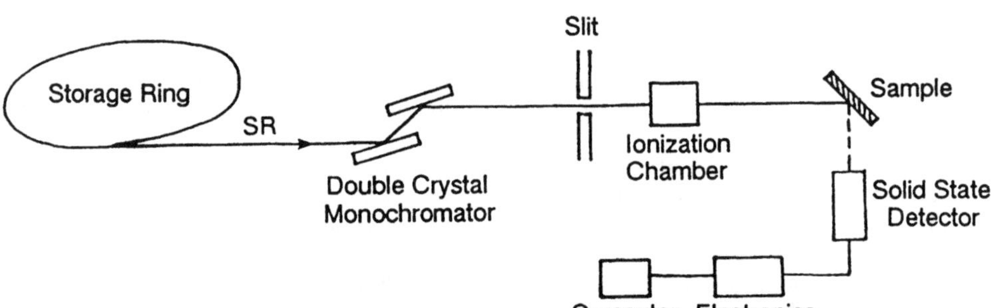

Fig. 1. Experimental arrangement for synchrotron radiation X-ray fluorescence analysis at beam line-4A at Photon Factory of the National Laboratory for High Energy Physics.

RESULTS AND DISCUSSION

The presence of trace elements in molluscan shells has been the subject of extensive studies [6]. Despite many investigations, the role of these elements on the process of shell formation is not known in detail. We believe we should investigate the problem to determine whether or not the trace elements consist merely as contaminations in the molluscan shells, and whether or not the micro-structure of shells is controlled by some trace elements contained in the structure.

Sea origin mineralized samples are not uniform and are normally rich in Na, Ca and Mg, which critically interfere in performing analysis of trace elements by currently available means. Though ICP (Inductively Coupled Plasma) -MS and ICP-AES are well recognized among analysts as a powerful modern tool and analytical technique for multi-element analysis, they need dry or wet ashing. Since characteristic X-rays from samples submitted quantitative information on elements, the transmission electron microscope, scanning electron microscope etc., have also been developed for the method of analysis for elemental quantities. However, excitation processes of the X-rays due to electrons, give abnormally high background, and the electron beam current covers the real characteristic X-ray peaks, resulting in a detection limit decreased to the % level. This level is considered insufficient in modern analytical methods for trace characterization of elements [7]. PIXE (Particle Induced X-ray Emission) analysis has been used because of its high sensitivity with simultaneous multi-element analysis and scanning analysis with a microprobe [8-10]. However, some difficulties such as maintaining protons as high flux for a long time with a stationary state, shortening of the analytical time, etc. occur during two dimensional data acquisition [7]. The Photon Factory of the National Laboratory for High Energy Physics has recently installed a new system of microbeam focussing a synchrotron monochromatized X-rays. This potential makes the two dimensional scanning possible for a small area of sub-millimeter level with a high photon flux beam. This seemed to be the possibility currently challenging the problems which tasked scientists concerned with the molecular bio-physics in organs and tissues for trace elemental characterization at the cellular level. Therefore, using SRXRF, we attempted to determine the detailed distribution of some elements concerned in the small area of the growth front of the oyster shell.

Fig.2 shows a photograph of samples which were observed just in the state of paper shell formation. The arrow shows the thin membrane-like shell materials. The process of the paper shell formation occurs on soft jelly-like materials. PS1 was cut out without the jelly-like materials (Fig.3a), and PS2 was with the jelly-like materials (Fig.3b). The analyses were made for the areas indicated by the frames in Fig.3a and Fig.3b. The results are given in Fig.4 indicating the Ca, Fe, Zn and Sr images of the PS1. The intensity was scaled with 15 densities. The Ca image in Fig.4a corresponds well with the distribution of white colored membrane-like shell materials in Fig.3a. Though the intensity is lower than that of Ca, the

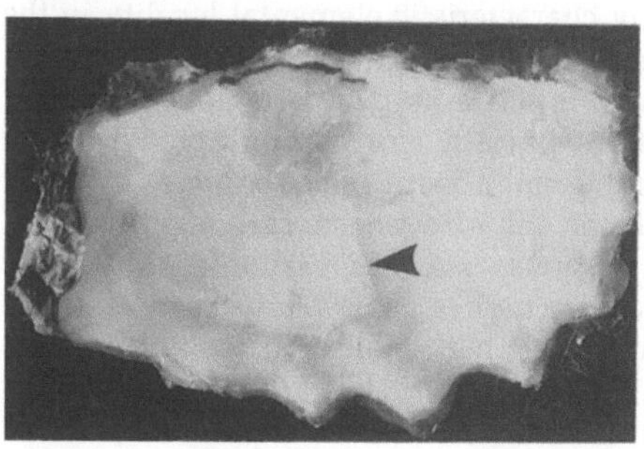

Fig. 2. A photograph of analyzed sample which was observed in a state of paper shell formation obtained from a chambered oyster, *Crassostrea gigas*. Arrow shows the thin membrane-like shell materials.

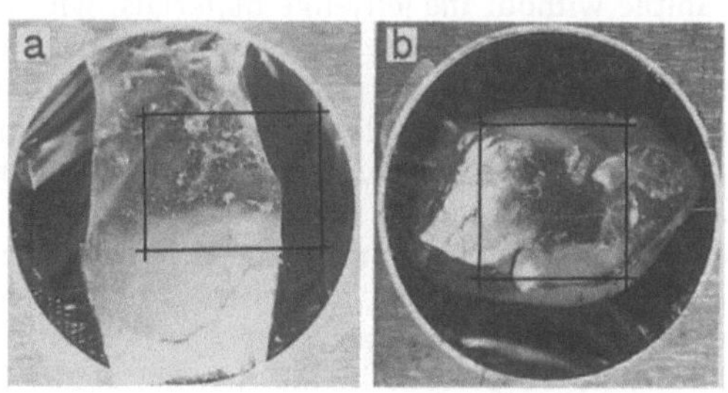

Fig. 3. Growth front of the paper shell. a: without jelly-like materials (PS1). b: with jelly-like materials (PS2). The analyzed areas are indicated by the frames.

Sr distribution pattern was similar to that for Ca (Fig.4d).Fe and Zn did not represent any characteristic elemental locality in the sample without the jelly-like materials except for some rather high accumulation at some sites (Figs.4b,4c).The results for chemical images of PS2 are shown in Fig.5. Ca was characteristic and its pattern was similar to that for Ca in the PS1(Fig.5a). The Sr distribution pattern was similar to that for Ca as shown in Fig.4 (Fig.5d). Particular interest is given to Zn, which was similar to that of Ca and Sr in the sample with the jelly-like materials (Fig.5c). The intensity of Fe is higher than that of PS1. Also a higher Fe accumulation was observed in some small areas of PS2 (Fig.5b).

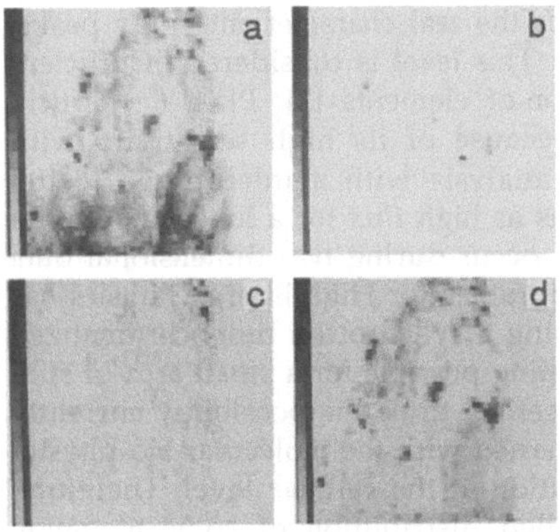

Fig. 4. X-ray fluorescence images of the PS1 shown in Fig.3a. a: Ca $K\alpha$ image. b: Fe $K\alpha$ image. c: Zn $K\alpha$ image. d: Sr $K\alpha$ image.

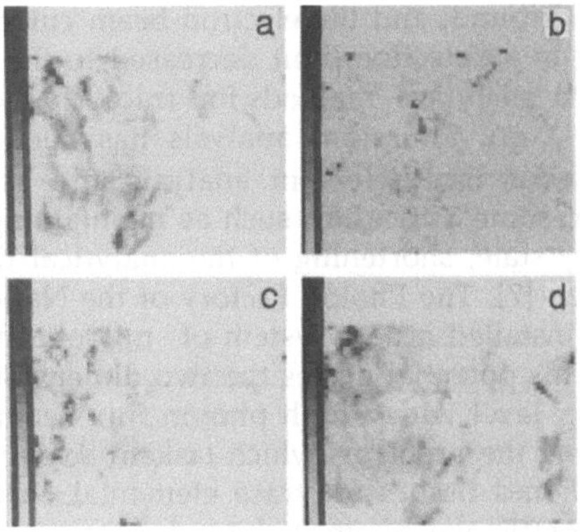

Fig. 5. X-ray fluorescence images of the PS2 shown in Fig.3b. a: Ca $K\alpha$ image. b: Fe $K\alpha$ image. c: Zn $K\alpha$ image. d: Sr $K\alpha$ image.

Whether the jelly-like materials are present or not, the distribution patterns of Ca correspond well with that of the thin membrane-like shell materials (Figs.4,5). As for Sr, the tendency is almost the same though the intensity is much lower than that of Ca. Therefore, it is considered that Ca and Sr are contained in the membrane-like shell materials as main component parts probably in some chemical states such as $CaCO_3$ and $SrCO_3$. On the other hand, Zn did not represent any characteristic elemental locality in the sample without the jelly-like materials, whereas the distribution pattern was similar to that for Ca and Sr with the jelly-like materials. The results indicate that Zn is mainly contained in the jelly-like materials. Also, the distribution pattern corresponding with the mineralized areas is clearly evident. Zn is an integral component of carbonic anhydrase, alcohol dehydrogenase, glutamate dehydrogenase, lactic dehydrogenase, carboxy-peptidase, alkaline phosphatase, etc.. The presence of zinc enzymes such as carbonic anhydrase, alkaline phosphatase and malic dehydrogenase in oyster tissues has already been reported [11]. Nearly all the zinc in oysters is bound either to soluble high-molecular weight proteins or to structural cellular components such as cell membranes [11]. Carbonic anhydrase, known to be present in the mantle, could function in two ways to increase the rate of proton and CO_2 removal [12] in the process of shell formation. From the facts previously described, the behavior of Zn in the jelly-like materials might probably relate to the process of initial mineralization concerned with shell chamber formation. Zn may also be present as an integral component of the concerned enzymes. Further research, such as histochemical analysis, is now in progress. These discussions will be reported in detail elsewhere. It is the subject for a future study concerning whether the higher accumulation of Fe observed in some small parts of the PS1 and PS2 is merely contaminations.

From the results of our experiments, it is considered that the sensitivity of the SRXRF system is more effective than conventional analytical tools. This system is effective for simultaneously obtaining the two dimensional information for main and trace elements presented in calcified samples. Because the beam of SR is truly nondestructive, it is also possible to undertake *in vivo* compositional analysis of shells of exoskeletonous organisms. Therefore, this analytical system is very promising in the field of biomineralization, and this method will be applied to the various biological fields in the near future.

ACKNOWLEDGEMENT

The authors are indebted to Dr. S. Hayakawa of University of Tokyo for providing them with the computer program for the experiment.

REFERENCES

1. Sweet RM and Woodhead AD (1989) Synchrotron Radiation in Structural Biology. Plenum New York, pp. 1-362
2. Iida A, Sakurai K, Matsushita T, Gohshi Y (1985) Nucl Instr and Meth 228: 556-563
3. Okoshi K, Mori K, Nomura T (1987) Aquaculture 67: 313-320
4. Okoshi K, Sugawara Y, Nomura T (1987) Venus (Jap J Malac) 46: 237-249
5. Okoshi K, Ishikawa M, Nomura T (1988) Nippon Suisan Gakkaishi (Bull Japan Soc Sci Fish) 54: 1213-1219
6. Simkiss K (1983) In: Westbroek P, Jong EW de (eds) Biomineralization and Biological Metal Accumulation. D. Reidel, Holland, pp.363-371
7. Ishikawa M, Iida A, Ishii T, Hayakawa S, Okoshi K. Nippon Suisan Gakkaishi (Bull Japan Soc Sci Fish) (in press)
8. Imaseki H, Ishikawa M, Kitao K (1982) J Radioanal Chem Letters 55: 49-56
9. Carriker MR, Swann CP, Ewart JW (1982) Mar Biol 69: 235-246
10. Ishikawa M, Okoshi K, Kitao K (1989) Proceedings of the 7th Symposium on Ion Beam Technology Hosei University: 173-178
11. Wolfe DA (1970) J Fish Res Bd Canada 27: 59-69
12. Wheeler AP (1975) Ph.D. Thesis, Duke University, Durham, North Carolina

CHAPTER 3.5
Crystallographic Behavior of Iron and Magnesium in Hydroxyapatite Crystals

M. OKAZAKI

Osaka University Faculty of Dentistry, Japan

Key words: Apatites, Iron, Magnesium, Crystallinity, Solubility

INTRODUCTION

Iron and magnesium, indispensable elements for human life, are also trace elements in tooth minerals. They have been implicated in the early stages of caries. Iron, an indispensable element for human life, forming for example, the core of hemoglobin molecules [1] is also a trace element in tooth minerals [2]. Iron accumulates in the outer surface of tooth enamel [2]. Analyses of total enamel have shown considerable variation in the iron content of individual human teeth, probably related to differences in iron ingestion and age. However, the form in which iron is deposited in the enamel and the mechanism of its acquisition are still not fully understood. Francis et al. [3] reported that Fe^{2+} ions are effective in reducing the subsequent acid dissolution rate of enamel. This result seems to be related to the incorporation of Fe^{2+} ions into apatite crystals.
On the other hand, magnesium is clinically focused because of its close relation to the metabolism and physiology of the human body [1]. Bone and teeth also contain cetain amounts of magnesium [2]. It may play an important role in the initial formation of apatites and have a significant effect on their physicochemical properties. For example, magnesium may control the rate of apatite formation in the early stages of mineralization in matrix vesicle [4]. It has been reported [5] that magnesium may be located on the surface of bone apatite crystallites. On the contrary, recent investigators [6,7] have reported that it can be incorporated to a limited extent in the apatite crystals. Since most of biological apatites except tooth enamel are poorly-crystallized and furthermore magnesium is a minor constituent of human bone and teeth, the nature of its association with the mineral phase, especially hydroxyapatite, is still not clear.
Synthetic apatite well-crystallized can sometimes give us a useful information on the physicochemical properties of the biological apatites, which are more difficult to observe themselves. In this study, the possibility of iron and magnesium substitution into apatite crystals was examined.

MATERIALS AND METHODS

Hydroxyapatites were synthesized at 80±1°C in the presence of both ferrous (Fe^{2+}) and ferric (Fe^{3+}) ions, by varying iron-calcium feed molar ratios $(Fe/Ca)_f$ in the range of 0 - 1. Precipitates were made by feeding 0.5 L of 100 mM $Ca(CH_3COO)_2H_2O$ solution containing $FeCl_2nH_2O$ or $FeCl_3 6H_2O$ and 0.5 L of 60 mM $NH_4H_2PO_4$ solution into 1 L of mechanically stirred 1.3 M CH_3COONH_4 solution. Both $FeCl_2nH_2O$ and $FeCl_3 6H_2O$ as Fe sources showed deliquescence, and n was determined as 2.6 by chemical analysis. The two solutions were supplied at 250 ml/h with a Tokyo Rikakikai microtube pump. The pH was maintained at 7.4±0.1 with occasional addition of concentrated NH_4OH solution. The suspension was stirred for 3 hr, then kept at 25 °C for 24 hr. The slurry was filtered, washed with distilled water, then dried at 110°C. Mg-containing hydroxyapatites were also synthesized by varying the magnesium feed concentration contained in the $Ca(CH_3COO)_2H_2O$ solution from 0 to 100 mM as $Mg(CH_3COO)_2 \cdot 4H_2O$.
X-ray diffraction was employed to identify precipitates and estimate their lattice constants and crystallinity. Measurements were made on a Rigaku Denki X-ray Diffractometer with graphite-monochromatized CuK_α radiation at 35 kV, 23 mA. The a-

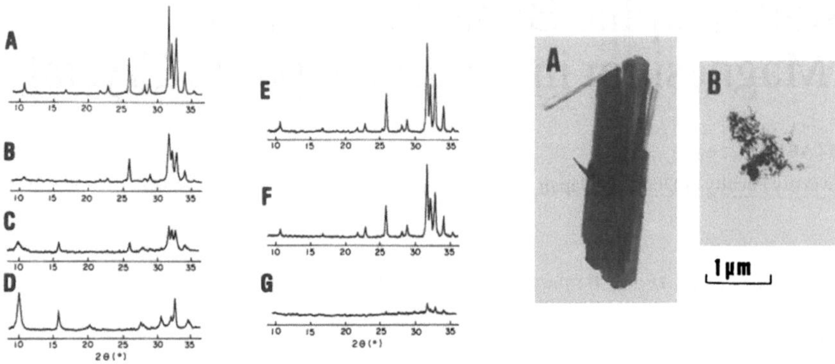

Fig. 1 X-ray diffraction patterns of Fe-containing hydroxyapatites, together with the transmission electron micrographs.
$(Fe^{2+}/Ca)_f$= 0 (A); 0.07 (B); 0.37 (C); 0.73 (D)
$(Fe^{3+}/Ca)_f$= 0.01 (E); 0.1 (F); 1.0 (G)

and c-axis dimensions were calculated from (300) and (002) reflections, respectively. To estimate crystallinity, the inverse of the half-value breadth was calculated from (300) and (002) reflections as representations of the a- and c-axes using silicone as a standard.
Calcium, iron and magnesium concentrations were determined by atomic absorption spectrophotometry. Total phosphate concentrations were determined by the UV spectrophotometric method of Eastoe [8].
To estimate solubility, samples were left to stand in 5 or 50 mL of 0.5 M acetate buffer (pH 4.0) at 37°C for 1 month.

RESULTS

Fe-containing Hydroxyapatites Figure 1 shows the X-ray diffraction patterns of Fe-containing hydroxyapatites. Hydroxyapatites synthesized using $FeCl_2{\cdot}nH_2O$ rapidly became poorly crystallized with the increase of ferrous ions in the solution (Fig.1A and 1B), and at higher concentration of ferrous ions, a different calcium phosphate in addition to apatite was formed, as shown in Fig. 1C and 1D. This was identified as $Ca_3Fe_4(PO_4)_4(OH)_6{\cdot}3H_2O$. In the hydroxyapatites synthesized using $FeCl_3{\cdot}6H_2O$, the degree of crystallization decreased more slowly with the increase in iron concentration than in the corresponding ferrous compounds (Fig.1E and 1F). Nevertheless, at higher concentrations, precipitates rapidly became amorphous, as shown in Fig. 1G, and their patterns differed completely from those of the iron sources. Transmission electron microphotographs of the precipitates are also added to Fig.1. Morphology of the crystal varied dramatically with the increase of iron content. The crystal growth of hydroxyapatites was extremely inhibited by the presence of iron.

Table 1. a- and c-Axis dimensions of Fe-containing hydroxyapatites.

$(Fe^{2+}/Ca)_f$	a-axis (Å)	c-axis (Å)
0	9.447_8	6.887_6
0.007	9.446_0	6.887_2
0.07	9.441_6	6.882_5
0.15	9.442_1	6.884_0
$(Fe^{3+}/Ca)_f$		
0.01	9.447_8	6.886_4
0.1	9.445_1	6.887_2

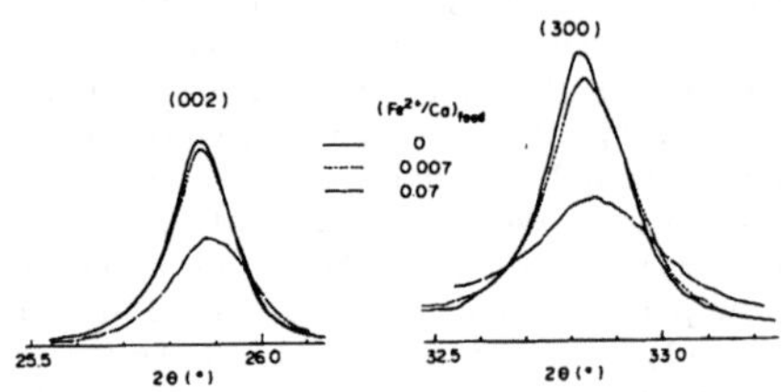

Fig. 2 Typical examples of the (300) and (002) reflections of Fe-containing hydroxyapatites.

Table 2. Chemical compositions of Fe-containing hydroxyapatites.

$(Fe^{2+}/Ca)_f$	Ca (mmol/g)	P (mmol/g)	Fe (mmol/g)
0	9.36	5.49	0
0.007	9.57	5.67	0.05
0.07	8.79	5.71	0.55
0.15	7.51	5.29	0.94

Table 3. Apparent solubility of Fe-containing hydroxyapatites.

$(Fe^{2+}/Ca)_f$	Ca (mM)	P (mM)
0	32.8	20.4
0.007	26.7	13.0
0.07	25.9	12.5
0.15	25.6	12.4
0.37*	19.3	8.64
0.73*	5.16	2.02

* Original samples contain other Fe-compounds besides apatites.

Ferrous (Fe^{2+}) ions seemed to be taken up into apatite crystals, contrary to ferric (Fe^{3+}) ions. The contraction of a- and c-axis dimensions of Fe-containing apatites with the increase of iron content were suggested by the data in Table 1, although the difference found were too small to be conclusive. The peak shifts of the expanded (300) and (002) reflections shown in Fig. 2 support this suggestion.
Calcium content decreased with the increase of the Fe content, as shown in Table 2, but phosphate content was almost independent of iron content below $(Fe^{2+}/Ca)_f=0.15$. These results suggest that Fe ions may substitute at Ca sites. The lower calcium and phosphate contents at higher Fe concentration may be due to the formation of different compounds, which remained partly insoluble even when the samples were dissolved in 0.1 N HCl solution for analysis of chemical composition.
Table 3 shows the apparent solubility of Fe-containing hydroxyapatites as the concentrations of calcium and phosphate ions dissolved in 0.5 M acetate buffer solution, pH 4.0, in one month. The solubility decreased dramatically with the increase of original iron content in spite of the decrease of crystallinity shown in Fig. 1. This decrease in the concentrations of soluble calcium and phosphate ions suggests that iron has the effect of preventing dissolution. Even if the relative decrease in calcium content by substitution of Fe^{2+} ions is considered, the apparent solubility decreases greatly, especially at very low Fe^{2+} content. This dissolution-preventive effect of iron may be due to crystal stabilization by Fe^{2+} ions substituted into the apatite crystal, and/or inhibition by insoluble precipitates of iron compounds formed at the crystal surface. The extreme decrease of the phosphate concentration in the solution with the increase of the iron content suggests the possible formation of different precipitates from apatites.
Mg-containing Hydroxyapatites With increasing magnesium-calcium feed molar ratio, $(Mg/Ca)_f$, calcium content decreased and magnesium content increased as shown in Table 4. Phosphate content showed no significant change below $(Mg/Ca)_f$ of 0.5.
Figure 3 shows X-ray diffraction patterns of precipitates, which were synthesized at 80°C and pH 7.4. These were typical of calcium apatites at lower magnesium-calcium

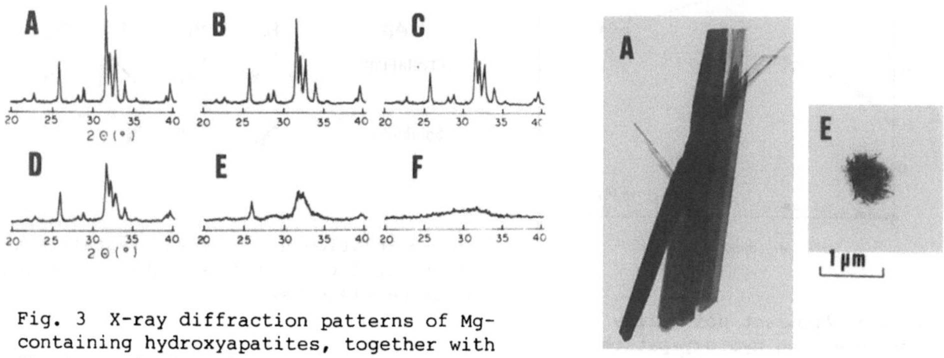

Fig. 3 X-ray diffraction patterns of Mg-containing hydroxyapatites, together with the transmission electron micrographs.
$(Mg/Ca)_f$= 0 (A); 0.01 (B); 0.05 (C); 0.1 (D); 0.5 (E); 1.0 (F)

Table 4. Chemical compositions of Mg-containing hydroxyapatites.

$(Mg^{2+}/Ca)_f$	Ca (mmol/g)	Mg (mmol/g)	P (mmol/g)	(Ca+Mg)/P
0	9.81	0	5.91	1.66
0.01	9.69	0.061	5.90	1.64
0.05	9.58	0.201	6.02	1.62
0.1	9.51	0.303	5.92	1.66
0.5	9.05	0.910	5.94	1.68
1.0	6.63	2.26	6.12	1.45

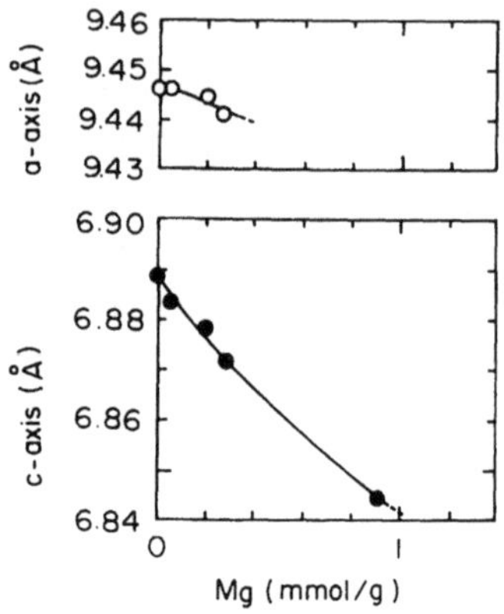

Fig. 4 a- and c-Axis dimensions of Mg-containing hydroxyapatites as a function of magnesium content.

feed molar ratio, $(Mg/Ca)_f$, but indicated poor crystallization at higher ratios and an amorphous structure at $(Mg/Ca)_f$=1. However, none of the samples gave a tricalcium phosphate pattern, even though the chemical composition of the sample of $(Mg/Ca)_f$=1 had a similar value 1.45 of (Ca+Mg)/P to tricalcium phosphate 1.5.

Transmission electron micrographs of the precipitates are also added in Fig. 3. The crystal size and shape altered dramatically with the increase of magnesium content. At low concentration of magnesium, the apatites seemed to be thin plate-like crystals, while samples synthesized at $(Mg/Ca)_f$=0.5 and 1 were small and coagulated. c-Axis dimensions decreased with the increase of magnesium content, as shown in Fig. 4. These results, taken together with those of chemical analysis, thus suggest that Mg^{2+} ions may substitute into the Ca^{2+} sites.

Figure 5 shows the apparent solubility of the precipitates as the concentrations of calcium, magnesium and phosphate dissolved in 0.5 M acetate buffer at 37°C and pH 4.0 in 1 month. The samples became more soluble with the increase of the original magnesium content. The residual sample with $(Mg/Ca)_f$=1 showed dicalcium phosphate dihydrate pattern.

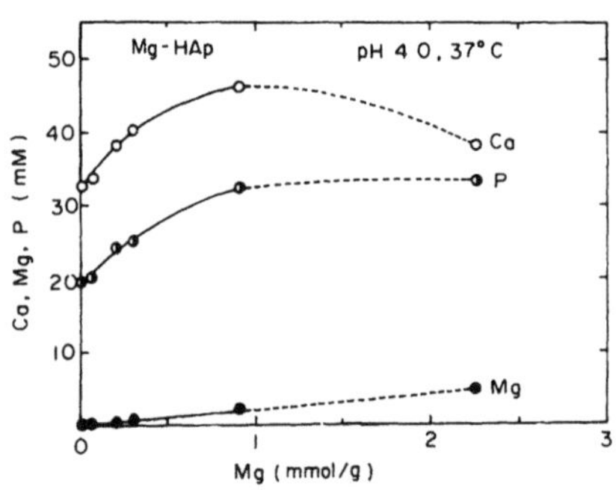

Fig. 5 Apparent solubility of Mg-containing hydroxyapatites.

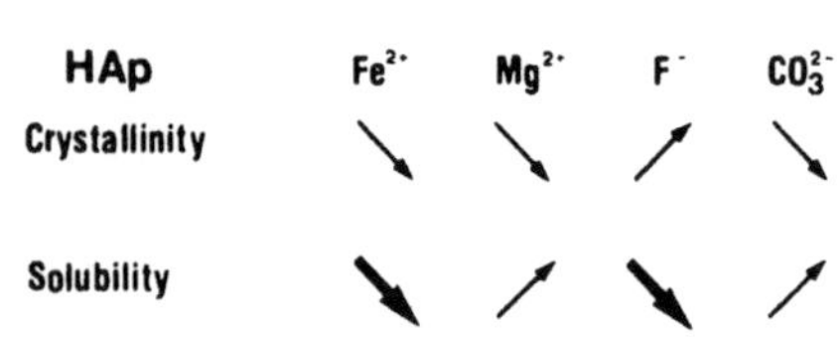

Fig. 6 Relationship between crystallinity and solubility of hydroxyapatite affected by trace elements.

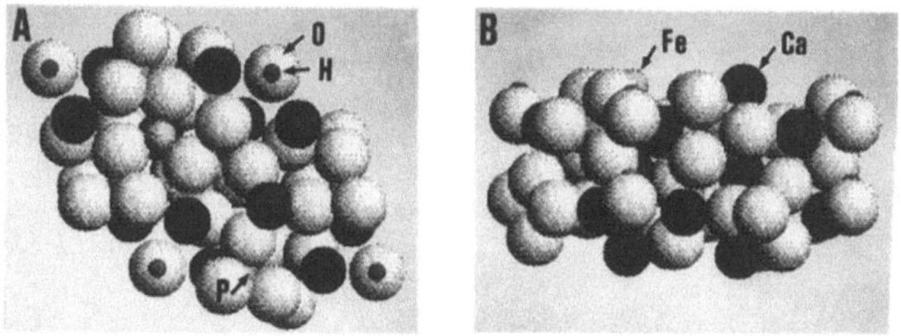

Fig. 7 Computer graphics of Fe-containing hydroxyapatites.
A: structure projected on the X-Y plane
B: rotated by 90° around the X coordinate

DISCUSSION

The ferrous ion (Fe^{2+}) has the same charge as the calcium ion (Ca^{2+}) but a smaller radius: 0.75 Å as opposed to 0.99 Å. Therefore, it seems reasonable that lattice dimensions would contract when Fe^{2+} ions substitute at Ca^{2+} positions, or suggested by the data in Table 1. Fe^{3+} ions would be less likely to substitute at Ca^{2+} positions than Fe^{2+} ions because of the loss of electroneutrality and their smaller ion radius (0.64 Å). Mg^{2+} ions were found to induce a clear contraction of c-axis dimensions. Mg^{2+} ions have a small radius of 0.65 Å, as opposed to Ca^{2+} ions. These cations such as Fe^{2+} and Mg^{2+} seem to show mainly a contraction in c-axis dimension, contrary to anions such as CO_3^{2-} and F^-, which can be substituted into PO_4^{3-} and OH^- positions, respectively. The solubility of Fe-containing hydroxyapatite decreased dramatically in spite of the decrease of crystallinity shown in Fig. 6. It is hard to imagine how an extremely small amount of iron compound on the crystal surface could bring about such a great decrease of the solubility without substitution into the apatite crystals.
Figure 7 shows an example of a computer graphics model of Fe (Mg)-containing hydroxyapatite, in which Fe^{2+} or Mg^{2+} ion substituted into columnar Ca^{2+} position. These computer graphics [9] may be helpful in understanding the crystallographic properties of apatites.

REFERENCES

1. Stryer L (1981) Biochemistry. W.H. Freeman and Co. San Francisco.
2. Miles AEW (1967) Structural and Chemical Organization of Teeth (II). Academic Press New York.
3. Francis MD, Gray JA, Griebstein WJ (1968) Arch. Oral Biol. 3: 83-120
4. Bourne GH (1976) The Biochemistry and Physiology of Bone. Academic Press New York.
5. Neuman MF, Mulryan BJ (1971) Calcif. Tissue Res. 7:133-138
6. LeGeros RZ (1984) Tooth Enamel IV. Elsevier Amsterdam pp.23-36.
7. Okazaki M, Takahashi J, Kimura H (1986) Caries Res. 20: 324-331
8. Eastoe JE (1965) Calcified Tissue. Université de Liège Liège.
9. Okazaki M, Sato M (in press) Biomaterials

CHAPTER 3.6
Magnesium in Normal and Pathological Calcifications

R.Z. LeGeros
New York University College of Dentistry, New York 10010, USA

Key words: Magnesium, Apatites, Whitlockites, Calcifications, Calculus

INTRODUCTION

Magnesium, Mg, is a minor but important constituent in some biological calcium phosphates, Ca-P (e.g., tricalcium phosphate, TCP or whitlockite, apatites (AP) and amorphous calcium phosphates, ACP) occuring in normal and pathological calcifications [1-5]. Mg, with carbonate, CO_3, was also associated with caries [6,7]. In synthetic systems, the presence of Mg in solution was shown to have the following effects: stabilizes octacalcium phosphate, OCP [8] and ACP [9,10] against transformation to AP; allows the formation of b-TCP [11,12] and promotes the formation of ACP [10,11]. The purpose of this paper is to present some of the findings from our past and current studies on the role of Mg on the formation of synthetic Ca-P and to discuss into its role in the formation of biological Ca-P.

MATERIALS AND METHODS

The effect of Mg on the formation of different types of Ca-P were investigated using precipitation methods: reaction temperatures - 25, 37, 60, 95°C; initial pH-5, 6.8, and 7.5. These conditions were based on the observed pH/temperature dependence [13] of the formation of dicalcium phosphate dihydrate, $CaHPO_4.2H_2O$ (DCPD); OCP, $Ca_8H_2(PO_4)_6.5H_2O$; Ca-deficient apatite; and ACP. Hydrolysis of DCPD (60°C) and of monetite, $CaHPO_4$, DCP (95°C) in solutions containing calcium, Ca^{2+} and Mg^{2+} ions, initial pH 5 and 7.5 was also employed [12,14]. The Mg/Ca molar ratio in solution, $[Mg/Ca]_s$, ranged from 0/1 to 1/1.

Biological materials analyzed included: samples of dental calculus [15,16 and this study]; pathological calcifications [4]; arrested dentin caries [14]; enamel, dentin and bone [5,11,17].

Biological and synthetic materials were characterized using x-ray diffraction (XRD), infrared (IR) spectroscopy, scanning (SEM) and transmission (TEM) electron microscopy, thermogravimetry (TGA), atomic absorption for the determination of Ca^{2+} and Mg^{2+} and colorimetry for phosphate analysis.

RESULTS

Biological Calcium Phosphates

(a) Human dental calculus. Highly mineralized samples of human dental calculus consisted of three types of Ca-P: CO_3-apatite, OCP and Mg-substituted b-TCP (b-TCMP) or whitlockite (Fig. 1). The Mg-for-Ca substitution in the b-TCP phase was ascertained from the known effect of Mg on contracting the a-axis dimension (Figs. 2) and the observed correlation of Mg content with Ca/P ratio ranging from 1.4 to 1.6 (Fig. 3). Presence of b-TCMP were also shown by SEM as rhombohedral crystals of varying sizes (Fig. 4A, 4B).

(b) Pathological calcifications: Non-visceral calcifications occuring in heart, lung and skeletal muscle were identified as ACP by XRD (Fig. 5A) with Mg/Ca of 1/5.7; visceral calcifications occurring in subcutaneous tissue and around tendons gave apatitic XRD patterns (Fig. 5C) and Mg/Ca of 1/30.3. Apatitic deposits showed carbonate, CO_3, in their IR spectra, amorphous deposits did not [4]. Ignition at 600°C caused the formation of b-TCMP from non-visceral Ca-P deposit and of well crystallized AP mixed with b-TCMP from non-visceral Ca-P deposits (Fig. 5B,5D).

(c) Dentin caries: Rhombohedral crystals associated with arrested human dentin caries (Fig. 4C) were identified as whitlockite by electron diffraction [14].

(d) Normal calcifications (enamel, dentin, bone). The biological apatites of enamel, dentin and bone of adult human differ in crystallinity, reflecting differences in their crystal sizes: enamel(E) dentin(D) bone(B) shown in Fig. 6. They also differ in the concentrations of minor elements, notably, Mg and CO_3 [2, 5, 17]: wt% Mg: 0.3 (E), 1.11 (D), and 0.48 (B); wt % CO_3: 3.5 (E), 5.8 (D) and 6.6 (B). Ignition at 800°C caused the formation of b-TCMP phase from enamel and dentin but not from bone, the b-TCMP/AP ratio being greater for dentin than for enamel [5,11,17].

Synthetic Calcium Phosphates.

(a) Effect of Mg on DCPD formation and transformation. Mg favored the formation of DCPD even at neutral and basic pH, but delayed and inhibited DCPD formation at low pH (Table 1). Under similar conditions in the absence of Mg, low pH normally favors the formation of DCPD; neutral or basic pH, the formation of apatitic Ca-P at 37°C [13, 17, 18]. When present in solution, Mg inhibited the transformation of DCPD to OCP and to AP but, in the simultaneous presence of Ca^{2+}, caused the transformation of DCPD to b-TCMP (Fig. 7).

Fig. 1: XRD of human dental calculus showing the presence of predominantly whitlockite, W, in (A); mixture of apatite, A, OCP, O, and W in (B); O and W in (C) [16].
Fig. 2: Mg-for-Ca substitution in whitlockite causes a contraction in the a-axes [15].
Fig. 3: Mg concentration in human dental calculus correlates with Ca/P ratio of 1.4 to 1.6 indicating the Mg in calculus is associated mostly with the whitlockite phase [16].
Fig. 4: SEM showing whitlockite, b-TCMP in dental calculus (A,B); in arrested dentin caries (C); from solutions with Mg/Ca = 0.25 (D). [14,16,18].
Fig. 5: XRD of non-visceral (A) and visceral (B) calcifications from uremic patients. The Ca-P deposits were characterized as ACP (A) and apatitic (B). Ignition (600°C) resulted in the formation of b-TCMP from (A), of CO_3-AP and b-TCMP from (B). [5].
Fig. 6: XRD of biological apatites from enamel, dentin, bone showing difference in crystallinity [5,11].
Fig. 7: XRD of b-TCMP obtained from synthetic systems and from dental calculus.
Fig. 8: IR of ACMP before (A) and after transformation to DCPD (B) after suspension in acidic solutions.
Fig. 9: Transformation of synthetic ACMP (A) to mostly b-TCMP (B); of AP (C) to CO_3-AP + b-TCMP (D) after heating (600°C).
Fig. 10: Effect of Mg on the formation of OCP.
Fig. 11: Effect of Mg on the crystallinity of AP and formation of ACP at 37°C [5,10,11].
Fig. 12: Effect of Mg on the b-TCMP/HA ratio of ignited 'AP' containing 0.2 (A), 0.3 (B) and 0.4 (C) wt% Mg, indicating the enhancing effect of Mg on HPO_4 incorporation.
Fig. 13: Formation of b-TCMP at the expense of AP with increasing Mg/Ca in solution (T= 95°C, pH =5).

1

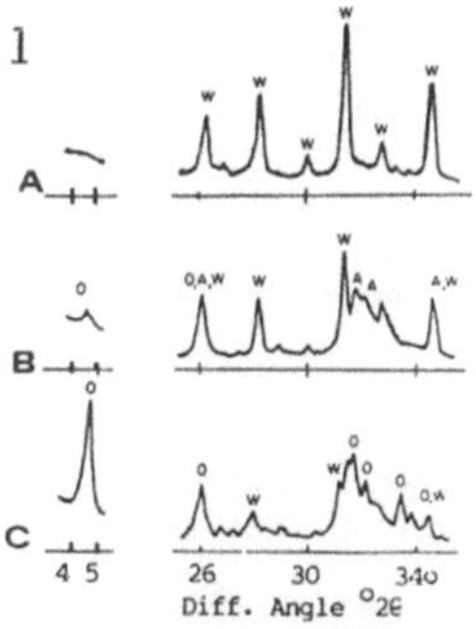

EFFECT OF MG SUBSTITUTION ON A-AXIS

2

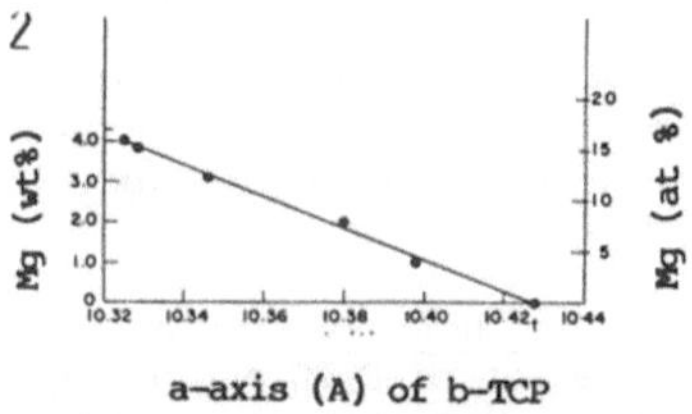

3

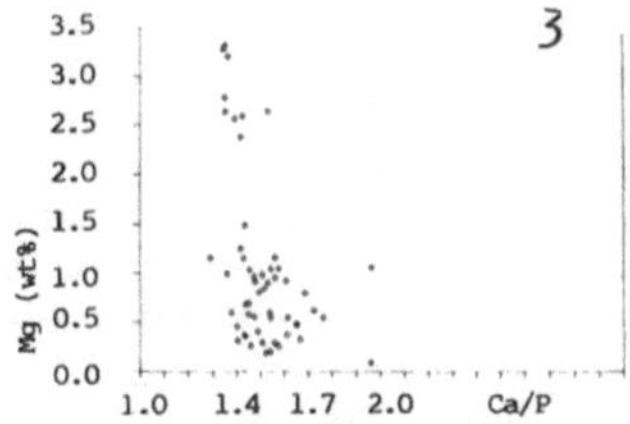

4A

4B

4C

4D

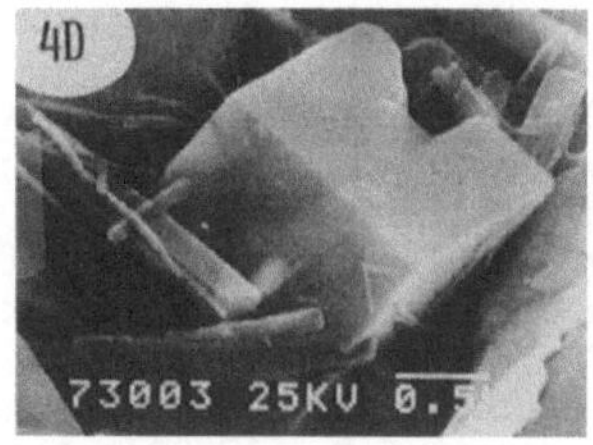

5

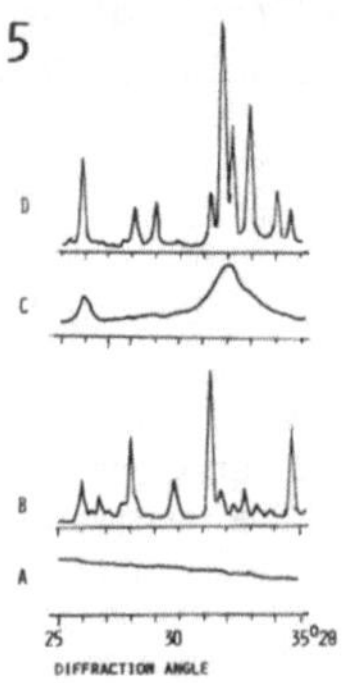

6

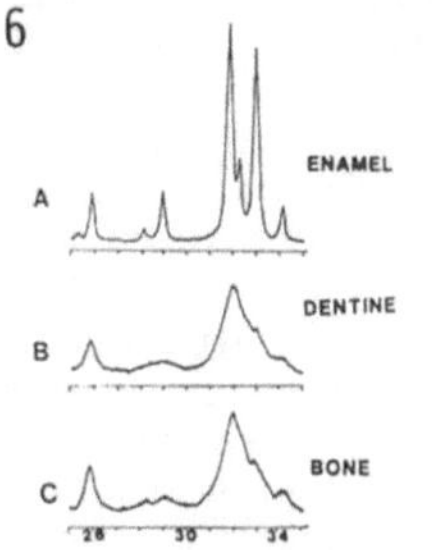

WT % Mg: 0.35 (E), 1.11 (D), 0.48 (B)
WT % CO_3: 3.5 (E), 5.8 (D), 6.6 (B)

7

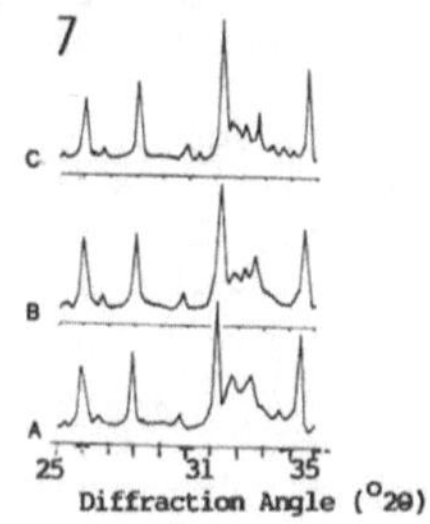

A,B - synthetic b-TCMP + AP
C - human dental calculus

8

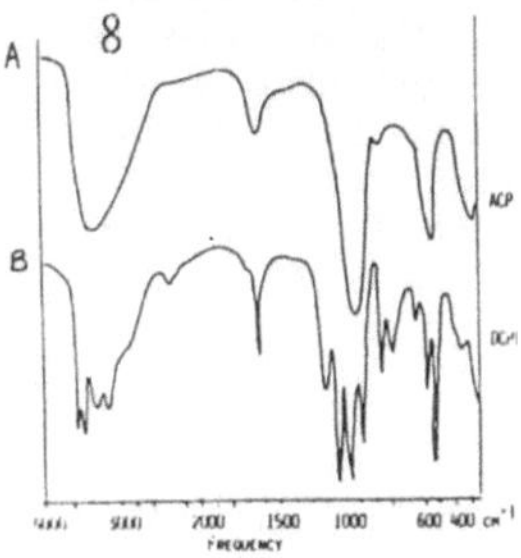

9

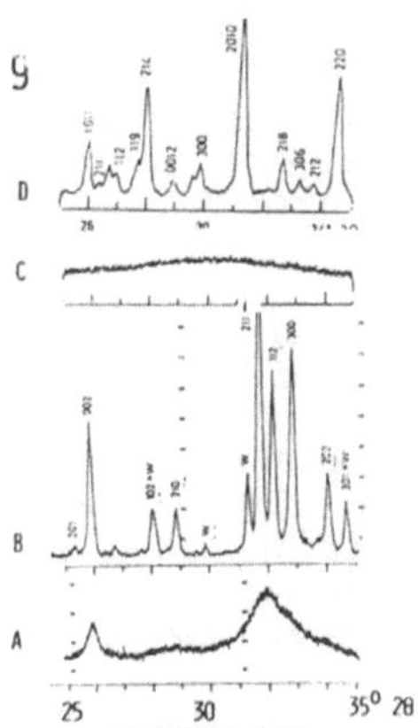

10

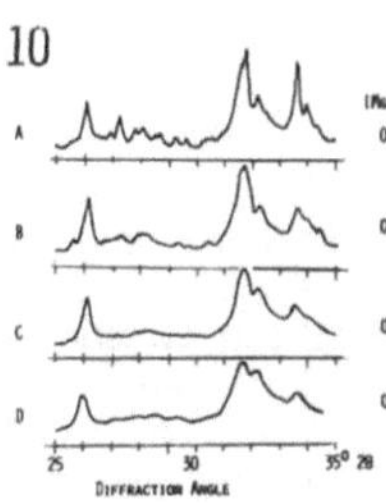

11

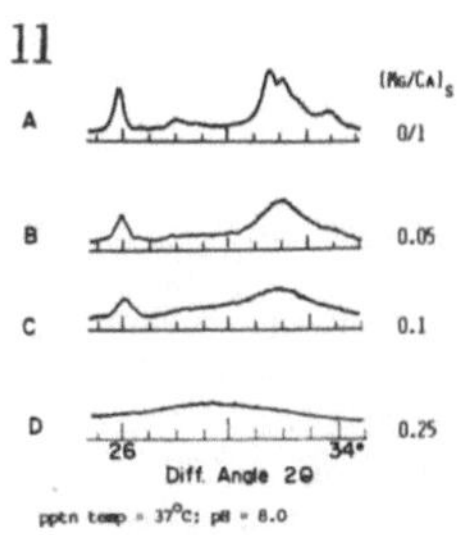

12

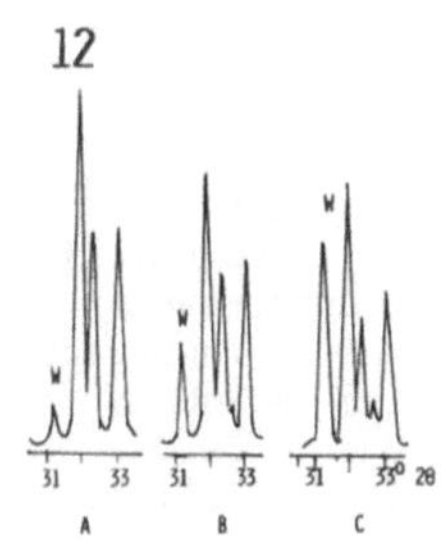

13

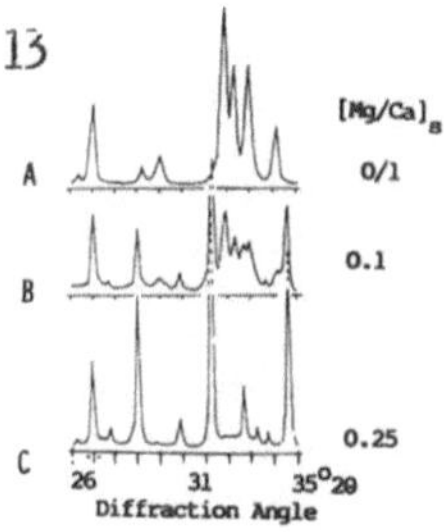

TABLE 1: EFFECT OF MG ON CALCIUM PHOSPHATE FORMATION

$[Mg/Ca]_S$	(A)	(B)	(C)	(D)
0/1	DCPD	OCP	'AP'	'AP'
0.03/1	DCPD	OCP	'AP'	'AP'
0.1/1	DCPD	DCPD + OCP	DCPD + W	DCPD + W
0.25/1	DCPD*	DCPD	DCPD + W	DCPD + ACMP
0.66/1	NO PPT	DCPD	ACMP	ACMP
1.0/1	NO PPT	DCPD	ACMP	ACMP

PPTN TEMP = 37°C; pH, 4.5 (A); 6.8 (B); 7.5 (C); 8.8 (D)
*APPEARANCE OF DCPD AFTER 3H
W = MG-SUBSTITUTED WHITLOCKITE, B-TCMP;
ACMP = MG-CONTAINING ACP

(b) Effect of Mg on ACP formation and transformation. High $[Mg/Ca]_s$ promoted the formation of Mg-containing amorphous calcium phosphate, ACMP (Table 1, Fig. 8A) which transformed to DCPD after 24h in solution maintained at pH 6.8 (Fig. 8B); remained amorphous even after more than 500 hrs at pH 8.4 when solution contains Mg^{2+}, but transformed to an Mg-containing apatitic precipitate after 400 hrs in Mg-free solution [19]. ACMP (Fig. 9C) transformed to b-TCMP after ignition at 600°C (Fig. 9D) similar to that observed after ignition of amorphous Ca-P deposit from non-visceral calcifications (Fig. 5).

(c) Effect of Mg on OCP formation and transformation. At 37°, pH 6.8, the presence of Mg promoted the formation of DCPD at the expense of OCP (Table 1); at 60°, pH 5, increasing $[Mg/Ca]_s$ disturbed the crystallization of OCP (Fig. 10). Mg inhibited the transformation of OCP to AP and to b-TCMP.

(d) Effect of Mg on AP formation. Mg becomes incorporated in the forming apatite to a very limited extent, maximum about 0.4 wt% [12], causing a decrease in the crystallinity of apatite (Fig. 11) and an increase in the incorporation of HPO_4^{2-}. The enhancing effect of Mg on the incorporation of HPO_4 is reflected in the greater b-TCMP/AP ratio in ignited (800°C) AP containing higher Mg content (Fig. 12). The disturbing effect of Mg on the crystallinity of apatite is enhanced by the simultaneous presence of CO_3 and minimized by the simultaneous presence of F^- [17-20]. Ignition at 600°C of apatitic Ca-P containing Mg and CO_3 resulted in the formation of b-TCMP and CO_3-AP (Fig. 9) similar to that observed upon ignition of amorphous and apatitic pathological Ca-P deposits (Fig. 5).

(e) Effect of Mg on the formation and transformation of b-TCMP. Presence of Mg in solution allows the formation of whitlockite from aqueous systems (Fig. 13). Although TCP formation is predicted from thermodynamic considerations [21], pure b-TCP, $Ca_3(PO_4)_2$ can only be obtained by solid state reactions at 1000°C [11,21] [11,21] or by ignition of AP with low Ca/P [17]. Mg-for-Ca substitution in b-TCMP, $(Ca,Mg)_3(PO_4)_2$ is limited to 16 atomic wt% or 3.8 wt% reflected in the lattice parameters [1,3,4,11,15,17,18] and in the IR spectra [18,19]. TGA analyses of b-TCMP shows that it contains HPO_4^{2-} [18]. The morphology of synthetic b-TCMP crystals (Fig. 4D) is similar to that observed in dental calculus (Fig. 4A,4B). The simultaneous presence of F^- or CO_3^{2-} with Mg^{2+} ions in solution suppresses the formation of b-TCMP, favoring instead the formation of (F,OH)-AP or CO_3-AP, respectively [17,18]. While unsubstituted b-TCP is easily hydrolyzed to AP, b-TCMP was not easily transformed to AP under similar conditions.

DISCUSSION

Our studies [4,5,10-20,23] demonstrated the importance of Mg/Ca molar ratio in solution, $[Mg/Ca]_s$, in promoting or inhibiting the in vitro formation and transformation of different types of Ca-P: ACP, DCPD, OCP, b-TCP, AP; modifying the pH/temperature requirements for their formation in the absence of Mg^{2+} ions in the solution [18]. Critical $[Mg/Ca]_s$ have also been shown to form struvite, $MgHPO_4.3H_2O$, instead of DCPD [24,25]; weddelite, $CaC_2O_4.2H_2O$ instead of whewellite, $CaC_2O_4.H_2O$ [25]; these compounds occur as crystalline components of some renal stones [25,26].

The observed effects of Mg on the formation and transformation of synthetic Ca-P may explain in part the formation and stability of biological Ca-P. For example, DCPD was reported present in 'young' (e.g., 3-day calculus) but not in 'older' (e.g. 30-day) calculus [27]. Synthetic systems showed that DCPD can hydrolyze to OCP, AP or b-TCMP depending on the pH and the $[Mg/Ca]_s$ in the oral fluid which fluctuates with stimulation, or Mg intake [28]. XRD pattern of b-TCMP obtained by transformation of DCPD or by direct precipitation is similar to that of some dental calculus specimens (Fig.7). The observed presence of rhombohedral crystals (whitlockite) in arrested dentin caries [29, 30] but not in enamel caries [30] may be due to the higher Mg content of dentin which upon partial dissolution during caries provides the critical $[Mg/Ca]_s$ which favors the subsequent formation of b-TCMP. The presence of different types of Ca-P in pathological but not in normal calcifications may be due to elevated $[Mg/Ca]_s$ in pathological conditions. Based on the effect of CO_3 on crystallinity of apatites [5, 11, 17, 31], the differences in crystallinity among the mineral phases of enamel, dentin and bone (Fig. 6) may be due in part to the synergistic effects of Mg and CO_3.

Acknowledgements. The author gratefully acknowledges the professional collaboration of: Drs. R. Kijkowska, G. Daculsi, I. Orly, J.P. LeGeros, P. Cheng, P. Morales, B. Kerebel, S. Contiguglia, A. Alfrey, T. Abergas; the technical assistance of: W. Torres, R. Zhang, M. Retino, D.J.LeGeros; and the support of NIH Research Grant Nos. DE 04123, DE 07223, S07 RR05332.

REFERENCES

1. Tovborg-Jensen A, Rowles SL: Nature 179: 912-913 (1957).
2. Zipkin I. In Schraer H (ed): Biological Calcification. Appleton-Century Crofts, New York, pp 69-104 (1970).
3. Rowles SL (1967): Bull Soc Chim (France) pp. 1797-1802 (1968).
4. LeGeros RZ, Contiguglia SR, Alfrey AC: Calc Tiss Res 13: 173-185 (1973).
5. LeGeros RZ: Prog Crystal Growth Charact 4: 1-45 (1981).
6. Verbeeck RMH: In: Tooth development and caries. vol. 1. Driessens FCM, Woltgens (eds). CRC Press, pp 95-156 (1986).
7. Hallsworth AS, Robinson C, Weatherell JA: Caries Res 6: 156-168 (1972).
8. Brown WE: Nature 196: 1048 (1962).
9. Boskey AL, Posner AS: Mater Res Bull 9: 907-916 (1974).
10. LeGeros RZ, Shirra WP, Miravite MA, LeGeros JP: In: Colloque Int CNRS No. 230: Physico-chimie et cristallographie des apatites d'interet biologique, Paris, 1973, pp. 105-115.
11. LeGeros RZ: Crystallographic studies of the carbonate substitution in the apatit structure. PhD Thesis, New York University (1967).
12. LeGeros RZ: In: Fearnead RW, Suga S (eds): Tooth Enamel IV, Amsterdam, Elsevier, pp. 32-36 (1984).
13. LeGeros, Kijkowska R, Abergas T, LeGeros JP: J Dent Res 65: 8, abstr 682 (1986).
14. Daculsi G, LeGeros RZ, Jean A, Kerebel A: J Dent Res 66: 1356-1359 (1987).
15. LeGeros RZ: J Dent Res 53: 45-50 (1974).
16. LeGeros RZ, Orly I, LeGeros JP. Scan Micr 2: 345-356 (1988).
17. LeGeros RZ: In: Myers H (ed): Monographs in Oral Sciences Vol. 15 (in press).
18. LeGeros RZ , Daculsi G, Kijkowska R, Kerebel B: In: Itokawa V and Durlach J (eds) Magnesium in Health and Disease: J Libbey & Co., pp. 11-19 (1988).
19. LeGeros RZ, Chang, X-F: J Dent Res 68: 215: abstr 268 (1989)
20. LeGeros RZ, Kijkowska R, Jia W, LeGeros JP: J Fluor Chem 41: 53-64 (1988)
21. Brown WE: In: Environment Phosphorus Handbook. J. Wiley & Co., New York (1973).
22. Terpstra RA, Driessens FCM, Schaeken HG: Z Anorg Allg Chem 507: 206-212 (1983).
23. Cheng P-T, Grabher JJ, LeGeros RZ: Magnesium 7: 123-132 (1988).
24. Abbona F, Madsen HEL, Boistelle R: J Crystal Growth 74: 581-590 (1986).
25. LeGeros RZ, Morales P: Invest Urol 11: 12-20 (1973).
26. Sutor DJ, Scheidt S: Br J Urol 40: 20-28 (1968).
27. Schroeder H: Formation and inhibition of dental calculus. Hans Hubert Publ, Vienna.
28. Dawes C. Caries Res 1: 333-342 (1967).
29. Takuma S, Ogiwara H, Suzuki H: Caries Res 9: 278-285 (1975).
30. Daculsi G, Kerebel G, LeCabellec MT, Kerebel LM: Caries Res 13: 190-202 (1979).
31. LeGeros RZ, LeGeros JP, Trautz OR, Shirra WP: Adv X-ray Anal 14: 57-66 (1971).

CHAPTER 3.7
Manganese Accumulation in Freshwater Bivalves

S. YOSHIOKA and M. TERAI
Aichi University of Education, Kariya, Japan and Tokyo Metropolitan Institute of Technology, Tokyo, Japan

Key words: *Hyriopsis schlegeli*, Manganese, Shell, Aragonite.

INTRODUCTION

Marine carbonate sediments consist of carbonate skeletons such as coral and molluscan shell. Since these organisms uptake various chemical elements from environmental seawater and form calcium carbonate skeleton, the relation between trace element content of calcareous skeletons and environmental condition, and the estimation for the chemical natures of ancient seawater have been studied [1]. And the distribution coefficients of many chemical elements between calcium carbonate and aqueous solution have been measured by many investigators [2]. But these skeletons are not formed directly from environmental water but from extrapallial fluid existing between the valve inner layer and mantle. Therefore, the study on the chemical composition of extrapallial fluid as well as skeleton and environmental water is needed for the understanding of a biogenic calcification mechanism. It was reported that many metal elements are concentrated from environmental water to extrapallial fluid. And the distribution coefficients of metals between extrapallial fluid and calcareous skeleton were shown [3].

Some marine bivalves are used as a biological indicator for metal pollution through the measurement of the uptake of metals into various organs of mussels [4]. The present authors have measured the trace element concentrations in several marine and freshwater bivalves by neutron activation analysis, and have noticed that large-shelled freshwater bivalves such as *Hyriopsis schlegeli* and *Anodonta woodiana* accumulate manganese, and further that different amounts of manganese are contained in different shell layers having the same mineral. It has been already reported that *Hyriopsis schlegeli* shows relatively a high manganese content [5].

The aim of this study is to analyse manganese to understand the manganese accumulation in freshwater bivalves.

MATERIALS AND METHODS

Samples Freshwater bivalves, *Hyriopsis schlegeli* and *Anodonta woodiana*, used in the present study were collected from Oe River, Gifu Prefecture. Collected bivalve was slightly opened with shell opener and extrapallial fluid was carefully sucked using a syringe. A part of the extrapallial fluid was filtered through a 0.45 μm membrane filter and then 10 $\mu\ell$ of conc HNO_3 was added into 1 ml of the filtrate. The soft part separated from shell was dissected into mantle surface, mantle margin and gill. These soft samples were completely

dried with diphosphorus pentaoxide in a desicator. The shell was washed with pure water and dried in air. It was cut into individual shell layers and the individual layer was ground.
Experimental Chemical analyses were carried out by NAA (neutron activation analysis) and ICP (inductivity coupled plasma atomic emission spectrometry). For NAA, an extrapallial fluid sample was dried in air after dropping its 100 $\mu\ell$ on a small No.6 filter paper. The dried filter paper was heat-sealed in a clean polyethylene bag. Dried soft tissues and powdered shell samples were weighed and heat-sealed in clean polyethylene bags. The samples were irradiated in a TRIGA MARK II reactor at Rikkyo University and the γ-ray spectra were measured. For ICP (LEEMAN LABS, INC), powdered shell samples were dissolved with 2N nitric acid, and dried soft tissue and extrapallial fluid were decomposed with sealed teflon vessels (Sanai Kagaku).
Thin sections prepared from lengthwise cross sections were analyzed for calcium, sodium and manganese by EPMA using a JEOL JXA-50A electron probe microanalyzer.

RESULTS AND DISCUSSION

The major constituents in seawater such as Na, K, Ca, Mg, Sr, Cl and Br are contained in extrapallial fluid and also shell of marine bivalve. These elements are selectively uptaken in freshwater bivalve in regardless of their low concentrations in environmental water. The different shell layers of *Hyriopsis schlegeli* and *Anodonta woodiana* are only aragonite. The

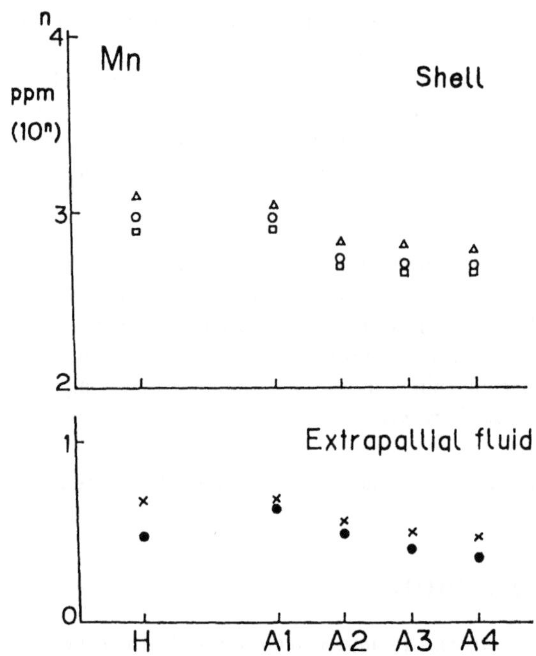

Fig. 1. Manganese concentrations in shell and extrapallial fluid.
H: *Hyriopsis schlegeli*; A: *Anodonta woodiana*
Shell length, A1:13.0cm; A2:17.5cm; A3:18.5cm; A4:21.5cm
△: inner layer; □: outer layer; ○: ventral margin
●: 0.45μm filtrate; ×: no treatment

sodium and strontium contents of aragonitic shell are very similar but the heavy metal contents are different among samples and the portions of the same sample. The high manganese content of freshwater bivalve is characterized.

Figure 1 shows the manganese concentrations in shell and extrapallial fluid of freshwater bivalve, *Hyriopsis schlegeli* and *Anodonta woodiana*. These manganese concentrations of *Hyriopsis schlegeli* were usually higher than that those of *Anodonta woodiana*. The shell of *Hyriopsis schlegeli* was observed to be very thicker and solider than that of *Anodonta woodiana*. The figure shows that the manganese content of inner shell layer is higher than that of outer layer, and the manganese content decreases slightly in larger bivalves.

TABLE 1. Heavy metal contents of soft tissues (μg/g in dry weight)

		Mn	Fe	Zn
freshwater bivalves				
Hyriopsis schlegeli	Gill	15000		
	Mantle	170 - 11000	1000 - 1200	98 - 290
Anodonta woodiana	Gill	7300 - 16000	1200 - 4400	410 - 840
	Mantle surface	330 - 6200	550 - 3300	64 - 320
	Mantle margin	160 - 3400	580 - 3900	41 - 320
marine bivalves				
Mytilus edulis	Gill	10 - 26	210 - 440	42 - 64
	Mantle surface	2.0 - 8.7	100 - 110	25 - 27
	Mantle margin	9.2 - 13	460 - 890	46 - 49
Saxidomus purpuratus	Gill	14	230	180
	Mantle surface	0.7	110	46
	Mantle margin	4.3	540	76

Table 1 shows the heavy metal contents of the soft tissues. They are very high in freshwater bivalves as compared with those in marine bivalves. Especially, the manganese concentrations are markedly high in gill and mantle surface by which an inner shell layer is secreted.

Figure 2 shows the distributions of Ca, Na and Mn along the cross section of *Hyriopsis schlegeli* shell. Various parts of shell were analyzed from inner to outer layer. In middle layer to outer layer, The sodium content is higher than the manganese content. It seems reasonable because of aragonitic shell layer. But from middle layer to inner layer, the manganese content increases markedly, whereas the sodium content is almost constant. The difference in the manganese content between outer and inner aragonitic shell layers suggests the difference in the mode of calcium carbonate secretion. A similar result has been reported for the strontium content[6]. The manganese accumulation in inner shell layer seems to be influenced by a biochemical control and associated with the behavior of phosphorus. The figure represented as D in Fig. 2 shows the decrease in specimen current at EPMA. It was reported that manganese in calcium carbonate resists against the destruction of calcium carbonate sample by electron beam (Kato, in JEOL Application Note). The decrease in specimen current at outer layer was larger than at inner layer. It is suggested that the shell material from middle layer to the outer layer is composed of pure aragonitic crystal but the inner shell layer contains some manganese impurity.

Figure 3 indicates the relationship between manganese and phosphorus contents of *Hyriopsis schlegeli* shell is positively leaner. Shells of sample numbers 1 to 4 are young. Sample No.1 is the smallest shell with 2 cm length. Sample No.6 was the largest shell with 22 cm length and showed two different appearances such as chalky and nacreous portions. In this figure, it is

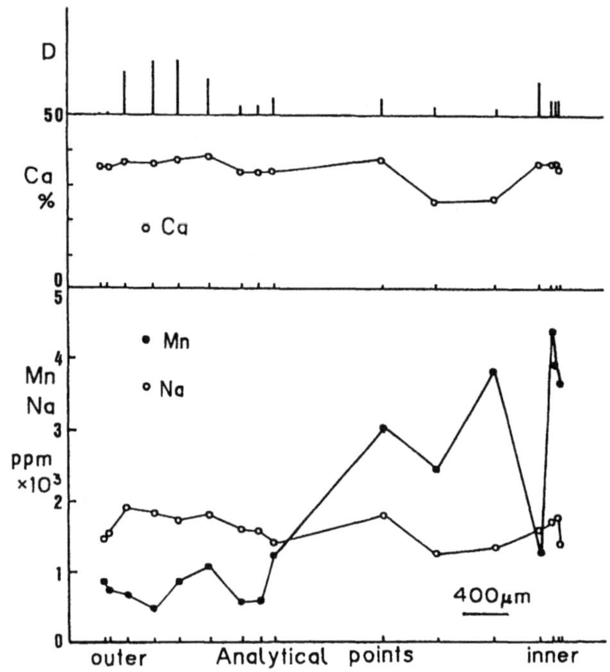

Fig. 2. Distribution of Ca, Na and Mn in cross section of *Hyriopsis schlegeli* shell.

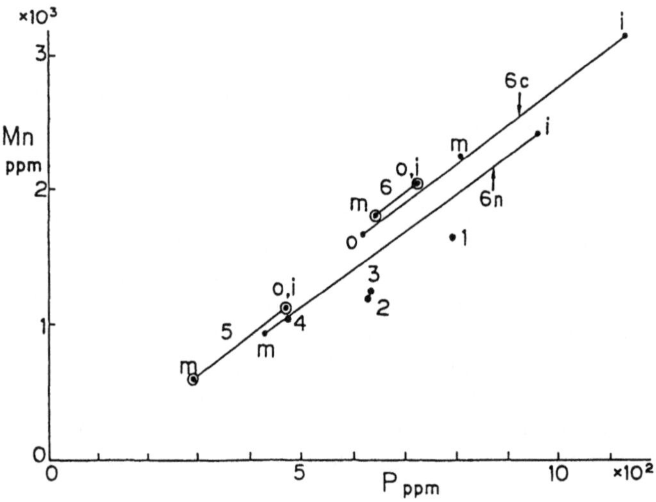

Fig. 3. Relationship between Mn and P contents of *Hyriopsis schlegeli* shell.
Shell length, 1:2.2cm; 2:3.0cm; 3:3.1cm; 4:4.2cm; 5:17cm; 6:22cm
i: inner layer; o: outer layer; m: ventral margin
c: chalky portion; n: nacreous portion

clear that manganese and phosphorus are accumulated in inner shell layer secreted by mantle surface than outer layer, and in chalky and thick-shelled portion than nacreous portion. The decrease in manganese and phosphorus contents was observed for growing individuals except for older shell (No.6). It was reported that a negative correlation between cadmium content and mussel size [4]. These results suggested that such elements might be uptaken at early stage of calcification and might be moved from soft tissue to shell.
It has been reported that manganese are strongly adsorbed on calcium carbonate [7]. Phosphate and organic materials are adsorbed on aragonitic calcium carbonate. A further experiment on the manganese uptake in calcareous skeleton is needed.

ACKNOWLEDGMENTS

The authors wish to thank Emer. Prof. Y. Kitano, Nagoya University, for his pertinent guidance throughout the present study and for his critical reading of this manuscript. Our thanks are also due to Dr. K. Suzuki, Nagoya University for technical help of EPMA.

REFERENCES

[1] Lowenstam HA (1961) Jour Geol **69**: 241-260
[2] Kitano Y, Kanamori N, Oomori T (1971) Geochem J **4**: 183-206
[3] Kitano Y, Kasai K, Wada K (1982) In: Assimilative Capacity of The Oceans for Man's Wastes. SCOPE/ICSU Academia Sinica, Taipei, Republic of China, pp. 232-237
[4] Coleman N, Mann TF, Mobley M, Hickman N (1986) Mar Biol **92**: 1-5
[5] Wada K, Fujinuki T (1974) Bull Natl Pearl Pes Lab **18**: 2085-2110
[6] Hallam A, Price NB (1968) Geochim Cosmochimi Acta **32**: 319-328
[7] McBride MB (1979) Soil Sci Soc Am J **43**: 693-698

CHAPTER 3.8

Scanning and Transmission Electron Microscopy, X-ray Microanalysis and Electron Diffraction of Pineal Concretions in Aging Rats

W. HUMBERT, M. MASSON-PEVENT, J.C. VOEGEL[1], J. HEMMERLE[1], and P. PEVET
U.L.P., CNRS-URA 1332 "Neurobiologie des fonctions rythmiques et saisonnières", Laboratoire de Zoologie, 12 rue de l'Université, 67000 Strasbourg, France [1]INSERMU157 ULP, 1 place de l'Hôpital, 67000 Strasbourg, France

Key words : Pineal gland - Concretions - Calcium - Aging - Rat (Wistar)

INTRODUCTION

The pineal gland of the rat is considered to play an important neuroendocrine role in chronobiology, circadian rhythmicity and transduction of certain external stimuli [1]. Several mammalian species, including man, contain calcified concretions (corpora arenacea, brain sand) in their pineal gland [2,3,4]. The aim of the present study was to examine these concretions in aging rats by means of SEM, TEM, X-ray microanalysis and electron diffraction. Deep-freezing followed by freeze-drying, is a means of tissue preparation that has been shown to be a satisfactory alternative for electron microscopic tissue preparation [5]. An attempt was made to study the mechanism of formation of these structures and to understand their possible role in pineal physiology.

MATERIALS AND METHODS

Animals 28-months-old male Wistar rats and young adult males (2-3 months old) were used. They were housed in plastic cages in a room under artificial photoperiod (12 L/12 D) with food and water at libitum. They were sacrificed by decapitation at 9 a.m.
Scanning electron microscopy - X-ray microanalysis. Pineal glands were quickly removed from the brain and frozen in a mixture of liquid propane (80%) and isopentane (20%) cooled to -196°C by liquid nitrogen (LN_2) [6]. Bulk fractures of the pineals were obtained by fracturing the tissue with chilled instruments on a LN_2 cooled brass. The tissue was then carried under LN_2 to a freeze-drying apparatus operating for 48 h under LN_2 and under the following conditions : vacuum, $3\text{-}10^{-5}$ torr ; temperature of the stage carrying the specimens, -90°C. At the end of the drying process, the stage was progressively heated to +30°C. Samples were then mounted on carbon planchettes using carbon glue (Leit C) and coated with carbon or gold/palladium. All observations were made with a Philips PSEM 501B scanning electron microscope. The elemental composition of the concretions was analysed using an energy-dispersive spectrometer (Link System series 2) attached to the scanning electron microscope (conditions of analyses : voltage, 20 KV ; beam current, 20µA). Analyses were exclusively performed on carbon-coated samples. Semi-quantitative results were expressed as counts/50 sec. X-ray microanalysis included the sum total of Ca emissions from 10 scanned areas ($20\mu m^2$) of intra- and extracellular tissue compartments of each fractured pineal. Areas with visible concretions or calcification foci were avoided in this analysis.
Transmission electron microscopy - X-ray microanalysis. Pineal glands were quickly removed from the brain, sliced into two parts and prepared for analysis by the following method : fixation with 3% glutaraldehyde in 0.1 M cacodylate buffer (1 h) and prepared for conventional electron microscopy. Samples destined for X-ray microanalysis were fixed unosmicated. They were dehydrated with a series of alcohol and embedded in araldite.
Unstained, non osmicated and carbon coated sections were analyzed with a transmission electron microanalyser CAMEBAX-TEM. To minimize background emission a graphite-beryllium holder was used. The probe was fitted with vertical wavelength dispersive spectrometers using a (PET) crystal for the analysis of calcium (accelerating voltage : 45 KV ; specimen current : 150 nA ; beam diameter : 100-200 nm ; counting time : 30 sec).
Electron Diffraction. Electron diffraction was performed using a JEOL 100 B electron

microscope working at 100 kV. References were obtained under the same conditions with the aid of grids coated with evaporated LiF. Measurements on the enlarged photographs were performed with the aid of a precision ruler (Siemens, type 941, Erlangen). The photographs were calibrated using the LiF rings. Indexing was performed using ASTM files.

RESULTS

Scanning electron microscopy. Freeze fractured pineals of young rats (2-3 months old) only showed concretions in 1 out of the 9 samples examined. All the old rats, on the other hand, had distinguishable concretions.

Most of the observed concretions were located in the periphery of the gland. Concretions appeared as grape-like structures built up from aggregated ovoid nodules believed to be calcification foci (o1-4μm) (Figs.1-3). Rhomboedric structures were often observed to be associated with these globules.

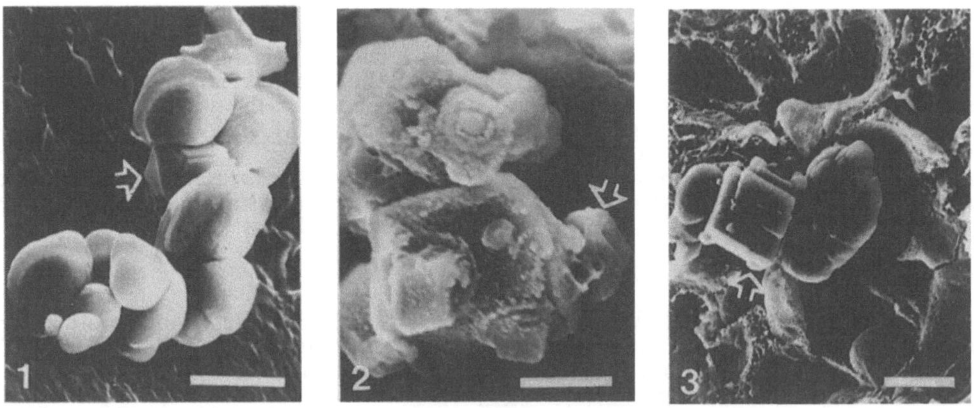

Fig. 1-3. SEM of pineal concretions observed on the surface of a frozen fractured pineal. Note rhomboedric structures (arrow) among ovoid globules ; Bar = 4μm.

Elemental analysis of pineal concretions. Calcium emission was measured on fractured pineals by means of X-ray microanalysis. Results, expressed as counts/50 sec, were a semi-quantitative representation of intra- and extracellular calcium (areas with visible concretions were avoided in this analysis). Figure 4 shows an increase of Ca X-ray emission in old rats (B) compared with young ones (A).

A typical X-ray spectrum of pineal tissue is represented in Figure 5a. In addition to large peaks of potassium, phosphorus and sulphur, smaller amounts of silicon, chlorine, calcium and traces of aluminium, magnesium, iron, copper and zinc were detected. Areas heavily loaded with calcium but without any visible concretions were analysed. Typical prominent calcium peaks associated with Al, Si, P, S, Cl, K, and traces of Mg, Fe, Cu and Zn are represented on Figure 5b. These areas may represent intracellular "calcification foci" where genesis of concretions may occur. Figure 5c represents the spectrum of a concretion similar of that represented in Fig. 1. A marked calcium peak is associated with Si, P, S, Cl and traces of K, Cu, Zn.

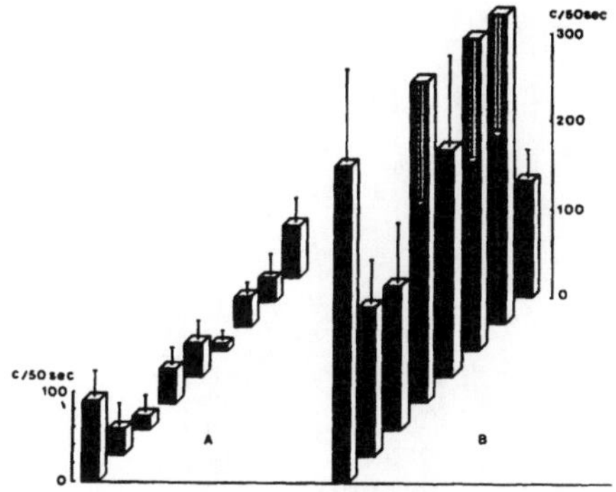

Fig. 4. Mean (+ SEM) of X-ray emissions (counts/50sec) of calcium on concretion-free areas (20µm^2) of freeze-fractured pineals from young (A) and old (B) rats. Each histogram represents the mean of 10 analyzed areas of one pineal gland.

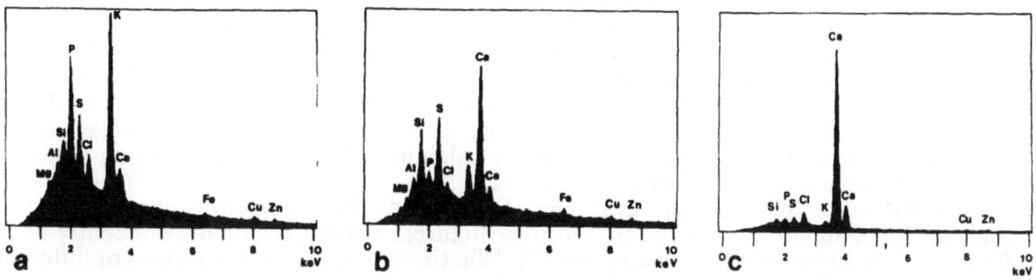

Fig. 5. Elemental energy dispersive spectra (0-10 keV) showing the composition of specific compartments obtained from fractured freeze-dried pineal glands of old rats.
a. X-ray spectrum from pineal tissue. Note the prominent peak of K, the noticeable peaks of P, S, Cl, Ca and small peaks of Mg, Al, Si, Fe, Cu, Zn.
b. X-ray spectrum of a tissue area rich in Ca. In addition to the large peak of Ca, note the presence of Mg, Al, Si, P, S, Cl, K, Fe, Cu, Zn.
c. X-ray spectrum of a concretion similar to that presented in Fig. 1. In addition to a large peak of Ca, smaller amounts of Si, P, S, Cl, and traces of K, Cu and Zn are detected.

Transmission electron microscopy. Morphologically there are two types of pinealocytes (light and dark) in the pineal gland. Concretions are built up in dark cells, which seem to be more numerous in old rats than in young ones. They are generally amorphous, with alternated electron dense and clear layers or crystalline, characterized by needle-like crystallites. Electron diffraction of an araldite embedded concretion demonstrates the presence of a typical hydroxyapatite phase (Fig. 6).

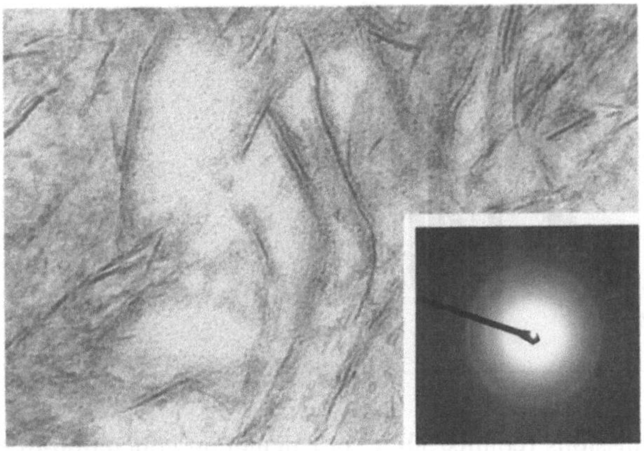

Fig. 6. Typical needle-like crystallites observed by TEM and electron diffraction pattern characteristic of hydroxyapatite.

DISCUSSION

The data presented here are the first to demonstrate that the presence of such concretions is an age-related problem ; they specifically show that old rats have a higher pineal calcium content than young rats. All the pineals of old rats (28 months old), for example, have concretions whereas only 1 out of 9 young rats (2-3 months old) do. In this study the high content of total calcium observed in the older animals suggests a relationship between the appearance of concretions and a calcium-rich environment which could in turn be responsible for the growth of concretions.
Mainly restricted to the subcapsular region of the gland, the concretions consist of ovoid globules often associated with rhomboedric structures. Similar forms have recently been described in the intestine of sea-water eels [7]. Whether these structures represent different crystalline forms of the same concretion remains an unanswered question.
Concretions have been reported to consist of both organic and inorganic substances. The organic substances have been demonstrated to be proteoglycans and glycoproteins [8]. Japha et al.[9] have reported that the matrix of the concretions is composed of carbohydrates, probably acid mucopolysaccharides complexed to a protein. This idea is supported by our analysis, which demonstrates the presence of sulphur in these concretions. In the present study we have also revealed the presence of inorganic substances by X-ray microanalysis. A noticeable Ca peak with smaller amounts of P, S, Cl, Si and traces of Cu and Zn have been observed. Likewise we have also found a high content of trace elements in uncalcified pineals of rats and of European hamsters (Humbert, unpublished results).
To date, only human and gerbil pineal concretions have been intensively studied. It has been shown [4,10,11,12] that pineal concretions contain P and Ca, and smaller amounts of Mg, S, Sr, which has led these authors to conclude that these concretions are mainly built up from hydroxyapatite and calcium carbonate. Moreover, traces of Si, Mg, Zn, Fe, Mn and Cu (review in [4]) have been described. Iron [13] has also been detected in these concretions. In the present study, we have detected all these trace elements with the exception of Mg, Sr and Mn. The concretions in the rat thus appear to be similar to those in gerbils and humans.
Data presented in this article show that the appearance of concretions in old rats is related to a general increase of calcium in the pineal gland. Is this increase of calcium and the appearance of concretions related to an increase in the intensity of metabolic activity in the gland as suggested in the mongolian gerbil [9,14] or is it related to a degenerative state ? This question remains unanswered for the moment and needs further investigation.
Concerning the origin of the concretions, several hypotheses have been put forward. It has been reported [9,14,15] that concretions in the Mongolian gerbil originate from vacuolated pinealocytes. The pineal cell degenerates, breaks down and the concretion is released into the extracellular space [14,16]. Our present results do not answer this

question but the possibility that pineal calcification in old rats is a different process from that observed in gerbils cannot be excluded.
In conclusion, our study demonstrates that pineal concretions in rats are dependent upon aging and upon the calcium content of the pineal gland. Thus, pineal calcifications in old rats appear to be an indicator of aging and/or a degeneration state. This can be related to the observations of Quay [17] who found calcifications of the human pineal gland in association with age and disease.

REFERENCES

[1] Wurtman RJ, Axelrod J, Kelly DE (1968) The pineal. Academic Press, New York

[2] Bargmann W (1943) Die Epiphysis cerebri. In : Möllendorff WV (ed), Handbuch der mikroskopische Anatomie des Menschen Bd VI/1/, Springer Berlin , pp 309-502

[3] Vollrath L (1981) The pineal organ. Springer, Berlin, Heidelberg, New-York

[4] Welsh MG (1985) Pineal calcification : structural and functional aspects. Pineal Res Rev 3:41-68

[5] Humbert W, Simmoneaux V, Kirsch R (1990) Scanning electron microscopy and X-ray microanalysis of freeze-dried digestive mucus in Anguilla anguilla L. : Modification of ion content during the early steps of secretion. Biol Cell 68 : 51-55

[6] Jehl B, Bauer R, Dörge A, Rick R (1981) The use of propane isopentane mixtures for rapid freezing of biological specimens. J Microsc 123:307-309

[7] Humbert W, Voegel JC, Kirsch, R. Simonneaux V (1989) Role of intestinal mucus in crystal biogenesis: an electron-microscopical diffraction and X-ray microanalytical study. Cell Tiss Res 255:575-583

[8] Palladini G, Alfei, L. Appicciutoli L (1965) Histochemical observations concerning the corpora arenacea of human epiphysis. Arch Ital Anat Embriol 70:253-270

[9] Japha JL, Eder TJ, Goldsmith ED (1976) Calcified inclusions in the superficial pineal gland of the Mongolian gerbil, Meriones unguiculatus. Acta Anat 94:533-544

[10] Krstic R (1976) A combined scanning and transmission electron microscopic study and electron probe microanalysis of human pineal acervuli. Cell Tissue Res 174:129-137

[11] Krstic R, Golaz (1977) Ultrastructural and X-ray microprobe comparison of gerbil and human pineal acervuli. Experientia 33:507-508

[12] Allen DJ, Allen JS, Didio LJA, McGrath JA (1981) Scanning electron microscopy and X-ray microanalysis of the human pineal body with emphasis on calcareous concretions. J Submicrosc Cytol 13 : 675-695

[13] Earle KM (1965) X-ray diffraction and other studies of the calcareous deposits in human pineal glands. J Neuropathol Exp Neurol 24:108-118

[14] Lukaszyk A, Reiter RJ (1975) Histophysiological evidence for the secretion for polypeptides by the pineal gland. Am J Anat 143:451-464

[15] Welsh MG, Reiter RJ (1978) The pineal gland of the gerbil Meriones unguiculatus I. An ultrastructural study. Cell Tissue Res 193:323-336

[16] Reiter RJ, Welsh MG, Vaughan MK (1976) Age related changes in the intact and sympathetically denervated gerbil pineal gland. Am J Anat 146:427-432

[17] Quay WB (1974) Pineal chemistry, Charles C Thomas, Springfield, pp 54-58

4 Evolution

CHAPTER 4.1

On the Initiation of the Hard Tissues of Animals During the Late Precambrian to the Early Cambrian

M. OMORI

Faculty of General Education, Azabu University, Kanagawa, 229 Japan

Key words: Precambrian, Hard Tissue, Animal, Ediacara, Glaciation,

INTRODUCTION

In recent years, scientific knowledge concerning the fossil metazoan fauna belonging to the late Precambrian and the earliest Cambrian have been accumulating remarkably. So there are many papers discussing the phylogenesis of the metazoa during these period. Lowenstam & Margulis [1] once wrote a paper showing the basic idea on the genesis of the hard tissues of metazoa during this period. In this paper, the writer wishes to summerize the recent advancement of the study of fossil metazoa and discuss the geological environments which affected the initiation of the hard tissues of metazoa during these period.

METAZOAN FOSSILS FROM THE LATE PRECAMBRIAN

Since the first discovery of the Ediacara fauna [2, 3], consisting of rich soft-bodied animals, various localities of the world yielded simialr fossil fauna as shown in Figure 1 [4, 5]. Among

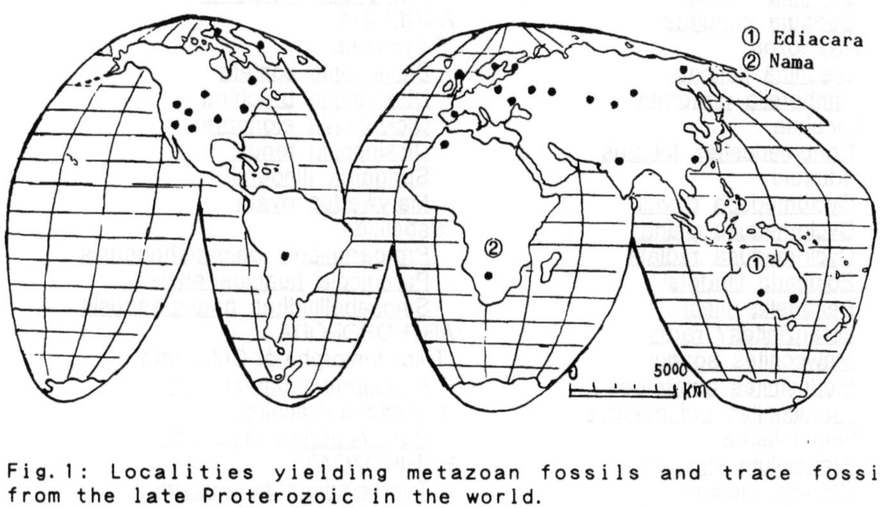

Fig. 1: Localities yielding metazoan fossils and trace fossils from the late Proterozoic in the world.

these localities, the Nama Group distributed in South Africa occurred also trace fossils made by metazoa [6], which are called Nama ichnofauna by the writer. Figure 1 includes the localities of fossil ichnofauna correlative to the Nama ichnofauna. The absolute age of the Ediacara fauna and that of the Nama ichnofauna have been considered to be 700 MaBP, however those of formations at the localities cited in Figure 1 are more or less variable ranging from 900 MaBP to 600 MaBP. Among them the oldest occurrence of metazoa is from Southeastern China [7], which yields worm-like animals such as Parenicola, Protoarenicola and Sinosabellitidites belonging to Sabellitida with algal fossils of Chuaria and Tawnia. The geological age of these formations are calculated to be 840 MaBP (Rb-Sr). Table 1 shows the main components of the Ediacara fauna and its correlatives and Table 2 cites the main trace fossils with the interpretation of habitat [8-12]. Beside these metazoan fossils, multicellular algal fossils such as Tawnia, Proterotawnia, Lanceforma, Grypania and Helminthoidichnites? are reported from the older horizon near to 1300 MaBP.

Recently, small shell fossils (SSF) are reported from the Precambrian and Cambrian boundary in various fields. SSF from the

Table I Main composition of the Ediacara fauna and its correlatives.

Taxic position and specific name	Taxic position and specific name
CNIDARIA	Charniodiscus arboreus
Hydorozoa Chondorophora	Charniodiscus oppositus
Chondroplon bilobatum	Charniodiscus concentricus
Ovatoscutum concentricum	Glaessenerina grandis
Eoporpita medusa	Glaessenerina longua
Spriggia wedae	Phyllozoon hansenni
Spriggia annulata	ANNELIDA
Scyphozoa	Polychaeta
Brachina delicata	Dickinsonia costata
Kimberella quadrata	Dickinsonia brachina
Conulata	Dickinsonia elongata
Conomedusites lobatus	Dickinsonia tenuis
Medusoid	Spriggina flindersi
Cyclomedusa davidi	Marywadea ovata
Cyclomedusa plana	Sabelitida
Cyclomedusa radiata	Protoarenicola baiguashanensis
Ediacaria flindersi	Parenicola huaiyuanensis
Beltanella gilesi	Sinosabellitidites huaiyuanensis
Lorenzinites? rarus	ARTHOROPODA
Mawsonites spriggi	Trilobitomorpha or Chlicerata
Medusinites asteroides	Praecambridium sigillum
Rugoconites enigmaticus	Crustacea Brachiopoda
Pennatulacea	Parvanchorina minchami
Pteridinium simplex	ECHINOZOA?
Charnia masoni	Tribrachidium heraldicum
Paracharnia dengyingensis	PROTOCONODONT or
	Spine of CHAETOGNATHA
	Protohertzina sp

Table 2 List of trace fossils implied in the Nama ichnofauna and its correlatives and habitat of their host animals.

Ichnogenus	Habitat of host animal	Ichnogenus	Habitat of host animal
Skolithos	Dwelling burrows (infauna)	Helminthoidichnus	Horizontal feeding burrows or trails (benthos-fragilis,epifauna)
Archaeichnium	Feeding burrows (benthos-fragilis)	Torrowangea	ditto (ditto)
Planolites	ditto (ditto)	Didymaulichnus	ditto (ditto)
Bunyerichnus	Feeding burrows or trails (benthos-fragilis)	Curvolithus	ditto (ditto)
Buchholzbrunnichnus	ditto (ditto)	Scolicia	ditto (ditto)

uppermost Precambrian (Vendian) consists of conoidal microfossils systema uncertain, most of which have phosphatic skeleton [13] and SSF from the lowermost Cambrian (Tommotian) consists of Brachiopoda, Porifera, Mollusca and Hyolitha beside the former microfossils. The skeleton of the latter SSF mainly consists of calcareous matter.

It is noticeable that a chitinozoan has been reported from the late Precambrian Chuar Group in Arizona, USA [14]. This fossil possibly belonged to heterotrophic protista or primitive metazoa. On the other hand, the oldest fossil records of most of the Protozoa, the likely ancestor of metazoa is known from the early Paleozoic, so the fact that a chitinozoan occurred from the late Proterozoic might show the linkage of metazoa from its ancestoral form. The absolute age of the Chuar Group is supposed to be 750 ± 100 MaBP.

GEOLOGICAL ENVIRONMENT DURING THIS PERIOD

According to the Hargrave's model [15] on the development of the ocean and the continent, Proterozoic (2500 to 600 MaBP) is a stage when the scale of plate interaction and intercontinental drift increased with the growing oceanic area, and land area increased as the sea became more shallowed until the continents of approximately the present-day dimensions emerged finally. He remarks on the following facts, which characterized the geological environment during the late Proterozoic (1600 to 600 MaBP) as shown in Fig. 2 [16].

(1) Siliceous rocks with microfossils widely distributed since 1600 MaBP, which might have resulted by progressive segregation of sial and hydrospher into continents and oceans as pointed out by Hargrave. So the zoning of these siliceous rock into three was possible.

(2) Stromatolites came to be extensively developed since the same age, and they show also four zones of a Riphean stage by the morphological change of stromatolites and microphyllites.

(3) Fine clastic sediments with microfossils came to be distinguishable as a result of the extensive raising of the continents, making the proper biotope for the development of metazoa to the close of Proteozoic.

(4) Glaciogenic sediments are recognizable during the late Proterozoic to the early Cambrian. From their distribution, four

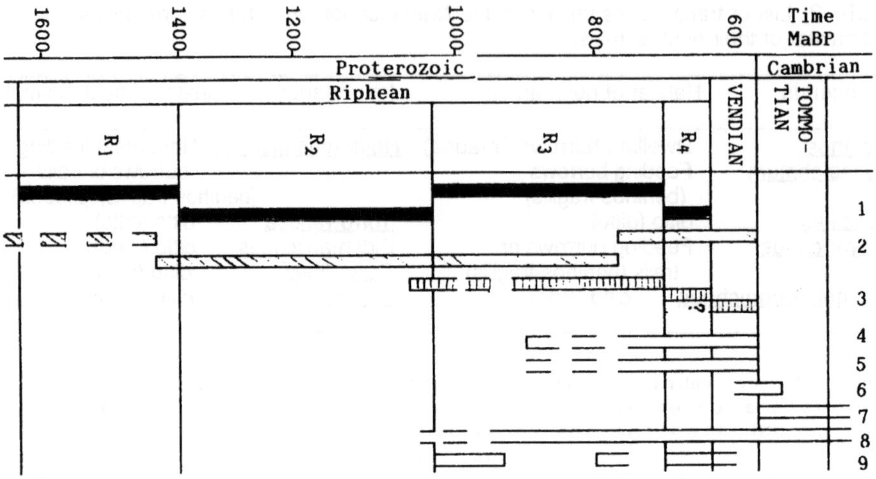

Fig. 2: Change of biotas and geological environments during the late Proterozoic and the early Cambrian. Modified from the work of Trompette.
1;Zones by stromatolites and microphytolites, 2;ditto by microbiotas from siliceous rocks, 3;ditto by microbiotas from fine clastic rocks, 4;Ediacara fauna, 5;Nama ichnofauna, 6;SSF, 7;Skeletal fauna, 8;Metaphyta, 9;Glaciogenic sediment

glaciations occurred with intervals of 50 to 100 Ma, the last of which was seemed to be the largest (Figure 2).
Considering the above-cited geological environments, he supposed that the extensive epicontinental shallow sea came to develop at the end of the Precambrian and the environment suitable for the life of benthic metazoa matured with rich nutrient salts.
These glaciations often set the earth in an ice-box like environment and the predominant dry climate of the terrain produced the enrichment of Ca^{2+} and HCO^- in the surface water. As a result of this, the wide distribution of stromatolites developed in the shallow waters.
On the other hand, in the deep sea bottom area, cold water mass with a low pH rich phosphate was produced. And when the sea-level rose after glaciation, the warm shallow sea water with relatively high pH precipitated forming apatite and the phosphorous sediments which accumulated on the sea-floor. Cook & Shergold [17] studied the genesis of the world-wide distribution of phosphorous sediments and phosphatite around the Precambrian and Cambrian boundary as shown in Figure 3. He concluded that these phosphorous sediments had been produced by the upwelling of deep sea water rich in phosphate in this way.

ON THE ORIGIN AND DEVELOPMENT OF METAZOA WITH SPECIAL REFERENCE TO THE INITIATION OF THEIR HARD TISSUES

The first appearance of eukaryota is considered to be 1400 MaBP, and metaphyta such as multi-cellular algae might have appeared near the time of 1300 MaBP as discussed before. So the evolving of metazoa would be later than the former. the writer assumes that it was between 1000 MaBP and 900 MaBP, because fine clastic sediments of terrigeneous origin rich in nutrient salts developed

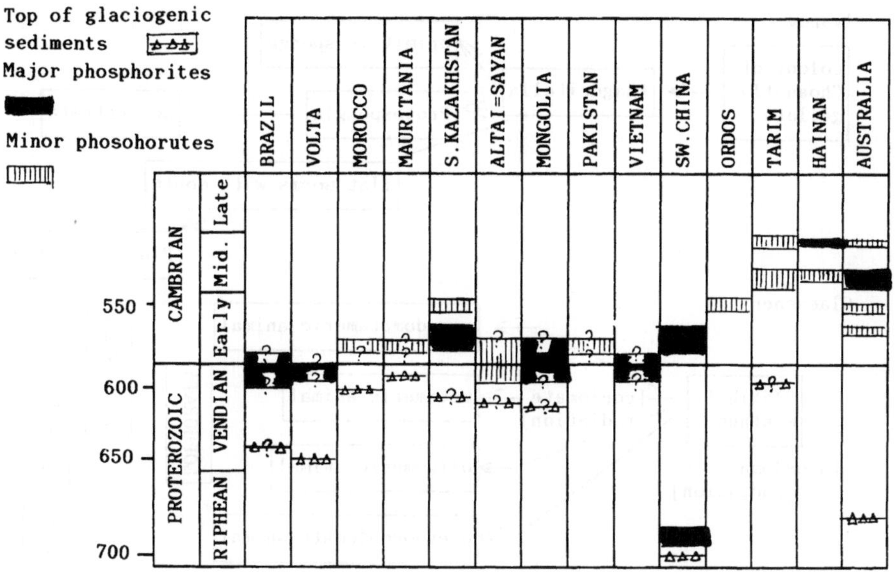

Fig. 3; Temporary distribution of phosphorous sediments and phosphatite around the Precambrian and Cambrian boundary.

since 1000 MaBP and the setting of the proper biotope for benthic animals commenced at that time. Moreove the oldest fossil record of metazoa is about 850 MaBP at present. Berbera [18] stated that the benthic biomass would have increased dramatically during the late Proterozoic (his meaning is 900 to 560 MaBP), which resulted by the shoaling of the sea waters. He supposed that pelagic planuloid might be ancestral metazoa, taking up benthic existence and evolved to Platihelminthes through flat-warm like animals. Beside the former, Ivanov [19] pointed out that the colony of choanoflagellata should be the origin of metazoa, which evolved as shown in Figure 4. Moreover, Gloessner [20] proposed another mode. The writer assumes that some kind of Zoomastigophora would be the ancester of metazoa, whose appearance was not before 1000 MaBP. Because the fine clastic sediments with nutri-

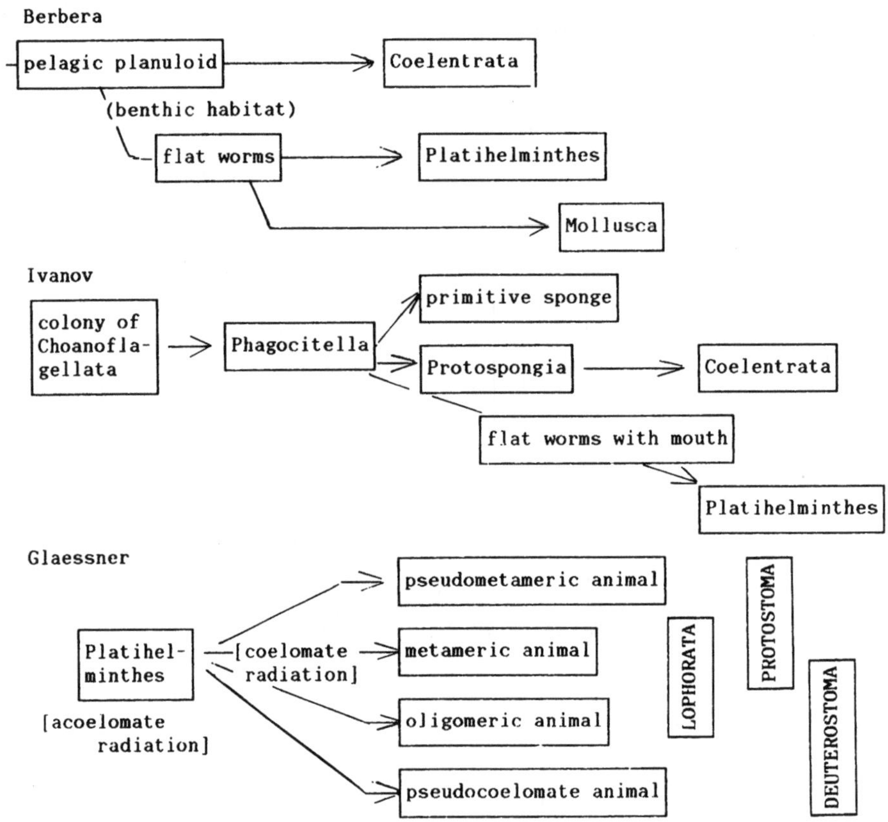

Fig. 4: Some models of the phylogenesis of metazoa proposed by other scientists.

ent salts developed since that time, and the oldest fossil record of metazoa is 840 MaBP. In addition to the former two reasons, the glaciation during the late Proterozoic setting the proper environment for benthic metazoa, commenced at the time of 1000 MaBP. The feeding habitat of the first benthic animals might be transfered from suspension feeding to mud feeding and scavenging to the development of phagocitella. Evolution of benthic life from the planktonic one might be the result of their escape from the destructive effect of UV rays and from the toxic effect of free oxygen produced by cyanophyta in the shallow sea area. Another of this adaptations was the development of their organization, to a coelomate radiation and a metameric one. Through these adaptation, some of them secured a burrowing life as infauna in the bottom sediments, and others realized a moving life on the sea floor as epifauna. These life-forms are clarified by the occurrence of trace fossils from the deposits during the late Proterozoic. Through the above adaptations, progress of the nervous and muscular systems was realized. It is probably the result of the modulation of Ca and calcium-binding protein such as troponin and myosin [1]. In certain groups, biochemical adaptation led to the development of cuticles or surface tissues as shown in the fossil record of Chitinozoa of this period. Then in some animals the cuticle and chitin combined with Ca excretion, which resulted in the production of calcareous skeletons. As shown by the fossil record cited before, utilization of phosphate was prior to the one of carbonate in the course of their mineralization. It is provably related to the appearance of phosphate rich shallow water resulting from the upwelling of deep water during the late Proterozoic. On the other hand, excess of phosphate in the shallow water caused the decrease the calcium excretion of Cyanophyta.

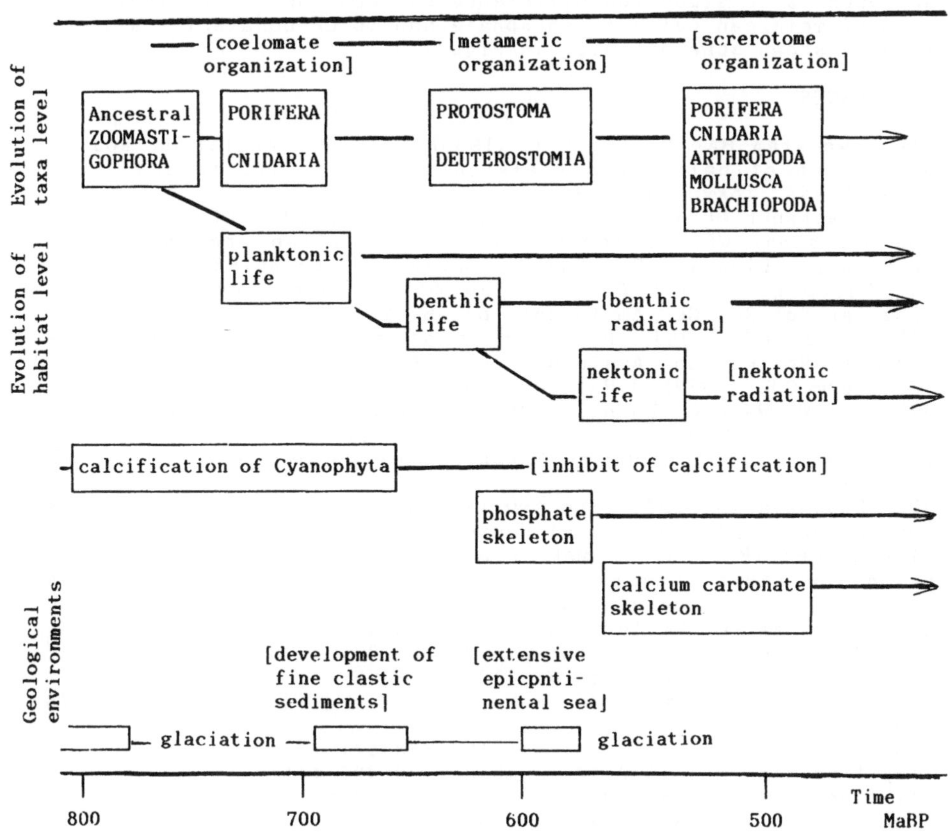

Fig. 5: Schematic presentation of the writer's view on the progress of metazoa during the late Proterozoic and the early Cambrian.

CONCLUSIONS

Genesis of metazoa are supposed by the writer to be a kind of Zoomsstigophora at the time of 1000 to 900 MaBP. Hard tissues of metazoa initiated around the Precambrian and Cambrian boundary, utilizing phosphates prior to that of carbonates, followed by a dominance of individual calcareous skeletons to the middle Cambrian. This change might be the result of the global change of the earth's environment, which commenced since the glaciation during the late Proterozoic.
The mechanism of the above cited substitution cannot be explained by the writer at present, however the following two approaches are needed for future study. The first approach for the solution is the analytical study of the geology and geochemistry of the sedimentary environment of stromatolites with special attension to the toxic behavior of metals and other compounds. The second approach is to study the difference between the biochemical reaction of metazoan cells to phosphate and to carbonate. The writer's idea is shown schematically in Figure 5.

REFERENCES

[1] Lowenstam HA, Margulis L (1980) Calcium regulation and the appearance of calcareous skeleton in the fossil record. Tokai Univ. Press, Tokyo, pp. 289-300.
[2] Glaessner MF (1971) Geol Soc Am Bull 82:509-514
[3] Glaessner MF, Walter MR (1981) Australian Precambrian paleontlogy. Elsevier Sci Pub Co, Amsterdam, pp. 361-396
[4] Salop LJ (1977) Precambrian of the northern hemisphere. Elsevier Sci Pub Co, Amsterdam, pp. 257-298
[5] Hunter DR (1981) Precambrian of the southern hemisphere. Elsevier Sci Pub Co, Amsterdam, pp. 1-882
[6] Germ GJB (1972) J Palaeont 46:864-870
[7] Sun W, Wang G, Zhon B (1986) Precamb Res 31:377-403
[8] Banks NL (1970) Trace fossils from the late Precambrian and lower Cambrian of Finnmark Norway, Seel House Press, Liverpool, pp. 19-24
[9] Fedonkin MA (1977) Precambrian Cambrian ichnocoenosis of the east European platform. Seel House Press, Liverpool, pp. 183-194
[10] Crimes TP, Legg I, Marcos A, Arborea M (1977) Late Precambrian low lower Cambrian trace fossils from Spain. Seel House Press. Liverpool, pp. 91-138
[11] Aceolaza FG, Miller H (1982) Precamb Res 17:133-146
[12] Glaessner MF (1969) Lethaia 2:369-393
[13] Brasier M (1986) Geol Mag 123:237-256
[14] Bloeser B, Schopf JW, Horodyski RJ (1977) Science 195:676-679
[15] Hargrave RB (1976) Science 193:363-371
[16] Trompette K (1982) Precamb Res 20:427-441
[17] Cook PJ, Shergold JH (1984) Nature 308:231-236
[18] Barbera ML (1978) Nature 273:22-25
[19] Ivanov AV (1970) Z Zool Syst Evolutionforsch 8:109-119
[20] Glaessner MF (1983) Precamb Res 20:427-441

CHAPTER 4.2
Nacre Formation in Bivalve and Gastropod Molluscs

H. NAKAHARA
Department of Oral Anatomy, Meikai University School of Dentistry, Saitama, 350-02, Japan

Key words: Nacreous layer, Ultrastructure, Mollusc shell.

Because of the simple and regular pattern of its organic and inorganic constituents, the nacreous structure of mollusc shell has attracted many investigators for a considerable period. This presentation deals with some observations on the fine construction of nacre, particularly the formation of the organic phase at the growing surface and the comparison of the pattern of nacre formation between two important mollusc groups, the gastropods and bivalves.

Technical remarks In order to retain the natural relationship among nacre, mantle and extrapallial fluid, careful handling was necessary. In the case of some bivalves, the shells were anesthetized before fixation [1]. However, in the case of gastropods, it is difficult to fix and embed the shell edge and mantle together without any disturbance. Therefore, we removed the mantle tissue from the shell before fixation. In so doing most structures related to the growth of nacre except mantle tissue were retained on the growth surface of the shell edge [2,3].

For observation of organic structures associated with the growing crystals of nacre surface, TEM in combination with ultrathin sectioning is necessary. So employed Araldite 502 for embedding and cut the shell with a diamond knife. For staining, the commonly used uranyl acetate-lead citrate double stain gave the best detail (Figs.4-6); however, this stain dissolves all of the $CaCO_3$ minerals. Therefore, when observation of the minerals was desired, lead citrate single stain was used (Fig.7). SEM was also employed to obtain growth patterns of surface crystals. A brief washing with water followed by drying was enough for preparation of bivalve materials (Fig.1); while for gastropod nacre, relatively tough organic structure, particularly the surface sheet, could not be removed by brief washing (Fig.8). To clean snail nacre, specimens were rinsed with 6% sodium hypochlorite

Figs.1-3. SEM of growth surface of nacre.
Fig.1: Pearl oyster, Pinctada radiata. Fig.2 and Fig.3: Archaeogastropod species Lunella coronata and Haliotis gigantea, respectively.

solution for 5 to 10 minutes before gentle wash with water (Figs.2,3). These materials for SEM were finally dried and coated with gold in an ion-sputtering apparatus.

Organic components Within the extrapallial fluid adjacent to the growing nacre surface, two organic components are formed as the first step of nacre formation. One is a flat "sheet" arranged parallel to the nacre surface [4], and the other is an "envelope" that surrounds the entire surface of each growing crystal [5]. The latter is assumed to have inducing and accelerating effect on crystal growth, while the former is involved in cessation and limitation of crystal growth [6]. The sheets contain glycine- and alanine-rich proteins, and the envelopes, aspartic acid-rich ones [7,8]. A structure similar to the envelope is observed in various shell types such as the prismatic layer of Pinctada (bivalve)[1], the hinge ligament of bivalves [9,10] and the crossed lamellar structure of Strombus gigas (gastropod)[11].

There are scattered, thread-like organic structures within the compartments formed between the parallel sheets; some of this substance may be incorporated into the envelope (Figs.11,12). Some of the envelope substances are probably become embedded in the aragonite tablets as mineral growth proceeds and turn into intracrystalline matrix.

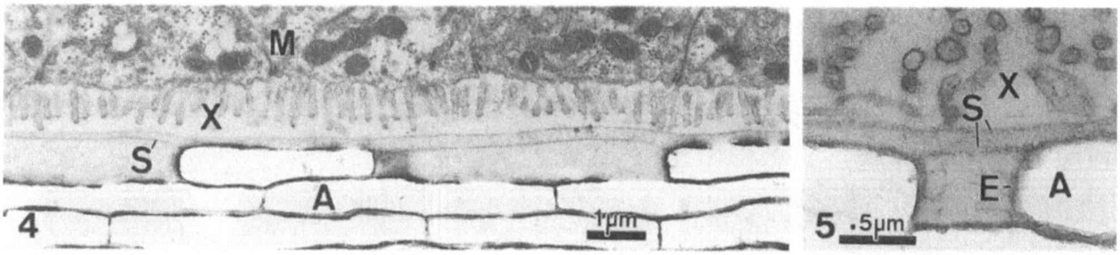

Figs. 4,5. TEM of section of bivalve (Pinctada radiata) nacre. Aragonite crystals (A) were dissolved due to the use of uranyl acetate-lead citrate double stain.
M: Mantle epithelium. X: Extrapallial space. S: Sheet. E: Envelope.

Though the basic morphology of the organic components are similar, the pattern of compartment formation of gastropod nacre differs from that of bivalves. In bivalve nacre, each sheet is formed independently within the extrapallial fluid (Figs.4,5); while in the gastropod a relatively thick surface sheet forms as a first deposit of organic structure, and then thin, ordinary sheets separate one by one from the surface sheet to form regularly spaced compartments (Figs.6,7, schematic drawings: Figs.9,10).

At higher magnification the sheets in gastropod nacre show an

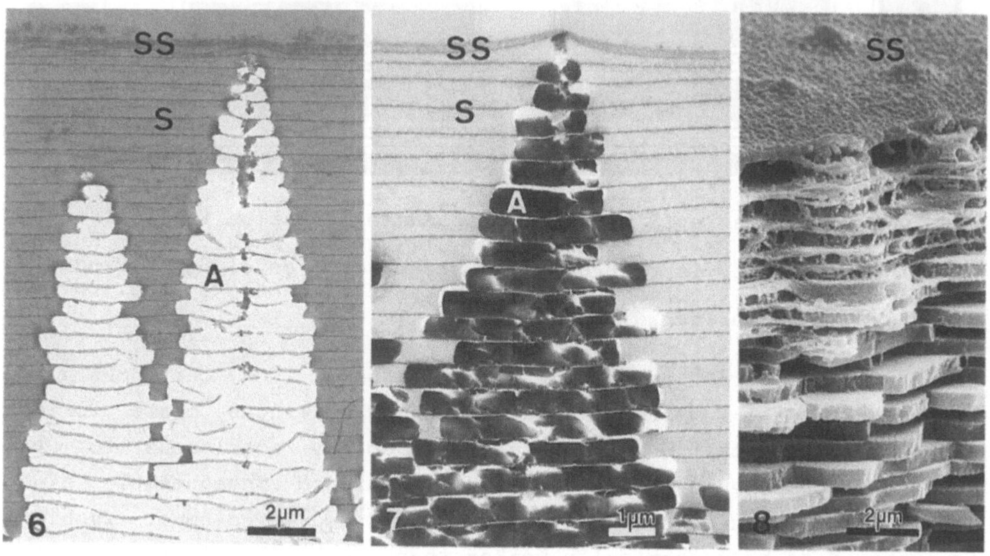

Figs.6,7. TEM of archaeogastropod nacre of Monodonta labio (double stain) and Tegula pfeifferi (lead citrate single stain). Surface sheets (SS) cover the top of aragonite stacks (A). S: Sheet.

Fig.8. SEM of Monodonta nacre. Washed with running tap water, then dried and fractured. Note the whole structure of the surface sheet (SS) is retained.

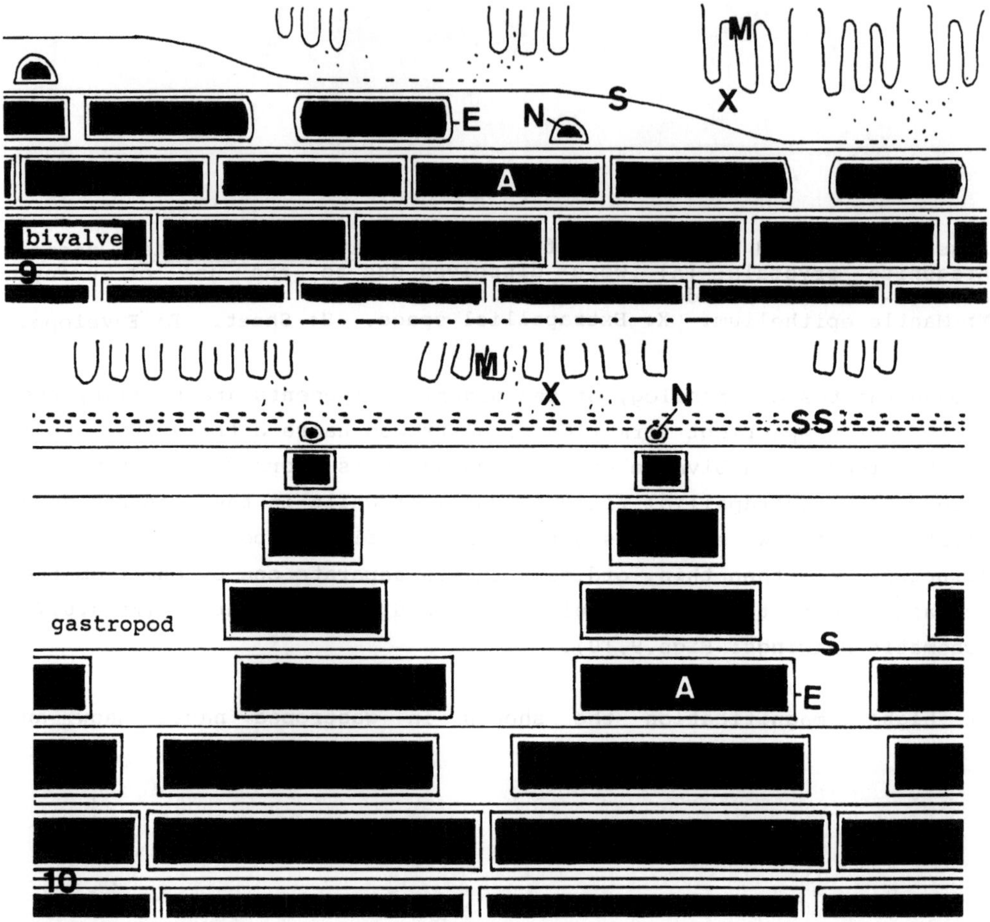

Figs.9,10. Schematic illustrations of nacre formation in bivalves and gastropods. A: Aragonite tablets. S: Organic sheet, SS: Surface sheet (exists only in gastropods). E: Organic envelope surrounding crystals. N. Newly formed crystal. M: Part of mantle epithelium. X. Extrapallial space (fluid).

electron-dense core that is a very thin (approx. 10nm) chitinous plate [7,12] (Fig. 11); while in bivalves, this chitinous plate is not obvious. This chitin structure is probably homologous to the chitinous components of the annelid or arthropod integument. Since snail nacre, which consists of crystals arranged in columnar fashion, is supposed to be mechanically weaker than the brick wall-type bivalve nacre, this chitin plate may be one means of compensation for the structural disadvantage.

Features of growth surface Since it does not show apparent damage after a wash in running water (Fig.8), the surface sheet of the

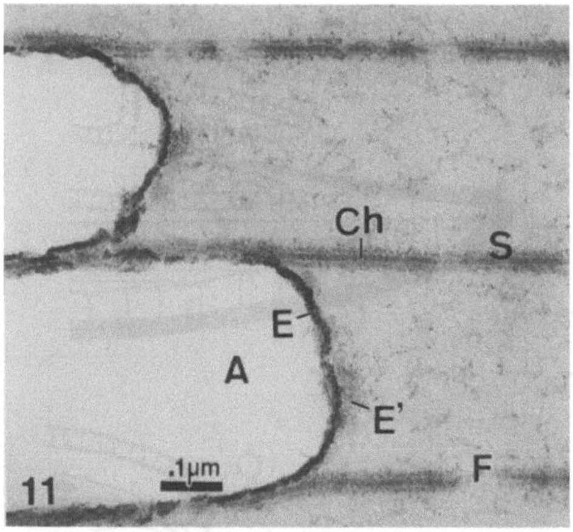

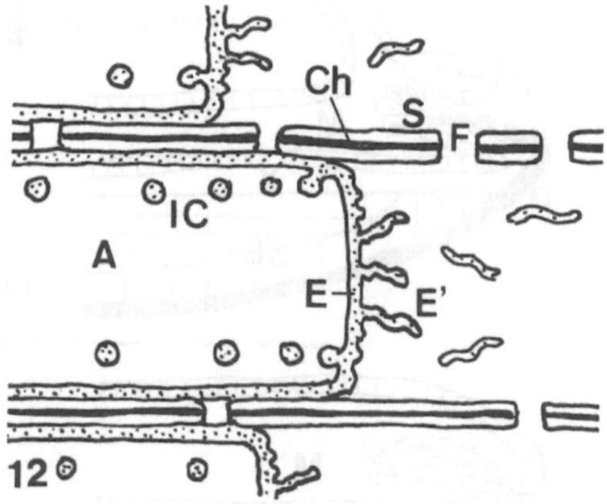

Fig.11. Part of a growing stack of gastropod (Haliotis rufescence) nacre. Fenestrations (F) and chitinous core (Ch) are seen in sheets (S), newly added envelope substance (E') is attached to envelope (E). A: Aragonite tablet (decalcified due to double stain). Fig.12. Schematic drawing of similar area shown in Fig. 11. IC: Intracrystalline matrix.

gastropod nacre (Figs.6,7) is considered to be a relatively tough structure [2,3]. This indicates the surface sheet probably serves as a protective shield covering the developing nacre surface. This protective activity may be particularly effective during exposure to the sea water when the mantle edge is retracted due to outside stimuli. In bivalves, on the other hand, the nacre surface is covered with a thin, stretched sheet of periostracum during mantle retraction [1,13], and the growing surface is therefore not exposed to the sea water (illustrated in Fig. 13).

The size of the area of nacre growth in bivalves is much larger than that in snails. In bivalves almost all of the inner shell surface is growth surface [14], whereas the growth surface of gastropod is limited to the relatively narrow aperture region [15]. In this situation, gastropod nacre needs to have faster growth rate than bivalve nacre. The pyramid-type construction seems to facilitate faster growth because it has more faces (lateral faces of the tablet) for growth than the step-type bivalve nacre.

The tablet arrangement of mature nacre also shows obvious differences caused by the different growth pattern: gastropod nacre exhibits the well- known columnar arrangement of mineral tablets while bivalve nacre shows the brick wall-type structure (with some exceptions such

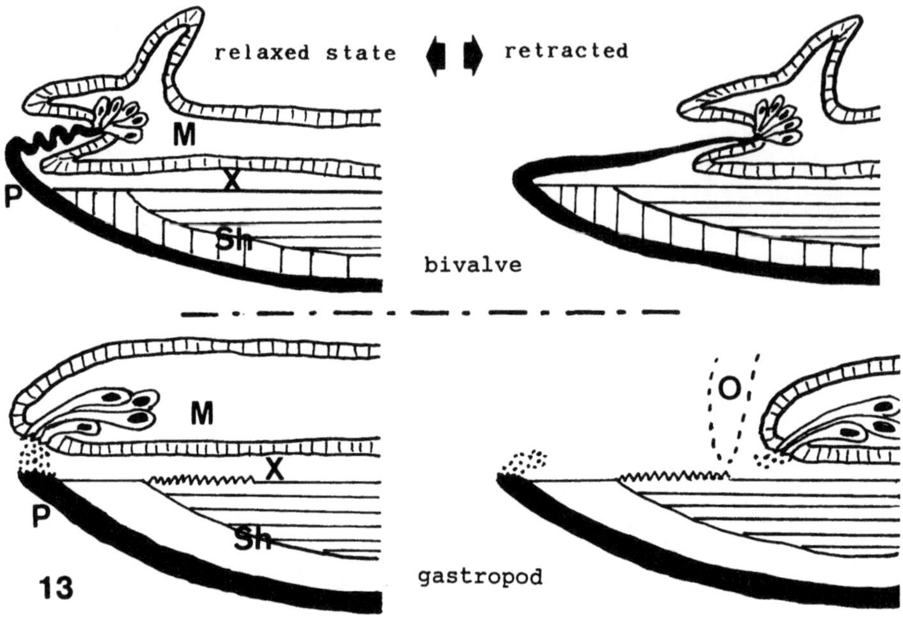

Fig.13. Schematic illustration of mantle, periostracum and shell relationship. M: Mantle. Sh: Shell. P: Periostracum. X: Extrapallial space. O: Operculum. Note marginal shell surface is exposed to sea water in gastropod when mantle retracts.

as Pinna). In the columnar-type nacre of gastropods, an interdigitating boundary apparently derived from the "staggered sequence" [15] of growing stack, is observed between neighboring columns. This irregular stacking of aragonite plates (Figs. 2,3,6,7,10) is probably caused by the uneven rate of crystal growth. However, these interdigitations may serve to increase the mechanical strength of snail nacre together with the chitin plate already

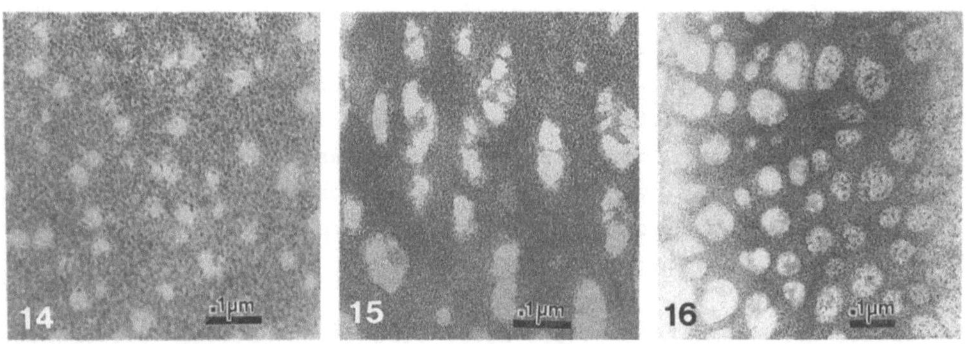

Figs.14,15. Section cut parallel to the interlamellar matrix of decalcified nacre. Fig.14: Bivalve (Pinctada), Fig.15: Gastropod (Haliotis gigantea), showing perforations. Fig.16. Section cut parallel to the organic "sheet" of Monodonta.

mentioned. Each pyramid on the growth surface contains usually 20 to 30, sometimes over 40, tablets.

A well-known feature of the organic phase of nacre is the perforations in the interlamellar sheets [16,17]. In gastropods, oval-shaped clearly defined holes in the sheets are observed in sections cut parallel to the interlamellar matrix of decalcified sheet (Fig.15) and in similarly cut uncalcified portions of organic sheet (Fig.16). The holes of gastropod sheet are also seen in cross section (Fig.11). The holes are also observed in the sheets of bivalves; however they are more obscurely edged than those in the gastropod (Fig.14). These perforations may facilitate passage of the substances moving between mantle epithelium and the compartment in which crystals grow. Perforations are not observed in the surface sheet; however, it exhibits a sponge-like appearance that suggests it is not a barrier to the movement of fluid substances. Since one of the known activities of the mantle epithelium is the ingestion of particulate substances from the extrapallial space [18], organic remnants not incorporated into calcified nacre may finally be moved to epithelial side and ingested by epithelial cells.

<u>Phylogenetic suggestions</u> The nacreous layer is generally considered as a rather primitive structure compared with the other major shell type, viz., crossed lamellar structure that is more widely distributed in both gastropod and bivalve groups [19,20]. The unit component of the crossed lamellar structure is a slender rod-shaped aragonite crystal found in bundles arranged in alternate directions. Each crystal is covered with an organic "envelope" that is apparently identical to the envelope observed in nacre [11]. However, this shell type does not have any organic "sheet" that forms the compartment as in the case of nacre. Since the only prominent organic structure of the crossed lamellar-type shell is the envelope, this shell actually contains a much smaller percentage of protein (0.1 to 0.01%)[21] than the nacreous shell (5 to 1%). These differences between the two major shell types indicate that the crossed lamellar type is constructed much more efficiently; in other words, it has progressed more evolutionally and is stabler than nacre. It can be concluded that the nacreous layer represents a moderately stable shell type that lies between ancestral chitinous integument and the more advanced shell types and that these evolutional changes have apparently occurred independently in the major molluc groups, Bivalvia, Gastropoda, and Cephalopoda as well.

Summary Two distinct organic structures are involved in nacre formation in both gastropods and bivalves. One is the organic "sheet", which forms parallel compartments in the extrapallial fluid; and the other is the "envelope", which intimately surrounds each growing crystal. The sheets are composed of glycine- and alanine-rich proteins, while the envelopes are rich in aspartic acid. In gastropods, crystal tablets grow within multilayered compartments and are arranged as a pyramid-shaped stack, while in bivalves the surface crystals appear to form a gently-sloped step within a one- or two-layered compartment formed at the growing surface of nacre. Only in gastropods does a thick and relatively tough surface sheet appear, which covers the top of the pyramids. This surface sheet may provide a barrier to protect against the disturbance caused by mantle movement.

REFERENCES

[1] Nakahara H, Bevelander G (1971) Calc Tiss Res 7:31-45
[2] Nakahara H (1979) Japan J Malac 38:205-211
[3] Nakahara H, Bevelander G, Kakei M (1982) Japan J Malac 41:33-46
[4] Bevelander G, Nakahara H (1969) Calc Tiss Res 3:84-92
[5] Erben HK, Watabe N (1974) Nature 284:128-130
[6] Bevelander G, Nakahara H (1980) In: Omori M, Watabe N (eds) The Mechanism of Biomineralization in Animals and Plants. Tokai University Press, Tokyo, pp.19-27
[7] Weiner S (1986) CRC Crit Rev Biochem 20:365-408
[8] Bevelander G, Nakahara H (1975) Earth Sci (J Assoc Geol Collab Japan) 29:87-91
[9] Bevelander G, Nakahara H (1969) Calc Tiss Res 4:101-112
[10] Marsh H, Hamilton G, Sass R (1978) Calc Tiss Res 25:45-51
[11] Nakahara H, Kakei M, Bevelander G (1981) Veliger 23:207-211
[12] Nakahara H (1983) In: Westbroek P, de Jong EW (eds) Biomineralization and Biological metal accumulation. Reidel, Dordrecht, pp.225-230
[13] Wada K (1968) Bull Natl Pearl Res Lab 13:1540-1560
[14] Nakahara H (1961) Bull Natl Pearl Res Lab 6:607-614
[15] Wise SW (1970) Eclogae Geol Helv 63:775-797
[16] Grégoire C (1957) J Biophys Biochem Cytol 3:797-808
[17] Mutvei H (1969) Stockholm Contr Geol 20:1-17
[18] Nakahara H, Bevelander G (1967) J Morph 122:139-146
[19] Bøggild OB (1930) K Danske Vidensk Selsk Copenhagen 2:233-326
[20] Kobayashi I (1969) Am Zoologist 9:663-672
[21] Hare PE, Aberson PH (1965) Carnegie Inst Wash Year Book 64:223-235

CHAPTER 4.3

Bacterial Mineralization in Cherts

K. TAZAKI[1], F.G. FERRIS[2], R.G. WIESE[3], and W.S. FYFE[4]

[1]Department of Geology, Shimane University, Matsue, Shimane, 690 Japan; [2]Nova Husky Research Corporation, NW, Calgary, Alberta, Canada T2E 7K; [3]Department of Geology, Mount Union College, Alliance, OH, USA 44601; and [4]Department of Geology, University of Western Ontario, London, Ontario, Canada N6A 5B7

Key words: Bacteria, Chert, Hematite, Graphite, Cell wall.

INTRODUCTION

Organo-metallic complexes are partly responsible for the partitioning of metals in aquatic sediments, and may play an important role in the formation of minerals during low-temperature diagenesis (1) (2). A number of studies have shown that substantial quantities of dissolved metals can be complexed by bacteria and their products (3) (4) (5). The cells served as centres for the formation of new mineral phases which are found in silicate and metalliferous sedimentary rocks (6) (7) (5).

The recognition of microfossils in ancient sedimentary rocks has confirmed that microorganisms which strongly resemble present-day procaryotes existed at least 3×10^9 years ago (8). In the present work we describe bacterial iron-mineralization and graphite formation in 2.0 Ga cherts from the Gunflint Formation, and show the preservation of intact cells by using electron micro-techniques.

MATERIALS AND METHODS

Basal gray and red stromatolitic cherts were collected from the 2.0 Ga age Gunflint Formation at the Mink Mountain and Schreiber localities, northern Ontario, Canada. Grains (<2 μm) of the samples were collected by hydraulic elutriation and pipetted onto specimen support micro-grids for morphological studies and selected-area electron diffraction studies by transmission electron microscopy (TEM), using a JEOL-EM 100C at 100 kv.

Au-coated polished-thin sections of samples containning bacterial cells were analyzed quantitatively using a JEOL JXA 8600 superprobe equipped with an energy dispersive analyzer, and operated at 15 kv.

X-ray diffraction studies of powdered samples were made using a Rigaku goniometer and Cu Kα radiation. Electron spectrochemical analyses (ESCA) were undertaken using an SSX-100 photoelectron spectometer equipped with a custom-designed vacuum system and sample treatment chamber. A monochromatized Al Kα X-ray exciting beam was used, and all binding energies were referenced to C (1s) at 285.0 ev.

RESULTS

X-ray powder diffraction patterns of the cherts shows the presence of large amounts of quartz with traces of lepidocrocite, calcite, hematite and graphite.

Optical microscopy of polished-thin sections of the Gunflint shows coccoid cyanobacterial communities in which individual cells are preserved by silicification and iron oxide mineralization. The high optical density and color of the cell walls suggest the presence of iron oxides (Fig. 1C). Fe Kα composition images of polished thin-sections indicate the remains of bacteria inside accumulations of iron (Fig. 1A and 1B). Quantitative analyses by electron microprobe of the coccoid shown in Fig.1A (arrow) indicate high concentrations of SiO_2(65.3%) and Fe_2O_3(6.7%), with MgO(0.21%) and Al_2O_3(0.11%). These components total only 72.32%, and it is likely that the remainder of the coccoid contains significant organic C, H_2O, or CO_2. The ESCA analysis of the same sample also shows high concentrations of Si and O in the chert, with traces of Fe, Na, Ca, Mg, C, and Al. The X-ray and ESCA results are averaged measurements from bulk samples. The cherts studied contain about 19% C (1s).

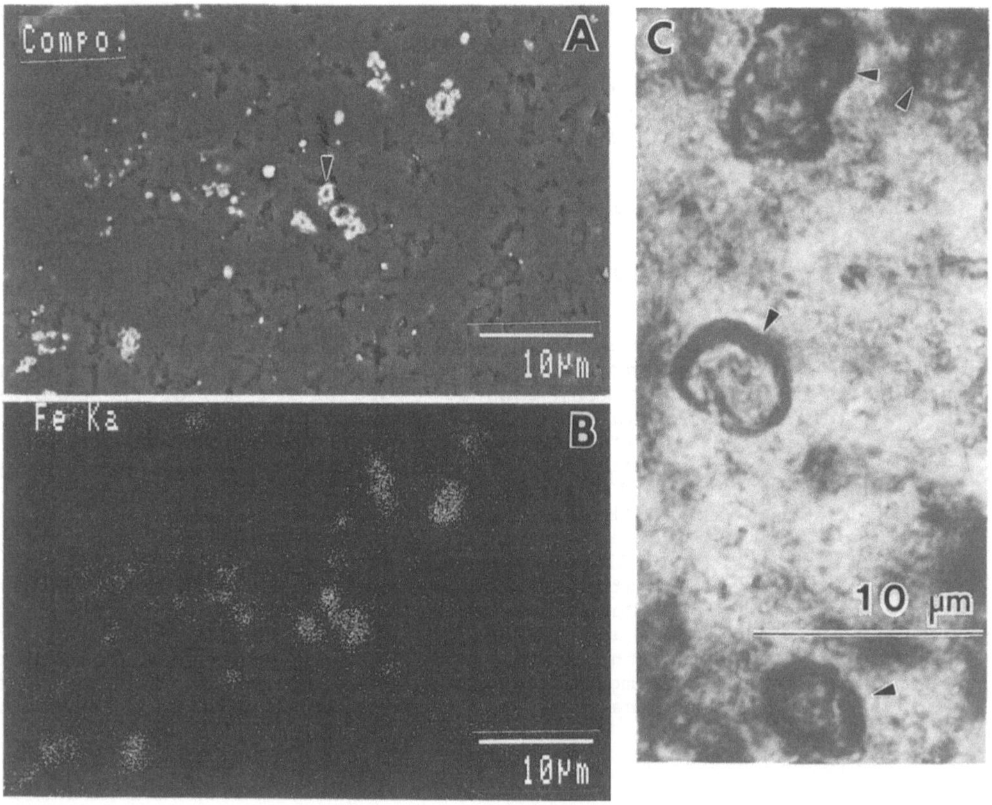

Fig.1 (A) Electron microprobe compositional image. (B) FeKα content map (C) optical micrograph of a thin section of the red chert showing a silicified coccoid. Arrows show iron-loaded cell walls. Note that Fe is concentrated in the coccoid communities.

Observation of iron-mineralization

TEM observations of microfossils in finely ground cherts shows progressive mineralization of iron-encrusted cellular fragments. Coccoid microfossils show concentrations of granular ferric particles on the cell walls. Various stages of iron mineralization are shown in Fig. 2. Area (1) shows the smooth surface of the cell wall of a fossil coccoid without a concentration of Fe. The incipient crystallization of iron in cell walls (Area 2) gives a weak electron diffraction of goethite. Where the cell walls are iron-loaded (Area 3), the electron diffraction pattern gives the characteristic d-spacings of hematite.

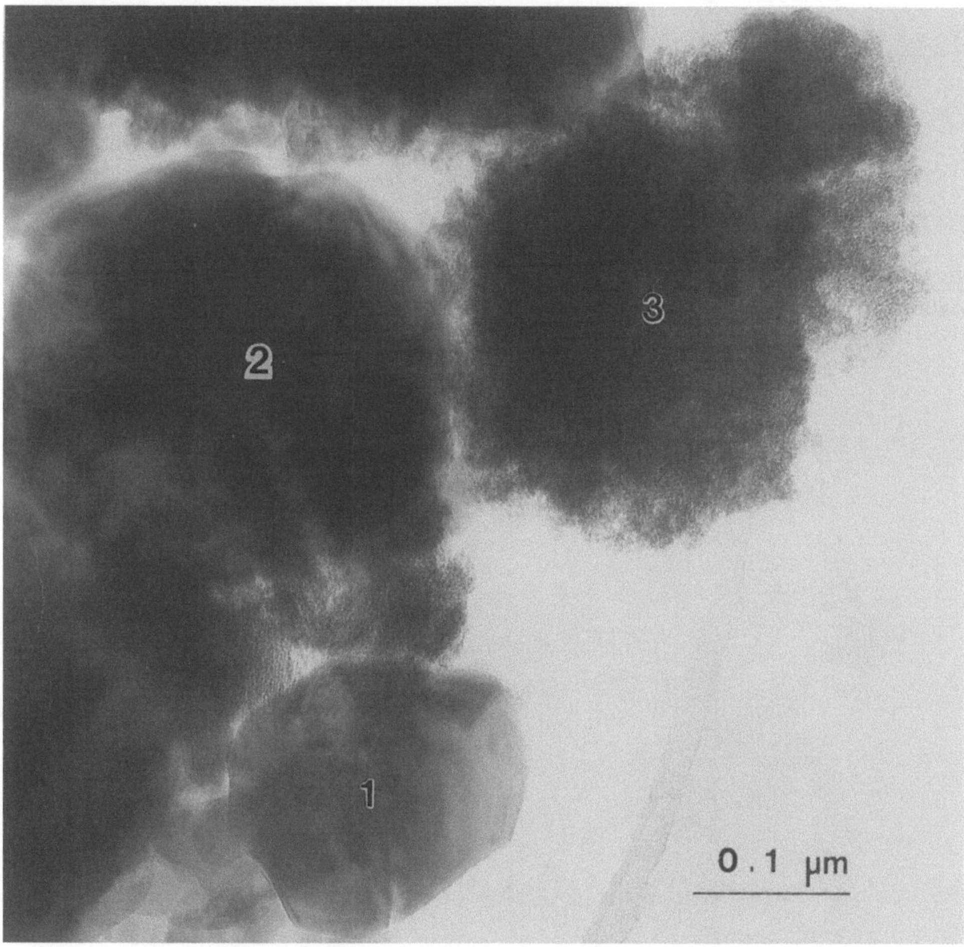

Fig.2 Transmission electron micrographs of iron-loaded coccoid. (1) non iron-loaded coccoid. (2) partly developed extracellular iron oxide. (3) Well-developed extracellular iron oxide.

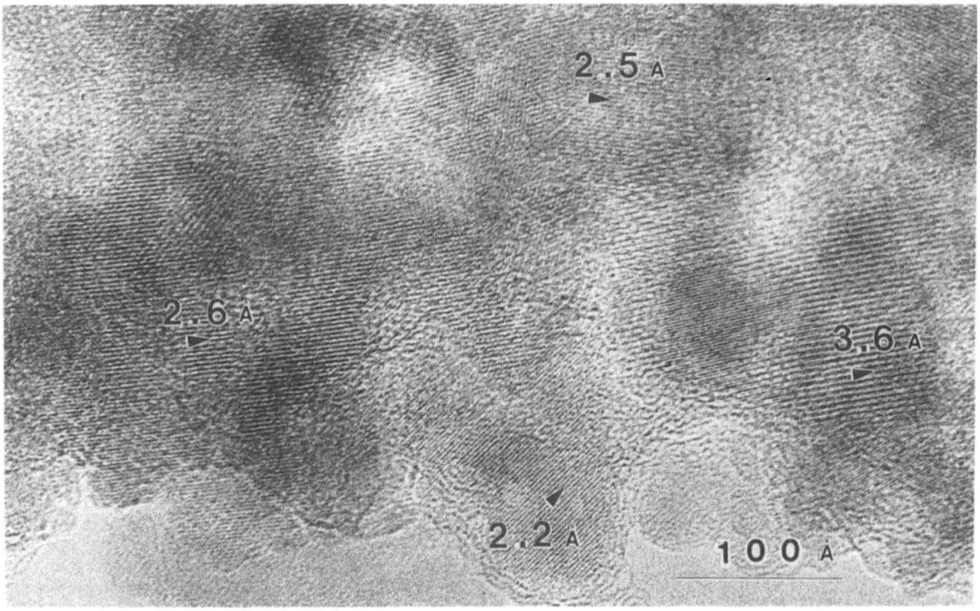

Fig.3 High resolution transmission electron micrograph of an iron-loaded bacterial cell wall showing complex mosaic structure with lattice images of hematite.

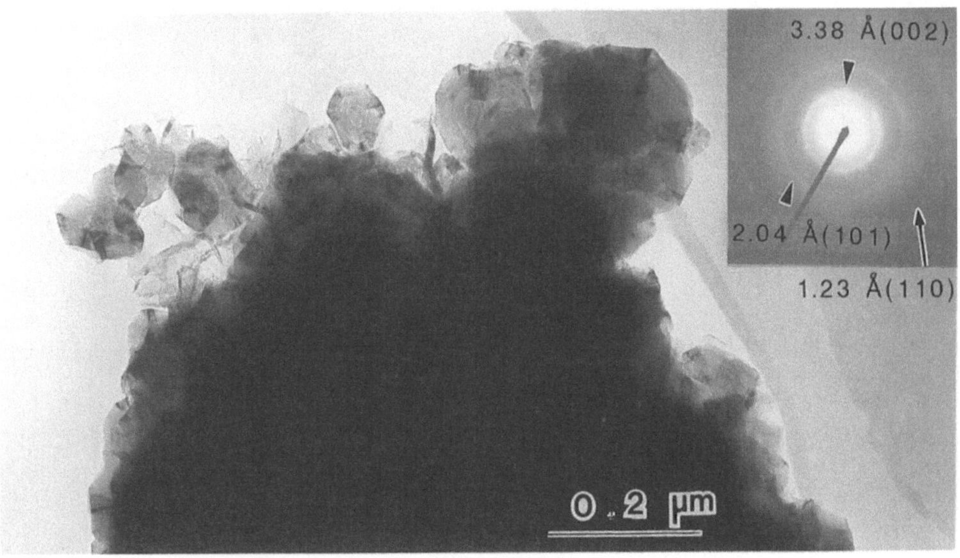

Fig.4 Transmission electron micrograph of graphite associated with bacterial cells. The graphite occurs as spiral and hexagonal crystals. The electron diffraction shows strong spots and rings at 3.38 Å(002), 2.04 Å(101) and 1.23 Å(110) of graphite reflections.

The occurrence of iron oxides associated with bacterial cells is clearly shown by high-resolutiion TEM (Fig. 3). The cell shown in Fig. 3 has 3.6-, 2.6-, 2.5-, and 2.2- Å lattice images on the sureface of its cell wall. These d-spacings suggest hematite in the mosaic of patches. The boundary of each patch consists partly of noncrystalline material. However, most patches are directly bound together.

Graphite

Organic materials in the cherts have also been studied and found to consist, at least in part, of graphite. The stages of formation of graphite range from poorly crystalline in some of the organic materials to well-crystallized graphite. Cellular debris containing is shown in Fig. 4. Selected area electron diffraction patterns gave d-spacings at 3.38, 2.04 and 1.23 Å (Fig. 4) indicative of graphite. Spiral and hexagonal graphite grains were noted to have a distinctly more graphitic stracture than do flat sheets.

DISCUSSION

Examination of fossil bacteria in the Gunflint Formation by electron micro-techniques has revealed not only the morphology of individual bacteria but also the occurrence of iron oxides and graphite associated with the bacteria. Clearly the microorganisms were instrumental in the formation of these minerals. Mineralization by these ancient microorganisms resembles mineralization and concentration of metal ions by present-day bacteria as reported by Nealson (9) and Mann et al. (10). Bacteria appear to have played a major role in the geochemical and biological clayes operative during the formation of the Gunflint sediments.

Acknowledgements

We thank the staff of JEOL in Tokyo for allowing us to use the electron microprobe JXA-8600 instrument.

REFERENCES

(1) Beveridge TJ, Meloche JD, Fyfe WS, Murray RGE (1983) Applied and Environmental Microbiology 45: 1094-1108
(2) Ferris FG, Fyfe WS, Beveridge TJ (1987) Chem Geol 63: 225-232
(3) Beveridge TJ, Fyfe WS (1984) Can J Earth Sci 22: 1893-1898
(4) Southgate PN (1986) Geology 14: 683-686
(5) Houot S, Cestre T, Berthelin J (1984) In: Proc. 11th Congr. Int.
(6) Irrigations and Drainage, Fort Collins, CO, pp.151-180.
(7) Robert M, Berrier J, Eyralde J (1983) Sci Geol 73: 95-103
(8) Ferris FG, Beveridge TJ, Fyfe WS (1986) Nature 320: 609-611
(9) Nealson KH (1983) In: Krumbein WE (ed) Microbial Geochemistry, Blackwell Scientific, Oxford, pp. 159-190.
(10) Mann H, Tazaki K, Fyfe WS, Beveridge TJ, Humphrey R (1987) Chem Geol 63: 39-43

CHAPTER 4.4

Biological Accumulation of Different Chemical Elements by Microorganisms from Yellowstone National Park, USA

H. MANN[1], W.S. FYFE, K. TAZAKI[3], and R. KERRICH[2]

[1]University of Western Ontario, London, Ontario, Canada N6A 5B7; [2]University of Saskatchewan, Saskatoon, Sk. S7N OWO, Canada; and [3]Shimane University, Matsue, Shimane, 690 Japan

Key Words: silica tubes, silicid acid, thermophilic algae

INTRODUCTION

Investigation of the silica budget for the Upper and Lower Geyser Basins of Yellowstone National Park by Truesdell et al. [1] suggest that the present fluxes of hotspring water and thermal energy may have been continuous for at least the past 10 000 yr. However, the record of any siliceous sinter deposited before 10 000 B.P. was destroyed by the Pinedale glaciation. Considering the presence of hydrothermally altered rocks incorporated within both early and late Pinedale glacial sediment indicates a long period of hydrothermal activity [2].

In a convecting hydrothermal system that discharges hot water at the surface of the earth and is recharged by cold surficial water, such as that at Yellowstone National Park, underground boiling temperatures are controlled by hydrostatic conditions [2]. At a given depth, the temperature of solution cannot exceed its boiling temperature, which is determined by the salinity and gas content of the solution and by hydrostatic pressure. The Abyss pool had a 8.65 pH and a temperature of 40.1°C.

The lowering water table and decreasing temperatures at given depths could induce thermal cracking of the quartz crystals and even hydrothermal explosions [3]. Decreasing temperature also causes a decrease in the solubility of quartz [4] which would result in overgrowths of silica on previously deposited quartz crystals during late Pinedale glaciation 45 000 - 14 000 B.P. Considering growth of thermophilic microorganisms in Yellowstone's hot-spring waters might be particularly favored during a period of thick ice accumulation because the increased hydrostatic pressure would prevent boiling throughout the system for long periods, and nutrient fluids might escape at a slower rate.

Except for oxygen, silicon is by far the most abundant of all the earth elements, accounting for as much as all the other 90 elements together, but in its soluble form as silicid acid [$Si(OH)_4$] (pH 7-8), it is relatively scarce in fresh and sea waters, ranging from undetectable levels to 140 μM in eutrophic Danish lakes and from detectable levels to 385 μM in sea water. But these values reflect the extremes, whereas typical ambient fresh water and sea water concentrations are approximately 100 μM and 10 μM, respectively.

MATERIALS AND METHODS

A number of thermophilic blue-green algae have been isolated from the geothermal environment from Abyss pool. During our investigation, specimens appeared to grow horizontally on site. General view at Abyss pool showed gray-white-dirty green color mats sitting about 10-15 cm below water level. In closer observation the mass consisted of longitudinal triable tube-like structures. The samples were dried at room temperature (18°C) and gold plated. SEM observation revealed tubular structures made of silica, with Cl and Ca traces (EDS) (Figs. 1-5). TEM was used for thin-section observation (Fig. 6). Blue-green filamentous algae could be

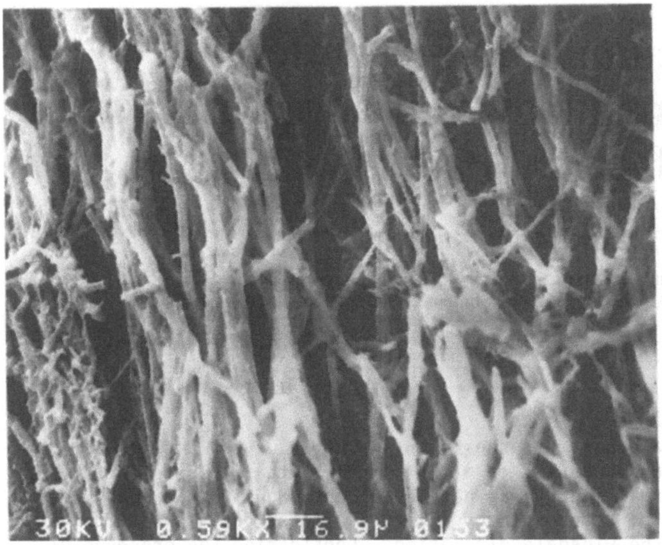

Fig. 1. SEM micrograph of filamentous microorganism encased in amorphous silica, specimens shown are in early stage of growth.

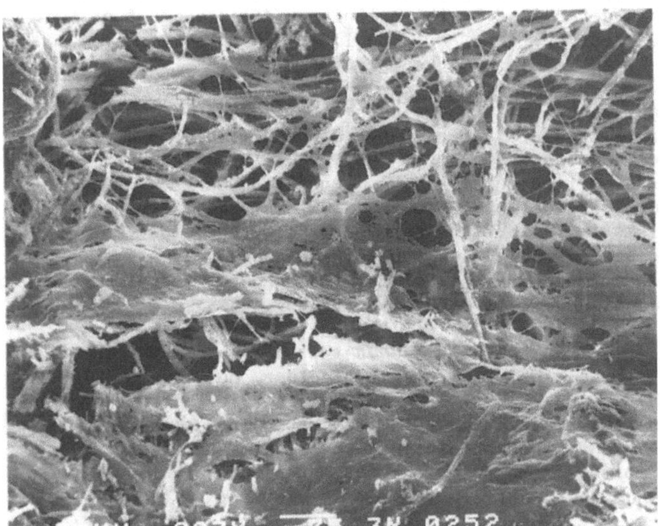

Fig. 2. SEM micrograph depicting later stage of cementation microorganisms now covered by "siliceous blanket".

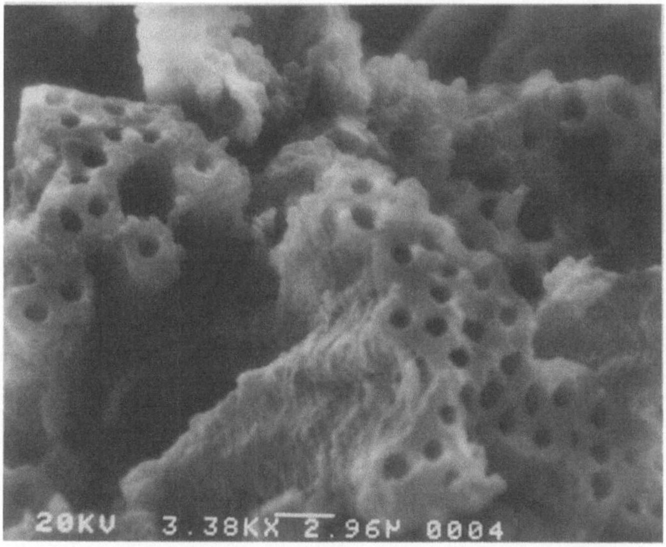

Fig. 3. SEM micrograph of silica tubes bound together in bunch-like formation.

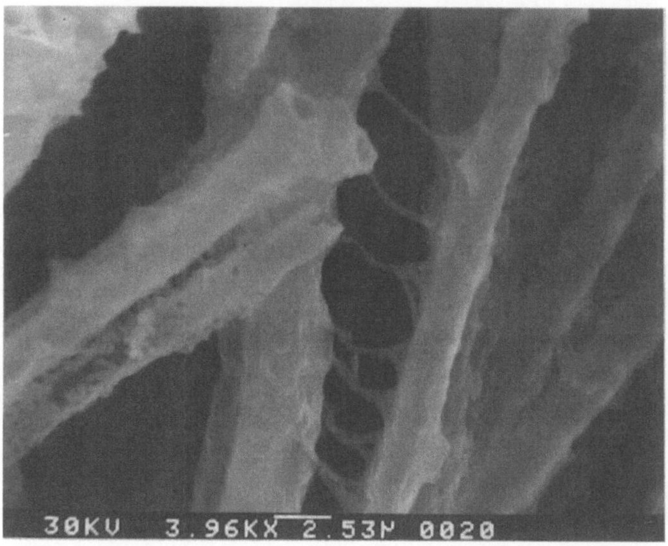

Fig. 4. SEM micrograph of two individual tubes in the process of cementing together.

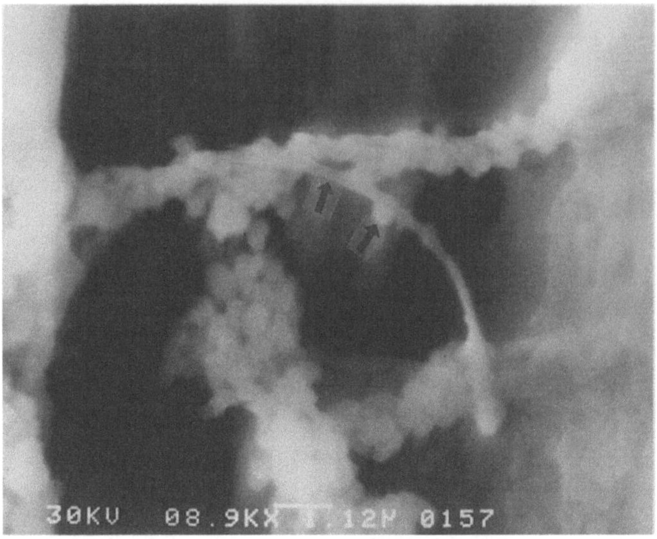

Fig. 5. Detail of two tubes cementing together. This process consists of a very thin string of amorphous silica droplets depositing on hair-like thin structure (arrows).

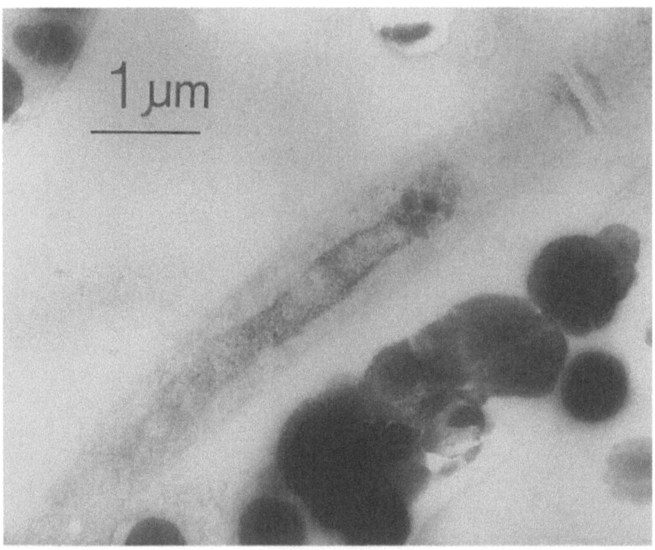

Fig. 6. TEM. Thin-section micrograph of one tube. Blue-green filamentous algae is present with right side of tube intact.

recognized inside of some siliceous tubes. Although further investigations are needed to better define the nature of these filamentous algae, their occurrence could indiate that the tubes are formed around them.

X-ray diffraction and electron microscopy has shown most silica in diatoms and other microorganisms is amorphous and occurs as a silica gel containing tightly bound water and many Si-OH units [4]. The general formula can be written $[Si_{n/2}(OH)_{4-n}]_m$ where n can be 0-4. The monomer is $Si(OH)_4$. However some crystalline silica is present along with amorphous silica in these tubes.

RESULTS

Most elements are at elevated levels in the warm waters. Elements concentrated in algae include Ca, Na, Mg, Al, Ga, Fe, Ti, Cr, Ni, Co, Se, Y, An, Pt, Ir with concentration factors of an order of magnitude or so. Some elements are not concentrated (e.g. B, Ge, Hg, Sb, see Table 1).

Table 1. Abundances of selected elements in algae and water from Abyss pool (µg/g),(ng/mL).

Elements	Algae	Water						
Ca	10,087	2,679	Li	188	-	Sc	30	30
Si	high	high	As		2,079	Y	2	0.3
Ge	34	70	Pb	6	9	Ta	0.2	<0.1
Sn	3	6	Cr	2,121	250	Au	1	0.7
Na	32,590	6,456	Rb	38	204	Bi	0.6	<0.1
Mg	246	34	Nb		0.2	La	4	6
K	-	-	V	-	-	Ce	9	73
Sr	6	4	Mo	8	81	Pr	1	0.1
Al	11,134	3,893	Ni	103	7	Nd	5	0.8
Ga	126	37	Ag	0.9	0.5	Sm	6	0.8
Tb	0.6	0.5	Co	3	0.4	Eu	0.5	
Fe	4,079	539	Rh	0.8	0.3	Gd	0.3	<0.1
Ru	9	0.37	Th	2	0.2	Tb	0.2	-
Os	2	0.2	Sb	2	112	Dy	0.9	-
Ba	31	21	Hg	3	4	Ho	0.2	<0.1
Zn	98	111	Se	169	24	Er	0.6	0.5
Br	183	1,813	Te	9	3	Tm	0.3	<0.1
I	5	28	U	0.6	0.2	Yb	1	-
P	-	-	Zr	55	43	Lu	0.2	-
B	8,007	8,148	W	12	892	Pt	1	0.2
Mn	167	7	Cs	153	309	Pd	5	1
Re	0.3	0.1	Cd	25	10	Ir	0.6	0.1
Cu	45	16	Be	11	13	Te	-	-
Ti	179	35	Hf	4	1			

REFERENCES

Truesdell, A.H., Fournier, R.O., White, D.E., and Muffler, L.J.P. (1968). Int. Geological Con., 23rd, Prague, Abstracts, p. 389.

Bargar, K.E., Fournier, R.O., and Theodore, T.G. (1985). Geology, 13, 483-486.

Muffler, L.J.P., White, D.E., and Truesdell, A.H. (1971). Geological Soc. of America Bull., 82, 723-740.

Pierce, K.L., Obradovich, J.D., and Friedman, I. (1976). Geological Soc. of America Bull., 87, 703-710.

Werner, D. (1977). In: D. Werner, Ed. The biology of diatoms. University of California Press, Berkeley, California, 1-17.

Mann, S., Perry, C.C., Williams, R.J.P., Fyfe, C.A., Gobbi, G.C., and Kennedy, G.J. (1983). J. Chem. Soc. Commun., 1314, 168-170.

Stober, W. (1967). In Equilibrium concepts in natural water systems. Adv. Chem. Ser., 67, 161-182.

Simpson, T.L., and Benjamin, E.V. (1981). Springer-Verlag, N.Y. 587.

CHAPTER 4.5

Genetic Characterization of a Magnetic Bacterium *Aquaspirillum* sp

T. MATSUNAGA, C. NAKAMURA, J.G. BURGESS, and T. SAKAGUCHI
Department of Biotechnology, Tokyo University of Agriculture and Technology, Koganei, Tokyo, 184 Japan

Key words: magnetite, Aquaspirillum, transposon mutagenesis, molecular cloning, iron uptake.

INTRODUCTION

Magnetic bacteria synthesize intracellular particles of magnetite (Fe_3O_4) which are surrounded by thin lipid films. These particles or magnetosomes are approximately 50-100nm in diameter and are aligned in chains of 30-40 units which appear to be asociated with the inner surface of the cytoplasmic membrane [1-4]. Bacteria which possess these structures have a permanent magnetic moment and may orient and navigate along geomagnetic field lines [5]. The mechanism of magnetite biomineralization remains unknown although models have been proposed involving transformation of hydrous ferrous oxide to magnetite within the magnetosome membrane [6]. The application of genetic engineering techniques to understanding the molecular mechanisms of magnetite biomineralization has been hampered by difficulties in growth, purification and colony formation of magnetic bacteria in the laboratory. We report here isolation and pure culture of a freshwater magnetic bacterium Aquaspirillum sp. capable of aerobic growth and colony formation on agar plates. Non magnetic mutants have been obtained by both chemical and genetic (transposon Tn5) mutagenesis. Several of these mutants have reduced iron uptake ability when compared with the wild type. This work represents the first example of gene transfer in a magnetic bacterium, and has allowed the cloning of genomic DNA sequences required for intracellular magnetite synthesis.

MATERIALS AND METHODS

Aquaspirillum sp. NK-1 was isolated from freshwater sediments obtained from ponds in Koganei, Tokyo using previously described purification procedures [7]. This strain was isolated as a pure culture and grown in the following medium which contained per litre of distilled water; 2 ml of Wolfe's mineral solution [8], 0.2g KH_2PO_4, 0.12g $NaNO_3$, 0.02g yeast extract, 0.02g malt extract, 0.05g of L-cysteine. HCl, H_2O, 10 μ M ferric galate, 0.5g biotin, pH 7.0. Nitrosoguanidine mutagenesis was carried out as shown in Figure 3. Iron concentrations were measured using ferrozine [9]. Transposon mutagenesis was carried out using plasmid pSUP1021 and E.coli S17-1 for conjugation [10]. Southern hydridisation was performed using the chemiprobe system according to manufacturers instructions [11].

RESULTS AND DISCUSSION

Figure 1 shows a transmission electron micrograph of the isolated magnetic Aquaspirillum strain NK-1. About 23 magnetosomes are present with an average diameter of 50nm. The growth curves of this strain are shown in Fig 2 where both aerobic and microaerophilic conditions were used. Aerobic

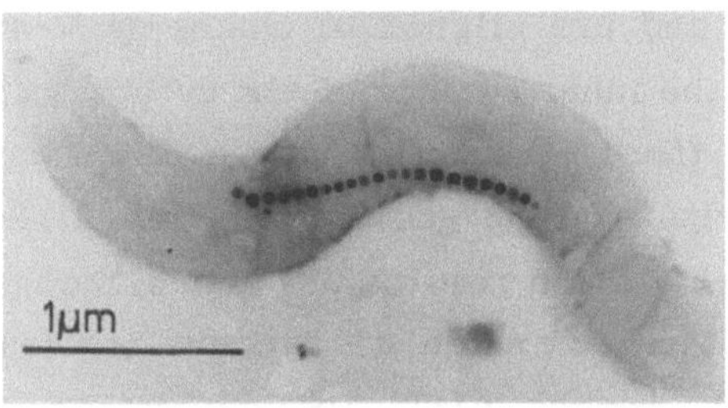

Fig. 1 Transmission electronmicrograph of isolated magnetic bacterium.

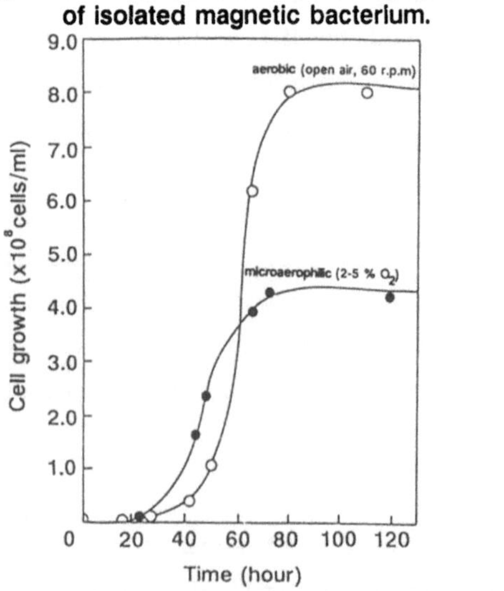

Fig. 2 Growth curves of isolated magnetic bacterium under aerobic and microaerophilic conditions.

Table 1 Colony formation ratio of isolated magnetic bacterium at various incubation time

Incubation time (day)	Colony number	Colony formation ratio (%)
1-5	0	0
6	59	23.6
7	164	65.6
8	247	98.8

Agar : 0.7%, Temp.: 26°C, 8 plates
Inoculation cell number : 250 cells/plate

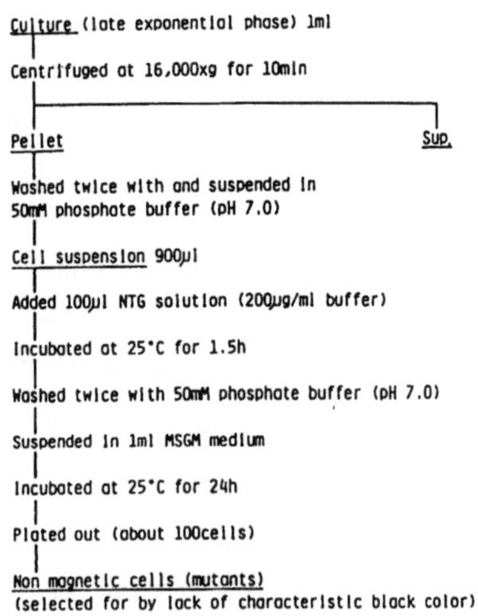

Fig. 3 Isolation of non magnetic bacteria using NTG mutagenesis.

growth is possible and the growth rate is 3 fold greater than that previously reported for Aquaspirillum magnetotacticum [5]. The ability to form colonies on 0.7% agar plates is shown in Table 1. Almost 99% of cells present formed colonies after 8 days growth. This phenomenon makes purification and genetic studies much easier than with strains which cannot form colonies. Initially nitrosoguanidine mutagenesis was used to obtained non magnetic mutants (Fig 3) some of which were completely non magnetic, others produced small amounts of magnetite (Table 2). The iron uptake properties of the wild type (Fig 4) were compared with those of an NTG generated nonmagnetic mutant (Fig 5). This data shows clearly the inability

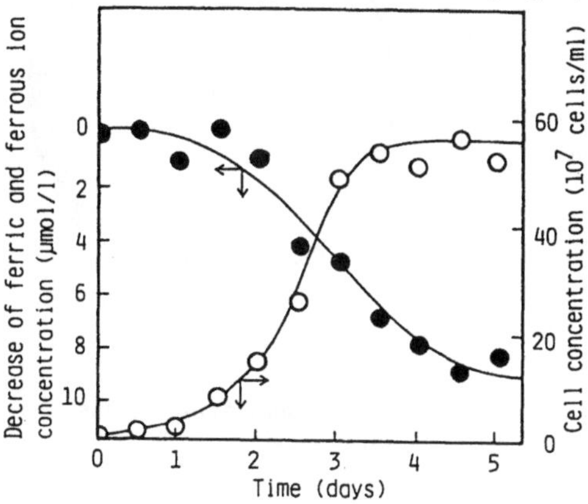

Fig. 4 Relationship between decrease of ferric and ferrous ion concentration and growth of wild type strain of magnetic bacteria.

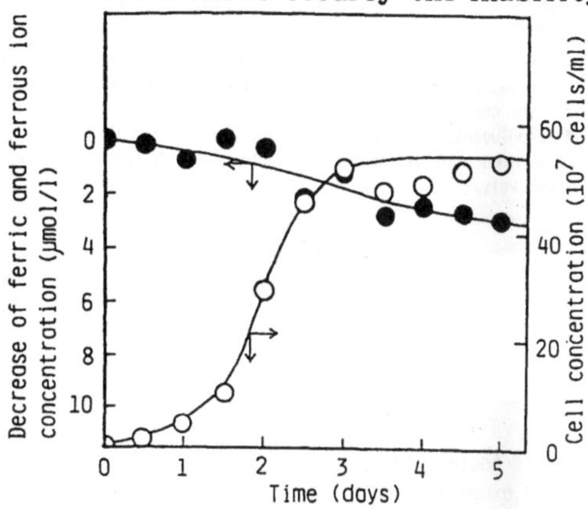

Fig. 5 Relationship between decrease of ferric and ferrous ion concentration and growth of mutant strain of magnetic bacteria.

of the mutant to remove iron from the medium to the same extent as the wild type. The plasmid pSUP1021 [10] shown in Fig 6 contains the transposon Tn5, which confers kanamycin resistance on cells containing transposon insertions in their chromosome. The plasmid was unable to replicate in Aquaspirillum sp. NK-1 (data not shown). Conjugation procedures were optimised as outlined in Fig 7. An important factor is the ratio of the E.coli

Table 2 NTG mutants of magnetic bacteria

Strain	Wild type	NTG-1	-2	-3	-4	-5	-6
Magnetite synthesis	++	-	-	-	-	+	-
Strain	-7	-8	-9	-10	-11	-12	
Magnetite synthesis	+	-	-	-	-	-	

Magnetite synthesis was estimated by extraction and quantification and quantification of magnetite. ++ : 10mg/g dry wt. cells ; + : 10mg/g dry wt. cells ; - : no magnetite
3% of mutagenised colonies were non magnetic.

to magnetic bacteria cells during conjugation. This is clear from Table 3 which demonstrates that a 1:10 ratio of donor to recipient gives the greatest number of transposon mutants. Southern blot analysis (Fig 8) shows that transposon insertions into the chromosome of the five nonmagnetic mutants

Table 3 Frequencies of Tn5 transfer to magnetic bacterial cells

Mating ratio (E.coli cells : Magnetic bacterial cells)	1 : 100	1 : 50	1 : 10	1 : 5
Frequency of transfer ($x10^{-5}$ per recipient)	3.3	5.0	19	9

Frequencies are expressed as the number of exconjugants per viable recipient cell.
118 exconjugants were isolated ,4.2% were non-magnetic.
Each non magnetic mutant was designated NM 1,NM 2,NM 3,NM 5 and NM 7 respectively.

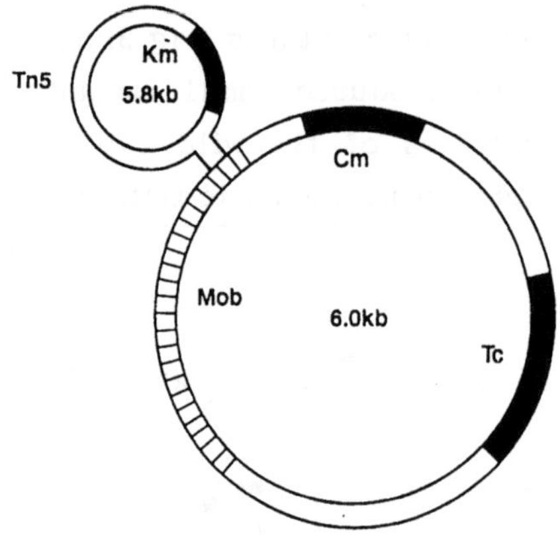

pSUP1021

Fig. 6 Plasmid pSUP1021 used for Tn5 mutagenesis.

Magnetic bacterial cells
(Middle exnential phase)

E.coli S17-1 fresh transformant
(pSUP1021)

Suspended in MSGM

Mixed together at differrent ratio

Cell mixture is spotted onto an MSGM plate

Incubated at 25°C for 6h (Mating)

Harvested and inoculated to MSGM plate
(Km 5μg/ml)

Km^r transconjugants containing nonmagnetic mutants

Fig. 7 Experimental procedure for Tn5 insertion mutagenesis

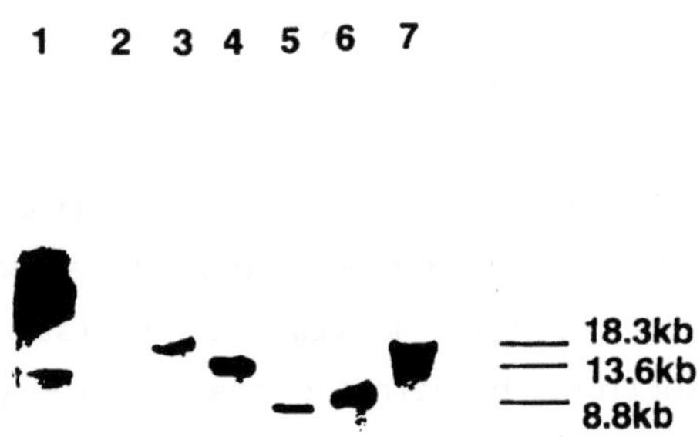

Fig. 8 Southern hybridization analysis of EcoRI digested DNA from non magnetic mutant strains using Tn5 probe.

Lane 1:pSUP1021,Lane 2:wild type, Lane 3:NM 1,Lane 4:NM 2,Lane 5:NM 3, Lane 6:NM 5,Lane 7:NM 7 has occurred. Mutants NM1 and NM7 have insertions on similar sized EcoR1 fragments and this may be the result insertion into the same gene. The hybridizing bands of the other three mutants are of different sizes suggesting insertion into 3 separate genes. Genomic fragments containing these sequences were shotgun cloned in to pUC19 and are currently under further characterization.

CONCLUSIONS

1. A new magnetic bacterium Aquaspirillum sp. capable of aerobic growth and colony formation has been isolated as a pure culture.
2. Nitrosoguanidine mutagenesis was used to produce ten nonmagnetic mutants several of which showed reduced iron uptake ability when compared to the wild type.
3. Transposon Tn5 mutagenesis was used to obtain 5 non-magnetic mutants with tranposon insertions into the chromosome.
4. Southern blot analysis has confirmed the presence of possibly four separate sequences required for magnetite synthesis.
5. These genomic DNA fragments have been cloned into E.coli and are currently being characterised in greater detail.

ACKNOWLEDGEMENTS

We thank Professor Puhler for the gift of pSUP1021 and E.coli S17-1. J.G.Burgess would like to thank the Japan Society for Promotion of Science (JSPS) for financial support.

REFERENCES

[1] Blakemore RP (1975) Science 190:377
[2] Balkwill DL, Maratea D, Blakemore RP (1980) J.Bacteriol. 141:1399
[3] Frankel RB, Blakemore RP, Wolfe RS (1979) Science 203:1355
[4] Towe K, Moench TT (1981) Earth Sci. Lett. 52:213
[5] Blakemore RP (1982) Ann. Rev. Microbiol. 36:217
[6] Mann S (1988) In: Kirschvink JL, Jones DS, MacFadden BJ (eds) Magnetite Biomineralisation and Magnetoreception in Organisms. A New Biomagnetism. Topics in Geobiology vol 5. Plenum Press, New York.
[7] Matsunaga T, Kamiya S (1987) Appl. Microbiol. Biotechnol. 26:328-332
[8] Wolin EA, Wolin MJ, Wolfe RS (1963) J. Biol. Chem. 238:2882-2886
[9] Stookey LL (1970) Anal. Chem. 42:779-781
[10] Simon R, Priefer U, and Puhler A (1983) Bio/Technology 1, 784
[11] Sambrook J, Fritsch EF, Maniatis T (1989) Molecular Cloning. A laboratory manual 2nd Edn. Cold Spring Harbor Laboratory Press. Cold Spring Harbor, New York.

CHAPTER 4.6

Coccolith Accumulation Rates: Cretaceous to Recent

A. PENTECOST

Division of Biosphere Sciences, King's College London, Campden Hill Road, London W8 7AH

Key Words: coccoliths, accumulation, sedimentation, productivity.

INTRODUCTION

Coccoliths, the minute calcite scales formed by marine phytoplankton belonging to the Prymnesiophyceae, have been responsible for most of the carbonate sedimentation in the oceans over the past 100 million years. The aim of this brief review is to compare rates of coccolith accumulation on the sea floor over time, and outline the factors thought to influence their rate of sedimentation. Recognition of Croll-Milankovitch cycles in the sedimentary record, and data from the Deep Sea Drilling Project (DSDP) and Ocean Drilling Program (ODP) are beginning to provide detailed information on rates of sedimentation [1,2,3].

Coccolith formation has been studied *in vitro* [5,6,7] and two 'types' are widely recognised: holococcoliths are composite structures formed of distinct, unmodified calcite rhombs; heterococcoliths are highly modified scales without discernable crystal faces. Most modern species, e.g. *Emiliania huxleyi* form heterococcoliths. Coccoliths have a long fossil history but have only become prominent since the Jurassic with accumulation reaching its acme in the Upper Cretaceous with the formation of extensive chalks [4].

Two major carbonate-depositing environments are recognised; carbonate platforms where rapid rates of carbonate production by benthic biota occur in warm, shallow (0-100 m) seas. Pelagic carbonates, deposited most rapidly at intermediate ocean depths (200 m - 3 km) consist mainly of coccoliths and foraminifera. The former make up 10-70% of the recognisable carbonate fraction [7], and today are considered a major carbonate sink in the oceans.

RESULTS AND DISCUSSION

The accumulation of coccoliths may be considered in four depth-dependent phases: a) carbonate production in the photic zone; b) transportation to the sea floor; c) surface accumulation and d) burial. Different factors affect each of these phases and water depth is of crucial importance. (a) The standing crop of cosmopolitan *E. huxleyi* varies enormously. During shelf-edge blooms, $CaCO_3$ production exceeds 40 g m^{-2}, possibly up to 400 g $m^{-2}a^{-1}$ [8] which far exceeds rates for the open ocean [9,10]. Since shelf sea primary production is about 30-40% global ocean production, the potential total shelf accumulation rate should be 5-8 times higher than the open ocean, assuming equivalent relative abundancies of coccolithophorids in these regions. Major factors influencing

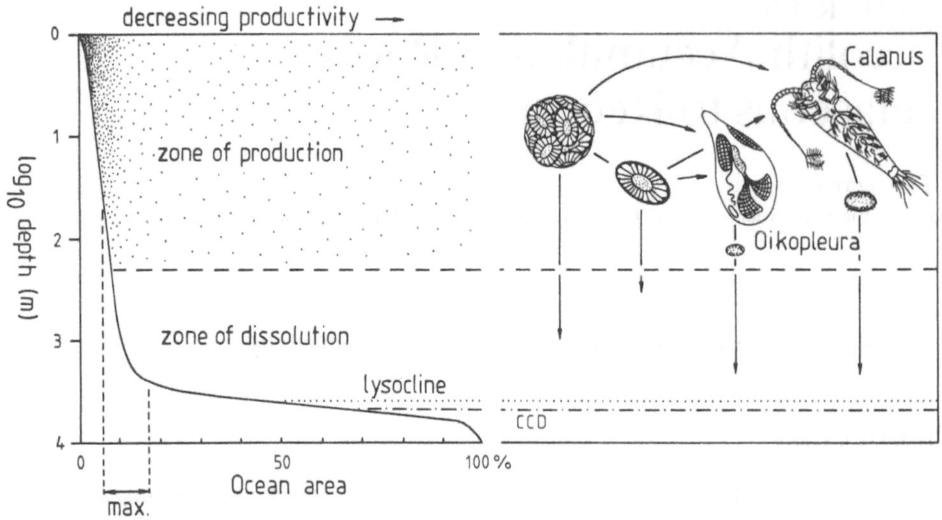

Fig. 1. Transport processes for coccoliths. The ocean hypsographic curve is plotted (left) showing the zone of pelagic photosynthesis (=carbonate production) declining in deeper water, and zone of dissolution. Region of potential high coccolith accumulation ('max') corresponds to deeper shelf seas and slopes and is flanked by deep water with lysocline, and shallow water (<40 m) where production is high but provided by other algae. Simplified pathway (right) of cells, single coccoliths and fecal pellets to sea floor with arrows indicating relative penetration of coccoliths into deep water.

coccolithophorid growth are the upwelling of essential nutrients, the depth of water and the grazing intensity (Fig.1). *E. huxleyi* contains comparatively low levels of chlorophyll and has a lower growth rate than many other phytoplankters [11].

(b) Coccoliths arrive at the sea floor by three routes: as scales, cells and fecal pellets. Stokes sinking rates for scales are about 10 cm d^{-1}, rates for cells are about 1 m d^{-1} [12] and rates for fecal pellets much higher, thus rates of sedimentation differ and fecal pellet transport is the most efficient [9]. Modern herbivorous zooplankton (Fig.1) feed on the cells [13] but during blooms, there is usually a grazing lag [11] so cells may sink directly. Coccoliths are readily detached from cells [8] but non-selective grazing would ensure that these too could be pelleted. A high but unknown proportion of coccoliths must sediment as pellets. Speckled limestones, interpreted as coccolith pellets, are known in the geological record [13]. Below the photic zone, seawater is generally aggressive towards calcite and dissolution is likely, though organic coatings on the scales give some protection. At a depth of c. 4 km dissolution increases rapidly as a response to increased CO_2 and hydraulic pressure (lysocline) and at c. 4.6 km reaches the calcite compensation depth (CCD, Fig.1) where dissolution is complete. Coccoliths can be found in deeper water, probably the result of mass-transport phenomena.

(c) Because sea floor accumulation is slow, coccoliths will remain effectively in contact with seawater for many years. Where oxygen levels are high, the benthic fauna will accelerate decay processes and cause bioturbation, but the effects of this on dissolution are not known. Under anoxic conditions, which exist locally today, but were more general in the past [14], decay would be slower, bioturbation reduced and preservation of coccoliths should be greater.

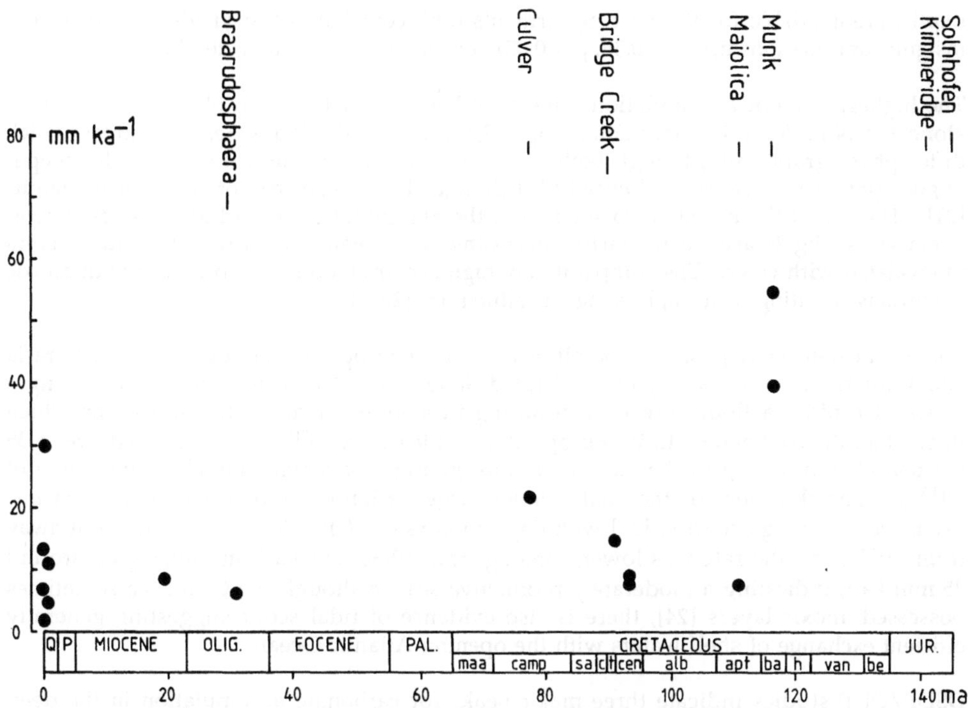

Fig. 2. Accumulation rate of coccoliths obtained/derived from [1-3, 9, 10, 15, 16, 20-3, 27, 28]. The rates assume accumulation of a pure coccolith limestone with 45% porosity (ave. for European chalk). Where data were unavailable, the porosity of Quaternary sediments was taken to be 70%. Abscissa shows geological epochs, with stages for the Cretaceous, and the position of some well known coccolith limestones.

(d) Sediments become dewatered and warmed on burial and further pressure-solution is likely. Little is known of this stage but burial in clay leads to better preservation than in pure limestone, due to retardation of pore water flow.

Rates of Accumulation

Accumulation rate data on coccolith-rich limestones are limited, and have been obtained mainly by dating either with biostratigraphic markers [15] or by the correlation of lithological periodicity with earth's orbital perturbations [3,16]. Some rates obtained by these methods are presented in Fig.2. These vary by less than two orders of magnitude, and no correlation between sediment age and accumulation rate is apparent. It has been argued that over a period of 20 ma following sedimentation, about half the accumulated pelagic carbonate would be removed by erosive processes, there being many hiatuses in the deep sea sediment record [17,18]. If this applies to coccolith-rich sediments, then rates prior to 20 ma BP would have been higher than shown by this amount. Even taking this into account, there is no significant correlation between rate and age ($r = +0.53$, $p > 0.05$).

Increasing water depth should decrease the rate of accumulation through the interplay of declining productivity and increasing dissolution. Although only approximate water depths are available for the past, Spearman's rank correlation test yielded a statistically insignificant relationship ($r = -0.38$, $p > 0.05$) between rate and water depth.

The highest modern accumulation rates have been estimated from blooms of *E. huxleyi* along the N.E. Atlantic continental slope [8,19] and in the Black Sea [20], areas which differ physiographically, though both are subject to high nutrient fluxes. In the deeper oligotrophic ocean, rates are lower [9,10] though this is also true of some shelf regions [21]. The coccolith laminae recovered from the Hauptblatterton and Munk beds (Lower Cretaceous, Fig.2) are particularly interesting, and seem to represent annual layers intercalated with clays. The comparatively high accumulation rate correlates with anoxic conditions resulting from high pelagic productivity [22,23].

The most extensive deposits of coccoliths occur in the upper Cretaceous chalks of Eurasia and Western Interior Seaway of the United States. The Chalk was deposited during a phase of rapid sea floor spreading resulting in a series of major transgressions which flooded all the continents. In W. Europe alone, at least 3×10^{20} g was deposited over a 35 ma period culminating in the late Cretaceous, giving an average annual accumulation of 10^{13} g. The thickness of the chalk shows large variations and in continental basins, extensive slumping occurred [24] with deposition rates of 100-200 mm ka^{-1} [25] but away from the basins, the rate was lower, ranging from 10-60 mm ka^{-1} and averaging around 25 mm ka^{-1}, indicating a moderately productive sea. Although the Chalk sea sometimes possessed anoxic layers [24], there is also evidence of tidal scour suggesting generally efficient exchange of shelf waters with the opening Atlantic Ocean.

DSDP/ODP studies indicate three major peaks for carbonate accumulation in the deep ocean over the past 100 ma. Global accumulation rates during the late Cretaceous are not known with precision but were certainly high [17]. It seems clear that during these periods, riverine input of calcium could not keep up with $CaCO_3$ deposition by pelagic biota. In the late Cretaceous, increased emanations of Ca and CO_2 [26] together with local recycling of deposited carbonate could offset the deficit. An early Oligocene minimum corresponds to a high sea level but is accompanied by a rapid fall in temperature. The late Oligocene maximum has been correlated with a rapid fall in sea level, exposing more continent, marginally increasing riverine input [18]. During major marine transgressions, it seems that shelf sea sedimentation increases at the expense of deep ocean deposition.

REFERENCES

1. Gartner S (1977) Mar. Micropaleont. 2: 1-25.

2. Fischer AG (1980) Spec. Publ. Geol. Soc. Am. 183: 93-104.

3. Hart MB (1987) Cret. Res. 8: 335-48.

4. Tappan H (1980) The Paleobiology of Plant Protists. W. Freeman, San Francisco.

5. Van der Val P, de Jong L, Westbroek P, de Bruijn WC, Mulder Stapel AA (1983) J. UJltrastr. Res. 85: 1139-58.

6. Westbroek P, van der Val P, van Emberg PR, de Vrind de Jong EW, de Bruijn WC (1986) In: Leadbeater BSC and Riding R (eds) Biomineralization in Lower Plants and Animals. Clarendon, Oxford, pp.189-203.

7. Bramlette MN (1958) Bull. Geol. Soc. Am. 69: 121-6.

8. Holligan PM, Violler M, Harbour DS, Camus P, Champagne-Phillipe M (1983) Nature 304: 339-42.

9. Honjo S (1976) Mar. Micropaleont. 1: 65-79.

10. Berger WH (1976) In Riley JP, Chester R (eds) Chemical Oceanography vol. 5, Academic Press, New York, pp.266-388.

11. Raymont JEG (1980) Plankton and Productivity in the Oceans 1, Phytoplankton, Pergamon, Oxford.

12. Bonin DJ, Droop MR, Maestrini SY, Bonin MC (1986) Cryptogam. Algol. 7: 23-84.

13. Hattin DE (1975) J. Sed. Petrol. 45: 686-96.

14. Fischer A, Arthur M (1975) Spec. Publs. Soc. econ. Paleont. Miner. Tulsa 25: 19-50.

15. Dean WE, Gardner JV, Cepek P (1981) Mar. Geol. 39: 81-101.

16. Schwarzacher W, Fischer AG (1982) In: Einsele G, Seilacher A (eds) Cyclic and Event Stratification. Springer, Berlin, pp.78-95.

17. Moore TC, Heath GR (1977) Earth Planet Sci. Lett. 37: 71-80.

18. Davies TA, Worsley JR (1981) Spec. Publs. Soc. Econ. Pet. Min. 32: 169-79.

19. Holligan PM, Groom SB (1986) Proc. Roy. Soc. Edinb. 88B: 239-63.

20. Ross DA, Degens ET (1974) Mem. Am. Ass. Pet. Geol. 20: 183-99.

21. Swift SA (1977) J. Geol. 85: 301-19.

22. Muttorlose J, Harding I (1987) Abh. geol. Bundesanst 39: 177-215.

23. Thomsen E (1989) Mar. Micropaleont. 15: 123-52.

24. Kennedy WJ (1987) Bull. Centres Rech. Expl. Prod. Elf Aquitaine 11: 91-126.

25. Häkansson E, Bromley RG, Perch-Nielsen K (1974) Spec. Publs. Ass. Sedimentol. 1: 211-34.

26. Mackenzie FT, Agegian CR (1989) In: Crick RE, (ed) Origin, Evolution and modern Aspects of Biomineralization in Plants and Animals. Plenum, New York, pp.11-27.

27. Geitzenauer KR, Roche MR, McIntyre A (1976) Mem. Geol. Soc. Am. 145: 423-48.

28. Maxwell AE, von Herzen RP, Andrews JE, Boyce RE, Milow ED, Hsu KJ, Percival SF, Saito T. (1970) Init. Repts. DSDP 3: 441-71.

CHAPTER 4.7
Amorphous Minerals and Theories of Biomineralization

K. SIMKISS

Pure & Applied Zoology, School of Animal and Microbial Sciences, University of Reading, Reading RG6 2AJ, England

Key Words, Glasses, Random networks, Hydration, Sol-Gel.

INTRODUCTION

Biomineralization is the process whereby organisms convert dissolved ions into solid minerals. These minerals are generally thought of as crystals i.e., solids with a regularly patterned geometrical arrangement of the structural components (atoms, ions or molecules). The smallest complete representation of this material is the "unit cell" and it is the regular repeating of this unit in space that gives a crystal its main characteristic feature i.e., long range order. Most of the commonly recognized forms of biomineralization are calcium salts such as $CaCO_3$ (as aragonite or calcite) and $Ca_{10}(PO_4)_6(OH)_2$ (bone mineral) and most theories of biomineralization are based upon suitable explanations for the deposition of crystals. There are, however, a number of condensed solids that are non-crystalline. They do not consist of regularly repeating structural units and are therefore referred to as amorphous i.e. possessing no long range order. They may, however, occur in large quantities. Of the 10^{13} kg of minerals that occur in the sea most exist as amorphous silica (SiO_2) formed by phytoplankton.

Table 1. Types and functions of amorphous minerals.

	Tissue	Function	Taxa
Precursor minerals			
Amorphous $CaCO_3$	Blood	Storage granules	Decapods
Amorphous $CaCO_3$	Connective tissue	Ions for blood	Gastropods
Amorphous $CaCO_3$	Gastroliths	Ions for blood	Crustacea
Amorphous $Ca_3(PO_4)_2$	Radula	Forms dahllite	Chitons
Amorphous $CaMgP_2O_7$	-	Phosphate storage	Protozoa
Structural minerals			
Amorphous $Ca_3(PO_4)_2$	Proboscis	Stabbing stylets	Nemerteans
Amorphous $Ca_3(PO_4)_2$	Muscle	Skeletal	Annelids
Amorphous SiO_2	Mandibles	Biting	Copepods
Amorphous SiO_2	-	Skeletal	Protoctista
Amorph $CaCO_3$&$Ca_3(PO_4)_2$	Epidermis	Spicules	Protochordata
Amorphous $Fe_3(PO_4)_2$	Radula	Grinding	Chitons
Amorphous CaF_2	Skin	Spicules	Gastropods
Deposits foreign ions			
Amorphous $Ca_3(PO_4)_2$	Malpighian tubule	Detoxification?	Diptera
Amorphous $Ca_3(PO_4)_2$	Hepatopancreas	Detoxification?	Crustacea
Amorphous $CaMgP_2O_7$	Hepatopancreas	Detoxification?	Gastropods

There is a tendency among biologists to regard amorphous minerals as little more than "imperfect crystals". The concept arises from the fact that foreign ions will often disrupt a crystal lattice and if sufficient disruptions occur the lattice structure is lost. In order to dispel this approach table 1 lists a number of biologically occurring amorphous minerals that have a relatively pure composition, although it is characteristic of amorphous solids that they may have a rather variable stoichiometry. Table 1 also indicates that many amorphous deposits can be associated with specific biological functions. The implication of this is that

the ability to form such solids is a product of natural selection. Amorphous minerals have properties that crystals do not possess and biological systems have exploited these. There seem to be at least 3 situations where amorphous minerals are superior to crystals. These are (i) a higher solubility making such deposits ideal precursors of either ions or other minerals (ii) structural properties in which network crosslinks provide a way of avoiding cleavage planes in the mineral and (iii) the ability to incorporate a variety of inclusions.

Biomineralization, therefore, produces two types of product, i.e. crystalline and amorphous solids. Crystals are thought to be formed in solutions that are kept supersaturated by the selective concentration of ions in regions where charged surfaces perform some nucleating role. Biologists usually invoke membrane pumps and matrix proteins to explain these activities. Amorphous solids do not fit easily into such schemes. Thus, they are typically more soluble than their crystalline equivalents and they have no regular repeating structure that would build on an equivalent template. This raises the question as to whether the processes of crystalline and amorphous biomineralization are similar or not.

THE STRUCTURE OF AMORPHOUS SOLIDS.

If we are to understand the biological processes involved in forming amorphous solids we clearly need to understand the structure of these minerals. Current models for the structure of inorganic amorphous solids derived originally from

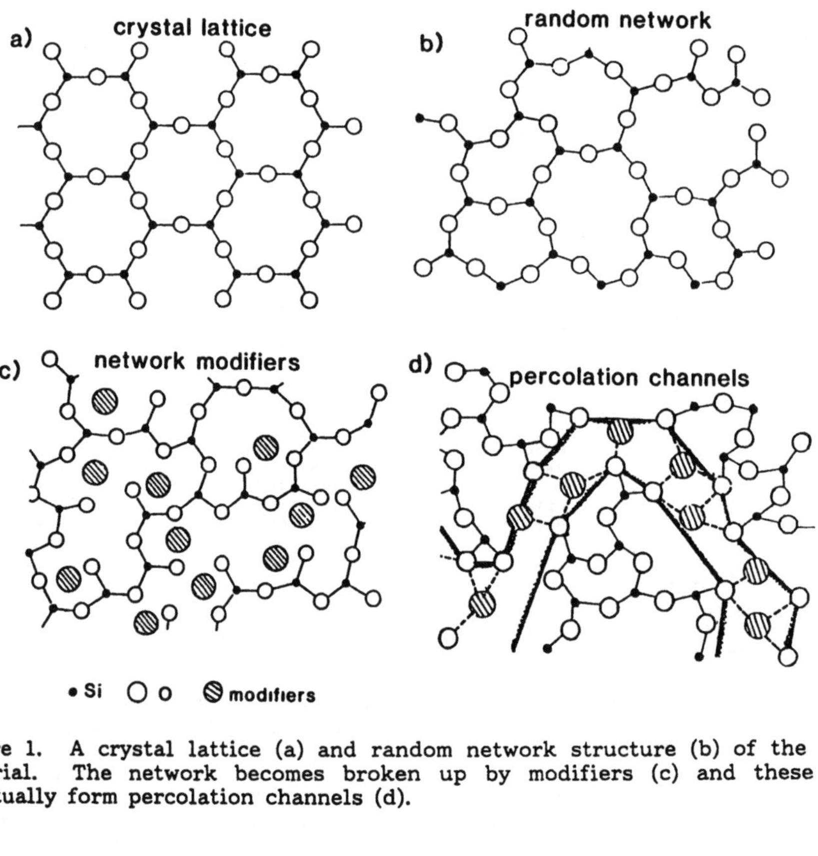

Figure 1. A crystal lattice (a) and random network structure (b) of the same material. The network becomes broken up by modifiers (c) and these may eventually form percolation channels (d).

the empirical observation that in glasses the ionic radii of the glass cation/anion are in the range of 0.2 to 0.4. Most crystals of this type have 4 anions around each cation and this has led to the concept that a tetrahedral arrangement is involved in most types of glass. In the first attempt to correlate glass formation with such a structural model Zachariasen [1] argued that the basic unit of a glass was probably the same as a crystal. Instead of a crystal network however he proposed that there should be an arrangement of these units in a random network. The essential feature of such three dimensional networks is that there is a variable set of bond angles, instead of the fixed crystalline arrangement, with the result that the structure has no periodicity (Figure 1). Such random networks are thought to form for kinetic reasons. Thus, a molten mass of glass contains the same basic structural units as those that occur in crystal formation. In glass formation, however, the cooling produces a large increase in viscosity that inhibits diffusion and this prevents these units from condensing into a regular lattice. There is, therefore, an energy barrier which stops the amorphous mineral from spontaneously transforming into a crystalline phase (Figure 2). The barrier to crystallization is determined by bond strength, and the number and type of crosslinks [2].

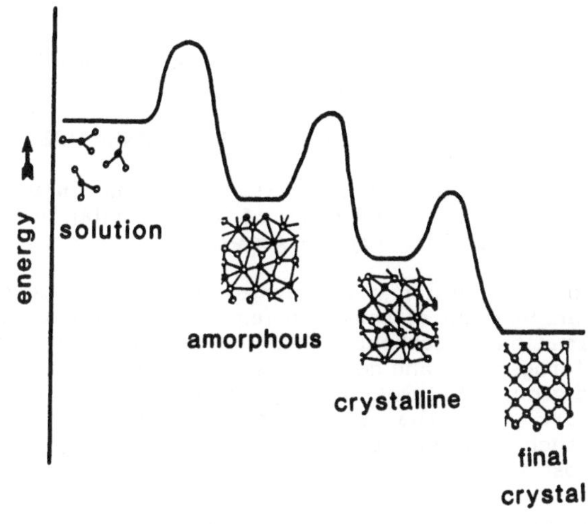

Figure 2. Energy barriers between materials in solution and their solid states.

The random network model is based on a number of rules that favour a non-crystalline structure. These are

1. each oxygen is common to only 2 cations
2. no more than 4 anions surround a cation
3. the anion polyhedra share corners but not edges or faces (Figures 1 & 3).
4. usually only one of the polyhedron corners may be unshared. [1]

Zachariasen's model was based on pure silica glasses but most commercial glasses contain soda (Na_2O) in addition to silica (SiO_2). With the advent of X ray analysis it was shown that in such soda glasses the sodium ions occupied the larger holes in the random network increasing the oxygen to silicon ratio and breaking up the random network (Fig 1). Such "modifiers" are used extensively in the glass industry since they lower the viscosity and melting point of the glasses making them easier to handle. Such a modified random network which is partly ionic and partly covalent is shown in figure 1. The structure is characterized by percolation channels through which ions may move. This property appears when the modifiers exceed about 16% of the total mass of the glass [3]. Most amorphous biominerals (e.g. SiO_2, P_2O_5 etc) appear to satisfy Zachariasen's rules and in order to make progress we should accept that the amorphous minerals shown in table 1 virtually all exist as random networks. The question of how such structures can be formed from biological fluids requires the further attention of chemists.

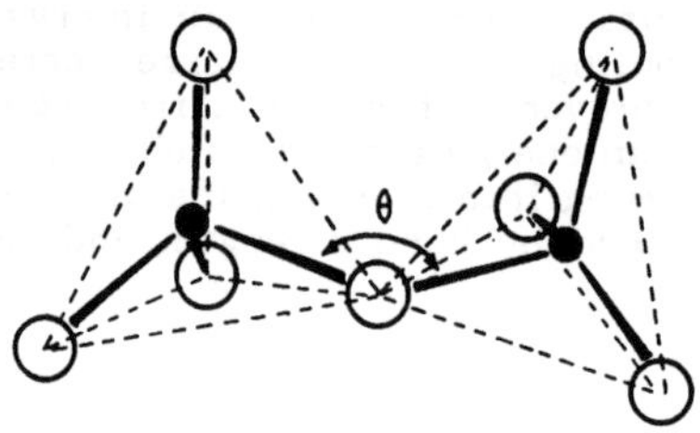

Figure 3. Tetrahedral structure showing a variable bond angle (θ) and the sharing of corners as they are thought to occur in a glass.

COMPARISONS OF CRYSTALLINE AND AMORPHOUS BIOMINERALIZATION

The basic steps in crystal formation are (i) an increase in ion-concentrations so as to exceed the solubility product of a particular mineral phase (ii) nucleation of a solid phase (iii) crystal growth.

In most forms of biomineralization it is assumed that (i) is achieved by ion pumps or the metabolic manipulation of some ion (e.g. the enzyme carbonic anhydrase facilitating the formation of carbonate ions). The process of nucleation (ii) is almost certainly heterogeneous and may involve not only a surface but specifically spaced and oriented groups that favour the grouping of particular ions in relation to a lattice structure. Ions that attach to this surface are involved in (iii) crystal growth and such ions are thought to move over the crystal surface to find the energetically most favourable site for this process (Figure 4a).

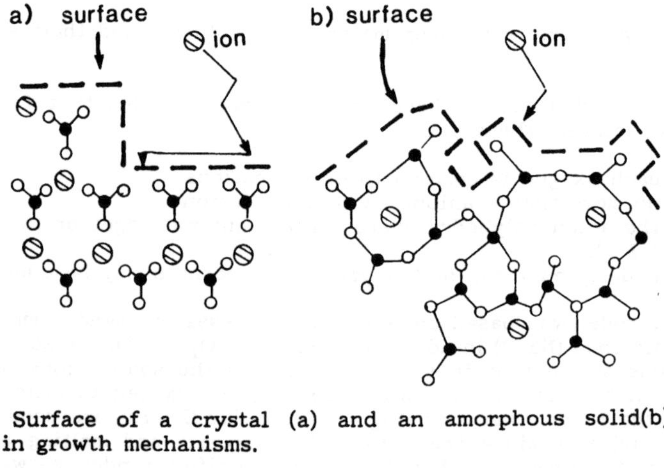

Figure 4. Surface of a crystal (a) and an amorphous solid(b) indicating differences in growth mechanisms.

If we now look at these processes in more detail we can identify their relationship with amorphous minerals. In biological systems the initial ion pairs and perhaps the first solid phases in most minerals are frequently hydrated since energy is required to remove their interaction with the aqueous solvent. For this reason such clusters are poorly organized until energy is available to dehydrate the ions. While they exist as hydrated ions they are often unable to pack very closely and thus form irregular structures (Figure 4b). Because such amorphous solids are heavily hydrated and because they lack the stabilizing forces of a crystal lattice they are usually highly soluble. This is not always the case however since for many salts the energy barriers to crystallization are sufficiently low that crystals form before the solubility product of the amorphous phase is exceeded [4]. An exception to this occurs if there is a very rapid increase in the concentration of ions and this is one of the factors known to favour amorphous mineral formation. A second factor that favours amorphous solids is when there is a large activation energy for breaking and reforming of bonds as there is in silica (Figure 2). Finally, factors that inhibit crystal growth clearly favour amorphous mineralization.

The formation of amorphous minerals is therefore favoured by a very rapid nucleation rate which inhibits the growth of larger crystals. Typically, amorphous biomineralization occurs in small vesicles. This may facilitate a very rapid change in ionic products because of the very large membrane surface that is available to transport ions in relation to the small volume of fluid that is present.

The nucleation of amorphous materials from aqueous solutions has received very little attention. Clearly the amount of disorder in the system favours amorphous structures and the presence of asymmetrical molecules is known to increase the tendency for glass formation. What clues are there for these influences in amorphous biomineralization? First, there is the tantalizing observation that amorphous biominerals often contain unexpected amounts of silica; that most common of glass forming ions [5,6]. It even occurs at the sites of bone formation [7] but the significance of these observations needs further consideration. Second amorphous biominerals contain an organic matrix.

It has become commonplace to attribute the formation of crystals in biomineralization to the crystal-orientating effect of an organic matrix. Such a matrix is variously envisaged as a phosphoprotein or glycoprotein rich in acidic groups. Typical analyses of matrix proteins for molluscan shell and echinoderm spicules are given in table 2. The only available analyses of the matrix proteins of the amorphous hydrous ferric phosphate from holothurian skin and the amorphous silica of sponge spicules are given for comparison. It is apparent that there are no major differences at this level of analysis in that both contain large amounts of dicarboxylic and other amino acids that are usually related to an epitaxic function.

Table 2. Main amino acids (residues/1000) in the soluble organic matrix of crystalline minerals from echinoderms and molluscs compared with the organic matrix from amorphous deposits in sponges and holothurians [8,9,10].

	Crystalline minerals		Amorphous minerals	
Amino Acid	Echinoderm	Mollusc	Sponge	Holothurian
Aspartate	110	148	140	124
Glutamate	121	72	90	108
Glycine	213	136	110	103
Alanine	83	61	68	75
Serine	125	73	63	57

One of the great surprises in biomineralization in the past decade has been the discovery that the soluble fraction of matrix proteins inhibits rather than facilitates mineralization [11]. It has, therefore, been suggested that the crystal inducing properties are only observed when the matrix protein is adsorbed onto a structural surface [12]. This raises the interesting possibility that soluble matrix may favour amorphous biomineralization by inducing rapid but uncontrolled nucleation. Soluble matrix plus structural protein on the other hand is sterically oriented and induces crystal deposition by facilitating lattice formation.
Once nuclei are formed their rate of growth, or stability, will depend on the rate at which ions can diffuse from solution onto the solid surface and then move to the site of crystal growth. One of the factors that favours glass formation from melts is the fact that cooling increases the viscosity of the liquid making it difficult for ions to assume preferred positions in the lattice. Organic molecules that attract ions and inhibit crystal growth will similarly favour amorphous biomineralization. This may be another feature of soluble as opposed to immobilized matrix proteins. Finally one needs to consider the actual process of crystal growth. It is generally accepted that it is unlikely that ions seek out and attach directly to the reactive site of a crystal surface. Instead the ion appears to attach to the surface and move over it in 2 dimensions until it reaches a kink along which it moves in 1 dimension until it attaches to the energetically most favourable site on the growing lattice (Figure 4). Clearly growth of an amorphous particle is different in that the surface of a random network provides a much more diverse set of variables about which we need more information.

Finally we should again emphasize that for kinetic reasons many crystals form by processes that initially favour the production of unstable minerals. Thus, calcite may initially form as an amorphous hydrated calcium carbonate that for thermodynamic reasons subsequently dissolves and reprecipitates as vaterite, aragonite and finally calcite (Figure 2). This is the Ostwald-Lussac sequence and it is of importance for two reasons. First, amorphous minerals may be stabilized in biological systems by the effects of ions such as magnesium or adenosine triphosphate which therefore delay the completion of the Ostwald-Lussac sequence. Second, it is difficult to understand the traditional explanation of the organic matrix as both a nucleator and controller of crystal form if different minerals are involved in an Ostwald-Lussac series.

BIOLOGICAL ASPECTS OF AMORPHOUS MINERALS

Recently there has been considerable interest in the possibility of the industrial production of glasses at room temperature. Instead of using melts or the direct precipitation of SiO_2, such schemes emphasize the hydrolysis of silicate esters, such as tetraethylorthosilicate ($(RO)_4Si$) to release units of silica hydroxides. These then condense as the centres of a polymerization reaction in a sol-gel reaction

$$\text{i.e.}\quad (RO)_3\,Si\,OR + H_2O \xrightarrow[\text{(hydrolysis)}]{} (RO)_3\,SiOH + ROH$$

$$(RO)_3\,Si{-}OR + (RO)_3\,SiOH \xrightarrow[\text{(polymerization)}]{} (RO)_3{-}Si{-}O{-}Si(OR)_3 + ROH + H_2O$$

The product is a silica gel that dries out to form a hard glass [13]. Much of the interest in the process centres upon the solvents, catalysts and gelation temperatures that are used to form the hydrated glass. Of particular interest is the fact that the sol-gel process can easily be modified to form composites and that both the hydrolysis and condensation reactions are particularly pH sensitive [14].

Model systems such as these present challenges to the study of silicification in biological systems. It has been known for some time that during silification in diatoms there are considerable changes in the soluble silicon pool. Much of this silicon exists as a trichloracetic acid-soluble form that is not silicic acid [15] and may represent a metabolic precursor. Clearly such molecules need to be identified.

The physical properties of amorphous minerals also differ from those of crystals, with a consequent effect on their biological uses. Perhaps the most exciting example of this occurs with the connective tissue cell of molluscs. This cell type is widely distributed in the mantle, foot, visceral complex and blood vessels of prosobranch and pulmonate molluscs. In the fully developed cell a mineral deposit occupies most of the cytoplasm with the organelles and nucleus displaced to the periphery. These granules are composed of amorphous calcium carbonate [16] deposited in an organic matrix. Under the influence of stimuli such as acidosis or calcium deprivation the membrane containing these deposits fuses with the plasm membrane to form a number of small pores. The amorphous deposit dissolves and the ions pass through these holes to enter the body fluids [17]. In this situation the amorphous mineral is clearly being used as a highly labile and soluble source of ions which, because of its amorphous structure, can be readily mobilized to repair more crystalline deposits such as a broken shell.

Finally it is worth noting that another amorphous mineral, $CaMgP_2O_7$ occurs intracellularly in the basophil cells of the molluscan hepatopancreas. These pyrophosphate deposits appear to react with a number of foreign ions in the blood and we have tried to relate these to the modified random network model of amorphous solids. Perhaps the most interesting of these reactions occurs if the snail absorbs large quantities of manganese ions. Under these circumstances the pyrophosphate granules become corroded within the cell apparently because the manganese ions act as Bronsted acids that presumably break down the modified random network of pyrophosphate ions near the surface of the granule (Fig 5). In the process they liberate pyrophosphate ions that reprecipitate at the surface of the granule as $MnMgP_2O_7$ and calcium ions that may lead to localized cell death [18]. Another equally revealing reaction occurs if the snail is exposed to zinc ions. In the presence of this ion the granule remains amorphous but the pyrophosphate becomes hydrolysed to orthophosphate. The same treatment in vitro results in the granules becoming crystalline (Fig 5) with a mixture of both pyrophosphate and orthophosphate anions [19]. This suggests that the organic matrix plays a role in stabilizing these amorphous deposits.

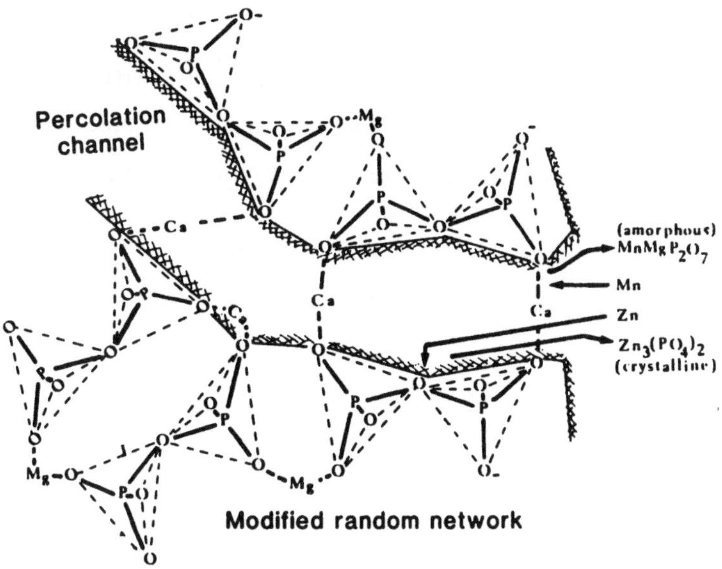

Figure 5. Reactions of Mn & Zn ions on the glass granules of a snail showing relationship to a modified random network structure.

Amorphous biomineralization has attracted relatively little attention but it is worth further study for the following reasons. First, it provides an interesting comparison for the usual theories of biomineralization. The formation of amorphous solids is favoured by high levels of supersaturation, rapid nucleation and restrictions on crystal growth, all of which involve processes that are also involved in crystal biomineralization. Second, although there is no clear distinction between amorphous and crystalline solids, the concept of a random network structure for amorphous minerals provides a theoretical basis for exploring some of the properties of these materials. These have probably been selected for in the evolution of these biomaterials. Finally the chemistry of amorphous mineral formation from aqueous solutions has been poorly explored and may provide major new insights into the biological pathways involved.

REFERENCES

[1] Zachariasen, WH, (1932) J. Amer. Chem. Soc. 54: 3841-3851.
[2] Ovshinsky, SR, (1985) In: Adler, D, Schwartz, BB, Steele, MC, (eds) Physical Properties of Amorphous Minerals. Plenum Press, N.Y. pp105-155.
[3] Greaves, GN, (1985) J. Non-Cryst. Solids, 71: 203-217.
[4] Williams, RJP, (1984) Phil. Trans. Roy. Soc. Lond. 304B: 411-424.
[5] Lowenstam, HA, Rossman, GR, (1975) Chem. Geol. 15: 15-51.
[6] Howard, B, Simkiss, K, unpublished.
[7] Carlisle, EM, (1986) In, Silicon Biochemistry: Ciba Sympos. 121, 123-136.
[8] Weiner, S, (1979) Calcif. Tiss. Int. 29: 163-1676.
[9] Sikes, CS, Wheeler, AP, (1986) Biol. Bull. 170: 494-505. Wheeler, AP, Rusenko, KW, Sikes, CS, (1988) In: Sikes, CS, Wheeler, AP. ('eds') Chemical Aspects of Regulation of Mineralization, University of South Alabama, Mobile, 9-13.
[10] Lowenstam, HA, (1972) Chem Geol. 9: 153-166.
[11] Wheeler, HAP, Sikes, CS, (1984) Amer. Zool. 24: 933-934.
[12] Addadi, L, Weiner, S, (1985) Proc. nat. Acad. Sci. USA, 82: 4110-4114.
[13] Ulrich, DR, (1988) Chem tech, 242-249.
[14] Brinker, CJ, (1988) J. Non-Cryst. Solids, 100: 31-50.
[15] Sullivan, CW, Volcani, BE, (1981) In: Simpson, TL, Volcani, BE (eds) Silicon & Siliceous Structures in Biological Systems, Springer-Verlag, N.Y. pp 15-42.
[16] Tompa, AS, Watabe, N, (1976) Calcif. Tiss. Res. 22: 159-172.
[17] Sminia, T, de With, ND, Bos, JL, Van Nieuwmegen, ME, Wittier, MP, Wondergen, J. (1977) Netherland, J. Zool, 27, 195-208.
[18] Taylor, MG, Simkiss, K, Greaves, GN, Harries, J, (1988) Proc. Royal Soc. Lond. 234B: 463-476.
[19] Taylor, MG, Greaves, GN, Simkiss, K, (1990) Europ. J. Biochem. (in press).

CHAPTER 4.8

Evolution of Enamel Proteins: a Paradigm for Mechanisms of Biomineralizaiton

H.C. SLAVKIN, R.J. KREJSA, A. FINCHAM, P. BRINGAS, JR., V. SANTOS, Y. SASSANO, M.L. SNEAD, and M. ZEICHNER-DAVID

Center For Craniofacial Molecular Biology, School of Dentistry, University of Southern California, Los Angeles, CA, USA

Key Words: Teeth, Enamel, Differentiation, Hagfish, Mouse

INTRODUCTION: DEVELOPMENT AND EVOLUTION

Knowledge of development provides a unique perspective into mechanisms of evolution, and a knowledge of evolution provides corresponding insights into development [1,2]. Several major questions are pertinent towards understanding the evolution of biomineralization:

(i) How do genes control development and how does this control contribute changes during the course of evolution?

(ii) What are the rules for timing mechanisms which operate during development and evolution?

(iii) Within the implied hierarchy of regulation, what are the levels of control beyond the regulatory and structural genes for vertebrate biomineralization?

a. Evolution of Enamel

One paradigm for development coupled with evolution is vertebrate tooth biomineralization; specifically, the tissues and their products associated with enameloid and enamel biomineralization [3-7]. Recent comparisons of macromolecules associated with the mineral phase of cell-mediated, tissue-specific, biomineralization, indicates that acidic glycoproteins and/or phospholipids are usually present [3-4].

The vertebrate tooth has a number of conserved as well as distinctive phenotypes which have changed during the course of vertebrate evolution. The recent advances from protein and immunochemistry, recombinant DNA technology, and experimental and comparative embryology provide new evidence to pursue mechanistic questions regarding the development and evolution of enameloid and/or enamel biomineralization [see critical discussions in 4-7].

We suggest that examination of enamel protein evolution may serve as a model towards understanding biomineralization. When and where were enamel gene products first expressed? Is there homology between very early vertebrate (e.g. 500 million years ago) and mammalian enamel gene products? Have the functions of enamel gene products been conserved during the evolution of vertebrate biomineralization?

b. Cyclostomes And Gnathostomes: Major Stems Of Vertebrates

In gnathostomous vertebrates, a series of sequential epithelial-mesenchymal interactions results in tissue-specific enameloid and enamel biomineralization (e.g. fish, sharks, frogs, reptiles and mammals) [5-9]. Indirect immunofluorescent microscopy has been used to demonstrate that rabbit anti-mammalian enamel polyclonal antibodies (i.e. anti-amelogenin) are cross-reactive with enamel extracellular matrices (ECM) from a number of disparate species within Chondrichthyes, Teleostei, Amphibia, Reptilia and Mammalia [8-10].

Several years ago our research group reported that anti-mouse amelogenin cross-reacted with Pacific hagfish (*Eptatretus stoutii*) [11]. Prior to that report, it was assumed that hagfish and lamprey teeth exclusively produced keratins [12-13]. Whereas anionic members of the keratin gene family have been demonstrated to participate in biomineralization [14], extant hagfish and lamprey teeth are not mineralized [12-13]. The cyclostomes comprise lampreys and hagfish which are eel-like extant aquatic vertebrates presumed to be the remaining survivors of a nearly extinct group of agnathan or jawless Paleozoic "fish" [11-15]. The hagfish possess a cartilaginous skeleton which is not mineralized. Therefore, the presence of enamel-like antigens or epitopes in the outer covering of hagfish teeth suggested considerable conservation within the enamel protein(s) structure during 500 million years of vertebrate dental evolution [3-12].

It is further evident that tooth development and biomineralization within mammals appears to be remarkably autonomous; isolated primitive tooth explants develop into complex and mineralized three-dimensional structures when cultured in serumless, chemically-defined medium [16-17]. The developmental program for enamel biomineralization can be expressed through short-range endogenous signals (e.g. growth factors, cytokines, hormone-like molecules), and does not appear to require long-range exogenous factors (e.g. steroid and polypeptide hormones) associated with neurogenic or vascular systems. Whereas this seems evident in embryonic mouse tooth tissues [16-17], it is not as yet known if this intrinsic and seemingly autonomous feature is also evident within hagfish, bony fish, shark, frog or reptilian tooth development.

We now hypothesize that early Paleozoic ancestors of the cyclostomous agnathan (e.g. extant Pacific hagfish) expressed genes related to enamel proteins. We predict that hagfish embryonic oral ectoderm gives rise to odontogenic epithelia which produce a number of different phenotypes (e.g. keratin producing cells) including the pokal cells. We suggest that pokal cells synthesize and secrete enamel-like gene products into the pyramid ECM covering material, and these cells are ancestral ameloblasts. Subsequent changes in maternal factors, cell cleavage patterns, cell lineage patterns, sequential gene expression patterns, and the relative timing of events contribute towards the evolution of ameloblasts and enamel biomineralization. Investigations of pokal cell development and evolution may provide new insights towards understanding the evolution of enamel biomineralization.

EXPERIMENTAL STRATEGY

The experimental strategy used in our investigations has been based upon the assumption that specific sequences of encoded information found within mammalian enamel genes represent highly conserved genetic information also found within hagfish tooth covering producing cells --- the pokal cells and their ECM products. The mouse amelogenin gene was isolated, cloned and sequenced [18], and the amino acid sequence deduced [18]. The deduced amino acid sequence for mouse is remarkably homologous to the bovine amelogenin protein sequence [19]. Based upon hydropathy analyses of the primary structure, we identified motifs appropriate for the production of synthetic oligopeptides to produce polyclonal antibodies specific for defined sequence regions [i.e. anti-peptide #1, -LPPHPGHPGYI-, amino acids 3-13].

This antibody was recently used to define when and where mouse amelogenin gene products were expressed during embryonic, fetal and neonatal stages of mouse molar tooth development [20]; antibodies were also used to define when and where "amelogenin-like" proteins were expressed during shark tooth development [21-22]. This antibody has now been used to define when and where these epitopes were expressed in hagfish tooth development. The data confirms our earlier studies using polyclonal rabbit anti-mouse amelogenin antibodies [11].

In tandem, studies were designed to establish the autonomy of primitive mouse embryonic cap stage tooth organs cultured in serumless-chemically-defined medium [16-17]. We also determined when and where amelogenin gene products were expressed during mouse molar tooth development *in vivo* [23], and we investigated the characteristic expression of amelogenin *in vitro* in association with biomineralization. Within 12 days *in vitro*, E15/16 mouse molars expressed biomineralization; by 21 days enamel biomineralization was readily apparent [16-17]. In replicate studies, we assayed for growth factor transcript expression using the RT-PCR (reverse transcriptase-polymerase chain reaction) technique [24]; β-actin was used as an internal control for the detection of low abundant mRNA. RT-PCR can detect 10 copies of specific transcript (e.g. EGF, TGF-α) per cell.

SEQUENTIAL EXPRESSION OF GROWTH FACTORS AND ENAMEL BIOMINERALIZATION

Since embryonic mouse tooth organs appear to contain an intrinsic developmental program for enamel biomineralization, an autonomous program independent of extrinsic controls, we designed experiments to determine if endogenous growth factor transcription could be detected in this simple *in vitro* model. Qualitative surveys for growth factor transcription products (e.g. EGF, TGF-α, TGF-β, IGF-I, IGF-II, bFGF) were made on E15 mouse molar tooth organs (cap stage) cultured for periods up to 31 days *in vitro* using serumless, chemically-defined medium.

Whereas relative levels of specific growth factor expression varied during progressive phases of enamel biomineralization, all cited growth factor mRNAs were detected at all stages of development. We argue that the changing patterns of expression, rather than a specific growth factor, is the key towards eventually discovering the metabolic code associated with tissue-specific biomineralization.

POKAL CELL DIFFERENTIATION, PYRAMID PRODUCTION AND THE ABSENCE OF BIOMINERALIZATION

Several caveats should be cited related to pursuit of studies using embryonic hagfish tissues. Only three embryos of *Myxine glutinosa* are known from the Atlantic, and from the Pacific only 34 sets of embryonic *Eptatretus stouti* remain from the original collections of Professor Bashford Dean founded in the last century (1898). We have obtained access to sectioned hagfish embryonic tissues, inclusive of the craniofacial regions, through the kind efforts of Professor Bo Fearnholm (Stockholm) and Ms. Elizabeth McLain (Boston). From these histological sections we have learned that hagfish form a dental lamina and the subsequent series of bud, cap and bell stages of tooth development characteristic of most vertebrates.

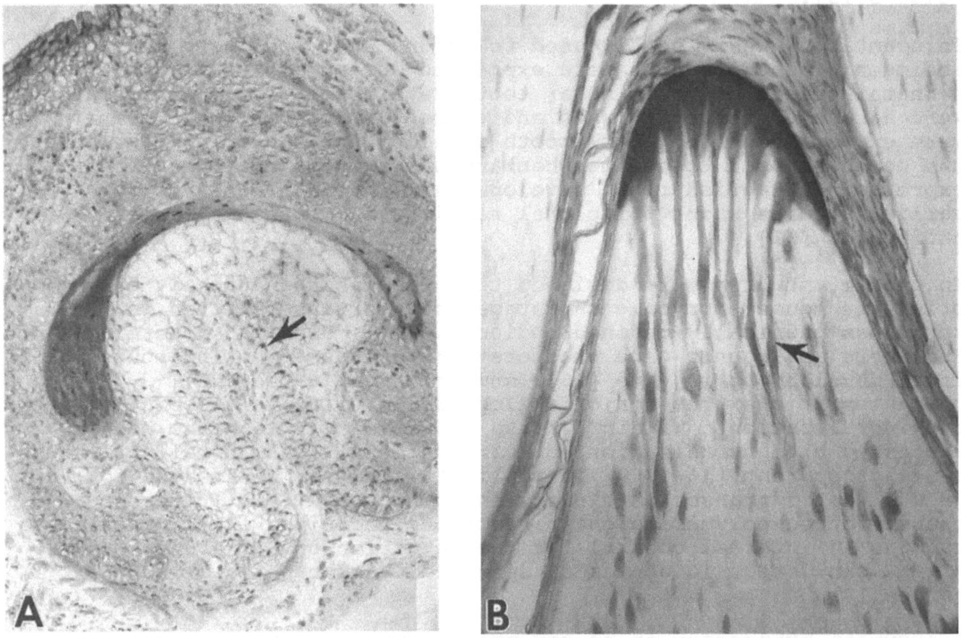

Fig. 1. (A) Hagfish cap stage development. Arrow indicates progenitor pokal cells. (B) Crown stage illustrating pokal cells producing the pyramid ECM covering the pokal cone. Pokal cells and pyramid ECM were immunostained with anti-peptide #1 antibody.

Curiously, the hagfish do not form inner enamel epithelia aligned in proximity and in juxtaposition to dental papilla mesenchyme; rather, enamel organ cells at the bud stage polarize, orient distal to the epithelial-mesenchymal interface, and appear to migrate, become pokal cells and are positioned into the forming pokal cone. During this process, specific epithelial pokal cell differentiation within the enamel organ epithelia, pokal cells and their secreted pyramidal ECM were immunostained with antibodies directed against the canonical amelogenin sequence (-LPPHPGHPGYI-, amino acids 3-13).

FROM POKAL CELL TO AMELOBLAST: CHANGING PATTERNS OF CELL LINEAGE

Examination of embryonic and adult stages of tooth development in numerous vertebrates, ranging from the Pacific hagfish to the mouse, indicate that in all vertebrates the oral ectoderm produces a discrete and sequential series of "enamel organ" forms (i.e. dental lamina, bud stage, cap stage). Within the enamel organ of cap stage hagfish, however, merocrine secretory cells do not form at the epithelial-mesenchymal interface as is the case for ameloblast differentiation in mammals, reptiles, amphibians, sharks and bony fish. Rather, inner enamel epithelia polarize in the opposite direction, 180 degree change in orientation, and align in linear arrays to produce the pyramid ECM covering for the pokal cone; pokal cells become polarized, non-dividing, merocrine-like secretory cells which synthesize and secrete products which contain epitopes shared with mouse amelogenin peptide 1 sequence (i.e. -LPPHPGHPGYI- corresponding to amino acids 3-13).

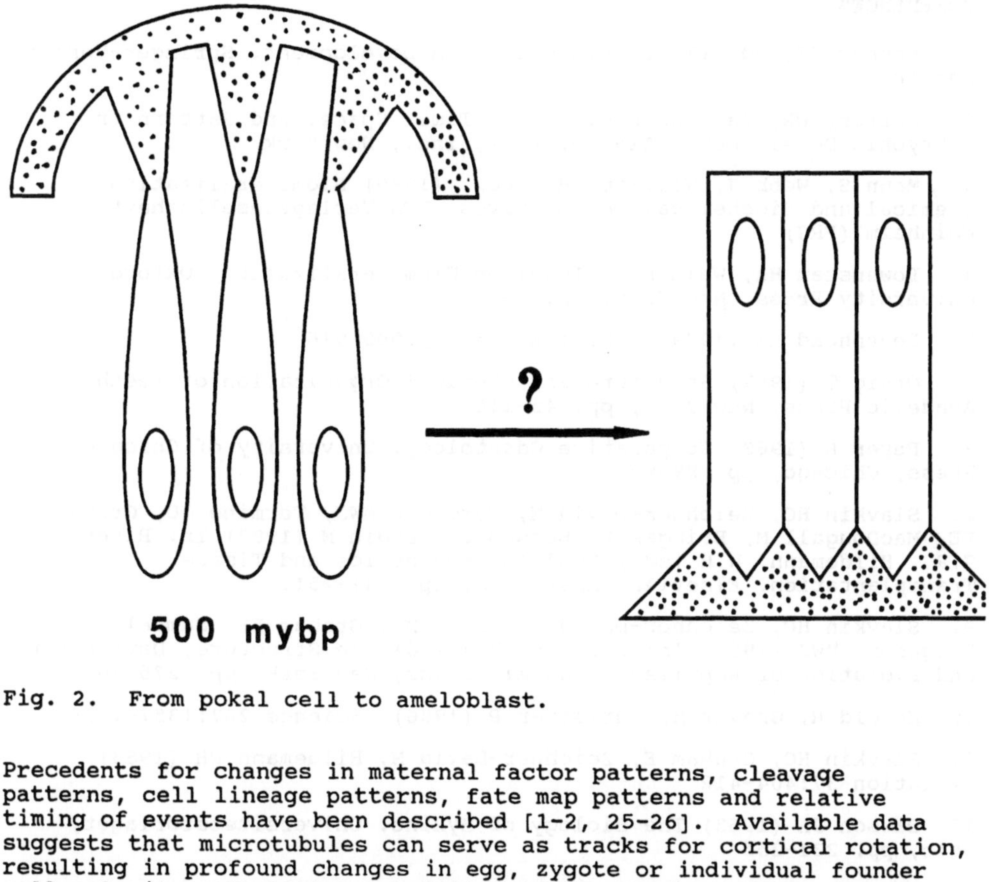

Fig. 2. From pokal cell to ameloblast.

Precedents for changes in maternal factor patterns, cleavage patterns, cell lineage patterns, fate map patterns and relative timing of events have been described [1-2, 25-26]. Available data suggests that microtubules can serve as tracks for cortical rotation, resulting in profound changes in egg, zygote or individual founder cell polarity. Such changes might provide partial explanations for the remarkable changes in polarity between (i) pokal cells secreting ECM constituents (i.e. sharing epitopes with the N-terminus of the mouse amelogenin) into the pyramidal tooth covering in the hagfish, and (ii) the inner enamel epithelia polarized towards a basal lamina in juxtaposition to adjacent ectomesenchyme and secreting enamel proteins. An explanation for how might natural selection influence the evolution from pokal cells to ameloblasts is not known. What is the explanation for the absence of biomineralization in the pokal cone and pyramidal ECM of the hagfish and lamprey? Hagfish are unique among vertebrates in having their body fluids isosmotic with sea water. What level of biological regulation controls the distinct patterns of differentiation characteristic of pokal cells versus inner enamel epithelia; both processes within the enamel organ?

Further, direct experimentation using hagfish embryos has not as yet been possible; animal husbandry for breeding hagfish has not as yet evolved. However, since hagfish produce multiple sets of teeth in the young and adult stages [27-28], microdissection, new culture strategies and new micromethods might provide a fruitful opportunity to pursue these studies of development and evolution.

REFERENCES

1. Bonner JT, ed. (1982) Evolution and Development. Springer-Verlag, Berlin

2. Jeffery WR, Raff RA eds. (1983) Time, Space, and Pattern in Embryonic Development. Alan R. Liss, Inc., New York

3. Mann S, Webb J, Williams RJP eds. (1989) Biomineralization: Chemical and Biochemical Perspectives. VCH Verlagsgesellschaft, Weinheim (FRG)

4. Lowenstam HA, Weiner S (1989) On Biomineralization. Oxford University Press, New York, pp. 1-324

5. Fearnhead RW (1979) J Dent Res 58(B):909-916

6. Orvig T (1967) Structure and Chemical Organization of Teeth. Academic Press, New York, pp. 45-110

7. Peyer B (1968) Comparative Odontology. University of Chicago Press, Chicago, pp. 89-99

8. Slavkin HC, Zeichner-David M, Ferguson MWJ, Termine JD, Graham EE, MacDougall M, Bringas P, Bessem C, Grodin M (1982) In: Rivere, G.R., Hildemann, W.H (eds) Oral Immunogenetics and Tissue Transplantation. Elsevier, Amsterdam, pp. 241-251.

9. Slavkin HC, Zeichner-David M, Snead ML, Graham EE, Samuel N, Ferguson, MWJ (1984) In: Ferguson MWJ (ed) The Structure, Development and Evolution of Reptiles. Academic Press, New York, pp. 275-304

10. Herold R, Graver H, Christner P (1980) Science 207:1357-1358

11. Slavkin HC, Graham E, Zeichner-David M, Hildemann WH (1983) Evolution 37:404-412

12. Dawson JA (1963) The Biology of Myxine. Universitetsforlaget, Oslo, pp. 231-255

13. Sognnaes RF, Lustig LL (1955) J Dent Res 34:132-143

14. Pautard FGE (1963) Nature (London) 199:531-535

15. Fernholm B (1969) Acta Zoologica (Stockholm) 50:169-177

16. Bringas P, Nakamura M, Nakamura E, Evans J, Slavkin HC (1987) Scanning Microsc. 1(3):1103-1108

17. Evans J, Bringas P, Nakamura M, Nakamura E, Santos V, Slavkin HC (1988) Calcif Tissue Int 42:1-12

18. Snead ML, Lau E, Zeichner-David M, Fincham A, Woo SLC, Slavkin HC (1985) Biochem Biophys Res Commun 12:812-818

19. Takagi T, Suzuki M, Baba T, Minegishi K, Sasaki S (1984) Biochem Biophys Res Commun 121:592-597

20. Slavkin HC, Bessem C, Bringas P Jr, Zeichner-David M, Nanci A, Snead ML (1988) Differentiation 37:26-39

21. Samuel NB, Bessem C, Bringas P, Slavkin H C (1987) J Craniofac Genet Dev Biol 4:371-386

22. Slavkin HC, Samuel N, Bringas P, Nanci A, Santos V (1983) J Craniofac Genet Dev Biol 3:43-52

23. Snead ML, Luo W, Lau E, Slavkin HC (1988) Development 104:77-85

24. Rappolee DA, Brenner, CA, Schultz R, Mark D, Werb Z (1987) Science 241:1823-1825

25. Wolpert L (1989) Development Supplement:3-12.

26. Gould SJ (1977) Ontogeny and Phylogeny. Harvard University Press, Cambridge.

27. Krejsa RJ, Bringas P Jr, Slavkin HC (1990) Courier Forschunginstitut Senckenberg 118:473-492

28. Krejsa RJ, Bringas P Jr, Slavkin HC (in press) Lethaia.

CHAPTER 4.9

Correlation Between the Size of Crystals and the Molecular Weight of Organic Fractions in the Soluble Matrices of Mollusc, Coral and Sponge Carbonate Skeletons

J.P. CUIF, P. GAUTRET, and F. MARIN

U.A. 723 du CNRS, Bât. 504 Géologie; Faculté des Sciences-F 91405-ORSAY

Key words : Invertebrates, carbonate crystals, soluble matrix, chromatography.

INTRODUCTION

The carbonate crystals which build calcified structures in Invertebrates show spectacular differences in the sizes, shape and optical properties [1]. In currently accepted models of biomineralization process [2], the steric characteristics of polypeptide assemblages determine the first step of crystal formation (calcite or aragonite), at the molecular level. But a major problem remains to be resolved before we can fully understand the biomineralization processes in various phyla : how the growth of the calcitic and aragonitic primary elements is controlled to give the required shape and orientation of the crystals.

Experiments reported in this paper analyse skeletons of molluscs, corals and calcifying sponges ; these forms have a very wide range of shapes and sizes of crystals. Correlation is made between the general features of these crystals and the results of high resolution chromatography of their soluble matrices.

MATERIALS AND METHOD :

Source of the illustrated data in this paper :

Pelecypods : prismatic and nacreous layers of *Pinna nobilis* from the Mediterranean see (biological station of Port-Cros, France); prismatic and nacreous layer of *Pinctada margaritifera* from Takapoto (atoll of french Polynesia); Cephalopod : pillars of the *Sepia.* skeleton; Madreporaria : *Caryophyllia ambrosia* from Atlantic Ocean (Gascogne gulf) ; calcified Sponges : *Astrosclera, Ceratoporella, Acanthochaetetes* and *Vaceletia* from Indian and Pacific ocean.

Method :

After the selected shell layers had been cleaned, quantities from 100 mg to 1 g have been powdered, and dissolved in acetic acid with permanent control of pH, by a Radiometer automatic titrator (pH fixed to 4). Insoluble components were removed in two steps : centrifugation (20 minutes, 7.OOO rpm) followed by filtration on Millipore 0,5 µm. Desalting was achieved using SEPHADEX G25 with 6g/l NaCl in Milli-Q water as the eluting medium. The resulting solution containing organic material was vacuum concentrated.

Gel permeation process :

Two types of chromatographic columns were used to separate soluble organic components of different molecular weights : 1 m long and 16 mm diameter SEPHACRYL S2OO or S 4OO columns, and SUPEROSE S6 and S12. Eluting media were : NaCl 6g/l in Milli-Q water or ammonium acetate 0,05 M or TRIS-HCl, pH : 7.5.

Detection was made by UV absorption at both 226 and 280 nm, on LKB-Pharmacia materials.

Complementary data concerning amino acid composition of separated fractions have been obtained by hydrochloric acid hydrolysis under standard conditions, followed by PITC derivation, and RP chromatography with a methanol/ammonium acetate gradient, in Milli-Q water, on a C18-25 cm long column.
U.V. detection of separated amino acids was achieved at 254 nm with a Beckman system.

CRYSTALLOGRAPHIC CHARACTERISTICS OF SELECTED MATERIALS

The crystallographic organisation of skeletal components is different in molluscs, coral and sponges. In order to compare the organic matrices, two major points of crystallography need to be clarified.

1° - Discrimination between monocyclic and polycyclic crystals.

This discrimination can be demonstrated by considering the shell structure of the bivalve mollusc *Pinna nobilis*. In this species, the external shell layer consists of the well known "simple prisms" (Fig. 1). Numerous successive growth stages within these microstructural units are easily demonstrated by physical and chemical etching of polished surfaces (Fig. 2). These mineral units are limited by organic sheets, (the compartment envelope illustrated elsewhere by Bevelander and Nakahara [3] and can be separated by pyrolytic heating under controled conditions. (Fig. 3).
However, radial thin sections through the prismatic layer clearly show, in polarized light, the perfect monocrystalline optical conformity of the prisms (Fig. 4). In each growth stage, the crystallographic axis of the newly produced calcium carbonate unit is in exact correspondence with the axis of the mineral unit previously deposited in the prism, in spite of the continuous organic sheet separating these units.

In contrast, in the nacreous layer of the shell of *Pinna nobilis*, crystal growth is limited to a lateral expansion, and does not have several biomineralization cycles. There is no continuity between the successive growth stages of the nacreous layer : the nacreous crystals are monocyclic, but the prisms of the external layer are polycyclic.

EXPLANATION OF PLATE

Fig. 1 : Enlarged view of external shell layer of *Pinna nobilis* , after dissolution of distal part of calcitic crystals. Scale : 40 µm
Fig. 2 : Result of proteolytic etching on polished longitudinal section of prisms of *Pinna nobilis*: the parallel envelopes of successive growth stages are clearly visible. Scale: 35 µm
Fig. 3 : Isolated calcitic units from prisms of *Pinna nobilis*, separated at temperature of 250°C during pyrolysis experiment under N2 gas flow. Scale : 40 µm
Fig. 4 : Longitudinal thin section in external shell layer of *Pinna nobilis* : thin section observed in polarized light. Each prism appears as a crystalline unit. Scale : 40µm
Fig. 5 : Mosaic structure of the calcitic crystal in a simple prism from the external layer of *Pinctada margaritifera*. Scale : 20 µm
Fig. 6 : Aragonitic prismatic structure in the skeleton of *Sepia* sp. . Scale : 60 µm
Fig. 7 : Aragonitic skeleton of *Caryophyllia ambrosia* (Scleratinian coral), after proteolytic etching : the organic laminae produced at each growth stage are visible. Scale : 30 µm
Fig. 8 and 9 : Skeletal structure of two calcified Demosponges : *Ceratoporella nicholsoni* (Fig. 8), built by aragonitic fibrous fascicles, and *Astrosclera willeyana* (Fig. 9), built by spherulitic aragonitic fibrous units. Scales : Fig. 8 : 35 µm, Fig. 9 : 12 µm.
Fig. 10 : Biomineralization unit in the skeleton of *Vaceletia crypta*. Scale : 100 µm.
Fig. 11 : Aragonitic microcrystals in biomineralization unit of*Vaceletia*. Scale : 0,5 µm.
Fig. 12 : Calcitic microcrystals in skeleton of *Acanthochaetetes wellsi* Scale : 2,5 µm.

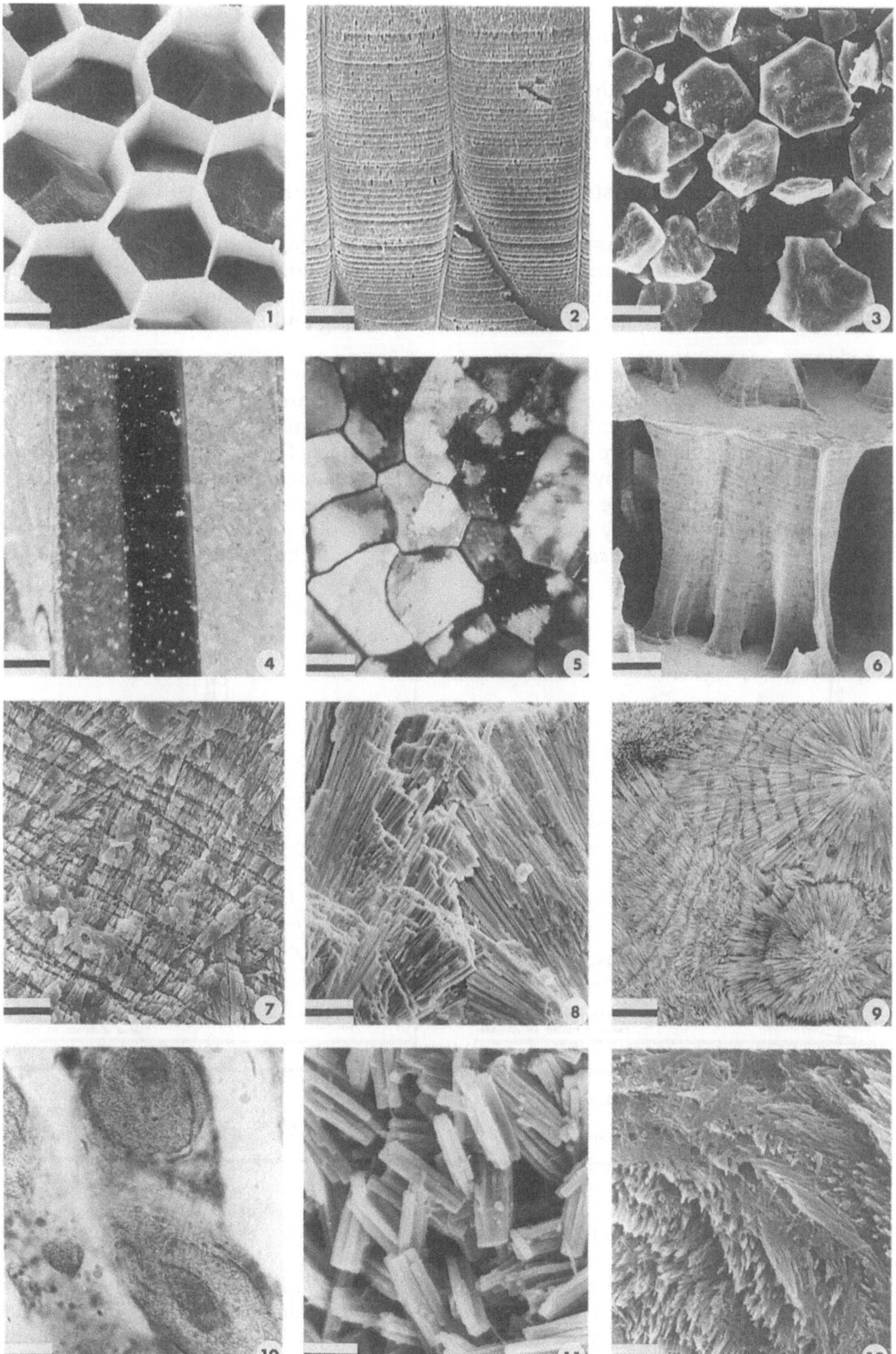
1
2
3
4
5
6
7
8
9
10
11
12

Aragonitic fibers building the Scleractinian corals (Fig. 7) or the skeletons of calcified Demosponges, like *Ceratoporella* (Fig. 8) or *Astrosclera* (Fig. 9), are also polycyclic crystalline units, in which the internal traces of organic sheets can be demontrated by proteolytic etching. Other species of Demosponges, like *Vaceletia* and *Acanthochaetetes* produce very minute crystals (Figs. 10, 11, 12)

2° - Distinction between crystalline units and microstructural units.

In many cases, this distinction is obvious. In the Demosponge *Vaceletia*, for instance, the microstructural unit are more or less spheroidal (80-130 µm in diameter), with a clearly definite central point (Fig.10), whereas the crystals are very minute fasciculate elements (Fig.11). This structure is also documented in other living and fossil Vaceletids [4].

On the other hand, the perfectly monocrystalline prisms of the external layer of *Pinna nobilis* are both crystalline and microstructural units. In all microstructural classifications, they are the model for the "simple prism" group.
However,"simple prisms" are not always crystalline units. The external layer of *Pinctada margaritifera* is also built by "simple prisms" , with a thick external organic envelope. They are very similar to the prisms of the *Pinna* from external point of view, but the calcite crystals exhibit a very distinct mosaic structure (Fig. 5). Consequently, the prisms of the external layer in *Pinctada* are microstructural units, but not true crystalline units.

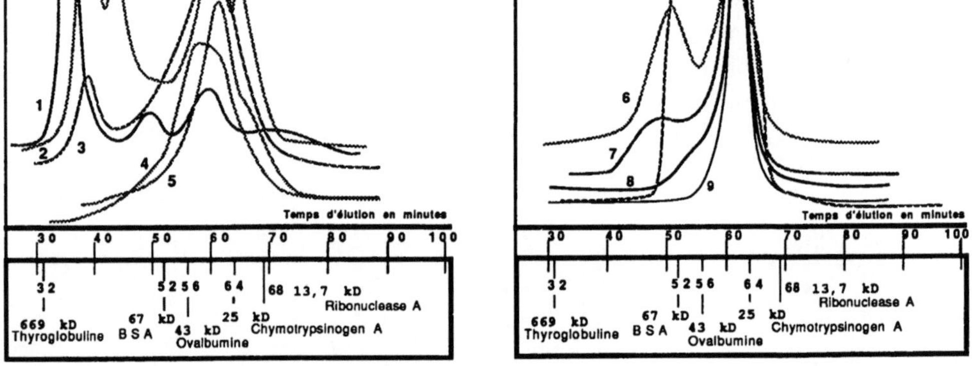

Fig. 13 : Gel-filtration chromatograms of soluble matrices in various biogenic carbonate structures.
Identification of chromatograms : prisms of *Pinna nobilis* (1), *Pinctada margaritifera* (2)*Sepia* (3); nacreous layers of *Pinna nobilis* (4), *Pinctada margaritifera* (5); fibrous structure of *Astrosclera willeyana* (6), *Ceratoporella nicholsoni* (7); microcrystalline units of *Vaceletia crypta* (8), *Acanthochaetetes wellsi* (9); fibrous skeleton of *Caryophyllia ambrosia* (10)

RESULTS OF THE CHROMATOGRAPHIC STUDY OF THE SOLUBLE MATRICES

Fig. 13, summarizing the results of this study, shows the very large range of molecular weights observed in the soluble matrices extracted from the representatives of the three phyla. The extreme values are situated at $4{,}5.10^5$ and 12.10^3 daltons, the major parts being located between 25 and 70 kD. In all cases, the elution curves indicate that a broad spectrum of continuously varying masses is present. Generally speaking, these results are in good agreement with those previously obtained [5-6]

All polycyclic crystals are distinct from monocyclic crystals in the presence of higher molecular weights.
This is clearly demonstrated in molluscs with bilaminar shells, in which a prismatic/nacreous succession occurs. The polycyclic crystals of the prismatic layer exhibit heavy organic components, the highest value being observed in the prisms of *Pinna nobilis* which are also, as previously noted, perfectly monocrystalline. In the same species, the high molecular weight component is not present in the soluble matrix of the nacreous crystals. This contrast is also clear in some haliolitid gastropods (e.g. *Haliotis rufescens*), in which a prismatic, entirely calcitic external layer is followed by aragonitic nacreous material [7]. Heavy organic components are also present in the aragonitic pillars of the *Sepia*, these corresponding to a prismatic layer [8]
Another example of calcitic "simple prisms" is found in *Pinctada* . But here, the crystalline elements exhibit a mosaic structure : the highest soluble fraction is lower in molecular weight than in the soluble matrix of the *Pinna* prisms.

The soluble organic matrix of calcified Demosponges provides evidences for the same relationship. The long aragonitic fibers of Ceratoporellids and Astrosclerids are produced by succession of numerous biomineralization cycles, and exhibit also a high mass soluble component, which is lacking in the microcrystalline elements of Vaceletids and Acanthochaetetids.

CONCLUSION

The results outlined here suggest that the development of polycyclic crystals in the carbonate skeletons produced in the three phyla studied could be governed by specific organic components of high molecular weight (possibly polymeric in structure), whrereas medium or low weight organic components are mainly devoted to the production of the primary carbonate units.
Amino acid analysis of these separate fractions, showing that acidic amino acids (i.e. aspartic acid) are preferentially located in the low or medium weigth organic components is clearly relevant with this hypothesis.

1 - Cuif JP, Dauphin Y, Denis A, Gautret P, Lawniczak A, Raguideau A (1987) Bull. Soc. géol. Fr., 8 -III : 269-288.
2 - Weiner S, Traub W, Lowenstam HA (1983) In : Westbroek P, De Young EW (eds) Biomineralization and biological metal accumulation. Riedel Publ. Comp., pp. 205-224.
3 - Bevelander G, Nakahara H (1980) In : Omori M, Watabe N (eds) Proc. Third Intern. Biomin. Symp. Tokai Univ. Press, pp. 19-27.
4 - Gautret P (1985) Geobios 18-5 : 553-562
5 - Crenshaw MA (1972) Biomineralization 6 : 6-11
6 - Wheeler AP, George JW, Evans CA (1981) Science 212 : 1397-1398
7 - Dauphin Y, Cuif JP, Mutvei H, Denis A (1989) Bull. Geological Institution of the University of Uppsala 15 : 7-38
8 - Barskov IS (1973) Paleont. J. 3:285-94

CHAPTER 4.10
Exoskeletal Adaptations for Life Environment in Neritacean Gastropods

S. SUZUKI[1], Y. TOGO[2], and S. UOZUMI[3]

[1]Department of Earth Sciences and Astronomy, Fukuoka University of Education, Munakata, 811-41 Japan; [2]Department of Earth Science, Hokkaido University of Education, Iwamizawa, 068 Japan; and [3]Department of Geology and Mineralogy, Hokkaido University, Sapporo, 060 Japan

Key words: Shell structure, Calcified operculum, Mineralogy, Habitat, Neritacean gastropod.

INTRODUCTION

The Gastropoda has been the most successful class in the Mollusca in adaptive radiation to different life environments. Although gastropods are largely marine, considerable numbers of species also live in non-marine environments such as brackish- and fresh-water, and terrestrial environments. Among Archeogastropoda the Neritacea is only superfamily including marine, brackish-water, fresh-water and terrestrial species.

Recent neritacean gastropods have two kinds of exoskeleton, shell and operculum, except for the family Phenacolepadidae, in which all species have lost the opercula. The neritacean shells and opercula are calcified without exception. These two calcified tissues must have played a role in the course of adaptation for several life environments, because their basic and most important function is the protection of soft bodies. It is, therefore, expected that their internal structures and components were modified along with the change of life mode.

It has remained unclear in previous studies [1] [2] [3] [4] [5] how the internal characteristics of the exoskeletons relate to environment due to an insufficient number of the species examined and to a lack of detailed division of life mode in this superfamily. We will show in this study the adaptive patterns of the structural and mineralogical varieties in the neritacean exoskeletons, and will discuss their implications for the evolutionary trends of the neritacean group.

MATERIALS AND METHODS

The following twenty-one species were used for shell structural study. Neritopsidae (marine): *Neritopsis radula*; Neritidae (marine): *Nerita (Theliostyla) albicilla, N. (T.) squamulata, N. (T.) planospira, N. (Amphinerita) polita, N. (A.) insculpta, N. (Heminerita) japonica, N. (Ritena) plicata, N. (R.) helicinoides laevilarbris, N. (R.) signata, N. (R.) undata*; Neritidae (brackish- to fresh-water): *Neritina (Dostia) violacea, N. (Vittina) ziczac, N. (V.) communis, Clithon sowerbianus, C. retropictus, C. colona, Neripteron granosus, Septaria porcellana*; Helicinidae (terrestrial): *Waldemaria japonica japonica*; Hydrocenidae (terrestrial): *Georissa japonica.* Internal structure of the operculum (opercular structure) was also examined in sixteen species except *Neritopsis radula* and four neritid species.

The internal structures were observed under phase-contrast light microscope (PLM) and/or scanning electron microscope (SEM). The specimens were embedded in an epoxy resin, sectioned transversely or axially, and were polished and etched with 0.5 mol EDTA for about fifteen minutes. Acetate-peel replicas of the sections were prepared for PLM observations. Some polished and etched sections were sputter-coated with gold for SEM observations. Minerals of the exoskeletons were determined by X-ray powder diffractometry or by staining with a Fiegl's solution.

RESULTS

Shell structure

Three types of shell construction, type I, II and III, defined by mineralogy, shell layering and microstructure were clearly distinguished in the neritacean shells examined (Fig. 1). Type I is of the families Neritopsidae, Helicinidae and Hydrocenidae, and the other two are of the Neritidae.

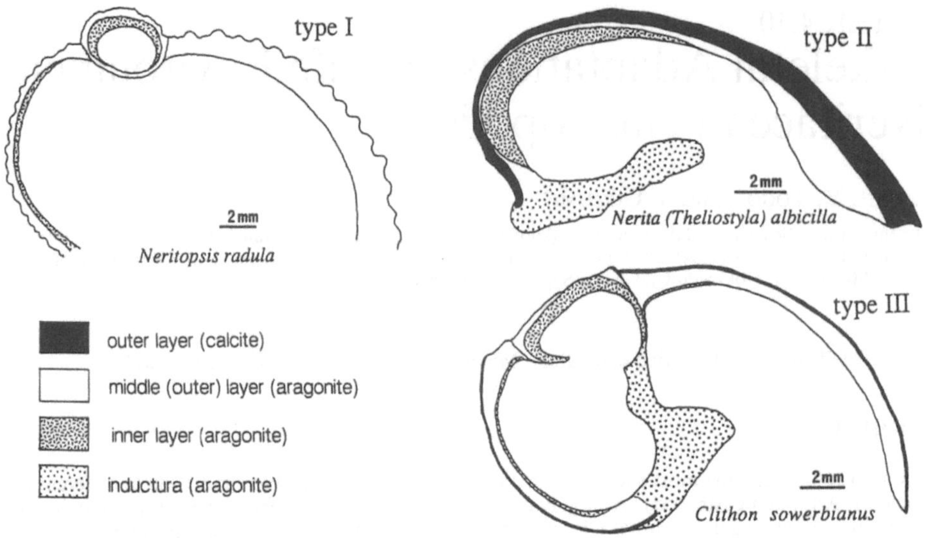

Fig. 1. Layering of neritacean shells: type I, axial section; type II and III, nearly transverse section.

Type I. — The shell is wholly aragonitic and consists of two shell layers. The outer layer is formed by the crossed lamellar structure; in the largest structural units (first order lamellas) the long axis is arranged almost perpendicularly to the direction of shell growth, that is, nearly parallel with the apertural margin of the shell. Structural units of first to third order in this microstructure are arranged more irregularly than those in the typical crossed lamellar ones reported by previous workers [6] [7] [8]. The inner layer of the type I corresponds with the innermost one of many other gastropod shells. This layer is formed mainly by the protocrossed lamellar structure (Fig. 2), an intermediate structure between the normal crossed lamellar and complex crossed lamellar structures [9]. The layer contains occasionally and partially the irregular prismatic [10], homogeneous (Fig. 2) and complex crossed lamellar structures; in *Georissa japonica*, which has a very small adult shell (about 2 mm high), the irregular prismatic structure is rather common than the protocrossed lamellar one.

Type II. — This type is found in the marine neritid species. The shell is composed of three layers, outer, middle and inner (Fig.3). The outer layer has a calcitic blocky structure [11] with a considerable thickness

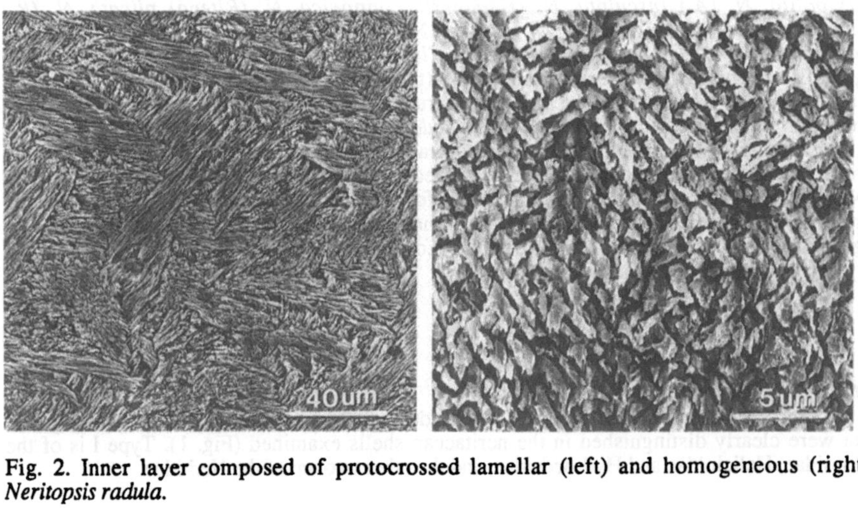

Fig. 2. Inner layer composed of protocrossed lamellar (left) and homogeneous (right) structures; *Neritopsis radula*.

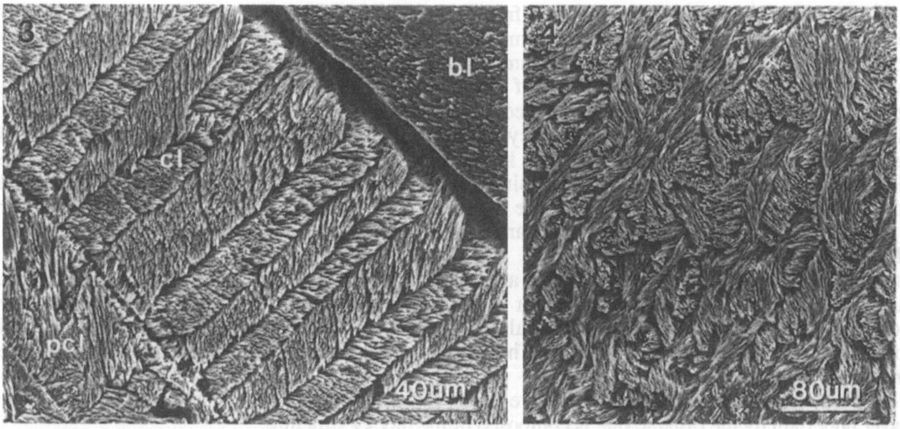

Figs. 3-4: 3, shell of type II showing outer blocky, middle crossed lamellar and inner protocrossed lamellar layers (*Nerita (Ritena) plicata*); 4, internal structure of inductura composed of protocrossed lamellar structure (*Nerita (Theliostyla) planospira*).

relative to that of type III (Fig. 1). The middle layer is formed by the crossed lamellar structure, in which the long axis of the first order lamellas is arranged normal to the growing direction of the shell. This microstructure shows a more highly ordered arrangement than that of type I. The inner layer is composed of the protocrossed lamellar structure with minor irregularly prismatic parts. A septum-like inductura on the inner apertural lip side, one of the most important characters of the shell of the Neritidae, is composed of a single calcified layer contiguous with the middle layer of the whorl shell. In spite of the continuity with the middle layer, the inductura is composed of the protocrossed lamellar structure (Fig. 4), which is not the microstructure in the middle layer but in the inner one of the whorl shell.

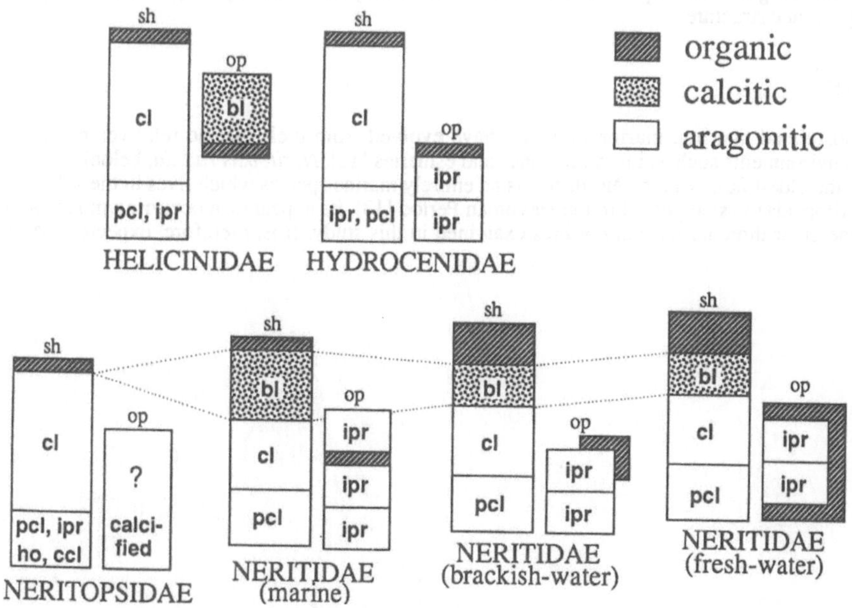

Fig. 5. Schematic columnar sections of the shell (sh) and operculum (op) in the Neritacea; dotted lines indicate certainty of correlation of calcitic outer layer: bl, blocky; ccl, complex crossed lamellar; cl, crossed lamellar; ho, homogeneous; ipr, irregular prismatic; pcl, protocrossed lamellar.

Type III. — This type belongs to brackish- and fresh-water neritids. The shell construction is basically the same as that of type II. However, there are some differences between the two types; type III has a very thin calcitic outer layer on a thick periostracum while type II has a thick calcitic layer and a thin periostracum. For example, in air-dried shells of *Neritina* (*Dostia*) *violacea* and *Neripteron granosus*, which belong to type III, the periostracum reduced by dehydration is still thicker than the calcified outer layer which may remain unreduced. This tendency towards a thick periostracum has never been observed in other shells examined. A difference also exists in the design of the apertural margin (Fig. 1), where the outer and middle layers are secreted by the epithelial cells near the mantle-edge. A microstructural difference is also recognized in the inductura; the inductura of type III is formed by the crossed lamellar structure, the same as in the middle layer, but the arrangement of the structure becomes more irregular.

Opercular structure

The neritacean opercula are composed of organic and calcified layers, the latter having a protrusion of one or two articulate apophyses into the foot except in *Waldemaria japonica japonica*. The opercula are classified into five types on the basis of mineralogy, layering of calcified parts, and the position and developmental manner of organic layer (Fig. 5). The following correlation between opercular type and life environment is recognized.

Type A. — *Waldemaria japonica japonica* (terrestrial). Construction: organic and calcitic layers. The calcitic layer underlain by the organic one is composed of the blocky structure.

Type B. — *Georissa japonica* (terrestrial). Construction: organic and two aragonitic layers. The aragonitic layers with an articulate apophysis have an irregular prismatic structure. The organic layer covers the outer surface of the calcified parts.

Type C. — *Nerita* (marine). Construction: organic and three aragonitic layers. The aragonitic layers have an irregular prismatic structure and are separated by the organic layer into two parts, outer and inner, the latter of which contains two aragonitic layers (Fig. 6). Two articulate apophyses protrude separately from the basal part of the inner layers into the foot.

Type D. — *Neritina* and *Clithon* (brackish- and fresh-water). Construction: organic and two aragonitic layers. This type closely resembles type B, but the organic layer does not cover the outer surface of the calcified layers, and occupies only a marginal area of the operculum along the outer lip of the shell. The aragonitic layers with two articulate apophyses have an irregular prismatic structure (Fig. 7).

Type E. — *Neripteron granosus* (fresh-water). Construction: two organic and two aragonitic layers. Each organic layer covers the outer or inner surface of the calcified part and is fused together at the margin of the operculum along the outer lip of the shell. The aragonitic layers with an apophysis are composed of the irregular prismatic structure.

DISCUSSION

It has been postulated that non-marine molluscs have evolved from their marine relatives by way of intermediate environments such as intertidal zones and estuaries [12]. *Neritopsis radula*, belonging to the Neritopsidae, the oldest family in the Neritacea, is an entirely marine species which lives in the sublittoral zone. The Neritopsidae first appeared in the Devonian Period [13], its appearance occurring much earlier than that of the other three neritacean families examined in this study. It is, therefore, expected that the

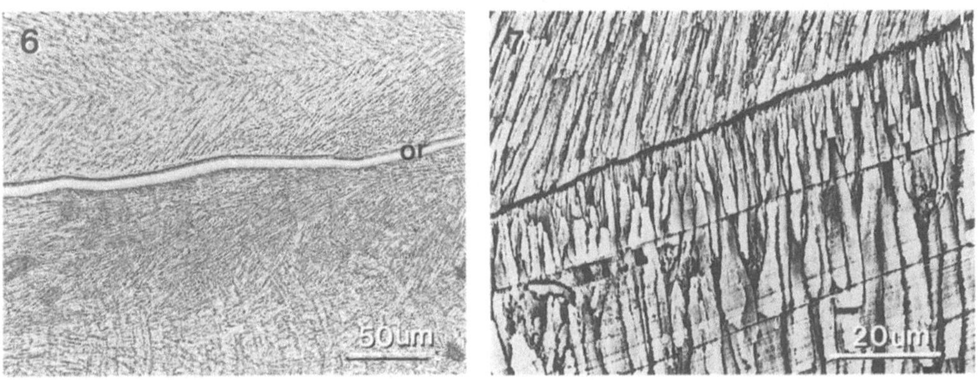

Figs. 6-7: opercular structure; 6, type C showing aragonitic layers separated by organic layer (*Nerita (Theliostyla) albicilla*; optical micrograph); 7, irregular prismatic layers of type D (*Neritina (Dostia) violacea*).

shell structure of *Neritopsis radula* remains in a primitive state. In fact, the crossed lamellar structure of the outer shell layer of this species has a less ordered arrangement than that of the corresponding layers in more advanced families such as the Neritidae. Variations in the microstructures of the inner (or innermost) layer of this species also suggest that the layer is still rather unstable in structural arrangement relative to that of other species in which the protocrossed lamellar structure is more stable.
Marine neritid shells are characterized by a calcitic outer layer which seems to be an added or descendent layer on aragonitic shell as seen, for example, in the shell of *Neritopsis radula*. These snails commonly live in the intertidal zone and they are required to be resistant against desiccation, spring tides, and varying temperature and salinity. The commonly accepted sequence of diagenetic alteration (magnesian calcite > aragonite > low-magnesian calcite) is only strictly observed in solutions near or about saturation for aragonite [14]. If such a saturation state is true in the case of the marine neritid habitat, it is likely that calcite will be deposited on the neritid aragonitic shell as a protection against dissolution. Although calcite deposition does not alone enhance shell strength or resistance to abrasion [15], bimineralic shell may increase the strength of some mechanical properties [16]. We have recognized that many intertidal prosobranch superfamilies such as Trochacea, Littorinacea, Muricacea, etc., which basically consist of species with wholly aragonitic shells, tend to produce the species with bimineralic shell [17]. A calcitic outer layer is also found on all shells of brackish- and fresh-water neritids, which have an extremely thin calcitic layer with thick periostracum. Such a shell construction probably results from the adaptation to life in waters of low salinity which tend to dissolve calcareous matters. The land species examined still preserve the basic neritacean shell, constructed by aragonite without a calcitic layer and a thick periostracum. It remains unknown whether or not the limy ground habitat of these species prevents the deposition of calcite in the shell.
Unfortunately, we have not been able to examine the operculum of *Neritopsis*; the only information we have obtained is that the operculum is calcified and accompanied with an articulate apophysis always protruding from calcified layer into the foot. It is, therefore, difficult to discuss here the primitive conditions of the neritacean operculum. However, the variety of opercular construction in living neritaceans may indicate that structural changes of opercula have occurred during adaptational radiations. Invasion of marine neritids into fresh-water habitats would cause their opercula to evolve into another form characterized by the loss of the outer calcified layer and by the development of an organic one. This tendency is somewhat similar to that of the shell. The helicinid operculum (calcitic) has a different construction from that of the hydrocenid one (aragonitic), which shows the same characteristics as that of some aquatic species. A construction like the helicinid operculum is rather common in the opercula of some land mesogastropods [18]. A helicinid operculum may be more advantageous than a hydrocenid one, but we are unable to explain why only epithelial cells of the foot have the function of calcite secretion in the species with wholly aragonitic shell layers.

REFERENCES

[1] Kessel E (1942) Z Morph Ökol Tiere 38: 197-250.
[2] Adegoke OS (1973) Malacologia 14: 39-46.
[3] Togo Y (1977) J Geol Soc Japan 83: 567-573.
[4] Vovelle J, Grasset M, Meunier F (1977) Malacologia 16: 279-283.
[5] Suzuki S, Togo Y (1988) Earth Sci. 42: 147-154.
[6] Taylor JD, Kennedy WJ, Hall A (1969) Bull Brit Mus Natl Hist Zool Suppl 3: 1-125.
[7] Uozumi S, Iwata K, Togo Y (1972) J Fac Sci Hokkaido Univ Ser IV 15: 447-478.
[8] Carter JG, Clark GR II (1985) In: Broadhead TW (ed) Mollusks, Note for a Short Course. Dep Geol Sci Univ of Tennessee, Knoxville, pp. 50-71.
[9] Togo Y (1974) J Geol Soc Japan 80: 369-380.
[10] Suzuki S, Uozumi S (1981) J Fac Sci Hokkaido Univ Ser IV 20: 7-20.
[11] Suzuki S (1983) J Geol Soc Japan 89: 433-442.
[12] Russell-Hunter WD (1979) A life of invertebrates. Macmillan, New York, pp. 1-650.
[13] Moore RC (ed) (1960) Treatise on invertebrate paleontology, Pt (I) Mollusca 1. Geol Soc Amer and University of Kansas Press, Lawrence, pp. 1-351.
[14] Walter LM (1985) In: Schneidermann N, Harris PM (eds) Carbonate Cement. Soc Economic Paleont Mineral, Tulsa, pp. 3-16.
[15] Taylor JD, Layman M (1972) Palaeontology 15: 73-87.
[16] Carter JG (1980) In : Rhoads DC, Lutz RA (eds) Skeletal Growth of Aquaitc Organisms. Plenum. New York and London, pp. 69-113.
[17] Togo Y, Suzuki S (1988) In: Omori M, Suga S, Goto M (eds) Biomineralozation and Phylogeny of Marine Organisms. Tokai University Press, Tokyo, pp. 113-134.
[18] Suzuki S, Togo Y (unpublished)

CHAPTER 4.11

Microstructural Organization of the Exoskeleton of Some Articulate Brachiopods (Terebratulida, Rhynchonellida) — The Importance of the Calcitic Granules and the Effects of Diagenesis

D. GASPARD

Université de Paris-Sud, Département des Sciences de la Terre, Bât. 504, F-91405 Orsay Cedex, France

Key-words: **Brachiopod exosekeleton, calcitic granules, growth stages, organic matrix, diagenesis.**

I. INTRODUCTION

In order to interpret the different microstructural patterns observed in fossil brachiopod shells, the palaeontologist must fully understand the biomineralization processes which are involved in forming Recent brachiopod shells. This information is essential for distinguishing original features from those which result from diagenesis.

The articulate brachiopod shell grows continuously with a general overlying rhythm which can, however, be disturbed by environmental events. But there also seem to be some breaks in the rhythm of growth which are independent of environmental factors.

The construction of the final calcitic shell is achieved by successive hierarchical stages, from the secretion of minute calcitic seeds or granules (the elementary units) to the elaboration of the constituent elements of the major layers.

Physicochemical modifications induced by different sources may result in small or significant diagenetic modifications of the shell ; some modifications can take place earlier and later obscure the results of the normal biomineralization processes. Characteristic information valuable for a specific identification in the fossil record is thus masked.

In this paper the impact of the environmental changes on shell growth, the effects of early post-mortem changes, and later changes during fossilization are highlighted.

II. MATERIAL AND METHODS

Recent Rhynchonellida and Terebratulida selected for study included: *Gryphus vitreus* (Born), Mediterranean; *Liothyrella notorcadensis* Jackson - off Palmer Peninsula, Antarctica ; *Campages furcifera* (Hedley.) - off Yaté, New Caledonia, *Chalcal* cruise; *Macandrevia africana* Cooper. - off Angola, *Walda* cruise; *Megerlia truncata* (Gmelin) - Mediterranean; *Waltonia inconspicua* (Sowerby) - New Zealand; *Terebratulina retusa* (Linné.) - Mediterranean, *Bracors* cruise and Firth of Lorne,Scotland; *T. hataiana* Cooper - off Phillipines, *Musorstom* cruise; and *Notosaria nigricans* (Sowerby) - Tikoraki Point, New Zealand.

Tertiary and Cretaceous brachiopods examined: *Gryphus minor* Phillipi, from the Lower Pleistocene, Capa dell Armi, Italy; *Rhynchora pectinata* (Linné), from the Maastrichtian of Scania, Sweden; *Rhynchonella vespertilio* d'Orbigny, from the Senonian of Charente, France; *Sellithyris cenomanensis* Gaspard, from the Cenomanian of Sarthe, France; *Gemmarcula menardi* (Lamarck), from the Middle Cenomanian of France; *Musculina sanctae-crucis* Catzigras, from the Hauterivian of the Jura region; *Cyclothyris compressa* (Lamarck) and *C. scaldisensis* (d'Archiach), from the Cenomanian of Gussignies, Belgium.

These shells were embedded in Araldite, cut, polished and then etched with a rapid decalcifier (RDO or RDC : 12.5% for a few seconds) and gold coated. They were then examined with a Jeol scanning electron microscope (SEM). Natural surfaces free of adherent material were also examined after gold coating. Microprobe analysis were made, as well as observations using cathodoluminescence.

III. GROWTH STAGES

The secretory regime of the Recent articulate brachiopods consist of different operations carried out by each of the epithelial cells on the mantle edge and originating from the generative zone [1],[2].These different

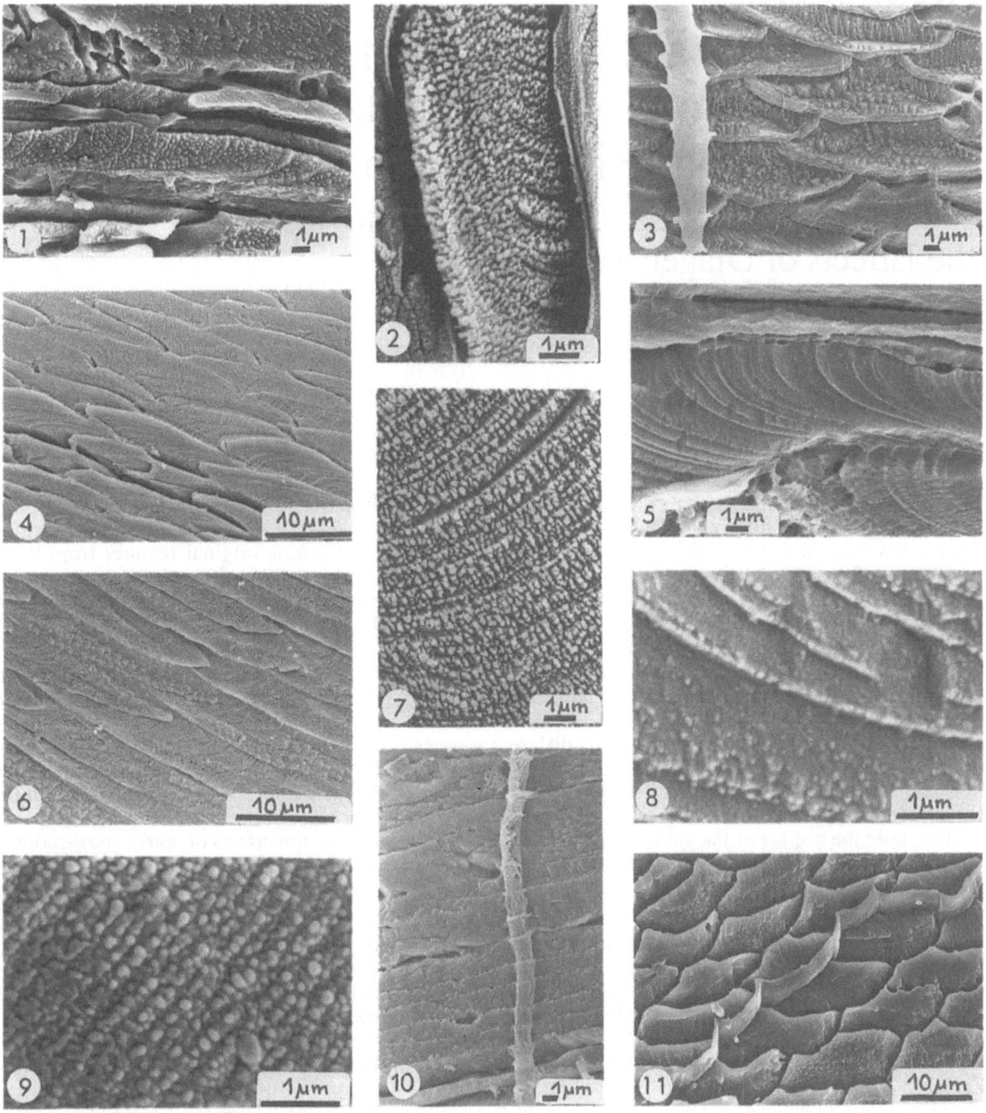

Fig.1: Longitudinal section of a *Macandrevia africana* Cooper shell, from off Angola coasts, showing growth lines and blocky crystallites in fibres in response to microboring organism activity. Fig.2: Granules and microgrowth increments near the terminal end of a fibre of *Notosaria nigricans* (Sowerby), New Zealand. Fig.3: Section in a shell of *Liothyrella notorcadensis* Jackson, off Palmer Peninsula, Antarctica, showing part of a caecal epithelium and transverse sections of fibres with blocky crystallites. Fig.4: Longitudinal section of a *Notosaria* shell exhibiting fibres with intermediate growth lines. Fig.5:Successive growth lines in fibres of a *Liothyrella* shell . Fig.6: Remains of granules in the new blocky crystallites formed in the shell of a bored shell of *Notosaria nigricans* (Sowerby) of the intertidal zone of New Zealand. Fig.7: Section of a shell of *Rhynchora pectinata* (Linné), Maastrichtian of Ignaberga, Scania, Sweden, showing well preserved granules and microgrowth increments. Fig.8: Detail at the border of growth lines of *Liothyrella* (cf. Fig.5) to show some remaining granules (compare with Fig.3). Fig.9: Well preserved granules in a shell of *Cyclothyris scaldisensis* (d'Archiac), Cenomanian of Belgium. Fig.10: Well preserved part of micropunctae crossing fibres partly recrystallized in a shell of *Gemmarcula menardi* (Lamarck), Middle Cenomanian, Sarthe (France). Fig.11: Remaining part of the organic matrix around fibres in a shell of *Rhynchonella vespertilio* d'Orbigny, Senonian of Charente, France.

operations contibute to a triple-layered exoskeleton: the outer organic periostracum, a calcareous primary layer, a calcareous-organic secondary layer, and a sporadically developed tertiary layer which is prismatic. It is interesting to highlight here how and why the results of biomineralization processes are modified by post-mortem diagenesis.

III.a.- Granules

Investigations of sections and natural surfaces of secondary and tertiary shell layers show that their constituents (fibres and prisms) are not homogeneous although they appear so crystallographically. In fact, in internal surfaces, their proximal ends (i.e. juvenile part) show granules or calcitic seeds embedded in an impersistent organic matrix [3]. This organic matrix is clearly visible in the first rows of fibres near the primary / secondary limit and disappears away from this zone and in the older part of the fibres as the granules become more numerous and coalesce. These granules differ in size between species (cf. *Macandrevia, Gryphus, Dallina* [3], *Campages* [4], *Terebratulina retusa* [5], *T. hataïana* [6], (Fig. 2)). They are identifiable in sections of Recent and fossil shells (Fig.9). The granules are better seen in juvenile fibres before the establishment of strong structures, because they are deposited on the first formed organic sheath by the somewhat microvillous active secreting surface of the outer epithelial cells. On this way organic arches, secreted by the anterior part of each outer epithelial cell of the external mantle lobe, separate the granules being secreted by the corresponding posterior part from their neighbours. This is the result of a differential secretion rate between the two parts of the active surface of the cell [4]. Following the patterns observed on fresh shells after removal of the mantle, I suggest that granules secreted are arranged in radiating rows [3].

I consider granules as the elementary units of the carbonate secretion and have considered the possible role of mitochondria in calcification [7].

Granules have been observed at the end of the fibres in an impersistent organic matrix, a soluble matrix which could be considered, by comparison with that of the molluscs, as a relative inhibitor of calcium carbonate deposition [8]. This matrix may act as a bath which isolates the granules from one another and reduces their growth.

III.b. Shell Elaboration

The shell is part of a dynamic system undergoing continuous changes and observation of granules during shell microstructural studies of Recent articulate brachiopods sheds light on the method of shell construction. This allows analysis of the final result of calcification during the process of the conveyor belt system of the mantle at several organisational levels: 1. secretion of granules by each cell of the mantle lobe to shape daily increments which are separated, in sectional view, by microgrowth lines [4],[9] (Figs. 2,7) ; 2. elementary increments coalesce to form bands, more or less important, separated by intermediate (e.g. lunar cycles) disturbances and major growth lines [5] (Figs. 1,4,5) ; 3. superimposed bands give rise to a fibre (in its organic sheath) or to a prismatic element.

Altogether, the cells of the outer mantle lobe, depending on their position, after the periostracum and the primary layer, elaborate the different elements of the secondary layer and of a tertiary layer when it is developed (cf. *Gryphus*).

Shell elaboration is dependent on physiological and environmental conditions and therefore reflects the results of normal growth and growth breaks [5] in the course of life. However soon after death and sometimes even during life, some modifications break the normal microstructure of the shell and intervene as results of physicochemical stresses.

IV. DIAGENETIC MODIFICATIONS

IV.a.- Early Diagenesis

Early stages of diagenesis can sometimes be seen in Recent shells, degraded by microboring organisms (algae, fungi) and bacteria [4],[10], as progessive modifications of the original calcitic granules into spiky or blocky crystallites. This phenomenon accelerates the process of maceration of shells which sometimes occurs [11].

The presence of the alga *Ostreobium queketti* has been detected in living *Gryphus vitreus* shells [10] and sections of *Frenulina sanguinolenta* and *Campages furcifera* shells from the Coral sea show hyphae of microboring organisms. Microorganisms, especially bacteria, use the organic matrix as a trophic resource, but their activity is not limited.

In other respects algae and fungi lead to a dissolution of calcitic surfaces in the form of the phenomenon of dissolution-recrystallization. Experiments in the laboratory [12],[13] show the role played by microboring organisms in carbonate diagenesis follows this scheme : 1. sparmicritization through dissolution; 2. production of spiky calcite through dissolution; 3. trapping and binding loose particles of calcite; 4. promoting growth of new crystals. All these microboring organisms involve a complex interaction of destruction and reconstruction of the substrate which here corresponds to the borders of the architectural elements of the brachiopod shell after the loss of the organic framework.

Spiky and blocky forms have been observed in Recent shells: *Campages* , *Frenulina* [10], *Liothyrella* and *Macandrevia* (Figs. 1,3), as well as in the fossil record: *Cyclothyris compressa* (Fig. 12), *Rhynchonella vespertilio, Rhynchora pectinata.* When the process of dissolution is not completely achieved (Fig.8) it is interesting to notice the presence of ghost shape of granules in the blocky crystallites this is the case for bored shells of *Notosaria nigricans* (Fig. 6).
All these sites of recrystallization can also be detected using cathodoluminescence.
Some siliceous epigenesis can occur in the presence of bacteria, other microorganisms and diatoms [6, 10]. This phenomenon will not be discussed here.

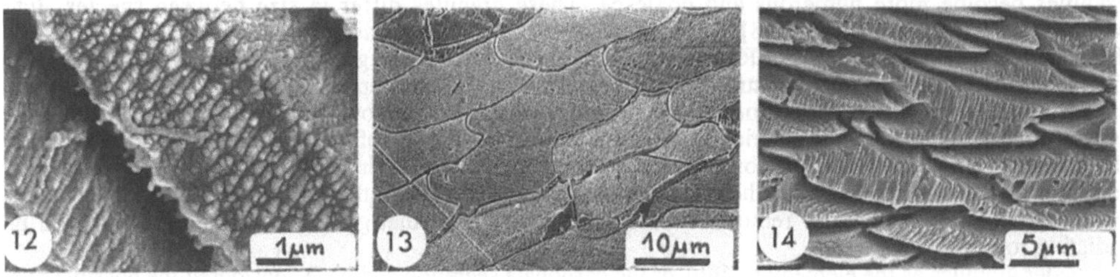

Figure 12: Phenomenon of dissolution-recrystallization observed in a longitudinal section of *Cyclothyris compressa* (Lamarck) from the Cenomanian of Sarthe, France. Fig.13: In part of a shell of *Cyclothyris scaldisensis* (d'Archiac) from the Cenomanian of Gussignies (Belgium), the organic matrix is nearly destroyed and coalescence occurs between some fibres. Fig.14: After the loss of their organic sheath, fibres of a *Gemmarcula menardi* (Lamarck) shell coalesce to form new blocks in the course of recrystallization.

IV.b.- Diagenetic Evolution
Progressive recrystallization of the shell during fossilization creates significant modifications in the architecture of the shell . Biochemical analysis of the organic matrix i.e the organic sheaths of fossils (cf. *Sellithyris cenomanensis* [14]) gives an idea of well preserved material, but this is not always the case and generally the matrix is poorly preserved or silicified (Fig. 11). In some cases after their use as trophic support by bacteria, fibres come into contact during progressive recrystallization and unite (Figs. 13,14).
During fossilization, micropunctae (when present), the organic matrix, growth bandings and granules are progressively masked according to the origin and evolution of the phenomenon. Thereby caecal epithelium is sometimes well preserved (Fig. 3), destroyed or recrystallized and punctae filled with crystallites. The pathways of micropunctae may be well preserved (Fig.10), traced out by recrystallized rods, or eliminated altogether. In this latter case it is impossible to recognize significant characters which might serve to distinguish species [14].
In granules, changes start just after death in early diagenesis and result, during the course of fossilization, in the total elimination of these fine structures (Fig. 10). Growth stage marks are also eliminated (Figs. 11,13). Continuous coalescence of growth banding and later of fibres or prisms then give rise to calcitic or siliceous nodules [10]. But there is no relationship between time and the extent and severity of the diagenetic modifications.

V. CONCLUSION

Patterns observed in fossil shells have been analyzed in the light of observations made on Recent shells, concerning the effects of the processes of biomineralization governed by organic material, environmental factors and on-off controls.
The SEM provides a valuable tool for the detection of alteration in fossil brachiopods for it allows the observation of varying degrees and stages of diagenesis of the shell during fossilization. These are best interpreted by reference to the original structures of Recent fresh shells and show that severity of alteration is not correlated with time but rather with environment and conditions of fossilization.
Analyses using cathodoluminescence revealed the location of alteration in the internal part of valves. Further studies using a microprobe should give complementary details from spot analyses of different parts of Recent shells and, later, information on the changes in shell composition during fossilization.

REFERENCES

[1] Williams A 1968 Lethaia 3: 268-287.
[2] Williams A 1971 Smithson Contrib Paleobiol 3: 47-67
[3] Gaspard D 1986 in: Racheboeuf P, Emig C (eds) First international Congress on Les brachiopodes fossiles et actuels, Brest: 77-83.
[4] Gaspard D 1990 in: Carter J G (ed) Atlas of biomineralization , in press.
[5] Gaspard D 1990 in: Balkema (ed) 2nd International Congress on Brachiopods, Dunedin, in press.
[6] Gaspard D 1990 Bull. Museum Hist nat, 1st National Congress on Palaeontology, Paris, in press.
[7] Lehninger A L 1983 in: Westbroek P, Jong E W de (eds) Biomineralization and biological metal accumulation: 107-121.
[8] Sikes C S, Wheeler A P 1983 in: Westbroek P, Jong E W de (eds) Biomineralization and biological metal accumulation: 285-299.
[9] Gaspard D 1989 in: Carter J G, K Bandel (eds) Skeletal biomineralization. Patterns processes and evolutionary trends 28th IGC, Washington DC, A.G.U. 5(2): 320-321.
[10] Gaspard D 1989 Bull Soc Geol Fr (8) V: 1207-1216.
[11] Alexandersson E T 1979 Sedimentology 26: 845-852.
[12] Jones B, Pemberton S G 1986 Jour Sed Petrology 57(4): 687-694.
[13] Jones B, Pemberton S G 1987 Can J Earth Sci 24: 903-914.
[14] Gaspard D 1988 Cahiers de Paléontologie CNRS (ed): 1-243.

CHAPTER 4.12

Amorphous and Crystalline Phases in Biominerals

M.G. TAYLOR and K. SIMKISS

School of Animal and Microbial Sciences, AMS Building, University of Reading, Reading RG6 2AJ, England

Keywords: PHOSPHATES GRAPHICS CALCIUM MAGNESIUM ALUMINIUM

INTRODUCTION

Many discussions of biominerals are concerned with the formation of calcium phosphate phases and their transformations from amorphous phases to a stable crystalline form, such as apatite with various lattice substitutions [1,2,3]. Other biominerals remain amorphous notably biogenic silica [4], some calcium carbonate deposits [5] and inorganic phosphate deposits found in many invertebrates [6]. The intracellular deposits found in the hepatopancreas of the snail Helix aspersa are an example. They are amorphous and have a mineral composition of calcium magnesium pyrophosphate with ca 18% water giving a formula of $CaMgP_2O_7.3H_2O$. Organic material, 5% w/w is also present [7]. A feature of these granules is their ready accumulation of dopant cations such as Mn, Fe, Co and Zn. The in vivo doped granules always remain amorphous to X-ray and electron diffraction techniques. The local atomic structures around Ca, Mn, Fe, Zn and P [8,9,10,11,12] have been determined by X-ray absorption spectroscopy using the Synchrotron Radiation Source at the SERC Daresbury Laboratory, UK. An open hydrated structure around calcium was deduced from EXAFS (Extended X-ray absorption fine structure) and density measurements [8] and it was proposed that the granules could be modelled by a modified continuous random network with metal ions cross linking the pyrophosphate chains with the water contributing to the interchain structure to allow the percolation and reaction of dopant cations [13]. At present the local structure around magnesium cannot be determined by EXAFS as there is no monochromator available for the appropriate wavelength. In order to test our model and investigate the cation interactions in the granules we are applying computer simulation techniques which have been used successfully in solid state chemistry to model perfect lattices and their defect properties [14]. Because of the uncertainties of modelling the amorphous states our approach is to study appropriate compounds with known crystal structures and to calculate the energies and lattice effects of substitution defects. Initially molecular graphics have been used to visualise and highlight features of the structures which may be significant for the movement of dopant cations and which suggest likely sites for substitution reactions. Structures of the dihydrates of calcium and magnesium pyrophosphates are illustrated as they are the molecules of the granules. By contrast, the structures of calcium phosphate as hydroxyapatite and aluminium phosphate are compared. It is considered that aluminium is not incorporated into the hydroxyapatite structure [15] but that it possibly affects the dissolution and growth of bone in aluminium associated osteomalacia by adsorption onto the apatite surface [16].

METHODS

Computer modelling was carried out on an IBM-PC-AT with a PC-CHEMMOD system. The connectivities and co-ordinates for graphics presentation were determined using programs written in the University of Reading (Inorganic Chemistry Modelling Group). The unit cell is built from the space group, lattice constants and the crystallographic co-ordinates obtained from crystal structure determinations in the literature or from the inorganic crystallographic database at Daresbury. The unit cell can then be translated in the x,y and z directions as required, within the limitations of the software which on the system used is 999 atoms. The image can be displayed as a stick model or space filled and manipulated by for example rotation around x, y and z axes to give views which highlight features of interest. Sizes of atoms for graphical representations were determined by considering the atomic charges and interpolating from ionic and covalent radii [17]. For example for $AlPO_4$ values of +1.4 for Al, +1.0 for P and -0.6 for O have been determined from charge density calculations [18]. Desk Top Molecular modeller [19] was used for some of these displays. Interatomic distances and bond angles were checked against literature values using CHEMMOD.

RESULTS AND DISCUSSION

Molecular graphics representations of the structures of the dihydrates of calcium and magnesium pyrophosphate are shown in Figure 1 (a) and (b). The calcium unit cell has been translated in the y direction and the magnesium structure is a fragment of the x translation. Two features are immediately obvious in both structures. Firstly the metal pyrophosphates form infinite layers and secondly the water molecules line channels separating the layers. The structures are remarkably similar. The $P\text{-}O_B\text{-}P$ angles are 123° for the calcium and 125° for the magnesium compound compared with 139° and 144° for the corresponding anhydrous compounds. Both cations are in two different sites and have a range of cation oxygen distances with mean distances of 2.09Å for magnesium which is 6 coordinate and 2.45Å for calcium which is 7 coordinate. Differences in the structures arise from the different ionic radii (Mg 65, Ca 99pm) and coordination of the two cations. The graphics appear to indicate that some of the cation sites may be more reactive to substitution than others. Experiments tend to support this as only in the in vitro reaction with zinc chloride, when the granule structure breaks down, is all the calcium replaced. [11] Cation substitution reactions require a degree of structural openness to allow the diffusion of foreign ions into and out of the structure. Molecular graphics of the crystalline pyrophosphate compounds qualitatively support the continuous random network model. The relative distributions of calcium and magnesium, however, require quantitative calculations. Whether or not there is a uniform distribution of calcium and magnesium throughout the structure or clusters of individual species can be estimated by calculations of the structural and energetic effects of substitution defects in the perfect lattice [14].

By contrast, structures of a calcium phosphate as hydroxyapatite and aluminium phosphate clearly show that the cation sites in these two compound are not related Figure 2 (a) and (b). Aluminium is tetrahedrally coordinated like phosphorus and is at a distance of 1.73Å from the phosphate oxygens. It is noted that aluminium can expand its coordination sphere to six coordinate. Calcium in hydroxyapatite is in two non-equivalent sites. In the first site calcium is surrounded by nine oxygens from phosphate groups with calcium oxygen distances ranging from 2.41 to 2.81Å. At the second site calcium is close to seven oxygens from phosphate groups at distances 2.34 to 2.71Å and also closely associated with the hydroxyl oxygen at a distance of 2.38Å. The hydroxyl groups can be clearly seen to lie in columns which can readily accommodate fluoride substitution. Substitution reactions to form new products are more likely to occur when there is a structural similarity between reactant and product and a minimum of structural reorganization.

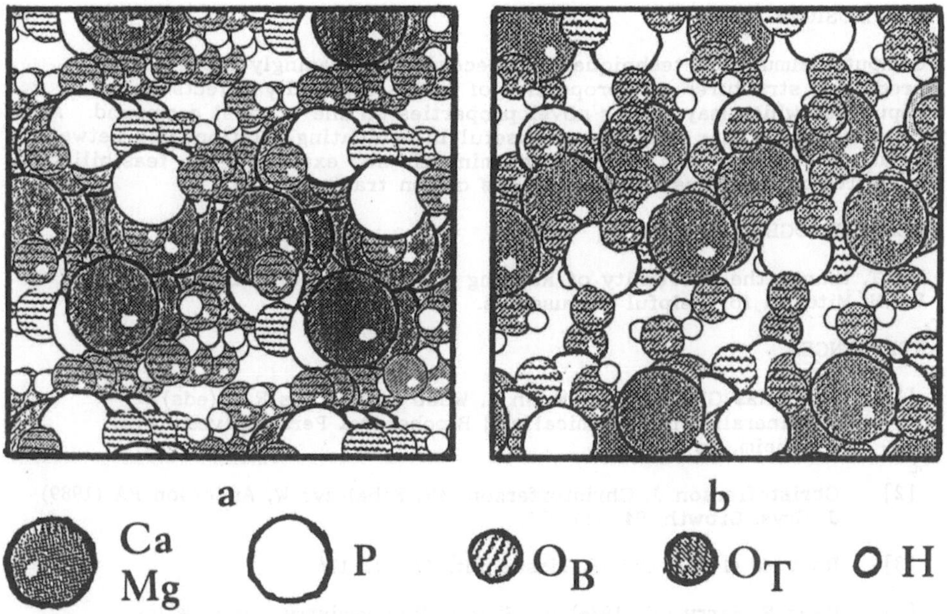

Figure 1. (a) Calcium pyrophosphate dihydrate [20] (b) Magnesium pyrophosphate dihydrate [21] from CHEMMOD.

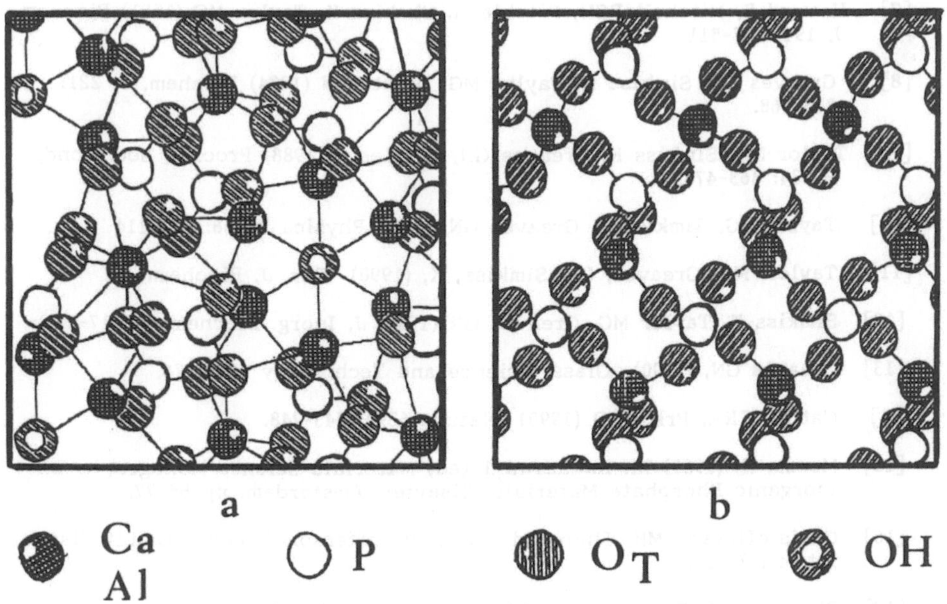

Figure 2. (a) Hydroxypatite [22] (b) Aluminium phosphate [18] from DTMM.

CONCLUSIONS

Computer simulation techniques are becoming increasingly important in predicting structures and properties of solids especially defects or impurities which may confer novel properties on the original compound. As a first step molecular graphics are useful in evaluating relationships between structures of materials including biominerals and examining the feasibility of substitution reactions and mechanisms of ion transport.

ACKNOWLEDGEMENTS

M.G.T. thanks the University of Reading for support for this work and Dr P.C.H. Mitchell for helpful discussions.

REFERENCES

[1] Nancollas GH (1989) In: Mann S, Webb J, Williams RJP (eds) Biomineralization Chemical and Biochemical Perspectives. VCH, Weinheim, pp 157-187.

[2] Christofferson J, Christofferson MR, Kibalczyc W, Anderson FA (1989) J. Crys. Growth. 94: 767-777.

[3] Root MJ (1990) Calcif. Tissue Int. 47: 112-116.

[4] Mann S, Perry CC (1986) In: Silicon Biochemistry. John Wiley, Chichester pp 40-58.

[5] Mason AZ, Nott JA (1981) Aquat. Toxicol. 1: 239-259.

[6] Taylor MG, Simkiss K, and Greaves GN (1986) Trans. Biochem. Soc. 14: 549-552.

[7] Howard B, Mitchell PCH, Ritchie A, Simkiss K, Taylor MG (1981) Biochem. J. 194: 507-511

[8] Greaves GN, Simkiss K, Taylor MG, Binsted N (1984) Biochem. J. 221: 855-868.

[9] Taylor MG, Simkiss K, Greaves GN, Harries J (1988) Proc. R. Soc. Lond. B 234: 463-476.

[10] Taylor MG, Simkiss K, Greaves GN (1989) Physica B 158: 112-114

[11] Taylor, MG, Greaves, GN, Simkiss, K, (1990) Eur. J. Biochem.

[12] Simkiss K, Taylor MG, Greaves GN (1990) J. Inorg. Biochem. 39: 17-23

[13] Greaves GN, (1990) Glass: Science and Technology 4B: 1-76.

[14] Catlow CRA, Price GD (1990) Nature 6290: 243-248.

[15] Monma H, (1989) In: Kanazawa T (ed) Materials Science Monographs on Inorganic Phosphate Materials. Elsevier, Amsterdam, pp 55-77.

[16] Christofferson MR, Thyregod HC, Christofferson J. (1987) Calcif Tissue Int. 41: 27-30.

[17] Shannon RD, Prewitt CT (1969) Acta Cryst. B25: 925-946

[18] Thong N, Schwarzenbach D (1979) Acta. Cryst. A35: 658-664.

[19] Crabbe MJC, Appleyard JR (1989) Desktop Molecular Mofdeller. Oxford University Press, Oxford.

[20] Mandel NS (1975) Acta. Cryst. B31: 1730-1734.

[21] Oka J, Kawahara A (1982) Acta. Cryst. B38: 3-5

[22] Kay MI, Young RA, Posner AS (1964) Nature 204: 1050-1052

CHAPTER 4.13
Evolutionary Trends of Shell Microstructure in Bivalve Molluscs

I. KOBAYASHI
Department of Geology and Mineralogy, Faculty of Science, Niigata University, Niigata, 950-21 Japan

Key words: Bivalve molluscs, Shell structure, Biocrystal, Phylogeny, Evolution

INTRODUCTION

During the evolutionary history of bivalve and gastropod molluscs, not only all the soft parts, but also the shells have changed considerably since the Cambrian period. Moreover, not only the outer shape of a shell, but also the internal shell microstructure has changed during this time. The internal shell microstructure is different in different species and is remarkably well related to bivalve classification. Therefore, shell structure is available for the phylogenetic study of bivalves and other molluscs. The internal shell microstructure expresses one face of molluscan evolution.
The molluscan shell microstructure has been examined from several points of view, such as shell formation, classification, functional morphology, ecology and evolution. The microscopic description of shell structure has been done since the latter half of the 19th century. The first great work on the molluscan shell structure was published by Bøggild in 1930 [1]. After that, works from the viewpoint of phylogenetic classification and evolutional change were done by Oberling [2], Kobayashi [3,4,5], Taylor *et al.*[6,7], Taylor [8], Uozumi·Suzuki [9] and Carter·Clark [10]. Moreover, there have been many investigations of the mechanisms of shell formation and shell regeneration [11,12,13], and studies of the organic matrices in shells. These are very important for research on the evolution of internal shell microstructure.
In this short article, some problems on the evolutionary trends of shell microstructure in bivalve molluscs are pointed out and considered on the basis of morphological and mineralogical data of recent and Quaternary specimens.

LEVEL OF SHELL STRUCTURE

Some levels of shell microstructure, such as layer structure, morphological type of shell structure, biomineral and organic matrix in descending order, are discriminated by means of microscopy, X-ray techniques and chemical methods. The evolutionary trends of shell microstructure have been found in each microstructural level and related to the other levels. For example, the simple prismatic structure of bivalve molluscs which is one of the morphological types of shell structure, is distributed only in the outer shell layer. On the other hand, the nacreous structure is only found in the inner shell layer. As another instance, two different kinds of organic materials, such as envelopes and sheets, which are associated with the growing or initiating calcium carbonate minerals [14], regulate the morphological type of shell structure.

LAYER STRUCTURE

The highest level of shell microstructure is layer structure which is observed under low magnification optical microscopy. Generally, there are two layers, namely outer and inner calcareous shell layers in addition to periostracal and myostracal layers. The outer shell layer is formed by the outer epithelium of the mantle edge and brings about the outward extension of the shell. On the other hand, the inner shell layer is formed by a part of the mantle epithelium further from the mantle edge than the portion forming the outer layer and brings about the increase of shell thickness. This two-layered system composed of outer and inner layers is widely distributed in bivalve molluscs. However, there are some species possessing more than three layers such as *Spondylus* and groups possessing one layer such as *Mizuhopecten*. In Pectinidae, the number of shell layers reduced from a three-layered system to an one-layered system during evolution [15]. Besides this example, the specialized shell layer is seen in the shell of epifaunal groups such as *Monia*, *Anomia* and *Placuna* in possession of foliated structure, and in the shell of deeply burrowing groups such as *Laternula* and *Panopea*.

The layer structure has probably changed during the evolution of bivalves. Three courses of the change of layer structure are discriminated as follows, 1) degeneration of an inner layer (e.g.in Pectinidae), 2) degeneration of an outer layer (e.g.in Mytilidae) and 3) differentiation of a layer (e.g.in Veneridae).

The change of layer system is mostly related to ecological condition. This phenomenon is also important for one of the changes of biomineralization system.

MORPHOLOGICAL TYPE OF SHELL STRUCTURE

In bivalve molluscs, several types of shell structure are distinguished on the basis of the microscopic observation of shell textures. These are called morphological types [4] and are classified into ten or more types [1,6,13]. Studies on the phylogeny and evolution of morphological type have been done by several investigators [4,5,8,9,16].

The shell of each species is composed of a constant combination of morphological types. A shell layer consists of one or more morphological types. There are a large number of combinations of morphological types in bivalve molluscs. From the result of a distributional survey of bivalves, main three groups, namely nacreous, foliated and crossed lamellar series are distinguished (Fig.1).

One of the most important relationships between morphological type and shell layer is the definite arrangement of morphological types from an outer to an inner layer. The representative arrangements from an outer to an inner surface are as follows [4,5].

1) simple prismatic st. or fibrous prismatic st. → nacreous st. → trans-prismatic st.
2) foliated st. → crossed lamellar st. → complex crossed lamellar st.
3) fibrous st. or composite prismatic st. → crossed lamellar st.and/or homogeneous st. → complex crossed lamellar st. and/or homogeneous st.

These definite arrangements of types have very significant meanings for the biomineralization and evolution of bivalve shells in relation to the evolution of secretory cells.

Fig.1 shows briefly the evolutionary trend of bivalve shell structure. Three main series are recognized as previously mentioned. Seven series of the evolutionary courses of the bivalvian shell structure were proposed by Taylor [8] and two main series also proposed by Uozumi · Suzuki [9]. It is inferred that each series has changed during bivalve history. Fibrous, crossed lamellar and composite prismatic structures made of aragonitic rods may form alternatively during shell formation [17]. The facts show that these have a very intimate relationship genetically.

There are two different opinions on the most primitive shell structure during the evolutionary courses of bivalves. One of them is a opinion [8] that the most primitive morphological type is nacreous and prismatic structures. The another opinion [9,17] is that it is primitive granular or simple structure.

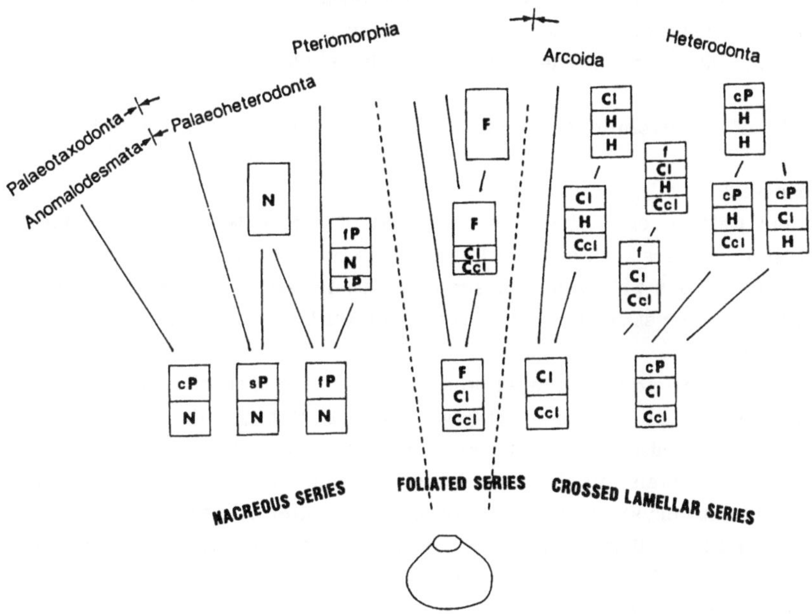

Fig.1: Evolutionary trends of bivalve shell structure
sP: Simple prismatic structure (Calcite, Aragonite)
fP: Fibrous prismatic structure (Calcite)
F: Foliated structure (Calcite)
tP: Trans-prismatic structure (Aragonite)
N: Nacreous structure (Aragonite)
cP: Composite prismatic structure (Aragonite)
Cl: Crossed lamellar structure (Aragonite)
Ccl: Complex crossed lamellar structure (Aragonite)
f: Fibrous structure (Aragonite)
H: Homogeneous structure (Aragonite)

BIOMINERAL

Form of Biocrystal:

The shape of biocrystal is different in each morphological type. Main shapes are tabulates, rectangular rods and grains which consist of calcite and aragonite crystals.

Nacreous structure: This structure is distributed in separated groups, such as Palaeotaxodonta, Filibranchia and Anomalodesmata. Almost all of these possess a sheet nacre type. There are some variations in shape of crystal during growth, but the biomineral formation will take place under similar mechanisms [14].

Crossed lamellar structure: This structure in Neotaxodonta shows the regular arrangement of first order lamellae which consist of aragonite crystals. On the other hand, the first order lamellae of crossed lamellar layer in some advanced bivalve molluscs are more intricate in arrangement and smaller in size. Some groups of Veneridae (advanced molluscs) possess complex and finer textures of crossed lamellar structure which is called finely or pseudo-crossed lamellar structure [4]. On the basis of the arrangement and shape of lamellae, crossed lamellar structure is subdivided into several types [10].

The diversity of the evolutionary trends of biomineralization will be made clear in the future.

Kind of Biocrystal:

The inorganic matters of a shell are composed of aragonite and calcite minerals. In general, bivalve molluscs with aragonitic shells are more abundant than ones with calcitic shells. Foliated, fibrous prismatic and simple prismatic structures of marine species are composed of calcite crystals. The other morphological types of shell structure consist of aragonite crystals.

The formation of a calcitic shell layer was initiated later than an aragonitic shell layer during the evolutionary course of bivalve molluscs. Ontogenetic data on the mineral formation [17,18] indicate that larval shells of *Pinctada* and *Mizuhopecten* are composed only of aragonite crystals, on the other hand, juvenile and adult shells are composed of calcite and aragonite crystals. *Pinctada* possesses an outer calcitic simple prismatic layer and *Mizuhopecten* has calcitic foliated structure. This phenomenon seems to indicate the change of physiological systems in relation to mineral deposition in the evolutionary trends of biomineralization.

ORGANIC MATRIX

Organic matrices of bivalve molluscs are examined by microscopes [19,20] and by organic chemical analyses [21,22,23].

The results show that there are several kinds of organic matrices which are composed of proteins and polyssacharids. Each morphological type of shell structure includes characteristic organic materials. The amino acid composition of protein matrices of crossed lamellar, complex crossed lamellar, homogeneous and composite prismatic structures is similar each other [23]. The curves of differential thermal analyses of them indicate the similar patterns [24] and the morphological features of them also are similar [20]. These facts mean that their types are closely related each other. Samata [23] pointed out that the amino acid composition of the water-soluble matrices of nacreous and prismatic structures, foliated and chalk structures resembles each other. The facts seem to correspond with the evolutionary series of morphological types.

Organic matrix secreted by epithelial cells is the most important matter in the formation of shell. Organic matrices control the mineralization and formation of shell structure. Moreover, under these regulated conditions the evolution of shell structure will occur. The evolution of shell is related with ecological, physiological, biochemical and genetical mechanisms.

I thank Dr.M.A.Whyte, the University of Sheffield, for helpful suggestions.

REFERENCES

[1] Bøggild OB (1930) Kgl Danske Videnskab Selsk Skr, Naturve ser.9 2:233-326
[2] Oberling JJ (1964) Mitt Nat Gesellschaft Bern 20:1-63
[3] Kobayashi I (1964) Earth Science 73:1-12
[4] Kobayashi I (1973a) Sci Rep Niigata Univ Ser.E 2:27-50
[5] Kobayashi I (1988) In:Omori M, Suga S, Goto M (eds) Biomineralization and Phylogeny of Marine Organisms.Tokai University Press,Tokyo.pp.97-112
[6] Taylor JD, Kennedy WJ, Hall A (1969) Bull Br Mus nat Hist,Zool,Suppl 3:1-125
[7] Taylor JD, Kennedy WJ, Hall A (1973) Bull Br Mus nat Hist,Zool,22:256-294
[8] Taylor JD (1973) Palaeontology 16:519-534
[9] Uozumi S, Suzuki S (1981) Habe T, Omori M (eds) Study of Molluscan Paleobiology, Prof.Omori,M Mem.Vol.pp.63-77
[10] Carter JG, Clark GR II (1985) In: Broadhead TW (ed) Mollusks, Notes for a Short Course, organized by Bottjer DJ, Hickman CS, Ward PD, University of Tennessee, Department of Geological Sciences Studies in Geology 13.pp.50-71
[11] Wilbur KM, Saleuddin ASM (1983) In: Saleuddin ASM, Wilbur KM (eds) The Mollusca. vol.4,Physiology,part 1.Academic Press.pp.235-287
[12] Watabe N (1983) In: Saleuddin ASM, Wilbur KM (eds) The Mollusca. vol.4, Physiology, part 1.Academic Press.pp.289-316
[13] Watabe N (1988) Trueman ER, Clarke MR (eds) The Mollusca.vol.11, Form and function. Academic Press.pp.69-104
[14] Bevelander G, Nakahara H (1980) In: Omori M, Watabe N (eds) The Mechanisms of Biomineralization in Animals and Plants.Tokai Univ Press,Tokyo.pp.19-27
[15] Waller TR (1972) 24th IGC,Sec.7,48-56
[16] Shimamoto M (1986) Tohoku Univ Sci Rep 2nd ser(Geol) 56:1-39
[17] Kobayashi I (1980) In: Omori M, Watabe N (eds) The Mechanisms of Biomineralization in Animals and Plants.Tokai Univ Press,Tokyo.pp.145-155
[18] Iwata K, Akamatsu M (1975) Bull His Mus Hokkaido 10:11-17
[19] Grégore C (1959) Inst Roy Sci Nat Belgique Bull,Tome 35:1-14
[20] Kobayashi I (1968) Jap Jour Malacology 27:111-122
[21] Hare PE, Abelson PH (1965) Carnegie Inst Wash Year Book 64:223-235
[22] Akiyama M (1966) Proc Japan Acad 42:800-805
[23] Samata T (1988) In: Omori M, Suga S, Goto M (eds) Biomineralization and Phylogeny of Marine Organisms.Tokai University Press,Tokyo.pp.183-202
[24] Kobayashi I (1973b) Jour Fossil Res 7:1-4

CHAPTER 4.14
Coexistence of Different Shell Microstructure Types in a Single Shell Layer of the Veneridae (Bivalvia, Mollusca)

M. SHIMAMOTO
Institute of Geology and Paleontology, Tohoku University, Sendai, 980 Japan

Key words: Shell Microstructure, Coexistence, Phylogenetic Implication, Veneridae, Bivalvia

ABSTRACT

For the purpose of understanding the relationship of component taxa of the family Veneridae (Bivalvia), 65 extant species have been observed for their shell microstructure on thin sections and acetylcellulose replicas with the aid of optical and scanning electron microscopy. The observation revealed that the microstructure of the inner shell layer shows intraspecific variation. The inner shell layer of venerids generally consists of either homogeneous or complex crossed lamellar structure. However, these two microstructure types coexist in some species (e.g., *Ruditapes philippinarum*, *R. variegata*, etc.) and even within a single shell of the species. In contrast, intraspecific variations are not recognized in the microstructure of the middle, outer, or outer-most shell layer, but replacement or accretion of some microstructures occurs in the middle shell layer of some species; e.g., the crossed lamellar layer accretes to the homogeneous layer during the later stage of shell growth (*Veremolpa mindanensis*, etc.), or the crossed lamellar structure replaces temporarily the homogeneous structure at growth breaks (*Mercenaria mercenaria*, *Protothaca (Notochione) jedoensis*, etc.). These lines of evidence imply that these shell microstructures have been developed concurrently in venerids, and give us important clues for understanding the phylogenetic relationships of the Veneridae.

INTRODUCTION

The shell microstructures of 65 extant species of the family Veneridae (Bivalvia, Mollusca) have been observed and five shell microstructure types were recognized. These types were composite prismatic, crossed lamellar, fine crossed lamellar, homogeneous, and complex crossed lamellar structures (Shimamoto, 1986)[1]. These microstructural types do not occur randomly in every shell layer; rather each type appears to be confined to a few layers. For example, composite prismatic structure is always secreted outside of other shell layers, complex crossed lamellar structure is restricted to the inner shell layer, and when both crossed lamellar and homogeneous structures coexist in a single shell, the former occurs exteriorly to the latter. Consequently the number of possible shell microstructural combinations in the Veneridae is strictly limited. Moreover, it has been said that a single shell layer consists of only one shell microstructure, and the intraspecific variation of shell microstructure has not been investigated. In order to consider the phylogenetic implications of shell microstructures, it is important to consider the intraspecific variation and coexistence of shell microstructures within a single shell layer. In this paper, these matters are discussed and the phylogenetic implications of shell microstructures are considered.

The term "homogeneous structure" is used here in a broad sense to include crossed acicular and/or fine complex crossed lamellar structures of Carter (1980)[2]. Bøggild (1930, p.245)[3] defined homogeneous structure on the basis of optical microscopy, and Carter (1980)[2] redefined in a narrow sense the homogeneous structure as the aggregation of grained crystals generally less than 5 μm in diameter. Homogeneous structure is accompanied with acicular and/or fine complex crossed lamellar structures in a single layer. However, it is difficult to distinguish the sublayer of acicular and/or fine complex crossed lamellar structures from the typical "homogeneous" sublayer under the optical microscope, and they do not occupy the major part of a single shell layer. Therefore, they are included in the homogeneous structure in this paper.

COEXISTENCE OF HOMOGENEOUS STRUCTURE AND COMPLEX CROSSED LAMELLAR STRUCTURE IN THE INNER SHELL LAYER

Shimamoto (1986)[1] examined the extent to which intraspecific variations occur in shell microstructures of the family Veneridae, and found that intraspecific variation is not recognized in shell microstructure outside the pellucid layer. In contrast, the microstructure of the inner shell layer showed intraspecific variation such that the inner layer of venerid shells commonly consists of either homogeneous structure or complex crossed lamellar structure, but these two microstructures coexist in some species (e.g., *Ruditapes philippinarum* (Adams & Reeve), *R. variegata* (Sowerby), etc.), and even within a single shell of the species.

The inner shell layer of *Ruditapes philippinarum* is commonly composed of homogeneous structure, but up to 20 percent of individuals have both homogeneous and complex crossed lamellar structures. Most shells of *R. variegata* or *Protothaca (Novathaca) euglypta* (Sowerby) have complex crossed lamellar structure in their inner shell layer, but up to 20 percent of all studied individuals have the inner shell layer in which both homogeneous and complex crossed lamellar structures coexist (Shimamoto, 1986)[1]. When these two mirostructures coexist, they constitute sublayers and alternate with each other. They may gradually transform from one microstructure to another laterally, and the boundary of each sublayer is usually obscure. Under the scanning electron microscope (SEM), not only typical granular crystals but also acicular and/or fine complex crossed lamellar ones can be observed in the inner shell layer. They are commonly recognizable near the area where homogeneous sublayer changes into complex crossed lamellar sublayer.

COEXISTENCE OF HOMOGENEOUS STRUCTURE AND CROSSED LAMELLAR STRUCTURE IN THE MIDDLE SHELL LAYER

The middle shell layer of venerids commonly consists of either homogeneous or crossed lamellar structure. However, the coexistence of homogeneous and crossed lamellar structures was reported in the middle shell layer of the subfamily Chioninae by Shimamoto (1988)[4]. Two types of coexistence were observed in the subfamily as follows.

First, the middle shell layer of *Veremolpa mindanensis* (Smith) consists only of homogeneous structure at the early stage of shell growth. Thereafter, crossed lamellar structure is secreted along the boundary between the middle homogeneous and the outer composite prismatic layers as the shell is growing older, which results in accretion of crossed lamellar layer between middle and outer layers. In this case, the homogeneous structure gradually changes into a crossed lamellar structure in the direction of the ventral margin, and the thickness of the crossed lamellar layer gradually increases as it replaces the homogeneous structure.

Second, the middle shell layer of *Mercenaria mercenaria* (Linnaeus), *Callithaca adamsi* (Reeve), *Protothaca (Novathaca) euglypta* (Sowerby), and *P. (Notochione) jedoensis* (Lischke) consists only of homogeneous structure early in growth, but crossed lamellar structure is secreted temporarily in later growth stages in a wedge manner near the boundary between middle and outer shell layers. The crossed lamellar structure, however, does not construct a distinct layer.

In both these cases, the change from homogeneous to crossed lamellar structure is gradual. It can be observed under the SEM that homogeneous layer is composed of irregular shaped granular crystals in early growth, but at the late stage of shell growth when crossed lamellar structure is secreted in the middle layer, not only is there an aggregation of granular crystals, but also acicular crystals occur in the transitional zone between homogeneous and crossed lamellar layers. The acicular crystals are arranged in two predominant dip directions, and intersect in a characteristic "cross-stitch" pattern visible in vertical sections.

COEXISTENCE OF CROSSED LAMELLAR STRUCTURE AND COMPOSITE PRISMATIC STRUCTURE IN THE OUTER SHELL LAYER

The outer shell layer of venerids is commonly composed of either crossed lamellar or composite prismatic structure, but in only one species, *Callista (Ezocallista) brevisiphonata* Carpenter, these two shell microstructures coexist in the outer layer.

The outer shell layer of *Callista (Ezocallista) brevisiphonata* ordinarily consists of composite prismatic structure, but the crossed lamellar structure is temporarily secreted as it replaces composite prismatic structure (Fig. 1). The boundary between these two microstructures is obvious, but an evident cessation of growth is not recognized near the boundary.

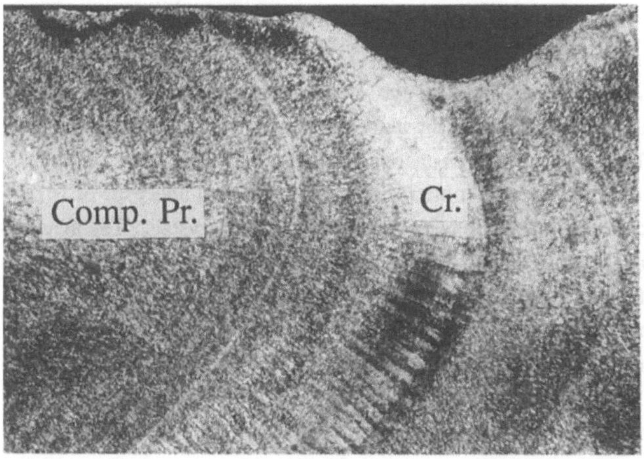

Fig. 1. Micrograph of a radial thin section of the outer shell layer of *Callista (Ezocallista) brevisiphonata* Carpenter. Comp. Pr.: composite prismatic structure, Cr.: crossed lamellar structure

In order to examine the relationship between the formation of crossed lamellar structure and growth rate of a shell, an analysis of growth lines was attempted (Fig.2). The crossed lamellar structure is recognized near the region where increment thickness is generally thinning, that is, the temporal secretion of crossed lamellar structure may be related to a low growth rate. Moreover, the frequency of secretion of crossed lamellar structure shows a tendency to increase during ontogeny (Fig. 2).

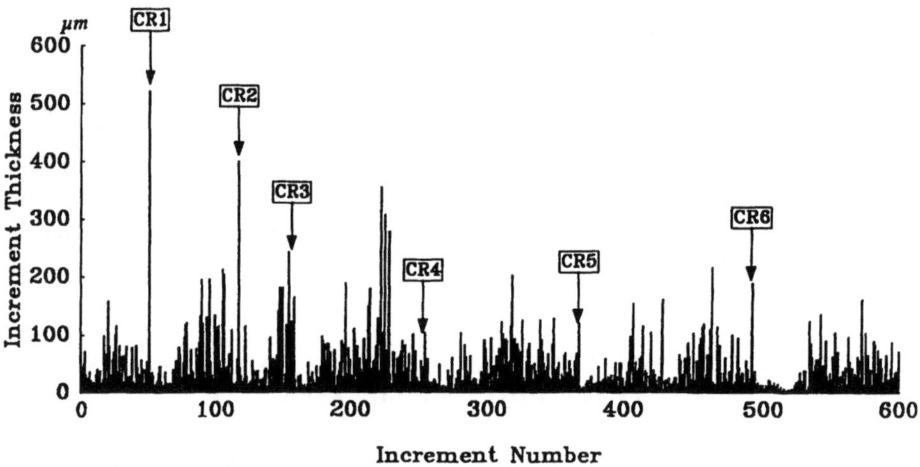

Fig. 2. Diagram showing the thickness of successive growth increments. The increment number is counted from the ventral margin toward the beak; the origin indicates the ventral margin. CR1-CR6 show the positions of crossed lamellar structures secreted temporarily. This examination suggests that the crossed lamellar structures are secreted immediately after decreasing in shell growth. Since growth lines are obscure in the position of crossed lamellar structures, thick increments indicated by arrows may include several increments.

DISCUSSION

As described above, homogeneous structure of shells in the Veneridae is secreted in very close relationship with the complex crossed lamellar structure in the inner shell layer, and with crossed lamellar structure in the middle shell layer. These microstructures change gradually into each other. When the transitional zone from homogeneous structure to another microstructure is observed under the SEM, the aggregation of granular crystals are commonly accompanied by one of acicular crystals. The acicular crystals cross in a characteristic "cross-stitch" pattern, and may correspond to crossed acicular or fine complex crossed lamellar structure of Carter (1980)[2]. As pointed out by Bøggild (1930)[3], Taylor *et al.* (1969)[5] and others, homogeneous structure may include traces of other fine-grained microstructural types. In the present study, the term "homogeneous structure" is used in a broad sense to include the crossed acicular and/or fine complex crossed lamellar structures of Carter (1980)[2]. In the Veneridae, homogeneous structure may include traces of fine-grained aggregations of crossed lamellar and complex crossed lamellar structures.

As demonstrated by Shimamoto (1986, 1988)[1,4], the inner shell layer of venerids shows intraspecific variation, that is, homogeneous structure coexists with complex crossed lamellar structure in some individuals. The position and relative proportions of these two shell microstructures are variable in each individual. Shimamoto (1986, 1988)[1,4] examined the geographic variations in the frequency of the two microstructure types. The results are as follows: it is difficult to define a clear topocline in the frequency of the two microstructures, but the dominant geographic distribution of each microstructure is different from the other: Species having homogeneous structure are predominantly distributed between 20°N and 45°N. Lat., in contrast, the number of species having complex crossed lamellar structure is greater to the south of 35°N. Lat., and decreases suddenly to the north of 35°N. Lat.. Therefore, the ratio of species having complex crossed lamellar structure is much higher to the south of 20°N. Lat..

To summarize these data, homogeneous and complex crossed lamellar structures appear to be interchangeable with each other and they may be regarded as equivalent shell microstructures.

In the middle shell layer, homogeneous structure coexists with crossed lamellar structure. However, the situation differs from the inner shell layer because intraspecific variation cannot be recognized in the middle shell layer. Crossed lamellar structure is secreted in a restricted area near the boundary between the middle and outer layers, and secretion is limited to the later stage of shell growth or related to breaks in shell growth. Therefore, the middle shell layer secretion may be more strongly controlled by genetic and phylogenetic factors than the inner shell layer secretion.

Another type of coexistence is recognized in the outer shell layer. The outer shell layer in venerids is composed of either crossed lamellar structure or composite prismatic structure, but in the outer shell layer of *Callista (Ezocallista) brevisiphonata* Carpenter these two microstructures coexist. The outer shell layer of the species ordinarily consists of composite prismatic structure, but crossed lamellar structure is temporarily secreted as it replaces composite prismatic structure. The timing of secretion of the crossed lamellar structure is not restricted during ontogeny, but it seems to be related to the time when shell growth rate is rather slow. Moreover, frequency of the secretion shows a tendency to increase during shell growth. Thus, the secretion of crossed lamellar structure may be related to growth rate.

The mechanism of coexistence of different shell microstructures in a single shell layer is unknown, but it is important that the same cell can secrete two kinds of crystal aggregates in a single shell layer. These lines of evidence imply that these shell microstructures have been developed concurrently in venerids.

Acknowledgments

The author would like to express his deep appreciation to Professors Kei Mori and Kenshiro Ogasawara of the Institute of Geology and Paleontology, Tohoku University for their encouragement, and critical reading of the manuscript. Sincere thanks are also due to Dr. Thomas M. Cronin of U. S. Geological Survey and Professor Kunihiro Ishizaki of the Institute of Geology and Paleontology, Tohoku University for their helpful comments on the manuscript. This research is partly supported by the Grant-in-Aid for Scientific Research of the Ministry of Education, Science and Culture, Japan (No. 02740390, 1990).

REFERENCES

[1] Shimamoto M(1986) Tohoku Univ, Sci Rep, 2nd ser(Geol) 56(1): 1-39.

[2] Carter JG(1980) In: Rhoads DC, Lutz RA (eds) Skeletal Growth of Aquatic Organisms: Biological Records of Environmental Change. Plenum Press, New York and London, pp.645-673.

[3] Bøggild OB(1930) D Kgl Danske Vidensk Selsk Skr Naturv, ser 9, 2: 232-326.

[4] Shimamoto M(1988) Saito Ho-on Kai Spec Pub(Prof T Kotaka Commem Vol): 239-245.

[5] Taylor JD, Kennedy WJ, Hall A(1969) Bull Brit Mus Nat Hist Zool, Suppl, 3: 1-125.

CHAPTER 4.15

Phosphate Gill Supports in Living and Fossil Bivalves

M.A. WHYTE

Earth Sciences Unit, University of Sheffield, Sheffield, S3 7HF, UK

Key Words: Bivalvia, gills, fossil, phosphate, amorphous.

INTRODUCTION

Specimens of fossil trigonioid bivalves have recently been discovered in which evidence of the gills and other soft tissues has been found preserved. Anatomical studies of these are still being carried out [1] but the traces of the gills are preserved as patches of a fibrous or bladed material which has often been disturbed or displaced during fossilisation. Comparison of these structures with the gills of the only extant trigonioid bivalve genus, Neotrigonia, indicates that what is being preserved is not the soft tissues of the gill filaments themselves but mineralised supports which run the length of each gill filament [2] and which in trigonioids reinforce the chitinous skeleton that is found in the gill filaments of other bivalves. Similar firm tissues, referred to as 'calcified rods' or 'baguettes calcifiées', also occur in one, and only one other, group of living bivalves, namely the various genera of the Order Unionoida [2,3,4], and studies of fossil unionoids have led to the discovery of specimens in which the gill supports have been preserved [1]. In this contribution the physical nature and chemistry of these gill supports in both living and fossil forms is investigated. Such studies can shed light on the preservation processes and on the preservation potential of these stuctures and hence on their usefulness to palaeontologists.

MATERIALS AND METHODS

Living bivalves. Excised gill tissues from specimens of the unionoid, Anodonta anatina (L.), were prepared, either chemically or physically, to remove their soft tissues and to release their inorganic components. Chemically this was done by macerating gill lamellae in a 10% solution of potassium hydroxide (c.f. [3]). Alternatively gills were subjected to low temperature plasma ashing using a Nanotech Plasmaprep 100 (Earth Sciences Unit, University of Sheffield). An aliquot of this plasma ashed material was further heated to 600°C for 24 hours and the gills of another specimen were similarly heat treated without previously being plasma ashed. X-ray diffraction studies of the residues were then carried out using a Philips X-ray Diffractometer (Earth Sciences Unit, Sheffield). A small quantity of quartz was added to some of the smear mounts as an internal standard. A sample of the plasma ashed residue was also embedded in Araldite resin and gill supports exposed on a polished surface of the mount were analysed using a Cambridge Instruments Microscan 9 (Earth Sciences Unit, Sheffield). The infrared spectrum of the plasma ashed material was recorded from a Nujol mull using a Perkin-Elmer Infrared

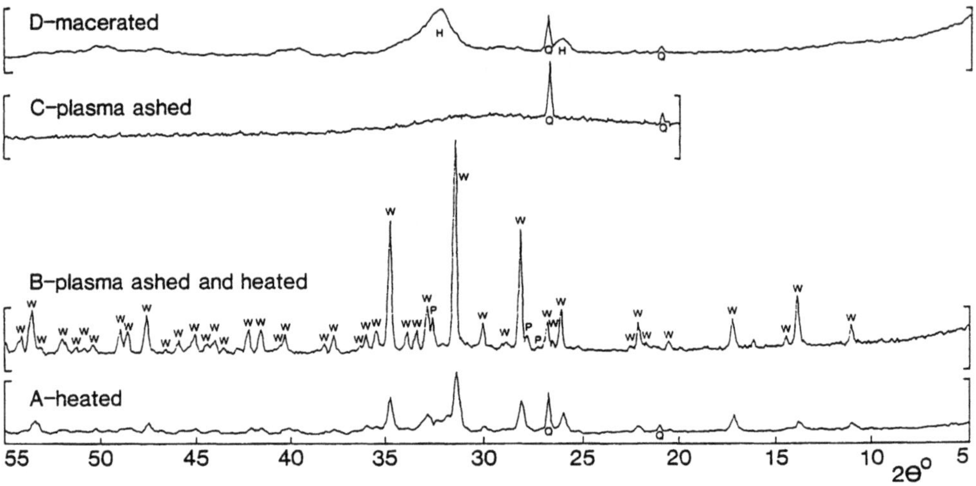

Fig.1 X-ray diffraction traces of residues from the gills of specimens of Anodonta anatina. Peak positions of whitlockite (W), calcium pyrophosphate (P), hydroxyapatite (H) and quartz (Q) are indicated. A - heated; B - plasma ashed and heated; C - plasma ashed; D - macerated in KOH.

Spectrophotometer (Department of Chemistry, Sheffield).
Specimens of the trigonioid, Neotrigonia margaritacea, were kindly supplied by the Museum of Victoria, Victoria, Australia and gill tissues were plasma ashed. X-ray studies were carried on this material both with and without supplementary heating. An infrared spectrum and electron microprobe analyses were also recorded.

Fossil bivalves. Gill supports of a specimen of the upper Jurassic (Portlandian) trigonioid species, Laevitrigonia gibbosa, were removed from the enveloping limestone matrix by etching in 10% acetic acid solution and x-ray difraction studies and electron microprobe analyses were carried out. Scarcity of material has so far hampered studies of fossil unionoids but an x-ray diffraction trace has been obtained for gill supports etched from a lower Cretaceous unionoid bivalve.

RESULTS

Living bivalves. The different x-ray diffraction traces of the gill supports of Anodonta show some interesting variations (Fig. 1). The simplest trace (Fig. 1C) is that of the plasma ashed material which apart from the peaks of the quartz standard, has only a single very broad and diffuse feature. This hump like form is characteristic of an amorphous calcium phosphate of the type which has been well documented in other biological materials and in synthetic preparations [5,6,7,8,9]. This is supported by the very complex trace produced by the plasma ashed material which has in addition been heated (Fig. 1B). The numerous well defined peaks in this trace are indicative of a mixture of a tricalcium phosphate, with a whitlockite structure, and minor amounts of calcium pyrophosphate. The unit cell dimensions of the tricalcium phosphate have been estimated as $a=10.343\pm.002$ and $c=37.011\pm.011$ which are smaller than those of pure whitlockite. Whitlockite is also evident in the trace of heated, but unashed, gill tissue (Fig. 1A) though here the peaks are less well defined and some

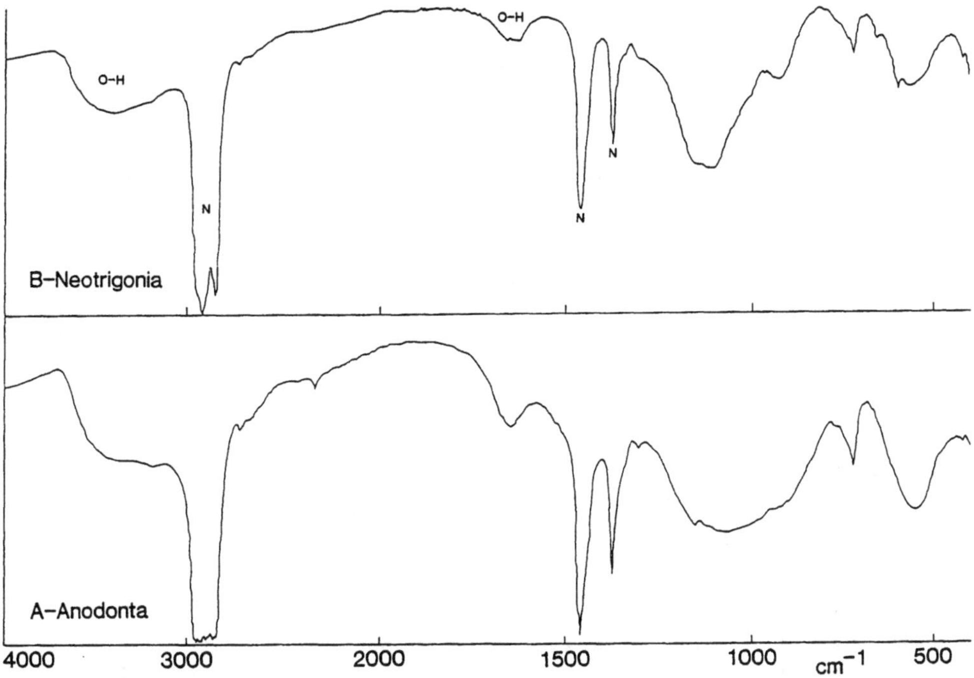

Fig. 2 Infrared spectra of gill supports in plasma ashed residues. Nujol (N) and hydroxyl (O-H) absorbance band positions are indicated. A - Anodonta anatina; B - Neotrigonia margaritacea.

apatite may also be present. Clearer evidence of an apatitic structure is seen in the trace of the macerated material (Fig. 1D) which is interpreted as a poorly crystalline hydroxyapatite [5,6]. In the case of Neotrigonia the trace of plasma ashed material also showed a broad hump suggestive of a calcium phosphate and, for material that had additionally been heated, the trace revealed the presence of calcium pyrophosphate with some minor amounts of whitlockite [1].

The amorphous nature of the phosphatic material of the gill supports of both Anodonta and Neotrigonia is supported by their infrared spectra which are poorly resolved (Fig. 2). The pattern of the Anodonta gill supports (Fig. 2A) is comparable to published spectra of amorphous calcium phosphate [7,10] while that of Neotrigonia (Fig. 2B) is most similar to the spectrum of an amorphous mixed calcium ortho-pyrophosphate [10].

The electron microprobe analyses (Table 1) gave low total counts and appear to have been affected by the small size and labile nature of the gill supports. Despite this the Ca/P ratio of Neotrigonia is lower than that of Anodonta, which is consistent with the differences detected in the x-ray and infra-red studies. A further obvious difference between the two genera is that Anodonta contains significant amounts of Mn while in Neotrigonia there are concentrations of Mg (Table 1). The reduced unit cell size of the whitlockite, produced on heating of the gill supports of Anodonta, probably at least partly reflects incorporation of the Mn into the crystal structure [11,12].

Fossil bivalves. In both the fossil trigonioid and the fossil unionoid the x-ray diffraction trace showed that the gill supports were preserved as the carbonate fluorapatite, francolite. The unit

TABLE 1

Average compositions (element weight %) of phosphatic gill supports in living (Anodonta and Neotrigonia) and fossil (Laevitrigonia) bivalves. Based on electron microprobe analyses.

Element	Anodonta	Neotrigonia	Laevitrigonia
Ca	25.76	10.10	34.08
Mg	0.26	4.23	0.06
Mn	5.41	0.02	0.02
Fe	0.71	0.17	1.64
P	14.85	9.39	13.30
S	0.14	5.16	0.37
F	0.76	0.70	2.81
Cl	0.05	0.19	0.10

cell dimensions were estimated as $a = 9.336 \pm .006$ $c = 6.893 \pm .005$ for the Laevitrigonia and $a = 9.337 \pm .005$ $c = 6.886 \pm .004$ for the Unio sp. The microprobe analyses of Laevitrigonia (Table 1) show a higher Ca/P ratio than either of the extant genera. The fossil material contains negligible amounts of both Mg and Mn.

DISCUSSION

The above results confirm Janssens' [4] conclusion that the inorganic deposits in the gill of Anodonta are a calcium phosphate and show in addition that this material is in an amorphous condition during life. They also confirm and amplify Lowenstam's [8] recognition of amorphous calcium phosphate in the gills of Neotrigonia.

In synthetic preparations [7] the stability of amorphous calcium phosphate is enhanced by the presence of either or both pyrophosphate and magnesium and thus the concentrations of these in Neotrigonia are consistent with the amorphous nature of its gill support material. Freshwater bivalves such as Anodonta are well known for concentrating manganese in gill and other tissues [13,14,15] and it is possible that in the unionoids Mn replaces Mg as a stabilising influence. This may in part reflect the higher ratio of Mn/Ca and lower ratio of Mg/Ca in freshwater compared to seawater [16,17] but a recent study of Anodonta [15] has suggested that the concentrations of Mn are due to the action of symbiotic bacteria in the gill tissues. This raises intriguing questions - firstly does Neotrigonia also have symbiotic bacteria in its gills and secondly are bacteria also involved in the production of the phosphate? A test culture from gill tissues of Anodonta failed to find any evidence of bacterial phosphate activity but the possibility deserves further investigation. The ability of trigonioid and unionoid bivalves to secrete phosphatic deposits and their role in the phosphorous cycle particularly in fresh waters perhaps deserves more attention than it has so far received. As filter feeders their extraction and accumulation of phosphate and of metals might be of importance particularly in polluted situations.

The maturation of amorphous calcium phosphate into crystalline hydroxyapatite is well known both in synthetic preparations and in some biological situations [5,6,7]. Though the amorphous material of the gill supports appears to be stable in life its varied responce to different treatments (Fig. 1) indicates that it will probably crystallise readily and rapidly in the diagenetic environment, provided that that is not too acidic [18]. It is to be expected that the amorphous phosphate will convert either directly, or through a hydroxyapatite intermediate stage, to the geologically stable francolite [19], which is the preserving medium of the recorded fossil occurrences. The microprobe analyses (Table 1) show that in the process the material becomes relatively richer in Ca and F and that the metals which act as crystal poisons are released. It is thus possible that Mg rich pore fluids might retard or inhibit the process but the preservation of gill supports would otherwise be a probable, and possibly common, phenomenon wherever trigonioid or unionoid shells containing soft tissues are being buried and becoming fossilised. Though they have gone almost unremarked by palaeontologists the preservation of gill supports is the molluscan equivalent of the better known preservation of spiral brachidia in the spiriferid brachiopods. As such they are of great potential importance to palaeontologists. Firstly, studies of their occurrence as fossils will shed light on the processes involved in the preservation of trigonioid and unionoid bivalves in particular and on the taphonomy of bivalves in general. Secondly as pointed out by Atkins [2] the similarity of the gill supports of trigonioid and unionoid bivalves is of phylogenetic importance since it concords with other evidence that the two groups may be closely related [20]. Detailed studies of the fossil gill supports will provide a unique opportunity to reconstruct the form and character of the gills of extinct bivalves, to trace the evolution of the gills in the trigonioid and unionoid bivalves and to elucidate the relationships between these two groups.

Acknowledgements. I am greatly indebted to the Museum of Victoria for providing the specimens of *N. margaritacea* and to Dr T Darragh and Ms S Boyd for their help in obtaining them. Prof DA Spears, Dr R Kanaris-Sotiriou and Dr FGF Gibb (Sheffield) gave help and advice with specimen preparation and analysis and Ms D Hall and Mr A Saxby gave technical assistance. Dr M Wainwright (Sheffield) carried out the bacterial cultures. JK Whyte, CL Whyte and A Whyte gave invaluable help with specimen collection. I am also grateful to Dr RA Hewitt for comments on and much valuable discussion of phosphates.

REFERENCES

1. Whyte MA in preparation.
2. Atkins D (1937) Q J Microscop Sci 79: 375-421.
3. Ridewood WG (1902) Phil Tran R Soc Lond B194: 147-284.
4. Janssens F (1893) La Cellule 9: 7-91.
5. Eanes ED Gillessen IH, Posner AS (1965) Nature 208: 365-367.
6. Glimcher MJ (1984) Phil Trans R Soc Lond B304: 479-508
7. LeGeros RZ, LeGeros JP (1984) In: Nriagu JO, Moore PB (eds) Phosphate Minerals. Springer-Verlag, Berlin, pp.351-385.
8. Lowenstam HA (1792) Chem Geol 9: 153-166.
9. Whyte MA (1988) Crustaceana 55: 219-224.
10. Termine JD, Lundy DR (1974) Calc Tissue Res 15: 55-70.
11. Nord AG (1983) N Jb Min Mh 1983: 489-497.
12. Roots M, Lindqvist B (1988) Geol For Stock Forh 110: 81-82
13. Bradley HC (1907) J Biol Chem 3: 151-157.
14. Dubuisson M, Van Heuverswyn J (1930) Arch Biol 41: 37-74.
15. Lautie N, Carru AM, Truchet M (1988) Malacologia 29: 405-417.
16. Culkin F, Cox RA (1966) Deep Sea Res 13: 789-804.
17. Turekian KK (1969) In: Wedepohl KH (ed) Handbook of Geochemistry Vol. 1. Springer-Verlag, Berlin, pp. 297-323.
18. Nancollas GH (1984) In: Nriagu JO, Moore PB (eds) Phosphate Minerals. Springer-Verlag, Berlin, pp.351-385.
19. McClellan GH (1980) J Geol Soc 137: 675-681.
20. Newell ND, Boyd DW (1975) Bull Am Mus Nat Hist 154: 57-162.

CHAPTER 4.16

Orientation of Apatite and the Organic Matrix in Lingula Shells

M. IIJIMA[1], Y. MORIWAKI[1], and Y. KUBOKI[2]

[1]Asahi University School of Dentistry, Gifu, 501-02 Japan and [2]Hokkaido University, School of Dentistry, Sapporo, 060 Japan

Key Words : Lingula shell, apatite, chitin, orientation , protein

INTRODUCTION

Lingula (Brachiopoda, Inarticulata) is a primitive animal which developed some 500 million years ago [1]. The shell is composed of alternating mineralized and the organic layers. The shell is unique, since apatite crystals grow in the organic matrix with a preferred orientation forming alternations of mineralized and the organic layers. $CaCO_3$ crystals grow into various morphology depending on the organic matrix in molluscan shells. Roles of the organic matrix on the regulation of the oriented and uniform growth of $CaCO_3$ crystals have been studied extensively [2-8]. In the case of Lingula shells, the mechanism of orientation of apatite crystals is still unclear. To clarify the mechanism, the structural relationship between apatite crystals and the organic matrix and the differences in the organic component between the mineralized and the organic layers are investigated.

MATERIALS AND METHOD

Lingula unguis and Lingula shantoungensis (Fig.1) were obtained live in Sumiyoshi Sea, Japan. Soft tissue was removed from the shells and the shells were air dried. Fracture of shells were observed by a scanning electron microscope [9].

<u>Analysis of the Orientation of Apatite and the Organic Matrix.</u>

The orientation of apatite and the organic matrix was analyzed by means of a micro-beam X-ray diffractometer using untreated and decalcified shells [10]. Air dried shells, transversal and vertical cross sections of the shells were used to get three dimensional information. X-ray diffraction photographs of the

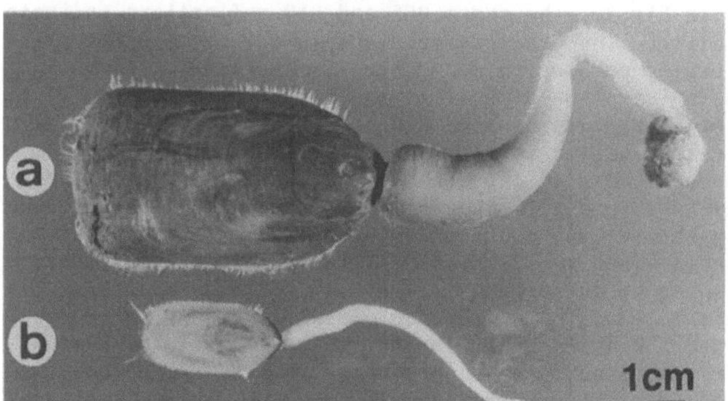

Fig.1 (a) Lingula shantoungensis and (b) Lingula unguis.

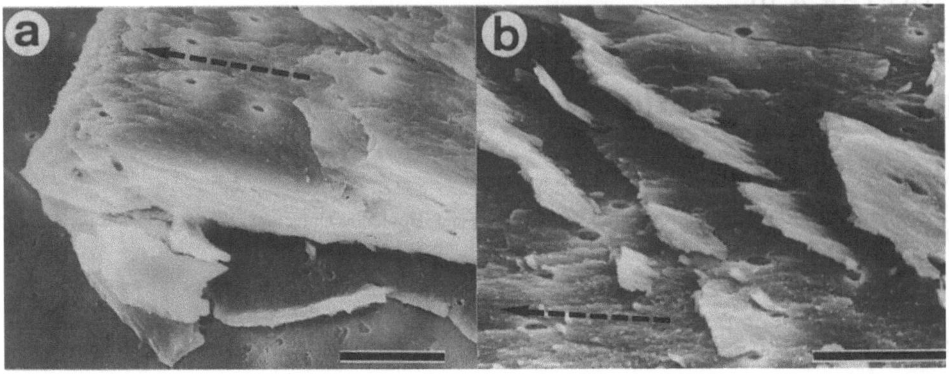

Fig.2 Fracture of Lingula unguis shell. (a) cross section and (b) surface of the mineralized layer. The broken lines indicate the growth direction of the shell. Scale line = 10μm.

mineralized and the organic layers of untreated and decalcified shells were taken using collimator with 30μm diameter.

Analysis of the Organic Components of the Mineralized and the Organic Layers.

The periostracum was removed from shells. The shells were ground to powder (< 75μm). The mineralized and the organic layers were separated by flotation in $CHBr_3+C_6H_6$; the mineralized layers = dense fraction (2.5 < d < 2.7) and the organic layers = light fraction (1.5 < d). The amino acid and hexosamine compositions were analyzed using a standard five-lithium-buffer system designed for physiological fluid analysis [11]. Proteins extracted by EDTA and guanidine (pH7.4) were analyzed by a gel filtration HPLC.

Results

Fractured shell of Lingula unguis showed a sheet-like texture (Fig.2), indicating that the mineralized layer is composed of oriented materials [9].

Figure 3 represents the X-ray diffraction patterns of the shell plane, transversal sections, and vertical sections of untreated and decalcified Lingula unguis shell [10]. The shell plane (Fig.3a) and the transversal section (Fig.3b) showed ark-like and intensified 002 and 300 reflections of apatite. While, the vertical section (Fig.3c) showed ring-like and intensified 300 reflection but weakened 002 reflection. These diffraction patterns indicate that the c-axis of apatite has a preferred orientation that arranges in the growth direction of the shell(Fig.4,[9]). Decalcified shells showed the oriented X-ray diffraction patterns of β-chitin (Fig.3d,e,and f). In Fig.3e, the sharp 002 and 013 reflections indicate the c-axis direction of β-chitin. The broad reflections on the equatorial line are 010 and 100. Chitin molecules combine to form fiber in the c-axis direction. The combined analysis of the orientation of apatite and β-chitin indicates that the c-axis of apatite and the fiber axis of β-chitin are parallel and both axes arrange along the growth direction of the shell.

Amino acid and hexosamine compositions of the mineralized layers were different from those of the organic layers (Table 1,[11]). Some differences between species were also observed. In both shells, the amounts of aspartic acid + asparagine, threonine, phosphoserine, serine, and glutamic acid+glutamine of the mineralized layers were larger than those of the organic layers. Lingula unguis shell contains a large amount of alanine and the difference of the glycine/alanine molar ratio between the mineralized and the organic layers is larger that of Lingula shantoungensis. The organic layers contains both glucosamine and galactosamine. Galactosamine was not detected in the mineralized layer. The molar ratio of glucosamine/amino acid of the mineralized layer was larger than that of the

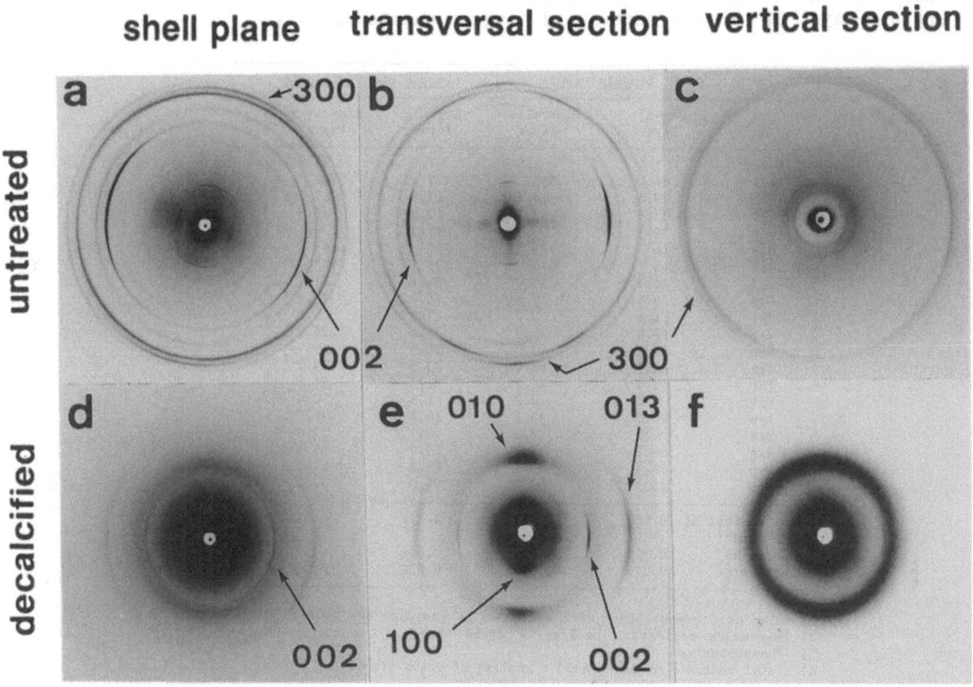

Fig.3 Micro-beam X-ray diffraction patterns of shell plane (a,d), transversal sections (b,e), and vertical sections (c,f) of untreated (a,b,c) and decalcified (d,e,f) shell of Lingula unguis.

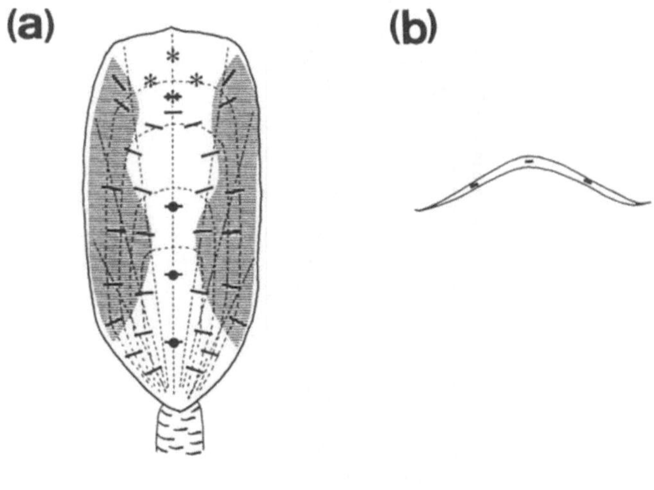

Fig.4 Orientation of apatite crystals in (a) shell plane and (b) cross section of Lingula unguis shell. ▬, ✱, and ● represent the direction of the c-axis of apatite, disoriented part, and the part with strong organic reflection, respectively. The hatched region is the highly oriented part.

Table 1 Amino acid[1)] and hexosamine[2)] compositions of the ineralized and the organic layer of Lingula unguis and Lingula shantoungensis.

	Lingula unguis		Lingula shantoungensis	
	Organic layer	Mineralized layer	Organic layer	Mineralized layer
P-ser[3)]	1.1	4.7	1.5	3.2
Asx	86.4	105	107	122
Thr	29.8	41.1	47.5	47.8
Ser	33.8	53.2	40.3	65.3
Glx	55.3	76.9	51.5	76.1
Gly	146	173	200	185
Ala	346	210	135	128
Val	46.1	15.9	50.1	21.9
Cys	9.1	18.5	14.1	13.9
Met	4.6	0.0	7.9	5.8
Ile	12.8	22.3	23.2	25.9
Leu	27.2	43.6	50.9	49.1
Tyr	25.9	28.0	23.8	23.2
Phe	18.9	28.9	34.3	36.6
Hyl	0.0	0.0	2.6	0.0
Lys	20.0	33.5	40.5	38.0
His	6.5	7.4	10.7	13.0
Arg	68.2	57.7	55.0	51.5
Hyp	16.9	20.8	27.2	21.1
Pro	45.4	60.7	77.0	72.0
Total content of amino acid[4)]	6.140	0.200	4.319	0.154
glucosamine	0.335	0.089	0.660	0.121
galactsamine	0.015	<0.001	0.032	<0.001

1) Amino acid compositions were expressed as residues per 1000 residues.
2) Hexosamine contents : μmol/mg of dried tissue.
3) Phosphoserine
4) Total contents of amino acid : μmol/mg of dried tissue.

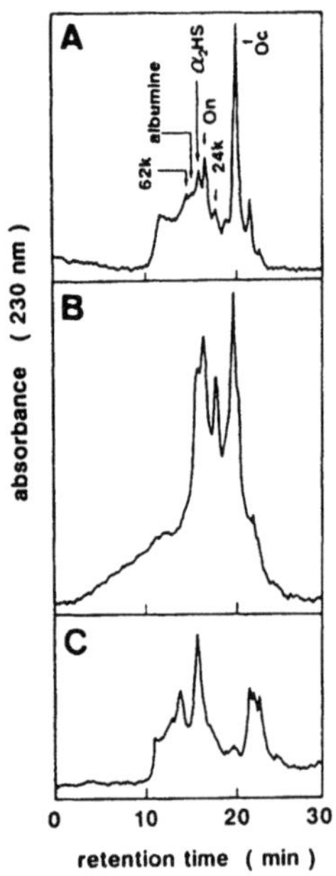

Fig.5 HPLC patterns of the G-extract of the mineralized layer (= C) and the organic layer (= B) of Lingula shantoungensis shell. The upper one (= A) represents the EN-extract of bovine bone.

organic layer in both shells.
Figure 5 represents the HPLC patterns of the G-extract of the Lingula shantoungensis shell. The HPLC pattern of the G-extract of the mineralized layer was different from that of the organic layer ; The G-extract of the mineralized layer was composed by proteins with higher molecular weight than that of the ohe organic layer.

Discussion

The present study demonstrates a relationship between the orientation of apatite and chitin ; the chitin fibers are almost parallel to the long axis of the needle-like apatite crystals (= c-axis direction) and the composition of the protein which bind to the insoluble organic components (chitin and insoluble protein) of the mineralized layers are different from that of the organic layers. The amino acid composition suggests that more acidic protein(s) distributes in the mineralized layer than in the organic layer. Since chitin is composed of glucosamine, the glucosamine/animo acid molar ratio suggests that the chitin/protein ratio of the mineralized layer is higher than that of the organic layer. In some of molluscan shells, there is a well-defined crystallographic relationship among chitin, oriented protein, and $CaCO_3$ crystal ; chitin fiber //a-axis of aragonite, protein chain//b-axis of aragonite [6-8]. These studies suggest that the insoluble organic matrix acts as an oriented template and the soluble protein(s) adsorbed on a rigid matrix acts as active nucleation sites [12-15]. Since chitin exists as a chitin-protein complex in most cases [16], chitin and acidic protein(s) are expected to assists the oriented growth of apatite in Lingula shells. Further analysis of the protein is now on going, however, the present study suggests a mechanism close to that of the mineralization of the $CaCO_3$ shells, i.e., 1) chitin acts as an oriented framework and influences the orientation of apatite crystals, and 2) acidic protein(s) which bind to chitin have some relation to the oriented precipitation of apatite. Galactosamine was not detected in the mineralized layers. This may in part explain the lack of mineralization in the organic layers.

Acknowledgment

The authors appreciate to Dr.Koji Wada and Mr.Yasuo Yamashita for collection of Lingula unguis and Lingula shantoungensis, and to Dr. Yuko Takita and Dr. Youko Oonuma for the biochemical analysis of the specimen.

References

1. Chapman F (1914) J R Micro Soc 5 : 28-31
2. Erben HK, Watabe N (1974) Nature 248 : 128-130
3. Weiner S, Hood L (1975) Science 190 : 987-988
4. Towe KM, Hamilton GH (1968) Calcif Tissue Res 1 : 306-318
5. Bevelander G, Nakahara H (1969) Calcif Tissue Res 3 : 84-92
6. Weiner S, Traub W (1980) Fed Eur Biochem Soc 111 : 311-316
7. Weiner S, Talmon Y, Traub W (1983) Int J Biol Macromol 5 : 325-328
8. Weiner S (1984) Am Zool 24 : 945-951
9. Iijima M, Moriwaki Y, Doi Y, Kuboki Y (1988) Jpn J Oral Biol 30 : 20-30
10. Iijima M, Moriwaki Y (1990) Calcif Tissue Int 47 : 237-243
11. Iijima M, Moriwaki Y, Takita Y, Kuboki Y (1990) Comp Biochem Physiol(in press)
12. Crenshaw MA (1972) Biomineral Res Rep 6 : 6-11
13. Weiner S (1979) Calcif Tissue Int 29 : 163-167
14. Greenfield EM, Wilson DC, Crenshaw MA (1984) Am Zool 24 : 925-923
15. Addadi L, Weiner S (1985) Proc Natl Acad Sci 82 : 4110-4114
16. Rudall KM (1963) Adv Insect Physiol 1 : 257-313

CHAPTER 4.17

Evolution of Fluoride and Iron Concentrations in the Enameloid of Fish Teeth

S. SUGA[1], Y. TAKI[2], K. WADA[3], and M. OGAWA[1]

Department of Pathology, The Nippon Dental University, Tokyo[1]; Tokyo University of Fisheries, Tokyo[2]; and National Institute of Aquaculture, Mie[3], Japan

Key words: Enameloid, Fluoride, Iron, Silicon, Evoution, Fish.

INTRODUCTION

It is well known that the fluoride has an ability to promote precipitation and growth of apatite crystal and to make its crystallinity higher due to its incorporation into crystal lattice. On the other hand, excess uptake of fluoride during tooth development interferes with the functions of the hard tissue forming cells, especially, the enamel forming cells, and, finally, induces hypomineralization of enamel (mottled enamel). Fluoride in the teeth, especially, in human enamel, under normal and various pathological conditions, has been examined by many investigators mainly in connection with dental caries [1, 2]. On the other hand, deposition of histologically detectable amount of iron in the enamel of amphibians, reptiles and some mammals and in the enameloid of some fishes has been known for many years. It has been considered that the iron minerals act as hardening agents on the surface of enamel and enameloid [3-5].

FLUORIDE AND IRON IN THE ENAMELOID OF BONY FISHES

The enameloid of fish teeth has been thought to be analogous to the enamel of teeth of tetrapods and is similarly very highly mineralized, although its organic matrix is composed of ectodermal protein (enamelin) and mesodermal collagen fibers, whereas the enamel of tetrapods is composed only of ectodermal proteins (amelogenin and enamelin). Our previous quantitative fluoride analysis made on the teeth of teleosts of bony fishes using the electron microprobe, which is an apparatus of non-destructive element analysis by measuring the characteristic X-rays emitted from very small region of specimens bombarded by the electron beam, about 1 - 1.5 µm in diameter, first pointed out that bony fishes can be divided into two groups; the fishes whose enameloid contains fluoride more than about 2.0% and those whose enameloid contains fluoride lower than about 0,4% (the dentin and bone of all the fishes examined contains less than 0.2% and 0.3%, respectively) [7]. In the human teeth, fluoride concentration is about 0.1% at the narrow surface layer and about 0.01% in the rest of enamel, whereas it is about 0.03% at the inner layer and about 0.38% at the pulpal layer, of the dentin [2]. Investigations made on the developing enameloid of the teleostean fishes by the same method showed that high fluoride concentration into the enameloid commences from the middle stage of mineralization and continues until the heavy mineralization is completed, whereas, in the adjacent dentin, its concentration is very low at the early stage and does not increase secondarily [8, 9].

According to many previous fluoride studies performed on the human enamel, it is generally considered that fluoride concentration in the sound enamel is closely related to that in the drinking water supplied during the tooth development. However, the fluoride concentration in the enameloid of bony fishes is independent of that in the environmental water [7], although marine water contains fluoride about 1.3 ppm over the world [10], whereas freshwater contains usually lower than 0.1 ppm. It is generally considered that fluoride higher than 1.0 ppm in drinking water has ability to induce hypomineralization of human enamel (so-called mottled enamel), when it is supplied during the tooth development continuously.

It is known that the teleostean fishes of the Perciformes contain a considerable number of freshwater species in addition to the predominant marine forms. Electron microprobe analysis indicated that all the 78 species collected taxonomically contains fluoride in the enameloid mainly more than 2.0%, whereas it is less than 0.6 % in the dentin and bone. The fluoride concentration in the enameloid showed no significant difference when related to the salinity of the environmental water of these fishes, nor between zoogeographical positions in the case of freshwater forms, i.e., primary, secondary, and vicarious freshwater species (Fig. 5a,b) [11].

Another investigation using the same method performed on the teeth of the marine teleosts of the order Tetraodontiformes which is a derivative of the order Perciformes indicated that the fluoride concentration in the enameloid of all the fishes examined from the primitive suborder Balistoidei was higher than 2.31% (Fig. 1-1A) whereas that in the enameloid of the fishes of the advanced suborder Tetraodontoidei was lower than 0.22%. Furthermore, the enameloid of the fishes of the Acanthuridae of the order Perciformes, from which the Tetraodontiformes may have been derived, contained fluoride higher than 3.85 per cent [12]. These facts seem to suggest that the fluoride concentration in the enameloid is related to the phylogeny of fishes rather than to the fluoride concentration in the environmental water and also suggest that a peculiar mechanism to concentrate the fluoride into the developing enameloid may have been lost in the course of fish evolution (Fig. 2).

There was also no correlation between the fluoride concentration in the enameloid and the shape and size of teeth which show diverse morphological adaptation for feeding habits. Therefore, it is speculated that the evolution of the mineral composition of fish enameloid is independent of that of morphology of teeth [8, 11]. Crystallographic investigation performed on the enameloid containing various amount of fluoride showed that fluoride incorporation resulted in increase in crystal size, a reduction in the carbonate content, a systemic decrease in the a-axis lattice parameter [12] and in the mode of crystal growth [23].

In the process of our fluoride studies, it was found that the high fluoride enameloid of a marine teleost fish, *Pagrus major* of the Perciforms, also contains high iron throughout the entire layer, more than 2.2% at the surface layer [9], whereas the low fluoride

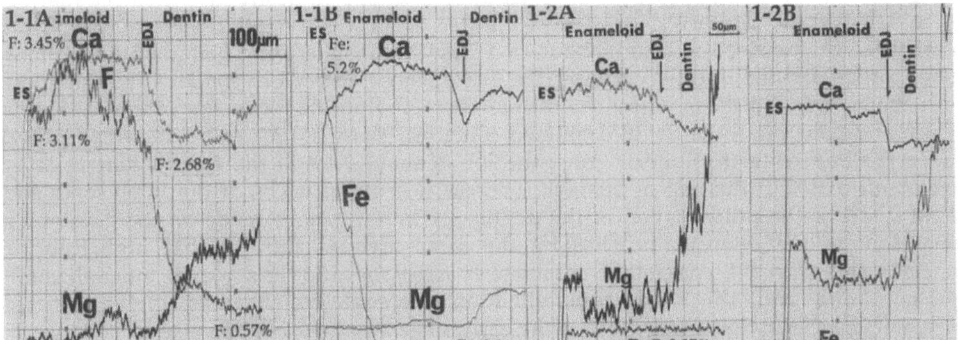

Fig. 1: The distributions of fluoride (F), iron (Fe), calcium (Ca) and magnesium (Mg) in the teeth of two teleosts, revealed by the line scan analyses of the electron microprobe. A shows the distributions of F, Ca and Mg, and B the distributions of Fe, Ca and Mg. (1) *Ostracion immaculatus* of the Balistoidei of the Tetraodontiformes (marine water), (2) *Carassius auratus* of the Cypriniformes (freshwater). ES-enameloid surface, EDJ- enameloid-dentin junction.

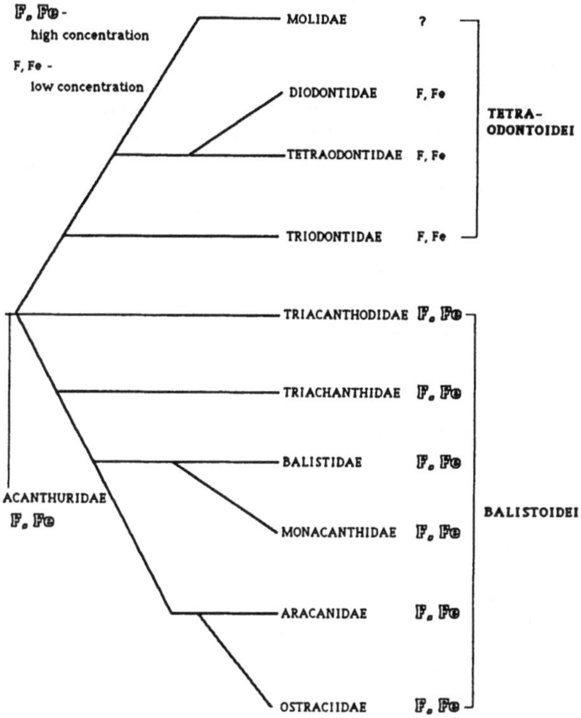

Fig, 2: The results of quantitative analyses of fluoride [12] and iron [15] by means of the electron microprobe performed on the enameloid of fishes of the Tetraodontiformes were plotted on the phylogentic tree proposed by Tyler [27] based upon osteological investigations .

enameloid of a freshwater teleost fish, *Cyprinus carpio* of the Cypriniformes, contains only about 0.2% at the very narrow surface layer [9] and human enamel contains about 0.02% [14]. Previous histochemical investigations indicated that the iron is deposited as a thin layer overlying the enameloid [3, 4] or superficially within the layer [5]. However, our investigation made using the electron microprobe indicated that, in the high iron enameloid, the iron is deposited throughout the entire layer and its highest concentration is observed at the surface or middle layers. Iron deposition to the developing enameloid commences from the surface layer towards the inner layer from the middle stage of mineralization although its timing of appearance and distribution pattern are not identical with those of fluoride [9].

Iron analysis made on the the enameloid of tetraodontiform fishes showed that the iron concentration ranges from 0,4% to 13,5% at the surface layer and is distributed through the entire layer, of the enameloid of the primitive suborder Balistoidei (Fig. 1-1B), whereas it is so low as could not be discriminated from the background value of the emission intensity in the enameloid of the advanced suborder Tetraodontoidei. The enameloid of fishes of the Acanthuridae also contains very high value of iron (2.7%- 3.9%). Thus it appears that the iron concentration in the enameloid is also related to the fish phylogeny and the mechanisms of iron concentration into the developing enameloid has been lost during evolution as is the case with fluoride [15] (Fig. 2). However, the biological significance of the concurrent deposition of fluoride and iron which have originally no chemical correlation is still wrapped in mystery. Reasonable biological and chemical interpretations are required for such peculiar phenomena appearing during heavy mineralization in the enameloid.

FLUORIDE AND IRON IN THE ENAMELOID OF CARTILAGINOUS FISHES.

It has been pointed out that the enameloid of sharks belonging to cartilaginous fishes is composed of fluorapatite which contains fluoride as high as about 3.5 % by X-ray diffraction [16], by X-ray diffraction and chemical analysis [17], by chemical analysis [18] and by the electron microprobe element analysis [7, 19]. Quantitative point and line scan analyses by the electron microprobe made on the developing shark enameloid showed that fluoride incorporation into the enameloid matrix progresses continuously from the very early stage until the late stage, of mineralization [19], whereas, in the teleost fish enameloid, it commences from the middle stage [8, 9].

However, a fluoride analysis of the enameloid of cartilaginous fishes in connection with the fish phylogeny and the environmental water has been not made yet. Furthermore, no investigation has been made on the iron in the enameloid of sharks and rays.

In the present study, the electron microprobe quantitative analysis of fluoride and iron

Table 1. Fluoride concentration in the teeth of cartilaginous fishes (%)

Species	Enameloid	Dentin
(SHARKS)		
Heterodontiformes		
Heterodontus japonicus	4,23 ± 0.21	0.42 ± 0.03
Hexanchiformes		
Chlamydoselachus anguineus	4.42 ± 0.20	0.32 ± 0.01
Heptranchias perlo	4.87 ± 0.11	0.51 ± 0.03
Lamniformes		
Caleocerdo cruvier	4.55 ± 0.25	0.52 ± 0.08
Prionace glauca	3.46 ± 0.12	0.21 ± 0.03
Carcharhinus japonicus	3.19 ± 0.13	0.31 ± 0.02
Carcharodon carcharias	6.30 ± 0.14	0.45 ± 0.12
Isurus oxyrinchus	4.89 ± 0.25	0.47 ± 0.05
Rhincodon typus	3.05 ± 0.12	0.52 ± 0.04
Scapanorhynchus owstoni	3.49 ± 0.23	0.25 ± 0.09
Squaliformes		
Centrophorus atromarginatus	3.96 ± 0.26	0.32 ± 0.11
Somniosus pacificus	5.76 ± 0.24	0.31 ± 0.02
Dalatias licha	4.22 ± 0.17	0.57 ± 0.08
(RAYS)		
Rajiformes		
Dasyatis akajei	2.75 ± 0.25	0.20 ± 0.03
Myliobatis tobiei	3.12 ± 0.07	0.27 ± 0.05
Disceus thayeri *	2.33 ± 0.15	0.19 ± 0.03
*Potamotrygon motro**	2.52 ± 0.31	0.21 ± 0.01
(CHIMAERAS)	(Pleromin)	
Chimaera phantasma	0.17 ± 0.03	0.19 ± 0.02
Hydrolagus alberti	0.16 ± 0.02	0.19 ± 0.01
Harriotta raleighana	0.18 ± 0.04	0.19 ± 0.02

* Freshwater stingrays obtained from Amazon basin near Manaus, Brazil (Fluoride concentration in the environmental water: 0.04 ppm).

1) Standard sample for quantitative fluoride analysis: Fluorapatite (Durango, F: 3.5%)

2) Cartilage of all the sharks and rays examined contains fluoride about 0.2%

was made on the teeth of 13 species of the shark, 2 species of the rays and 3 species of the chimaeras, obtained from marine water (Table. 1). The chimaeras possess very highly mineralized mesodermal tissues, so-called pleromin, in the axis of the tooth plate which is made of bone, whereas sharks and rays posses a highly mineralized enameloid layer in the coronal region of the teeth.

The quantitative point and line scan analyses indicate that the enameloid of all the sharks examined belonging from the primitive Heterodontiformes and Hexanchiformes to the advanced Squaliformes contains fluoride ranging from 3.05% to 6.30% and the rays ranging from 2.75% to 3.12%, at the surface layer, whereas the dentin contains fluoride only ranging from 0.20% to 0.57%. However, the pleromin and bone of the tooth plates of the chimaeras contains only from 0.16% to 0.18% and about 0.19% respectively (Table 1). In the shark and ray enameloid, the fluoride concentration which is highest at the surface layer decreases towards the enameloid-dentin junction (Fig. 4). Excess concentration of fluoride in the shark enameloid as compared with theoretical value of fluoride in fluorapatite seems due to deposits in a form of calcium fluoride or the absorption on the surface of fluorapatite crystals.

The teeth of two freshwater stingrays, *Disceus thayeri* and *Potamotrygon motoro* obtained from Amazon basin near Manaus, Brazil (the environmental water contains only 0.04 ppmF), were also examined. They contain fluoride 2.33% and 2.52% at the surface layer of the enameloid and 0.19% and 0.21% in the dentin, respectively. There is no significant difference between these results and that obtained from a marine water stingray *Dasyatis akajei* which is very close taxonomically to these freshwater stingrays (Table 1) (Fig. 4a, b). Present results suggest that the fluoride concentration in the enameloid of sharks and rays is also independent of the environmental water as observed in the bony (teleostean) fishes.

According to the iron analysis performed on the same specimens indicate that the iron concentration is very low as its emission intensity can not be discriminated from the background value in the enameloid, dentin and bone of all the sharks and rays and in the pleromin and bone of chimaeras.

In the bony fishes, as far as we examined already, the enameloid containing high fluoride is always accompanied by a relatively high iron concentration. However, the high fluoride

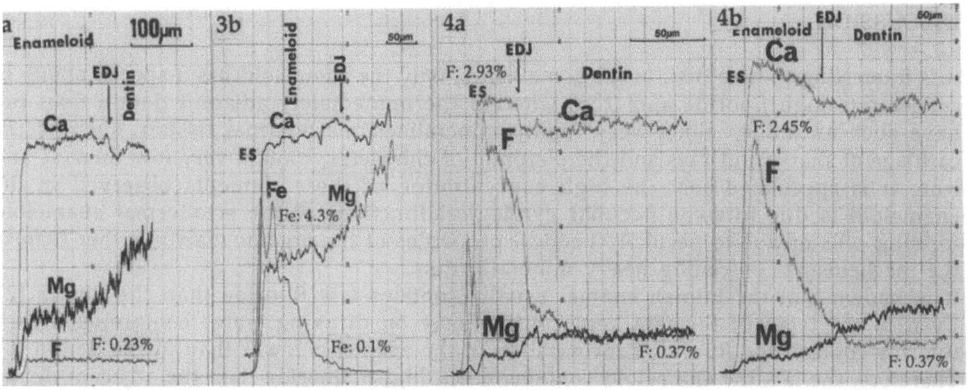

Fig. 3: The distribution of fluoride (F), iron (Fe), calcium (Ca) and magnesium (Mg) in the teeth of a fresh water teleost, *Leporinus friderici*, of the Characiformes, revealed by the line scan analysis by means of the electron microprobe. **a** shows the distribution of F, Ca and Mg, **b** the distribution of Fe, Ca and Mg. ES- enamelod surface, EDJ- enameloid-dentin junction.

Fig. 4: Comparison of the distribution of fluoride (F), calcium (Ca) and magnesium (Mg) in the teeth between the marine water and fresh water stingrays. **a** A marine water stingray, *Dasyatis akajei*, **b** a fresh water stingray, *Potamotrygon motro*. ES- enameloid surface, EDJ- enameloid-dentin junction.

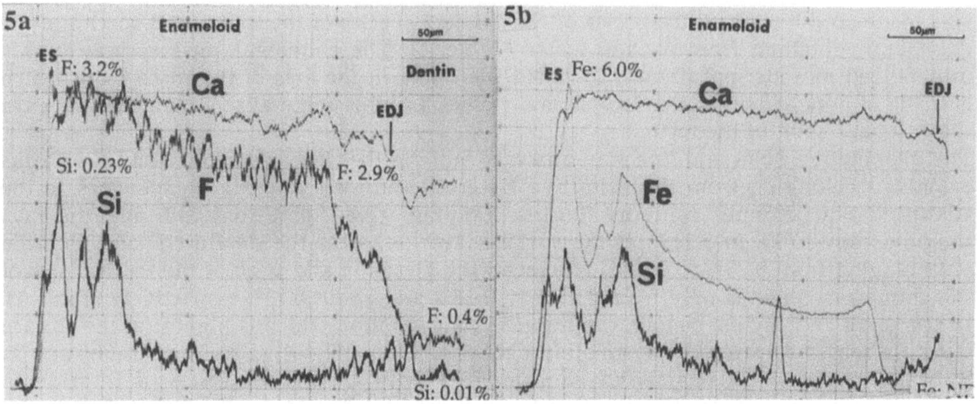

Fig. 5: The distribution of fluoride (F), iron (Fe), calcium (Ca) and magnesium (Mg) in the teeth of a freshwater teleost, *Oreochromis niloticus*, of the Cichlidae of the Perciformes, revealed by the line scan analysis by means of the electron microprobe. **a** shows the distribution of F, Si and Ca, **b** the distribution of Fe, Si and Ca. ES- enameloid surface, EDJ- enameloid-dentin junction.

enameloid of the sharks and rays is not accompanied by a high iron concentration. These facts seem to suggest that the deposition of fluoride and iron into the developing enameloid are conducted by the different biological mechanisms between the bony and cartilaginous fishes.

SILICON IN THE ENAMELOID OF BONY FISHES

Preliminary investigation [28] indicated that, in the teeth of some fresh water teleosts of the Cichlidae of the Perciformes, silicon is also more highly concentrated in the enameloid (about 0.3%) than that in the dentin of the same teeth (about 0.03%) and human enamel (about 0.02%) [29], in a distribution pattern similar to the iron concentration (Fig. 5).

DISCUSSION

It is interesting to note that, in the some fishes, only the enameloid has a special ability to concentrate high fluoride and iron, whereas the immediately adjacent dentin does not have such an ability. Since very highly mineralized mesodermal tissues, such as the cartilage of sharks and rays and the pleromin of chimaeras, contain very low fluoride and iron, it is speculated that the high concentration of these elements observed in the enameloid is due to some peculiar cytological functions of the ectodermal enameloid forming cells or due to peculiar chemical properties of the organic matrix, rather than to the mechanisms controlling heavy mineralization.

It is known that the human enamel usually contains less fluoride than the dentin [2]. Even in the teeth of the rats kept continuously on drinking water containing a high fluoride content, the fluoride concentration in the enamel is lower than in the dentin [8]. Therefore, the fact the enameloid contains much higher fluoride than the adjacent dentin suggests that the enameloid plays some unknown special biological role.

The enamel and enameloid have been generally thought to be very highly mineralized to protect the teeth from the very strong mechanical stresses which occur when the teeth are used as a tool for mastication. Since the fluoride incorporated into the crystal lattice of apatite makes crystallinity of apatite higher [13], it is supposed that the high concentration of fluoride in the enameloid is a sort of adaptation for feeding habits. On the other hand, the iron deposition has been also thought to harden the enameloid or enamel and to reduces abrasion and cracking [3], so that its concentration in the enameloid was considered to be related to feeding habits of fishes [4]. Crystallographic investigations indicated that the presence of $Fe3+$ induces the formation of apatite of high crystallinity [20]. The Mossbauer effect study made on the enamel and dentin showed that iron is incorporated mainly as $Fe3+$, in a form similar to ferritin or iron (III) hydroxide oxide (FeOOH), and it is mainly attached to the surfaces of apatite crystals [21]. So that, the iron is considered to be deposited partly in the crystal lattice of apatite and mainly in the microspaces in the tissue, such as the enameloid tubules, whereas the fluoride is included into the crystal lattice of apatite.

However, our studies indicated that the concentration of fluoride and iron in the enameloid is not related to the feeding habits of fishes. The fishes of Perciformes whose enameloid contains a high fluoride and iron have diverse feeding habits, some fishes subsist on mollusks, crustaceans and echinoderms and some on vegetable foods, such as, algae and plankton [8]. On the other hand, all the fishes of the Tetraodntiformes subsist on almost the same foods, such as, crustacean and mollusks [22]. So that, a more reasonable interpretation on the significance of high fluoride and iron concentration into the enameloid is required.

Dual high concentration of these elements has been observed at the site of heavy mineralization in other biological systems such as radular teeth of chiton (mollusks) whose coronal portion is made by the epithelial cells [9, 24]. Furthermore, there is another interesting fact that radular teeth of the limpet (mollusks) contains a very high iron and silicon [30]. These facts suggest that concurrent high concentration of fluoride and iron or fluoride, iron and silicon in the enameloid has a special biological significance which is probably related to the evolution of the mechanisms of the body fluid control, that originated from nonvertebrates and is still present in the some vertebrates.

It is known that, in certain epithelial tissues (kidney, digestive glands and hepatopancreas) of mollusks, there appear highly insoluble intracellular granules containing calcium, magnesium, phosphate and carbonate ions and also small amounts of various other ions which include iron and silicon. There was no evidence for their recycling. Such a biomineralization has been thought as a cellular detoxification mechanism [25].

The very highly mineralized enameloid seemed to provide a suitable site where excess amounts of fluoride and iron in the body fluids are excreted in order to detoxify the fishes [9]. It is speculated that the enameloid forming epithelial cells (ameloblasts), especially during the mineralization stage, act also as an excretion gland as observed in the epithelial tissues of mollusks. Furthermore, the enameloid is considered to be a suitable tissue to fix more firmly the elements excreted, as compared with the dentin and bone, because the fully mineralized enameloid is chemically very stable and does not have contact with living cells after eruption. Fish teeth are continuously being replaced throughout the life span of the fish (polyphyodont). It is considered, therefore, that this peculiar property of enameloid enable the fishes to remove excess fluoride, iron and silicon continuously into their surrounding waters as stable chemical compounds which do not pollute their environment [9].

During our project of quantitative analyses of fluoride and iron in the enameloid of various teleost fishes, it was found that there are different combinations of their concentrations, namely, high fluoride and high iron (Figs. 1-1A, B, 5a, b), low fluoride and high iron (Fig. 3a, b), and low fluoride and low iron (Fig. 1-2A, B), however, the combination of high fluoride and low iron has been not found so far. The combination of high fluoride and low iron concentrations was observed only in the enameloid of sharks and rays. These facts seem to indicate that there is an order in the disappearance of concentration mechanisms of fluoride and iron into the developing enameloid in the course of teleostean evolution and there is a difference in the concentration mechanisms of these elements into the enameloid between bony fishes and cartilaginous fishes (sharks and rays).

Our preliminary studies made on the primitive bony fishes showed that high concentrations of iron in the enameloid are observable in the polypterids, gars and bowfin whereas a very high concentration of fluoride throughout the entire layer of enameloid, as seen in many teleosteans, is found only in the bowfin. In other words, the mechanism of high iron concentration can be traced back to the cladistian stage of actinopterygian evolution, while that of high fluoride concentration to the neopterygian stage. These facts indicate that the two mechanisms are independent of each other with different evolutionary backgrounds [31, 32].

Comparative fluoride analysis made on the many species of fishes, amphibians, reptiles and mammals using the electron microprobe indicate that a high concentration of fluoride as observed in the enameloid of some fishes was not found in the enamel of all the tetrapods examined. On the other hand, high depositions of iron were observed in the enamel of some amphibians, reptiles and lower mammals [26].

Therefore, it is considered that the high concentration of fluoride and iron (and also silicon ?) in fish enameloid is related to ancient body fluid control mechanisms which have become modified in the course of evolution of some fishes.

REFERENCES

[1] Brudevold F. and Soremark R. (1967): In: Miles AEW (ed) Structural and Chemical Organization of Teeth, Academic Press. Vol. 2, pp 247-277

[2] Weatherell J A, Deutsch D, Robinson C and Hallsworth AS (1977): Caries Res 11 (suppl): 85-115

[3] Shellis RP and Berkovitz B K B(1976): J Zool 180: 69-84

[4] Motta P J (1987): Can J Zool 65: 106-112

[5] Schmidt W J (1969): Z Zellforsch Mikrosk Anat 93: 447-450

[6] Lowenstam H A and Kirschvink J L (1985): In: Kirschvink J L, Jones D S and Mocfadden E J (eds) Magnetite Biomineralization and Magnetoreception in Organisms. Plenum Publishing. pp 3-15.

[7] Suga S, Wada K. and Ogawa M. (1980): In: Omori M. and Watabe N (eds) The Mechanisms of Biomineralization in Animals and Plants. Tokai Univ Press. Tokyo, pp 229-240

[8] Suga S, Wada K and Ogawa M (1981): In: Binder K and Hohenegger M (eds) Fluoride Metabolism. Verlag Whilhelm Maudrich, Wien, pp 79-88

[9] Suga S (1984): In: Fearnhead RW and Suga S (eds) Tooth Enamel. Elsevier Science Publishers BV, Amsterdam, pp 472-477

[10] Greenhalgh R and Riley J P (1963): Nature 197: 371-372

[11] Suga S, Taki Y and Wada K (1983): Japan J Ichthyol 30: 81-93

[12] Suga S, Wada K and Ogawa M (1981): Japan J Ichthyol 28: 304-312

[13] LeGeros R Z and Suga S (1980): Calcif Tissue Int 32: 169-174

[14] Torell P (1957): Odont Tidskr 65: 20-23

[15] Suga S, Wada K, Taki Y and Ogawa M (1989): J Dent Res 68; 1115- 1123

[16] Trautz DR, Klein E. and Addelston HK (1952): J Dent Res 31: 472-473 (Abst.)

[17] Glas J.E (1962): Odont. Revy 13: 315-326

[18] Buttner W (1966): Adv Fluorine Res 4: 193-200

[19] Suga S, Wada K and Ogawa M(1978): Japan J Oral Biol 20: 67-81 (in Japanese).

[20] LeGeros R Z, Taheri M H, Quirolgico C B and LeGeros J P (1980): Proc 2nd International Congress on Phosphorus Compounds, Boston. pp 89-103

[21] Bauminger E, Ofer S, Gedalia I, Horowitz G and Mayer I(1985): Calcif Tissue Int 37: 386- 389

[22] Hiatt R W and Strasburg D W (1960): Ecological Monographs 30: 65-127

[23] Aoba, T., Moreno, E. C., Shimoda, S., Miake, Y., Prostak, K. and Suga, S. (1990): This volume.

[24] Lowenstam H A (1967): Science 156: 1373-1374

[25] Simkiss K and Mason AZ (1983): In: Hochachka PW (ed) The Mollusca Vol 2. Environmental Biochemistry and Physiology. Academic Press. New York, pp. 101-164

[26] Suga S and Ogawa M (1989): In: Fearnhead RW (ed) Tooth Enamel V. Florence Publishers, Yokohama, pp. 353-357

[27] Tyler J C (1980): U.S. Dept. Commerce Nat Ocean Atomes Admin, Nat Mar Fish Serv, NOAA Tech Rep, NMFS Circ. 434, 422 pp

[28] Suga S, and Ogawa M (1990) J Dent Res 69: Special Issue: Abst. No. 927

[29] Curzon MEJ and Cutress T W (1983) Trace elements and dental disease. John Wright PSG Inc. Boston, pp. 38, 70, 80

[30] Runham NW, Thornton PR, Shaw DA and Wayte RC (1961): Z Zellforsch 99: 608-626

[31] Suga S and Taki, Y (1981): J Dent Res 60:Special Issue A Abst. No. 715

[32] Suga S, Taki Y and Ogawa M (1989): J Dent Res 68: Special Issue Abst. No.479

CHAPTER 4.18
Evolutionary Trends of the Tooth Structure in Chondrichthyes

M. Goto
Department of Anatomy, School of Dental Medicine, Tsurumi University, Yokohama, 230 Japan

Key words: Teeth, Dentine, Enameloid, Sharks, Chondrichthyes.

INTRODUCTION

The history of Chondrichthyes spans more than 370 million years, from their appearance in the Devonian seas to the present varied groups. Their tooth morphology and structure have been changing in various ways. The purpose of this study is to compare the morphology and structure of fossil and extant teeth in Chondrichthyes and examine their evolution and adaptation in relation to their phylogenies and food habits.

TOOTH STRUCTURE AND DEVELOPMENT OF SHARKS

A basic tooth type of Chondrichthyes is observed in the frilled shark, Chlamydoselachus anguineus (Fig.1). The tooth is composed of three layers, that is, the outer enameloid layer, the inner dentine layer and the basal osseous tissue layer. The tooth support is by a fibrous attachment. The basal osseous tissue of the tooth is connected with the lamina propia mucosae of the oral mucous membrane by the bundles of collagenous fibres [1,2].
In a labio-lingual section of the jaw of sharks (Fig.2), the dental lamina invaginates from the oral epithelium toward a groove in the jaw cartilage, and tooth germs are formed at distal end. The growing tooth germ moves to the labial side, erupts at the position which the dental lamina invaginates and then becomes a functional tooth [3,4].
The hard tissue formation of shark teeth occurs in the following order: enameloid matrix formation, predentine formation, calcification of enameloid and dentine, and formation of

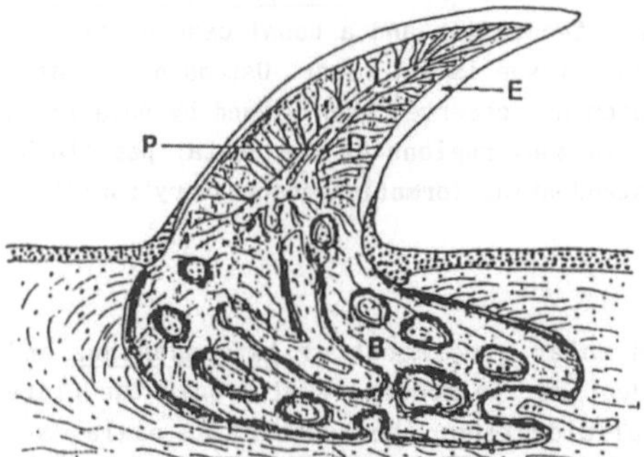

Fig.1 Labio-lingual section of a tooth of Chlamydoselachus anguineus. B:basal osseous tissue, D:dentine, E:enameloid, P:pulp

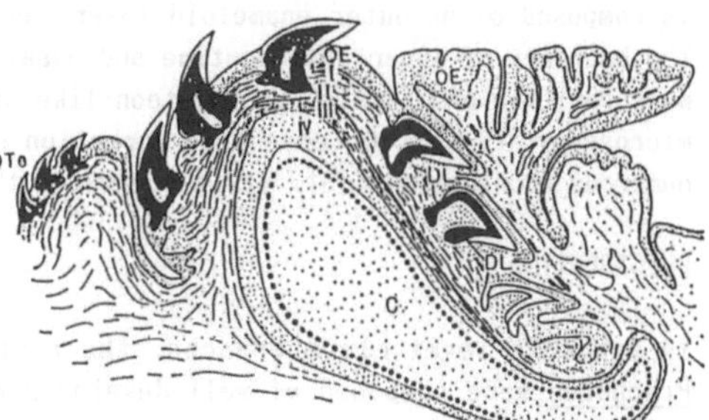

Fig.2 Labio-lingual section of a jaw of Triakis scyllia. C:jaw cartilage, DE:dental lamina, DTe:dermal teeth,OE:oral epithelium, I-IV:lamina propia mucosae

a basal osseous tissue. The enameloid is a highly mineralized tissue which is formed by both the epithelial and mesenchymal cells of the tooth germ [3,4].

ORTHODENTINE

The outer layer of the tooth of most cladodont level sharks was composed of orthodentine. The tooth of xenacanthoids, one of the Paleozoic cladodont level sharks, was composed of well developed orthodentine (Fig.4). In the modern level sharks, carcharhinoids, one of the meat-eating groups, have teeth which are composed of well developed orthodentine (Fig.3). Numerous dentinal tubules radiate from a central pulp cavity and growth lines are observed in the orthodentine. The boundary of the basal osseous tissue and the inner dentine layer is very clear in those shark teeth which are composed of well developed orthodentine.

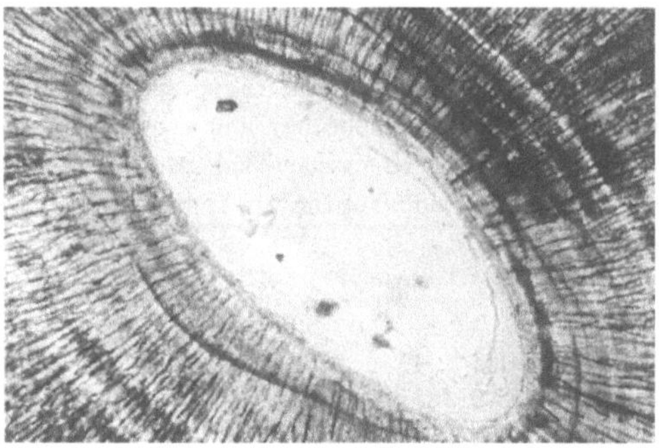

Fig.3 Horizontal section of a fossil tooth of Carcharhinus sp., Miocene, USA. ×58

Fig.4 Horizontal section of a fossil tooth of Xenachanthus texensis, lower Permian, USA. ×22

OSTEODENTINE

The inner layer of the dentine of some cladodont shark teeth was composed of osteodentine. Most of hybodont sharks had teeth which were composed of well developed osteodentine. In modern level sharks, only the tooth of lamnoid sharks, another one of meat-eating groups, is composed of osteodentine (Fig.5,6). The tooth of great white shark, Carcharodon carcharias, is composed of an outer enameloid layer, an inner osteodentine and a basal osseous tissue. The boundary of inner osteodentine and basal osseous tissue is not clear. Osteodentine has many cavities and canals. Many osteon-like structures are observed when viewed by polarizing microscopy (Fig.6). Osteodentine formation occurs in many regions of the dental papilla by numerous odontoblasts [5]. It is suggested that osteodentine formation occurs very rapidly.

PLICIDENTINE

In hybodont level elasmobranchs, the plate and molar type teeth of Asteracanthus and Ptychodus were composed of well developed plicidentine. In modern levels, only the plate type tooth of myliobatoid rays is composed of plicidentine. The tooth of an eagle ray, Myliobatis, is composed of an outer thin layer of enameloid, an inner plicidentine and a basal osseous tissue. Plicidentine is composed of many denteons, or dentinal osteons (Fig.7). Each denteon is a small area of orthodentine which has many dentinal tubules radiating from a central small pulp cavity (Fig.8). Plicidentine is formed by the folding and convolutions of the dental pulp.

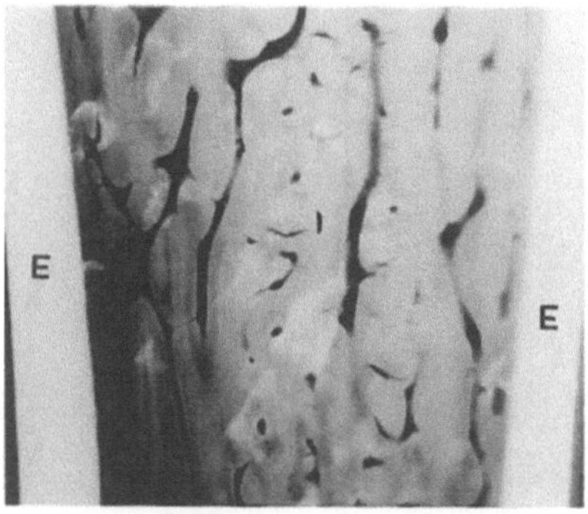

Fig.5 Microradiogram of labio-lingual section of a tooth of Carcharodon carcharias. E:enameloid, D:osteodentine. ×26

Fig.6 Horizontal section of a tooth of mako shark, Isurus oxyrinchus. Many osteon-like structures are observed in polarized light. ×60

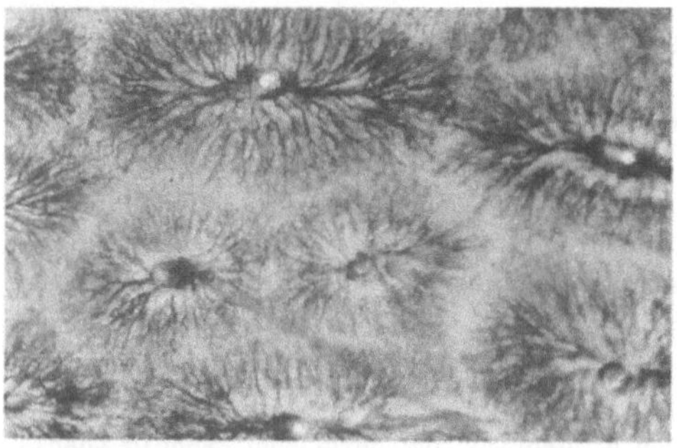

Fig.7 Horizontal section of a fossil tooth Myliobatis sp., Miocene, USA. ×68

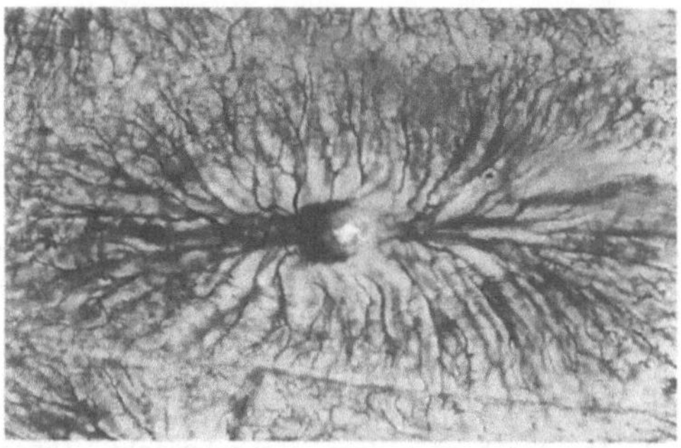

Fig.8 A denteon which composes plicidentine. High magnification of Fig.7. ×135

TOOTH PLATE OF COCHLIODONTS

Holocephali include many fossil chondrichthyan fish groups which have various tooth types. However, the basic tooth type of holocephalians was the same as elasmobranchs. The tooth of some holocephalian groups had a thin layer of enameloid but the other groups had no enameloid layer. The inner layer of the holocephalian tooth was composed of orthodentine, osteodentine and plicidentine similar to elasmobranchs [6].

The tooth plate of extant chimaeroids is composed of osteodentine and pleromin. Pleromin consists of whitelockite [7].

Cochliodonts were one of the Paleozoic holocephalian groups which had some large tooth plates like as chimaeroids. The tooth plate was continuously growing and composed of enameloid, dentine and osseous tissue.

In a labio-lingual section of a cocliodont tooth plate collected from the middle Permian of Japan (Fig.9), three layers, that is, the outer white enameloid-like layer, the middle black and basal gray osseous tissue layers were observed.

In another section of a cochliodont tooth plate from the same locality (Fig.10), only one layer which is composed of primitive plicidentine was observed. The primitive plicidentine had many canals from which dentinal tubules radiate.

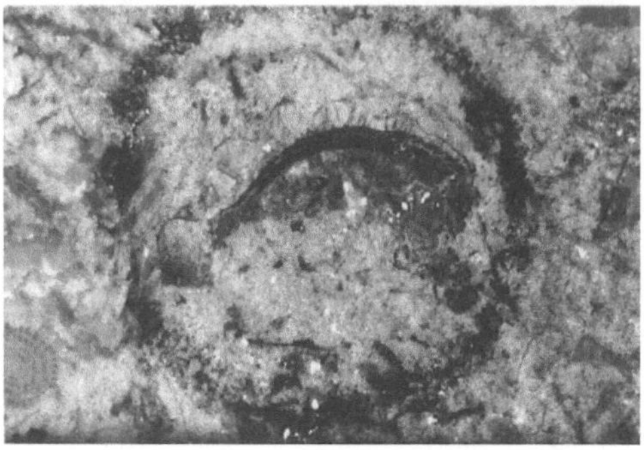

Fig.9 Labio-lingual section of a cochliodont tooth plate, middle Permian, Japan. ×4

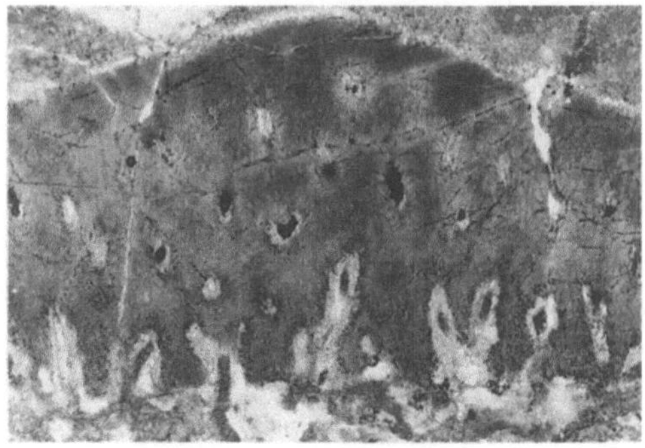

Fig.10 Section of a cochliodont tooth plate, middle Permian, Japan. ×16

EVOLUTIONARY TRENDS OF CHONDRICHTHYAN TOOTH STRUCTURE

A hypothesis on the evolutionary trends of tooth morphology and structure in elasmobranchs are summarized in Fig.11 [8]. The evolution of elasmobranchs is divided into three levels; the Paleozoic cladodont level, the Mesozoic hybodont level and the Cenozoic modern level.
Cladodont level sharks had multicuspid teeth which were composed of an outermost thin enameloid layer, an outer orthodentine layer and an inner osteodentine layer. The tooth of xenacanthoid sharks was composed of a thin enameloid layer and a well developed orthodentine. Hybodont sharks had multicuspid or molar and plate type teeth. Hybodus had teeth which were was composed of thin enameloid and orthodentine layers and a well developed osteodentine. The plate type tooth of Asteracanthus and the molar type tooth of Ptychodus were composed of plicidentine, while the tooth of Polyacrodus consists of orthodentine with a pulp cavity. In the teeth of modern elasmobranchs, three types of dentine are observed. Lamnoid sharks, one of the meat-eating group, have sharp triangular teeth which are composed of enameloid and osteodentine. Carcharhinoid sharks, another one of the meat-eating group, also have sharp triangular teeth, but their teeth are composed of enameloid and well developed orthodentine. Squaloid sharks and most rays also have teeth which are composed of enameloid and orthodentine, but, myliobatoid rays have plate type teeth which are composed of enameloid and well developed plicidentine.
Holocephalians have many types of teeth, some groups have large tooth rows which consist of many succesional teeth and some groups have several tooth plates. However, the basic tooth structure of holocephalians is regarded as the same as elasmobranchs.
Orthodentine is a basic type of dentine which was derived from the dermal tubercles of heterostracans. But, well developed orthodentine in carcharhinoid and xenacanthoid sharks can be regarded as an adaptation for meat-eating.
Osteodentine has been considered as a primitive type of dentine. But, the osteodentine in lamnoid shark tooth can be regarded as a structure whose formation occurs rapidly.
Plicidentine which is composed of many denteons is a complicated type of orthodentine. This structure is suitable for crushing the hard shells of invertebrates.
Therefore, it can be considered that the tooth structure of Chondorichthyes is not dependent on their evolutionary levels, but the result of an adaptation to their food habits.

ACKNOWLEDGEMENT

The author wish to thank Mr.M.Okura and Mr.H.Ogawa for their generous supplies of fossil tooth specimens and Prof.R.W.Fearnhead for his critical reading of this manuscript.

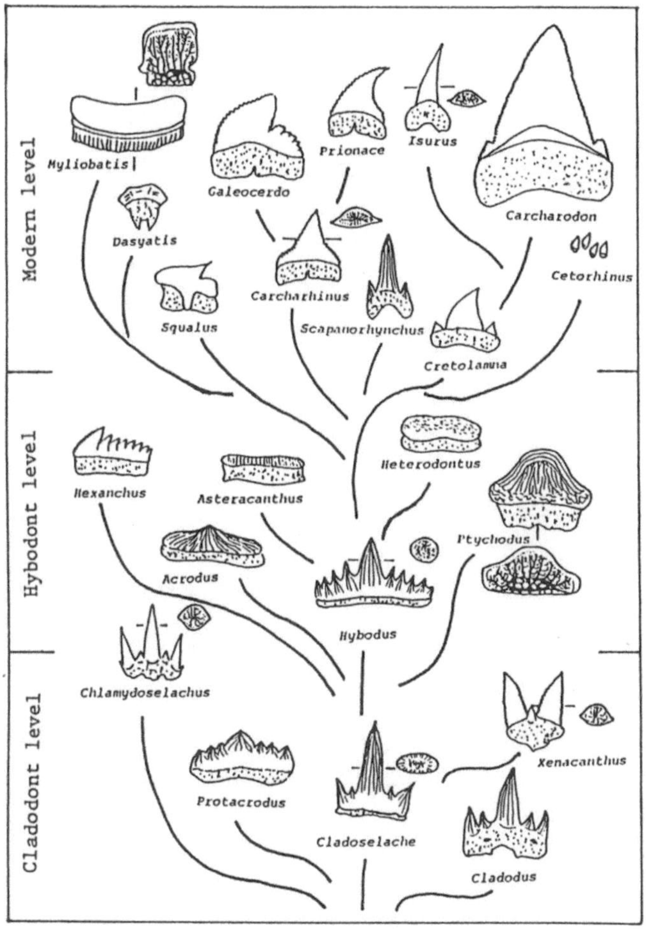

Fig.11 A hypothesis on the evolutionionary trends of the tooth morphology and structure in elasmobranchs.

This study was supported by Grant-in-Aid for Scientific Research (C) from the Ministry of Education, Science and Culture of Japan (No.01540647).

REFERENCES

1. GOTO M, HASHIMOTO I (1976) Jap J oral Biol 18: 362-377.
2. GOTO M (1978) Tsurumi Univ Dent J 4: 85-104.
3. GOTO M (1976) Earth Science (Chikyu Kagaku) 30: 206-221.
4. GOTO M (1978) J Stomatol Soc Jpn 45: 527-584.
5. KAKIZAWA Y (1984) Nihon Univ dent J 58: 59-69.
6. ZANGERL R (1981) Hand Book of Paleoichthyology. Vol.3A Chondrichthyes I. Gustav Fisher Verlag. Stuttgart, pp.1-115.
7. ISHIYAMA M, SASAGAWA I, AKAI J (1984) Arch histol jap 47: 89-94.
8. GOTO M (1985) In GOTO M, TAKAHASHI M, KIMURA M, HORIKAWA H (eds) Monograph 30: Evolution and Adaptation of Marine Vertebrates. The Association for the Geological Collaboration in Japan, Tokyo, pp.19-35.

CHAPTER 4.19

Ultrastructure of Pleromin, a Highly Mineralized Tissue Comprizing Crystalline Calcium Phosphate Known as Whitlockite, in Holocephalian Tooth Plates

M. Ishiyama, S. Yoshie, Y. Teraki, and E.W.T. Cooper[1]

Department of Histology, The Nippon Dental University, Niigata, Japan and [1]Vancouver Public Aquarium, Vancouver, Canada

Key words: Whitlockite, Pleromin, Tooth plate, Chimaerids, Ultrastructure

INTRODUCTION

The gross structure and microstructure of holocephalian tooth plates were initially investigated by Bargmann (1,2) and Brettnacher (3). Ørvig (4,5,6) subsequently reported in the detail microstructural characteristics of both extant and fossil holocephalian tooth plates. These studies indicate the holocephalian tooth plate to consist of bony tissue and hypermineralized pleromin (4), with the hardness of pleromin comparable to that of enamel and enameloid. Ishiyama et al. (7) found the crystalline material of pleromin to be not hydroxyapatite but whitlockite by a powder X-ray diffractometry, and pleromin to be the sole sound dental tissue consisting of whitlockite in vertebrates.

Compared to the gross structure and microstructure of pleromin, the inferior ultrastructure and development of pleromin have been elucidated. The ultrastructure and mineralogical characteristics of pleromin were thus examined by contact microradiography, light and electron microscopy and microfocus X-ray diffractometry, with special attention to the initial mineralization of pleromin.

MATERIALS AND METHODS

Two live chimaerids, Chimaera phantasma and Hydrolagus collei measuring 70 cm and 50 cm in total length, respectively, were used in this study. Their tooth plates were extracted from the head following decapitation, fractured and cut into small pieces and fixed in 2.5% glutaraldehyde + 2.5% paraformaldehyde mixture buffered to pH 7.4 with 0.1 M sodium cacodylate for 12h.

For transmission electron microscopy, the non-demineralized specimens were post-fixed in 1% osmium tetroxide for 2 h, and embedded in epoxy resin. Ultrathin sections were stained with lead citrate, and examined by a transmission electron microscope. For light microscopy, the specimens demineralized in formic acid and sodium citrate mixture were embedded in paraffin wax. Serial sections were cut, stained with hematoxylin and eosin or with azan stain, and examined with an light microscope. For scanning electron microscopy, some non-demineralized specimens were critically point dried and observed with a scanning electron microscope, after mounting and gold coating.

The remainning specimens were embedded in a polyester resin and cut into sections. The sections were ground and polished down to a thickness of 70 μm for contact microradiography and microfocus X-ray diffractometry.

RESULTS AND DISCUSSION

The tooth plates of the holocephalans consisted of bony tissue and pleromin (Fig.1). The pleromin appeared highly mineralized due to the radio-dense appearance in the contact microradiography, and was distributed intratrabecularly in basal and surface portions of the plates. The crystallites in mature pleromin were large in size and granular and appeared to differ from hydroxyapatite and fluorapatite showing a column that was hexagonal in appearance (Fig.2). Microfocus X-ray diffractometry showed typical peaks at diffraction angles (2Θ°) of 17.0°,27.8°,31.1°,34.4° (Fig.3). These values were extrapolated to Bragg's formula ($2d \sin\Theta = n\lambda$). The d-values corresponding to each peak were 5.21,3.21,2.88,2.60, respectively. These values were examined by the JCPDS Card, thus, the crystals of pleromin were identified as whitlockite. Ishiyama et al. (7) found the crystalline component of pleromin not to be hydroxyapatite but whitlockite by the powder X-ray diffractometry. The present study thus confirms the previous investigation (7) on the crystal component of pleromin. Whitlockite is known as a major constituent of dental calculus (8). The present study, however, indicated whitlockite to be an essential element in normal dental tissues of holocephalans.

Pleromin is produced by pleromoblasts, a type of odontoblast differentiated from osteoblast-like cells, without epithelial cell activity. Light microscopy indicated pleromoblasts to be high columnar clear cells having a penetration of cytoplasmic processes for the pleromin matrix (Fig.4). Transmission electron microscopy showed an immature matrix of pleromin prior to mineralization to be sparse tissue comprised of (a)scattered reticular collagen filaments 26-130 nm in diameter, (b)myelinated structures 300 nm in diameter and (c)tubular saccules 25-250 nm in diameter (Fig.5). In these

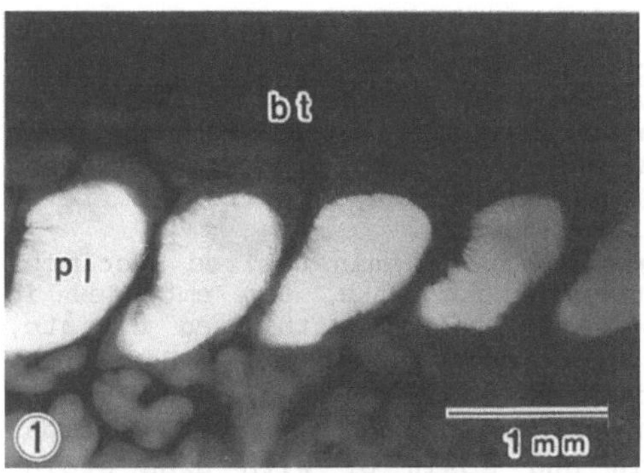

Fig.1: Contact microradiograph of the sagittal section of the tooth plate. Pleromin is found to be highly mineralized because of the radio-dense appearance. pl:pleromin, bt:bony tissue.

Fig.2: Scanning electron micrograph of the mature pleromin. Crystals of pleromin are granular and large in size.

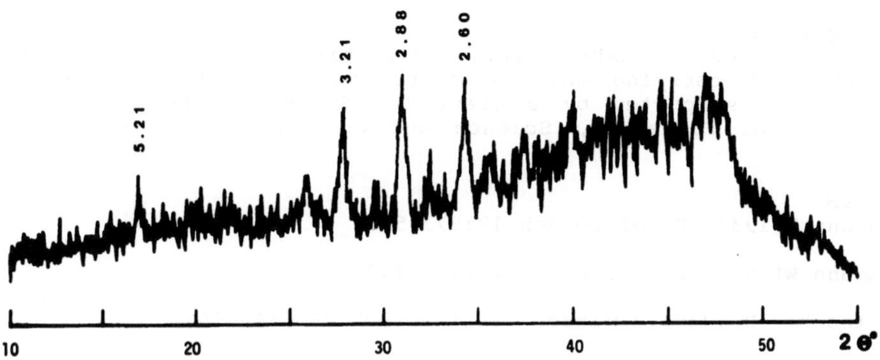

Fig.3: X-ray diffraction pattern of the mature pleromin. The d-values corresponding to each peak are 5.21,3.21,2.88, and 2.60.

components, tubular saccules were the major constituent of the immature pleromin matrix. Both tubular saccules and myelinated structures may possibly originated from pleromoblasts.
Pleromoblasts produced a matrix containing abundant collagen filaments along the peripheral portion of pleromin, as the final products of the pleromin matrix (Fig.4). Mineralization of pleromin did not begin until matrix formation have been completed. The mineralization of pleromin was initiated by a matrix vesicle-like structure in the peripheral portion of pleromin. Fine needle-like substances, possibly initial hydroxyapatite crystals, first appeared in the matrix vesicle-like structure, and, then mineralization followed along the collagen filament. The mineralization along collagen filaments could not propagate for the main portion of pleromin (Fig.6). Few collagen filaments in the main portion of pleromin appeared to hinder propagation of the mineralization along collagen filaments from the peripheral portion to the main portion of pleromin. Mineralization of the main portion started after completion of that in the peripheral portion. The initial crystals constituting the bulk of pleromin, i.e.whitlockite, were extensively present. These crystals, however, were deposited specifically within tubular saccules of cellular origin (Fig.7).
Garant(9) noted initial crystals of hydroxyapatite in shark (Squalus) enameloid to be produced specifically within small saccules 6-7 nm in diameter, as was also noted by Sasagawa (10) for Mustelus (shark), and considers that tubular saccules may possibly originate from odontoblasts. The tubular saccules of holocephalian pleromin may also originate from pleromoblasts, or odontoblasts. The present study indicates tubular saccules in holocephalian pleromin matrix to be larger than those in shark enameloid matrix as reported by Garant (9) and Sasagawa (10), and that the crystal components for pleromin and enameloid differ considerably. However, as a common feature, that the mineralization of both pleromin and shark enameloid is initiated by tubular saccules originating from mesenchymal scleroblasts, appears reasonable based on this study. The mineralization of mesenchymal hard tissue as pleromin, enameloid, dentine and bone is initiated by extracellular organells of scleroblastic origin, such matrix vesicles and tubular saccules.

Acknowledgements
We thank to Dr.Toshiro Sakae, Nihon University School of Dentistry at Matsudo, for carrying out the microfocus X-ray diffractometry. This study was supported by a Grant for Scientific Research from the Ministry of Education, Science and Culture, Japan (Grant-in-Aid,62771446).

REFERENCES

(1)Bargmann W(1933) Z Zellforsch 19:573-561

(2)Bargmann W(1937) Z Zellforsch 27:492-499

(3)Brettnacher H(1939) Z mikrosk anat Forsch 46:584-616

(4)Ørvig T(1976) Zool Scr 4:35-47

(5)Ørvig T(1980) Zool Scr 9:219-239

(6)Ørvig T(1985) Zool Scr 14:55-79

(7)Ishiyama M, Sasagawa I, Akai J(1984) Arch histol jap 47:89-94

(8)Jensen AT, Rowles SL(1957) Acta odont scand 16:121-139

(9)Garant PR(1970) J Ultrastr Res 30:441-449

(10)Sasagawa I(1989) J Anat 164:175-187

Fig.4: Light micrograph of the semithin (1 μm) section showing the pleromin formation. Note that the collagen-rich matrix (*) is produced as a final product of the pleromoblast (pb). ce:central portion.

Figs.5-7: Transmission electron micrographs of the pleromin matrix.

Fig.5: Immature matrix of pleromin, prior to mineralization, is sparse tissue. Loose reticular collagen filaments, myelinated structures and tubular saccules are found in the matrix.

Fig.6: The boundary between peripheral portion (pe) and central portion (ce) of pleromin. The mineralization of collagen filaments does not propagate for the central portion because of the few collagen filament in the central portion.

Fig.7: The early mineralization of the central portion of pleromin. Whitlockite crystals occur specifically within the tubular saccules.

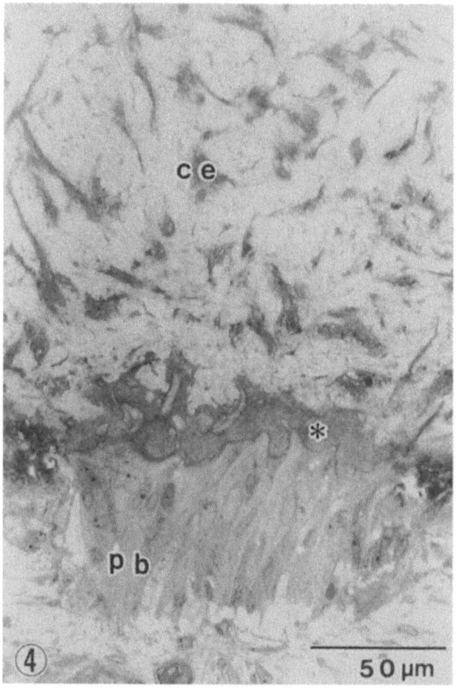
ce
*
pb
④
50 μm

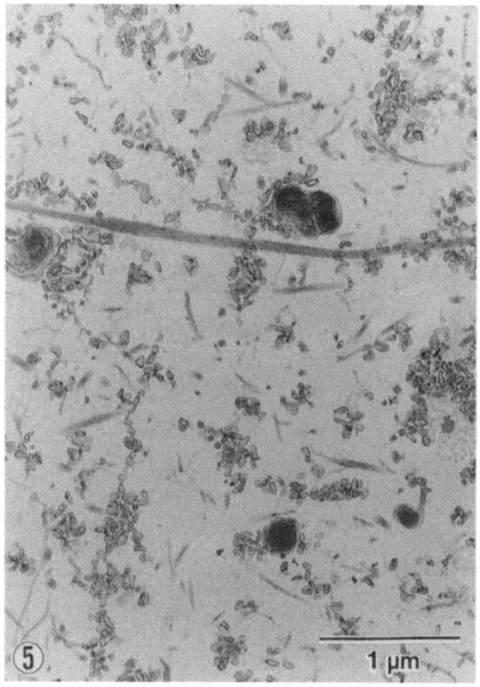
⑤
1 μm

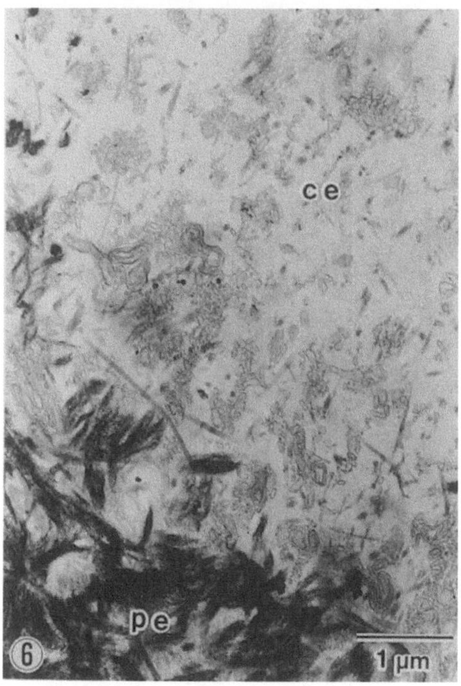
ce
pe
⑥
1 μm

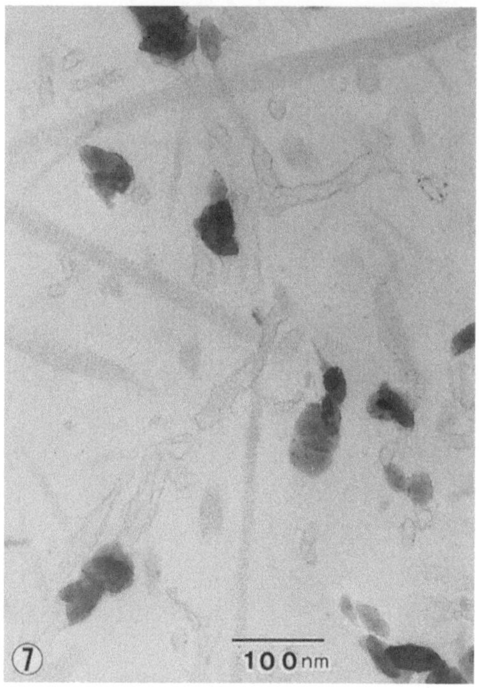
⑦
100 nm

CHAPTER 4.20

Dental Apatites in in Vertebrate Species: Morphology and Chemical Properties

T. AOBA, Y. MIAKE, S. SHIMODA, K. PROSTAK, E.C. MORENO, and S. SUGA[1]

Forsyth Dental Center, MA 02115, USA and [1]Nippon Dental University, Tokyo, Japan

Key words: Enameloid, Enamel, Fluoride, Carbonatoapatites, Mineralization

INTRODUCTION

The understanding of the basic mechanisms by which biological systems exert a precise control over the nature and properties (size, morphology, structural defects, and orientation) of the crystals in hard tissues, is greatly enhanced by comparative studies of mineralization processes in various species and their similarities in the evolutionary scale. It is known that carbonate and fluoride are common constituents of enameloid and enamel in vertebrate teeth [1-3] and the fluoride incorporation into fish enameloid does not reflect the fluoride concentration in environmental water and feeding habits but is related to the phylogeny of fish species [4,5]. However, little is known about the mechanism of species-dependent carbonation and fluoridation of the forming tooth mineral. The objectives of the present study were to investigate the nucleation and crystal growth of carbonated calcium apatites in various animal species and to gain an insight into the regulatory mechanism of in vivo precipitation processes, particularly in relation to fluoride.

MATERIALS AND METHODS

The biominerals under investigation were obtained from 1) a shell of Lingula unguis (the formation of calcium phosphates in non-vertebrate species), 2) the enameloid of shark Isurus glaucus and skate Raja erinacae (marine elasmobranch, the high-fluoride group), 3) the enameloid of red seabream Pagrus major (marine teleost, the high-fluoride group), 4) the enameloid of puffer Takifugu porphyreus (marine teleost, the low-fluoride group), 5) the enameloid of carp Cyprinus carpio (fresh-water teleost, the low-fluoride group), and 6) the enamel of pig (developing) and human (erupted) teeth. For comparative purposes, carbonatoapatites (dahllite- and francolite-types) were prepared in aqueous systems [5]. The mineral phase of all these samples was characterized using Fourier transform infrared spectroscopy (FTIR) and x-ray diffraction. The fluoride contents of the biominerals and synthetic apatites were determined using an ion-selective electrode or electron probe microanalysis (EPMA); their carbonate contents were determined chemically (Conway's microdiffusion) or using FTIR. For the observation by microradiography, FTIR microscopy, and electron microscopy, ground sections (about 60 um in thickness) were prepared of the biological specimens embedded in resin. On the basis of the findings obtained by microradiography and FTIR microscopy, electron microscopic observations were made of regions displaying typical features of the particular mineralization in each specimen. Ultra-thin sections were prepared with a diamond knife. The ultrastructure of the hard tissues was observed in a JEM 1200EX electron microscope at 120 kV; high resolution images of crystallites were obtained using the JEM-2000FXII and 2010 electron microscopes at 200 kV.

RESULTS

Tables 1 shows the composition and crystallographic parameters of the biominerals (and synthetic apatites). All the crystallized phases were most adequately described as carbonated calcium apatites but the degree of fluoridation and carbonation of the apatite lattice varied among species and, within a species, with developmental stages. Of special interest are the CO_3 mu_2 and PO_4 mu_1 bands which in a carbonated hydroxyapatite are at 873 and 961 cm^{-1}, respectively, while they are shifted to 866 and 966 cm^{-1} in a francolite-type mineral [6]. Lingula shell consisted of fluoridated carbonatoapatites containing various metal ions, e.g., Mg, Na, Mn, and Sr (data not included). In agreement with other reports [1,2], the shark enameloid contained high levels of fluoride (2.5 %wt or more) at the beginning of precipitation and its mineral phase was characterized as francolite by FTIR and x-ray diffraction. In teleost enameloid, the F contents were low (0.1 to 0.3 %wt F) in the early stage of mineralization. Interestingly, the fluoride content of seabream enameloid increased drastically (one order in magnitude) up to 3 % wt or higher with advancing mineralization; it was also verified by the shifts of FTIR bands, that a change from a dahllite-type to a francolite-type mineral structure took place with the fluoride accretion in the tissue. In contrast to seabream enameloid, puffer and carp enameloid exhibited only modest increases in the fluoride content through the developing stages. It is notable that the fluoride contents of enameloid in the low-fluoride group (0.1 to 0.3 %wt) are still higher by one order magnitude than those (0.01-0.03 %wt) of developing enamel in mammals. The results shown in the Table indicated that there is an inverse relationship between the fluoridation and carbonation of the bioapatites formed in the various species.

Figures A through H show electron micrograms of the crystallites in the various species. The mineralization in Lingula shell (Fig. A) was characterized by the precipitation of acicular crystals, which were scattered in aggregated forms. A similar feature of precipitation was reproduced at 37°C and pH = 7.3 in dilute supersaturated solutions containing 5-10 ppm F^- (Fig. B). The mineralization in the elasmobranch enameloid was characterized by the initial formation and

Table 1. Results of chemical and crystalline-structural analyses

Samples		F and CO_3 contents % wt		Wave numbers cm^{-1}		Lattice parameters Å	
		F	CO_3	$CO_3^{2-}mu_2$	$PO_4^{3-}mu_1$	c	a
Shell							
Lingula unguis		1.8	1.6	871	968	6.849	9.380
Enameloid							
Shark	(immature)	2.5	-	866	965	-	-
	(erupted)	4	0.5	866	966	6.878	9.384
Seabream	(immature)	0.1	-	872	962	-	-
	(erupted)	3	0.8	866	965	6.877	9.388
Puffer	(immature)	0.1	2.3	873	961	6.881	9.422
	(erupted)	0.2	2.0	873	961	6.880	9.423
Enamel							
Pig	(secretory)	0.01	4.0	874	961	6.886	9.437
	(maturing)	0.01	3.5	879 874	961	6.882	9.443
Human	(erupted)	<0.03	3.0	879 873	961	6.881	9.443
CO_3-OH-apatite		0	3.0	880 873	961	6.889	9.433
CO_3-F-apatite		4	3.0	873	966	6.896	9.350

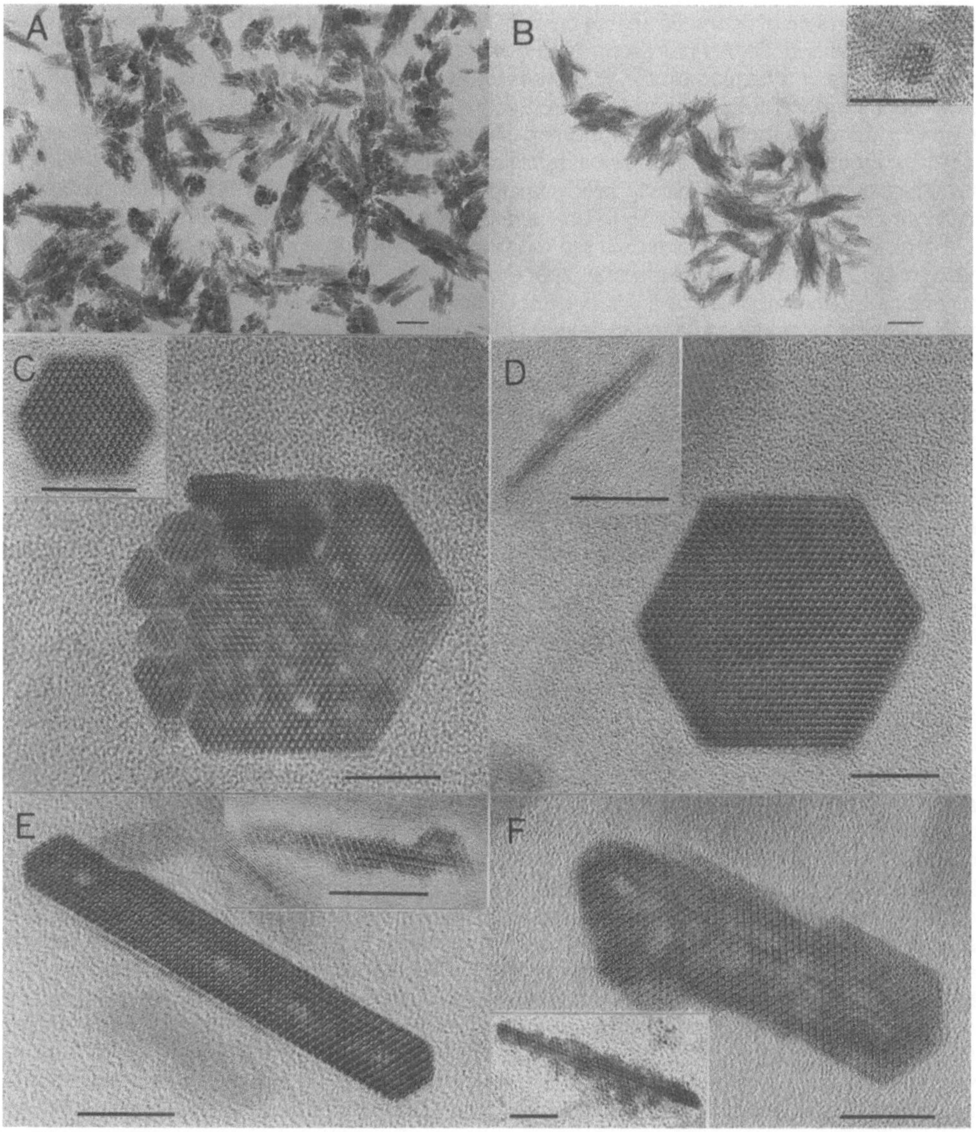

Figs. A through F. Electron micrograms of biominerals and synthetic apatites. (A) Lingula shell showing deposits of acicular crystals. (B) A similar type of apatite to that in Lingula shell obtained by precipitation in dilute supersaturated solutions ([Ca] = 1.0 mM; [PO_4] = 3 mM; [NaCl] = 160 mM; [F] = 5 ppm) at 37°C. (C) Prismatic apatite formation in the elasmobranch enameloid: note the growth of projections on the C-plane. (D) Seabream enameloid: the initial precipitation of thin-ribbons and its growth into francolite crystals having equilateral hexagonal cross section. (E) The crystal growth in puffer enameloid: note the initial precipitation of thin ribbons and the non-uniform accretion of apatitic unit cells on the prism planes (insert). The grown crystallites display flatten hexagonal cross section. (F) The crystal growth in pig enamel: note the similarity of the initial growth to that of puffer enameloid shown in Fig. E but, as a conspicuous difference, the presence of a central dark line in the grown crystallite. Thin and thick horizontal bars represent 100 nm and 10 nm, respectively.

subsequent growth of prismatic apatite crystals having hexagonal (frequently equilateral) cross-sectional areas (Fig. C). An interesting finding was that the growth along the c-axis of bioapatite crystals occurs in a fashion similar to that reported in vitro [7,8]; i.e., elongation and fusion of projections perpendicularly to the C-plane. Each elongating sub-domain had the lattice planes matching perfectly those of the supporting single crystal having an isodiametric hexagonal outline, suggesting that these individual apatitic projections consolidate into a single crystal. In the teleost enameloid, the initially precipitating crystallites were thin-ribbons in morphology, having thicknesses corresponding to only a few unit cells of the apatite lattice (Figs. D and E). In the case of seabream enameloid, the morphology of the crystallites changed from flatten-hexagonal to equilateral hexagonal shape in cross section with the advancement of mineralization accompanied by the fluoride accretion. In the low-fluoride group of fish enameloid, their crystallites maintained flattened hexagonal cross sections even when the tissue was highly mineralized. Their initial growth process was characterized by a non-uniform accretion of mineral on the lateral planes of a thin-ribbon (Fig. E, insert). In later stages, the thickening of the crystallites progressed mostly by a layer-type accretion of apatitic unit cells on the side prism planes; a step having the height corresponding to one unit-cell thickness of calcium apatite was frequently observed on the lateral planes. The enamel mineralization in mammalian species (Fig. F) was similar in some respects (the initial precipitation of thin-ribbons; the non-uniform accretion of apatite unit cells, and, in later stages, the formation of flattened hexagonal crystals in cross-sectional images) to the enameloid mineralization in the low fluoride group. However, an obvious difference was that the so-called central dark lines were commonly found in the crystallites of developing porcine enamel but these defects were not discernible or not as frequent in occurrence in the enameloid crystallites [9].

DISCUSSION

High fluoride content in a biomineral suggests that high concentrations of that ion were present at the time of mineralization. The present results indicate that fluoride ions at relatively high concentrations (although not yet determined in vivo) facilitate the de novo precipitation of apatite crystals, the elongation of the crystallites along the c-axis, and the growth on the prism planes. In this fashion, fluoride gives rise to the formation of prismatic apatite crystallites having equilateral cross sections as typically observed for fish enameloid in the high-fluoride group. The species-specific variations in crystal morphology and structure found within the teleost enameloid, are also explained in part by the effect of fluoride on the kinetics of precipitation, i.e., by facilitating hydrolysis of possible precursors and subsequent growth of apatite crystals. It is reasonable to consider that the absence, or less frequent occurrence, of central dark lines in enameloid crystallites of bony fish, in spite of the formation of thin ribbons in the preceding stage, is related to a relatively high accessibility of fluoride throughout the mineral accretion in these fish species, as compared to the limited incorporation of fluoride in mammalian species.

In biological media, it is most likely that other factors, such as potent inhibitors of crystal growth, may modulate significantly the effect of fluoride. Among regulators of biomineralization, much attention has been given to the role of organic matrix-mineral interaction. In considering the involvement of high fluoride in the initial crystallization of both elasmobranch enameloid and Lingula shell, a distinct feature is that precipitating apatite crystals in the former are relatively few in number and dispersed in matrix tissues; in contrast, the precipitation in Lingula shell (or in vitro), is indicated by a greater number of acicular crystals precipitated in an aggregated fashion. Further evidence of the importance of organic matrix in nucleation of crystals was obtained from studies using the teleost enameloid, showing that the initial precipitation of thin ribbons is mostly discerned along collagen-like fibers [9]. On the basis of such matrix-mineral interactions, it was proposed [10] that the crystals grow into preformed spaces created by the organic sheath material. However, the modes of crystal growth in either fish enameloid or mammalian enamel, i.e., the formation of projections on the basal plane or the non-uniform deposition on the prism planes, cast doubt on this concept. It is improbable that any matrix structures could be preformed to determine such modes of crystal growth. Nevertheless, very possibly organic components regulate the kinetics of crystal growth and prevent secondary nucleation, as advocated in enamel mineralization [11]. In this connection, it is a finding of importance that the fluoridation of apatite crystals enhances the adsorption affinity of enamel matrix proteins onto the crystal surface [12], since the adsorption of matrix proteins appears to be a prerequisite for exerting their inhibitory activities on crystal growth of bioapatites [13,14].

In relation to evolutionary aspects of calcium phosphate crystallization in vivo, the present results supports the idea that the regulatory mechanism of fluoride accretion in hard tissues is developed phylogenetically. This expression, in turn, seems to determine the increase of carbonation or defects in the crystal lattice. The formation of carbonated apatites in physiological media (the CO_3 in enamel fluid was determined to be about 10 mM [11,15]) is understandable since carbonatoapatites are the stable mineral phases in the presence of carbonate in the media [16]. Although our knowledge about the nature and properties of mineral in the vertebrate teeth is still limited, it is conceivable that the mechanism of carbonatoapatite formation in evolutionary processes is modified by possible changes in the microenvironments where the precipitation occurs. Questions remaining to be elucidated are how the driving force for precipitation of carbonated apatites is controlled in vivo and what is the significance of the phylogenetic changes in the mineral structure (and its stability) in relation to the highly organized tooth structures. Additional work is being conducted in our laboratories seeking answers for these questions.

ACKNOWLEDGEMENTS: This work was supported by grants DE08670, DE03187, and DE07677 from the National Institute of Dental Research.

REFERENCES

1. Glas JE (1962) Odont Rev 13:315-32631
2. LeGeros RZ and Suga S (1980) Calcif Tissue Int 32:169-174
3. Aoba T, Moreno EC (1990) Calcif Tissue Int (in press)
4. Suga S, Wada K and Ogawa M (1981) In: Binder K and Hohenegger M (eds) Fluoride Metabolism. Verlag Wilhelm Maudrich, Wien, pp 79-88
5. Suga S, Taki Y, Wada K and Ogawa M (1990) (in this volume)
6. Shimoda S, Aoba T, Miake Y, Moreno EC (1990) J Dent Res (in press)
7. McClellan GH and Lehr JR (1969) Amer Min 54:1374-1391
8. Doi Y and Eanes ED (1984) Calcif Tissue Int 36:39-47
9. Aoba T, Yoshioka C, Yagi T, Moreno EC (1984) J Dent Res 63:1348-1354
10. Miake Y, Aoba T, Shimoda S, K Prostak, Moreno EC, Suga S (1990) Calcif Tissue Int (in press)
11. Lowenstam HA and Weiner S (1989) On Biomineralization, Oxford univ Press, New York & Oxford, pp 175-188.
12. Aoba T, Moreno EC (1989) In: Fearnhead RW (ed) Tooth Enamel V. Florence, Yokohama, Japan, pp.163-167
13. Tanabe T, Aoba T., Moreno EC (1988) J Dent Res 67:536-542
14. Aoba T, Fukae M, Tanabe T, Shimizu M, Moreno EC (1987) Calcif Tissue Int 41:281-289
15. Aoba T, Tanabe T, Moreno EC (1987) Adv Dent Res 1:252-260
16. Aoba T, Moreno EC (1987) Calcif Tissue Int 41:86-94
17. Moreno EC, Aoba T (1990) Calcif Tissue Int (in press)

CHAPTER 4.21

Tooth Matrix Formation and Mineralization in Extant Fishes

K. PROSTAK, P. SEIFERT, and Z. SKOBE

Forsyth Dental Center, Boston, MA, USA

Key words: Teeth, enameloid, ultrastructure, mineralization, fish

INTRODUCTION

Three highly mineralized apatitic tissues cover dentin of fish teeth: (A) Osteichthian enameloid with matrix containing collagen [1] that, in some teleosts, is secreted by inner dental epithelial cells [2] and is degraded during the mineralization process [3, 4]. Therefore, this enameloid is different from enamel, dentin and bone. (B) Chondrichthian enameloid with matrix that is a unique non-collagenous product of odontoblasts to which IDE cells secrete carbohydrate [5,6]. (C) A non-collagenous, ectodermally derived, enamel-like matrix which is found on crossopterygian [7], dipnoan [8] and holostean [9] teeth. We research tooth development in diverse fish species to elucidate homology with mammalian odontogenesis and to characterize structural features of tooth formation which modulate the levels of fluoride (F) in the enameloid [10]. Here we report ultrastructural and biochemical parameters of several fish species illustrating various strategies used to form highly mineralized tooth structures.

MATERIALS AND METHODS

Species representing families of chondricthyes and osteichthyes were: skates (Rajidae); tautog (Labridae); parrotfish (Scaridae); pufferfish (Tetraodontidae); gar (Lepisosteidae); and goldfish (Cyprinidae). Control and colchicine injected (0.1mg/10g body weight) fish were routinely processed for TEM [3,4]. Other specimens were freeze-dried, polished and assayed for F by electron microprobe. F in developing enameloid and serum were measured by microdiffusion [11]. Secretory stage rat enamel matrix was extracted with guanidine-EDTA at 4°C, then used as an immunogen in rabbits to produce an antibody probe. Bovine enamel and fish enameloid matrices were extracted in 0.5 M acetic acid, dialyzed, separated on 10% acrylamide gels, then silver stained. Proteins were also transfered to Immobilon P membranes, incubated with antibody, then developed. Tissues for protein-A gold TEM immuno-labelling were fixed in 1% formaldehyde, 1% glutaraldehyde, 1% acrolein, post-fixed in 1% OsO_4 and embedded in Epon. Mammalian and fish teeth were immune-labelled according to Bendayan [12].

RESULTS

The mature enameloid tissues were categorized into two F groups. One group had a high F content [skate (3.05%), shark (3.86%), tautog (3.1%), parrotfish (2.72%)]; the other had low F levels [pufferfish (0.11%), gar (0.05%), goldfish (not detected)]. F content of developing parrotfish enameloid increased with progressive mineralization, and was 10-fold higher than that of pufferfish. In all fish, dentin contained less than 0.7% F and serum F levels were below 0.05 ug F/ml. Bony fish enameloid matrix was mainly comprised of collagen fibers. In colchicine-treated bony fish, IDE cells accumulated procollagen granules (PG) in the cytoplasm, 2 hrs post-injection (Fig 1), and at later time intervals collagen fibers similar to those in enameloid matrix formed between IDE cells. Accumulations of procollagen granules in odontoblasts of puffer- (low F) and parrot-fish (high F),

also indicated collagen secretion into the enameloid matrix.
In all bony fish, nascent enameloid crystallites formed parallel to the enameloid collagen fibers (C, Fig 2) near the dentin enameloid junction. At later stages, closely packed crystallites obliterated the enameloid collagen structure (Fig 3) and extended to the enameloid surface. However, at later stages, parrotfish crystallites (20 nm dia) were widely spaced and had no intervening enameloid collagen fibers (Fig 4). Fusion of numerous small crystallites observed along the enameloid collagen formed the large crystallites of parrotfish. Interpallisade crystallites (IP) of parrotfish enameloid also nucleate later than the bulk of enameloid (Fig 4). Goldfish enameloid collagen fibers (C, Fig 5) are wider and more highly striated than those of other enameloid matrices. Resulting mature goldfish enameloid crystallites were 11 nm wide (Fig 6).

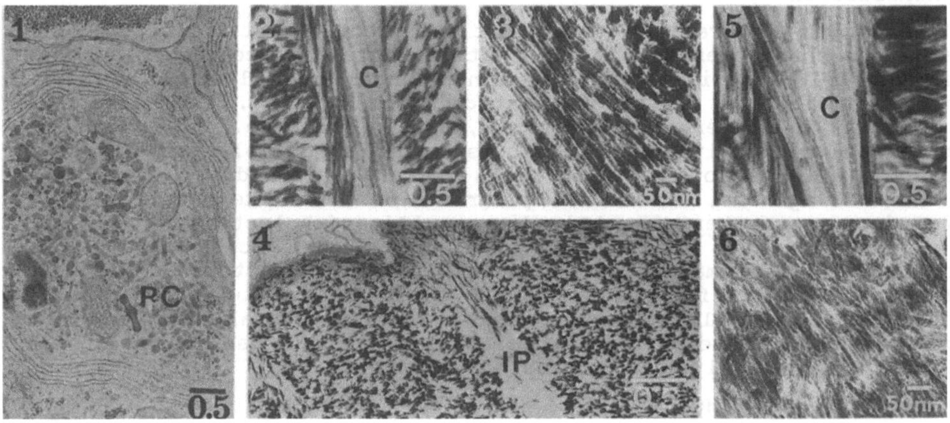

An amorphous layer (AM) devoid of collagen fibers was secreted on the enameloid surface by puffer- and parrot-fish (Fig 7) IDE. Initially, this material contained no crystallites, then at later stages, randomly oriented crystallites formed within the matrix (Fig 7). With subsequent growth, these surface crystallites were indistinguishable from the underlying enameloid crystallites.
Gar IDE cells also secreted a crystal containing matrix on the tooth shaft dentin and cervical enameloid [9]. This ganoid/enamel (E) had crystallites oriented normal to the surface, and in acid-etched SEM samples, showed incremental lines (Fig 8).

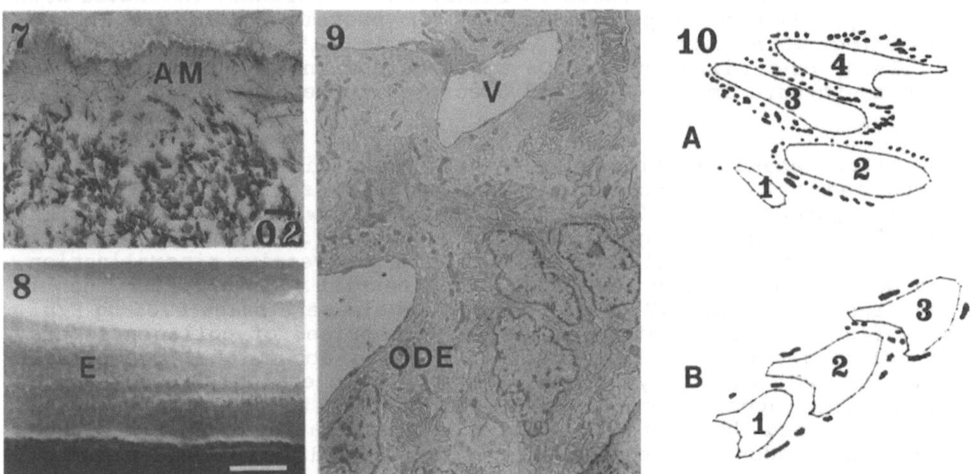

Distinctive features were observed in bony fish having high F content in enameloid. At the enameloid mineralization stage, the plasma membranes of outer dental epithelial cells (ODE) became highly infolded and the cells nearly surrounded numerous capillaries (V) (Fig 9, 10A). Although pufferfish (low F) ODE cells showed similar membrane configurations, very few capillaries were observed (Fig 10B). In contrast, tooth organs of gar and goldfish (low F) had neither specialized ODE cells nor numerous capillaries at any stage of enameloid formation.

Skate enameloid matrix morphology and formation was unique. The matrix consisted of 'giant' fibers having 17.5 nm periodic cross-striations, and unstriated, 'unit' fibers (25 nm dia) [5,6]. The 'giant' fibers (F) disaggregated to form 'unit' fibers (E, Fig 11). The IDE cells contained copious amounts of glycogen, but few strands of granular endoplasmic reticulum. Colchicine-treated skate IDE cells accumulated vesicles containing glycogen (v), some of which exocytosed along the lateral and apical plasma membrane (Fig 12). In both normal and treated skates, procollagen granules were notably absent from both the IDE and odontoblasts associated with enameloid matrix. Instead, odontoblast Golgi cisternae were filled with deeply stained material (M, Fig 13), indicating that odontoblasts secrete non-collagenous proteins in enameloid matrix.

At the mineralization stage, skate (high F) enameloid organs had neither specialized ODE cells nor any specific association with the vasculature. Nascent skate enameloid crystallites formed throughout the enameloid matrix but were not observed within 'giant' fiber material. Although the C-axis of some crystallites were oriented parallel to the 'giant' fibers, no prefered orientation to either 'giant' or 'unit' fibers was noted (Fig 11).

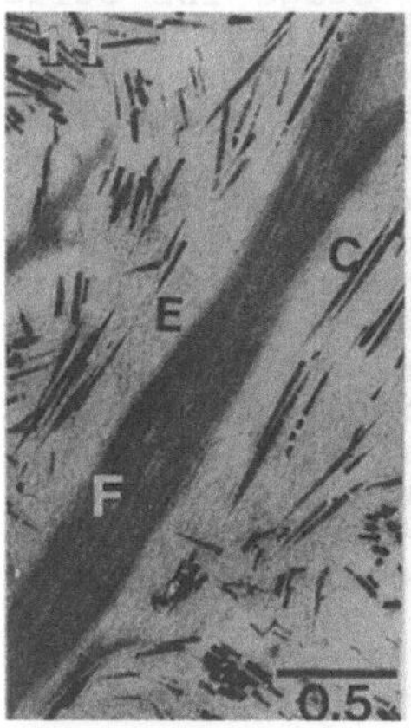

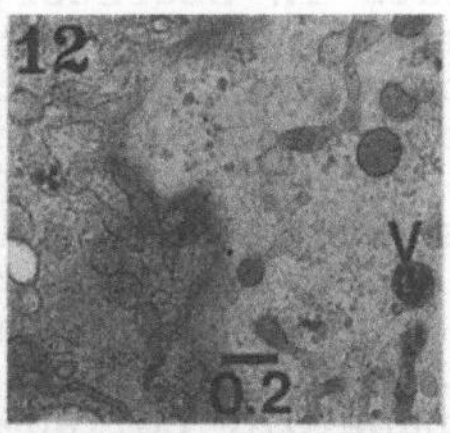

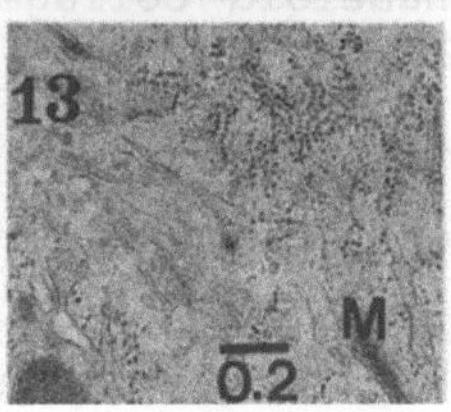

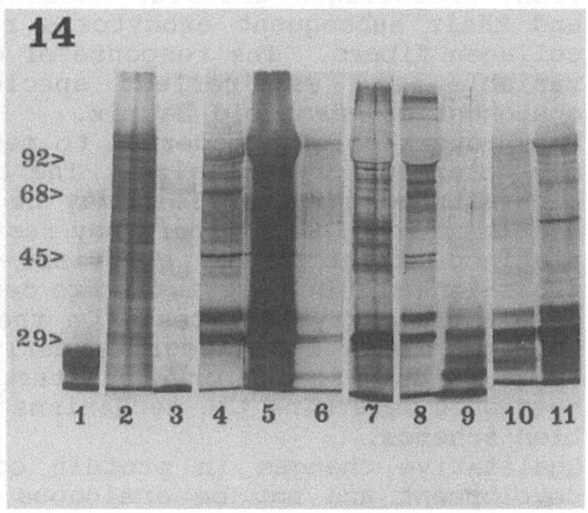

PAGE showed numerous enameloid matrix proteins of similar MW in pufferfish, tautog, and parrotfish (Fig 14; Lane 1=Bovine; 2=developing skate; 3=mature skate teeth; 4-6=Pufferfish I, rows 1-3; 7=Tautog; 8-9=Pufferfish II, rows 1-2; 10-11=Parrotfish, rows 1-2). All fish had high MW components (100 KD) suggestive of collagen. In bony fish, major proteins occured at 33 and 29 KD in early stage (cheese-like) enameloid, whereas in later stages (chalk-like), lower MW proteins (25, 23 KD) were enriched. Pufferfish enameloid matrix (low F) also had prominant bands at 85, 69, 47 and 44 KD, whereas a 56 KD band was highly visible in enameloid matrix of tautog and parrotfish (high F). No single band was predominant in developing skate teeth (Fig 14, lane 2), but the 100 and 33 KD bands were similar to those of bony fish proteins. Although developing skate teeth had adherent predentin, few bands were similar to mature teeth (Fig 14, lane 3).

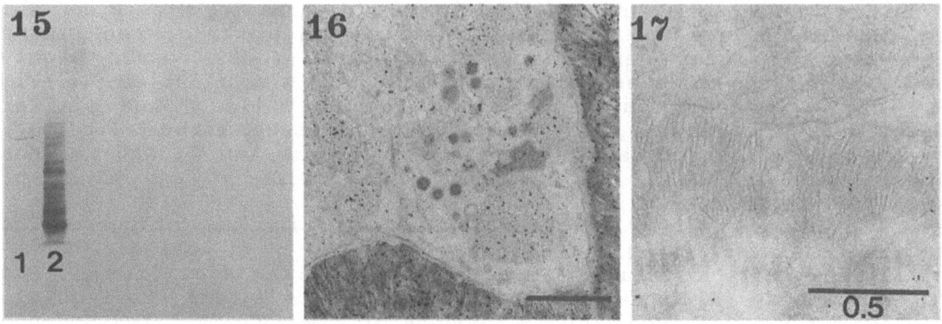

Western blots of bovine enamel matrix demonstrated its homology with rat enamel matrix (Fig 15), although a slight reaction with bovine albumin was sometimes observed. However, none of the fish enameloid proteins (skate, pufferfish, parrotfish, tautog) were recognized with this antibody. Electron microscopic immune-labelling of ferret secretory stage ameloblasts with this probe showed positive labelling in the enamel matrix, secretory granules and large vesicles (Fig 16). Identically prepared specimens of gar, parrotfish and pufferfish tooth buds showed no labelling in either the enameloid, amorphous surface layers, or ganoid/enamel (Fig 17).

DISCUSSION

Our results confirm the idea that osteichthian enameloid matrix collagen is mainly derived from the IDE [4]. In colchicine-treated fish, procollagen granules accumulate in secretory stage IDE cells and their subsequent exocytosis results in intercellular enameloid collagen fibers. The response of odontoblasts to colchicine is more variable, and may reflect species variations in the mesodermal component of enameloid matrix.

The proportion of ectodermal to mesodermal enameloid matrix collagen may determine crystal size. The highly striated collagen fibers of mineralizing gar enameloid may be mesodermal in origin [9]. As in dentin or bone, such fibers may restrict crystal size. Similarly, the distinct striations of goldfish enameloid collagen may indicate a cross-linked fiber, resistant to degradation, resulting in less growth of enameloid crystallites. On the other hand, parrotfish enameloid collagen loses its integrity early in mineralization, resulting in large, widely spaced crystallites. Further comparative studies are needed to determine the mechanisms of these unusual collagen degradation schemes.

Qualitative changes in protein composition occur during enameloid development and may be analogous to changes observed in mammalian enamel mineralization [13]. With increased mineralization, the spectrum of high MW enameloid proteins decrease with concomitant increases in lower MW peptides. Although no lower MW enameloid proteins were antigenically similar to mammalian enamel proteins, they may have similar functions in controlling crystallite growth. Proteolytic activity, observed in the 68 KD band (unpublished data) of tautog, puffer- and parrot-fish enameloid matrix, may also degrade enameloid matrix to facilitate its removal during mineralization.

Our ultrastructural and biochemical results indicate that skate enameloid formation is a unique process. Immune-labelling with enamel matrix antisera showed no affinity for skate enameloid matrix. Additionally, unlike bony fish odontogenic cells, neither skate IDE cells nor odontoblasts produce procollagen granules during enameloid matrix formation. Instead, colchicine-treated IDE cells were active in glycogen synthesis and secretion whereas odontoblasts synthesized material which formed the 'giant' fibers of the enameloid matrix [5,6]. Nascent skate crystallites were sometimes oriented along 'giant' fibers, but had no preferred orientation to 'unit' fibers. In contrast, all nascent crystallites in bony fish enameloid were oriented parallel to the enameloid collagen fibers.

The ganoid/enamel of the gar and the mineralizing, amorphous surface layers deposited on parrot- and pufferfish enameloid are not enameloid. During development of these layers, IDE cells do not contain procollagen granules, nor are collagen fibers observed within the amorphous matrix. Instead, IDE cells contain granules having either dense (in puffer- and parrot-fish) or 'stippled' (in gar) contents. Granules with similar morphology are also observed in mammalian secretory ameloblasts. Although our antibody did not cross-react with the spherical granules, amorphous material, or the ganoid/enamel, we cannot eliminate the possibility that antigens similar to the 'enamelin' class of enamel proteins [14] are present in these fish structures. The 'enamelin' proteins are suggested to nucleate apatite crystals, whereas the 'amelogenin' proteins may affect crystallite orientation and growth rate [15]. However, an interesting dichotomy is that gar ganoid/enamel has well ordered crystalites and incremental lines, whereas the amorphous surface layers of parrot- and pufferfish teeth, do not. Perhaps the geometry of the IDE apical membrane associated with these amorphous materials influences the microenvironment which, in turn, affects crystallite orientation.

The F content in enameloid is not directly related to serum or environmental F levels. However, there is a strong association of specialized ODE cells with numerous vascular elements in osteichthian species having high levels of F in their enameloid. These specialized ODE cells are remarkably similar to chloride transporting cells of gill epithelia [16] and, here, are suggested to transport F toward the enameloid. The relative paucity of vasculuature in pufferfish as well as the lack of specialized ODE cells in gar and goldfish support this contention. The observation that skate tooth buds have few features in common with bony fish enameloid formation suggests a unique mode of F incorporation into the enameloid mineral also exists. Possibly, F incorporation into elasmobranch enameloid is a matrix-mediated event, reflected in its unique biochemical footprint. Further investigations of enameloid formation in various species should clarify these unresolved issues.

REFERENCES

1. Poole DFG (1967) Structural and Chemical Organization of Teeth. Academic Press, New York, p 111.
2. Prostak K, Skobe Z (1985) J Craniofac Gen Devl Biol 5:75-88.
3. Shellis PR, Miles AEW (1976) Proc R Soc Lond B 194:253-269.
4. Prostak K, Skobe Z (1986) Arch oral Biol 31:73-85.
5. Prostak K, Skobe Z (1988) Amer J Anat 182:59-72.
6. Prostak K, Seifert P and Skobe A (1990) Amer J Anat (in press).
7. Miller W (1969) Nature 221:1244.
8. Ishiyama M, Terakai Y (1989) Tooth Enamel V. Florence Publishers, Tokyo, pp. 131-137.
9. Prostak K, Seifert P, Skobe Z (1989) Tooth Enamel V. Florence Publishers, Tokyo, pp. 188-195.
10. Suga S (1983) Japan J Oral Biol 25:419-436.
11. Whitford GM Reynolds KE (1979) J Dent Res 58:2058-2065.
12. Bendayan M (1984) J Electron Microsc Tech 1:243-270.
13. Robinson C, Lowe N, Weatherell (1977) Calcif Tissue Res 23:19.
14. Deutsch D (1989) Anat Rec 224:189-210.
15. Samuel N, Bessem C, Bringas P, Slavkin H (1987) J Craniofac Genet Devel Biol 7:371-386.
16. Pisam M, Boeuf G, Prunet P, Rambourg A (1990) Amer J Anat 187:21-31.

CHAPTER 4.22

Some Evolutional Tendencies of Mammalian Tooth Structure

Y. KOZAWA, H. MISHIMA, and K. SUZUKI

Department of Anatomy, Nihon University, School of Dentistry at Matsudo, Chiba, 271 Japan

Key words; Enamel, Enamel prism, Hunter-Schreger band, Evolution

INTRODUCTION

The pattern of Hunter-Schreger bands in the mammalian enamel was reported by Kawaii [5], whose classification followed the scheme of the proboscidean evolution proposed by Ijiri [3]. Yamakawa [7] also applied the Ijiri's idea to the rodent special enamel structure. On the other hand, Shobusawa [6] classified the morphology of mammalian enamel prism into 5 types, each of which is characteristic for each order. Boyde [1] reported 4 different patterns of arrangement of the enamel prism. The purpose of this study is to investigate: 1) the changes of width and form in the enamel prism, and 2) the pattern of Hunter-Schreger bands exhibited in an evolutionary series of several fossils and living mammalian specimens.

MATERIALS AND METHODS

The orders observed were; Insectivora, Primates, Lagomorpha, Rodentia, Cetacea, Carnivora, Codylarthra, Notoungulata, Proboscidea, Hyracoidea, Sirenia, Desmostylus, Pressiodactyla, and Artiodactyla. The specimens were cut sagitally, horizontally and tangentially, and polished. The polished surfaces were etched with 0.5% HCl and coated with C-Gold. The specimens were observed by the light-reflection microscope and the scanning electron microscope.

RESULTS AND DISCUSSION

The change in width of the enamel prism during the course of evolution were classified into; 1) no-chang (many species), 2) increased width (Proboscidea, Odobenus, etc.), 3) decreased width (Ungulata, Desmostylia, etc.), 4) variable width (Cetacea etc.) and 5) no-prisms.

On the Proboscidea, the enamel prisms changed from the arched type of about 5 μm to the ginko-leaf type of about 7 μm. However, many ungulates decreased the width of enamel prisms (e.g. *Hyracotherium* had arched type about 5 μm in diameter, but *Hipparion* and recent *Equus* have oval form prisms about 3 μm in diameter. The arrangement of enamel prisms also changed from alternate to parallel. The most typical example of the decreased pattern was observed in *Desmostylus* that had enamel prisms about 1.5-2.0 μm in diameter in the middle enamel layer. The ancestor of *Desmostylus*, *Behemotopus*, had enamel prisms about 3-4 μm in diameter.

The prism were arranged parallel with one another in these species. The form and size of the enamel prisms of the living cetacean were quite variable, some were prismless and others had prismatic enamel [4]. However, the ancestor of Cetacea, *Prozeugrodon* and *Squalodon*, had the hexagonal and oval pattern of enamel prisms (5 μm diameter) respectively. Both had the Hunter-Schreger bands. These size variations may be closely related to the type of prism arrangement, and the size of ameloblasts.

The pattern of the Hunter-Schreger bands were related to the molar form and to their masticatory function, and developed from a simple to more complex form. However, some species, e.g. Cetacea, Hyracoidea and Sirenia, whose ancestors had Hunter-Schreger bands, lack these bands in the extant teeth. The development pattern of Hunter-Schreger bands were classified into; 1) irregular, 2) horizontal, 3) vertical, 4) reduced or absent, and 5) special type (Fig. 1). The horizontal type were observed in many species. The ancestral proboscidean, *Palaeomastodon*, had also this pattern (Fig. 2). However, the evolved proboscidea, such as *Stegodon* and *Elephas*, had the horizontal type only in the outer enamel, but the inner half layer was the irregular type (Fig. 3). That pattern was found in the Hippomorpha molar enamel. The irregular type of the inner half layer of the enamel of *Proboscidea* and *Hippomorpha*, is considered to be the new character Hunter-Schreger bands (Fig. 4). Since this character was absent in the ancestor and appeared during the course of evolution. The irregular type was found in one of the ancestral Proboscidea, *Barytheium* (fig. 5). These irregular bands developed into a highly complex form in *Deinotherium*, which is the special types of Proboscidea (Fig. 6). These irregular type suggests that the original bands might be formed by the grouped movement of he ameloblasts(Fig. 7). Their grouping and dancing movement of ameloblast may have become more complex, which may have caused their fusion or differentiation to these three types of Hunter-Schreger bands (Fig. 1). The vertical type already discussed in Rimnocerous by Forterius [2]. Some species reduced and lost the Hunter-Schreger bands, e.g. Sirenia, Hyracoidea and Cetacea.

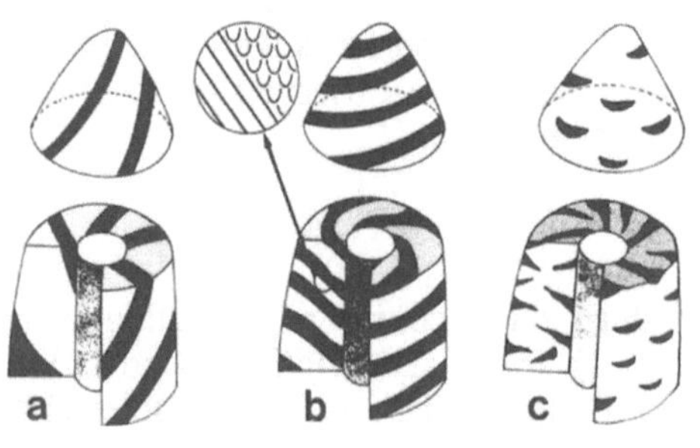

Fig. 1. Type of Hunter Schreger band (a) vertical, (b) horizontal, (c) irregular

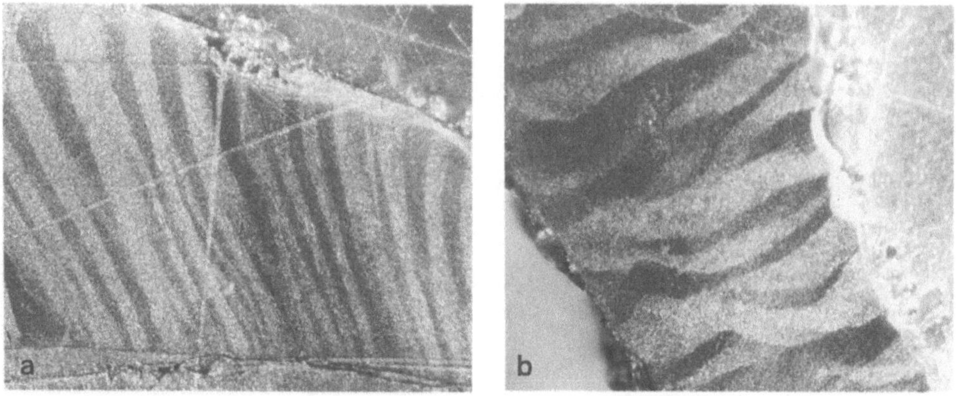

Fig. 2. Hunter-Schreger band of *Palaeomastodon.* (a) sagittal section, (b) tangential section (cusps is upper). Horizontal types of Hunter-Schreger bands are clearly observed on the sagittal section. Many mammalian species have similar type.

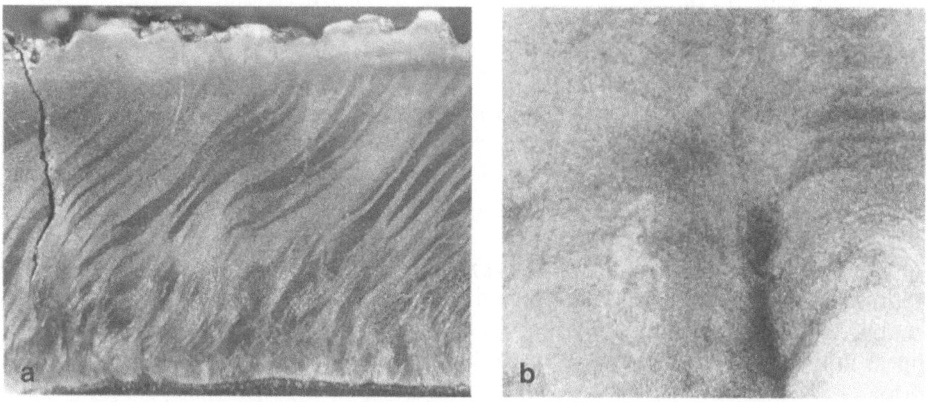

Fig. 3. Hunter-Schreger band of *Proboscidea.* (a) sagittal section (*Loxonta africana*), (b) tangential section (*Palaeoloxodon naumanni*). The inner half layer of enamel shows quite irregular pattern, but the outer half shows relatively regularly on the sagittal section. On the tangential section, former is irregular and latter is horizontal.

Fig. 4. Schema of Hunter-Schreger band in Elephas and Equus. The irregular pattern is observed in the inner half, which may be a new character appeared in their evolutionary course.

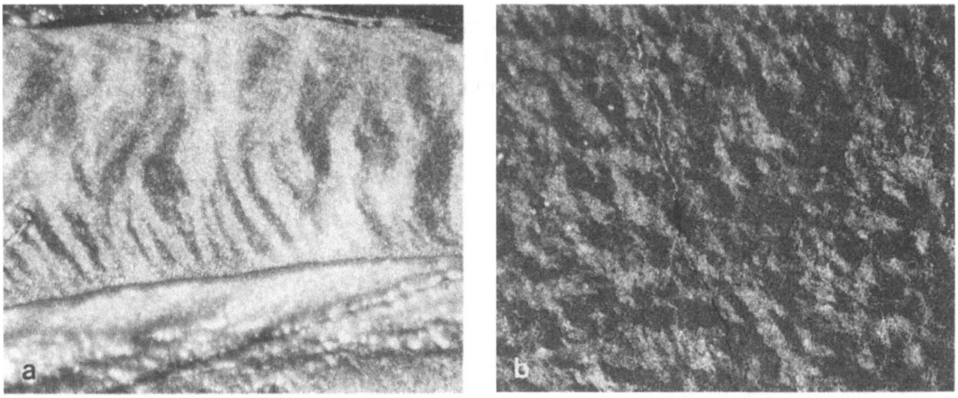

Fig. 5. Hunter-Schreger band of *Barytherium.* Both sagital (a) and tangential (b) section show the irregular type. Island-like bands are observed on the tangential section, which suggests that the grouping of ameloblast in the enamel organ may appear in the early evolutionary stage of the Hunter-Schreger band.

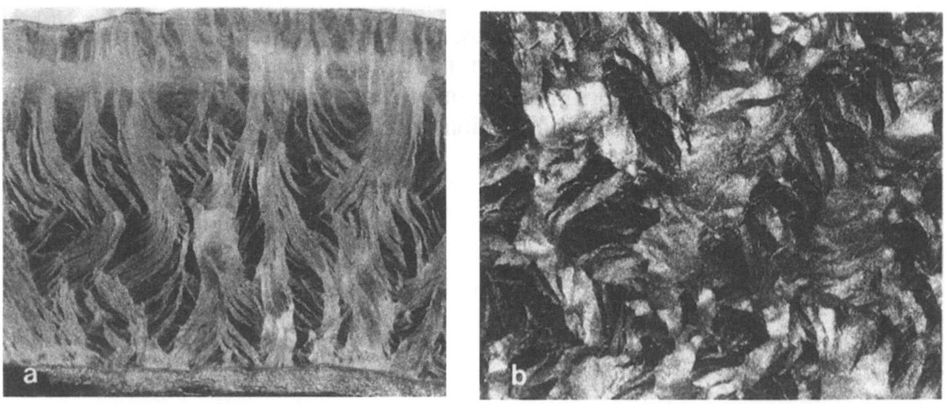

Fig. 6. Hunter-Schreger bands of *Deinotherium.* Highly complexed Hunter-Schreger bands are observed in both sagittal (a) and tangential (b) sections. This type is classified into the irregular type.

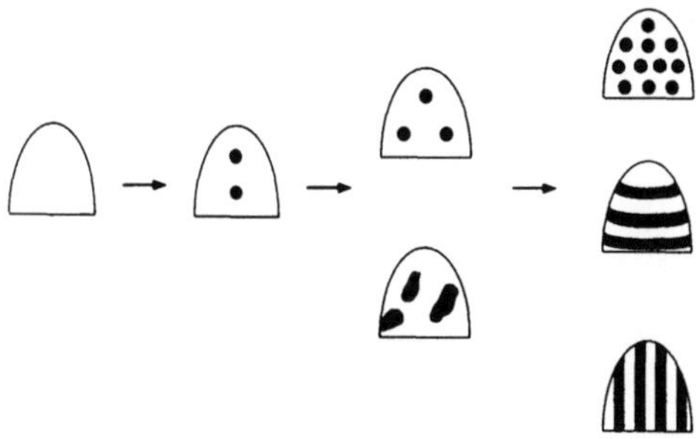

Fig. 7. Development of Hunter-Schreger band. The Hunter-Schreger bands cannot be observed in early mammalia (left). Next stage, ameloblasts may grouped and form several islands (center). Grouped ameloblasts may fuse or differentiate into three pattern (right).

CONCLUSION

The Hunter-Schreger bands developed from a simple to more and more complex pattern in each evolutionary course of mammalia except several species. These are classified into 5 types relative to the enamel prism patterns. These types are related to the molar form and to their mastication. The origin of the Hunter-Schreger bands may be found in the irregular bands of *Barytherium*, which shows the island-like grouping of ameloblast in the enamel organ. This island-like types of bands may have differentiated into the horizontal, vertical and irregular type. On the Proboscidea and Hippomorpha, the new character of enamel structure may have appeared in the early formed deep enamel layer.

REFERENCES

[1] Boyde A. (1964) Univ. London Thesis.
[2] Forterius M. (1984) In: Fearnhead R.W. & Suga S. (eds), Tooth Enamel IV, Elsevier, New York, pp.427-431.
[3] Ijiri S., Kawai N. (1948) Bull. Tokyo Sci. Mus., 23:1-6, Figs. 1-15.
[4] Ishiyama M. (1984) Jpn. J. Oral Biol., 26:1054-1071.
[5] Kawaii N. (1955) Okajimas Folia. Anat. Jap., 27:115-131, Pl. 1-7.
[6] Shobusawa M. (1952) Okajimas Folia. Anat. Jap., 24:371-392, Pl. 1-4.
[7] Yamakawa K. (1959) Acta Anat. Nipponica, 6:852-866.

CHAPTER 4.23

Proboscidea Fossil Teeth Suggest the Evolution of Enamel Crystals

T. Sakae, H. Mishima, and Y. Kozawa

Nihon University School of Dentistry at Matsudo, Chiba, 271 Japan

Key Words: Elephant, Apatite, Carbonate, X-ray diffraction

INTRODUCTION

Tooth enamel is a highly calcified tissue composed of well crystallized apatite in comparison with that in bone and dentin. Parker and Toots [1] considered that among these three hard tissues, tooth enamel showed the least degree of change in its chemical composition in fossil samples. The aims of the present study were to clarify the crystallographic characteristics of fossil tooth enamel, to reveal its stable and unstable chemical components, and to examine any differences existing among fossil enamels from animals belonging to the Proboscidea.

MATERIALS AND METHODS

Materials used in this study were enamels from molar teeth of Proboscidea which were alive between the recent and Eocene periods (Table 1). The names of fossil animals used in this study were those on the labels of the specimens, and did not follow the modern nomenclature in order to prevent confusion. For comparison, human enamel was also examined at the same time. Enamel was removed from each tooth using a dental disc and powdered using an agate mortar for subsequent analysis.

X-ray diffraction (XRD) was carried out for identification of crystal components, measurement of crystallinity, and determination of lattice parameters of the enamel crystals using a Rigaku X-ray diffractometer. Crystallite size was calculated using Scherrer's equation. XRD study of enamel crystals was also carried out after mixing with corundum and ignition at 200 ^{o}C, 600 ^{o}C, 1000 ^{o}C and 1200 ^{o}C for 2 h. Lattice parameters of enamel crystals were calculated by the least squares method. Infrared absorption spectrometry (IR) was carried out using a JASCO IRA-2 spectrometer to determine the carbonate content of enamel apatite and to examine minor components which could not be detected by XRD. The carbonate content of enamel apatite was calculated from the absorbance peak intensity ratio of CO_3 at 1415 cm^{-1} to PO_4 at 595 cm^{-1} following the method of LeGeros. Thermogravimetric analysis (TGA) and differential thermal analysis (DTA) were carried out to determine the thermal stability and decomposition rate of enamel crystals using a Rigaku TAS-100 thermal analysis system.

RESULTS

Figure 1 shows a typical XRD pattern of Proboscidea enamel. All the enamel samples showed XRD patterns similar to this one and to the XRD pattern of human enamel. Table 1 lists the a-axis length and crystallite size values for the enamel samples together with those of other animal species. *Gomphotherium* enamel showed the shortest a-axis length and smallest crystallite size among the listed data. These values suggest that *Gomphotherium* enamel has a large carbonate content. Enamels of the two geologically oldest Proboscidea, *Palaeomastodon* and *Moeritherium*, had shorter a-axis lengths than those of the other geologically younger Proboscidea, the latter showing similar but slightly shorter a-axis lengths than human enamel. The crystallite sizes for the enamel samples were mostly in the same range, around 230 A, except for *Palaeoloxodon* and *Gomphotherium* enamels. This uniformity of crystallite size suggests that the enamel samples had not suffered severe chemical attack.

Figure 2 shows the crystal components of the enamel samples after ignition at 1200 ^{o}C. These could be assigned to three groups according to the main crystal component: 1) Whitlockite; *Loxodonta, Elephas, Mammuthus, Palaoloxodon* and *Stegodon*, 2) a-TCP; *Stegolophodon* and *Dinotherium*, and 3) Apatite; *Gomphotherium, Palaeomastodon* and *Moeritherium*. Human enamel was included in Group 1, high temperature synthesized hydroxyapatite was in Group 2, and mineral fluorapatite was in Group 3. These groups seemed to indicate the thermal stability of apatite.

The decrease in a-axis length for Group 1 enamels upon heat treatment was similar to that seen for human enamel (Fig. 3a), and reduction of the a-axis length over a temperature range of 200 ^{o}C to 600 ^{o}C was less than 0.01 A whereas that for Group 2 was over 0.01 A (Fig. 3b). The reduction of the a-axis length for Group 2 over a temperature range from room temp. to 1000 ^{o}C was large, 0.02 A whereas that for Group 1 was within 0.016 A, except for *Stegodon* enamel, which showed a shrinkage as large as 0.02 A. The changes in a-axis length for Group 3 were not uniform (Fig. 3c). That for *Gomphotherium* was unique because after heating at 200 ^{o}C the a-axis length did not decrease but rather increased. *Gomphotherium* and *Palaomastodon* enamels showed an increase in a-axis length over a temperature range of 600 ^{o}C to 1000 ^{o}C. This unique feature may be related to ionic substitutions in the enamel crystals.

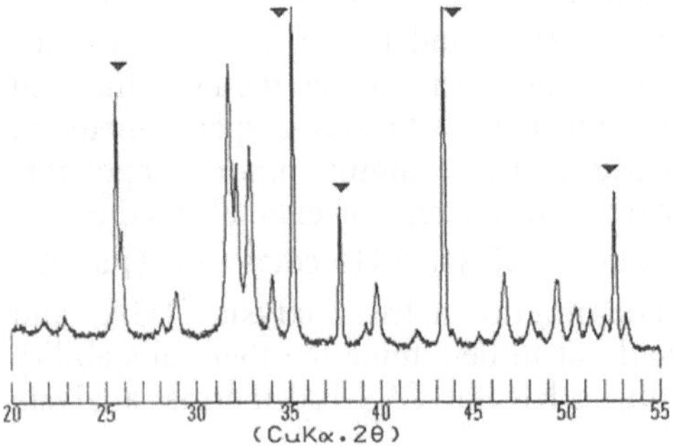

Fig.1. X-ray diffraction pattern for *Moeritherium* enamel. Arrow heads indicate the peaks of corundum.

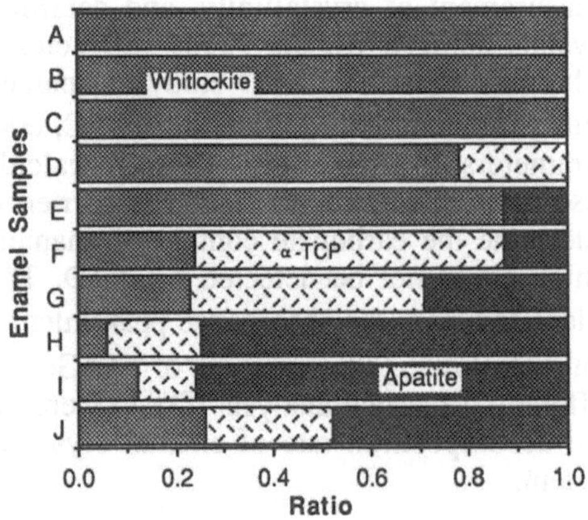

Fig.2. Crystal components of enamel after heating at 1200 ^{o}C.

TGA patterns showed that the weight loss curves of the enamel samples were mostly similar to that of human enamel. However, total weight losses of the enamel samples gradually decreased as the geological age of the samples decreased (Fig. 4).

The carbonate contents of the enamel samples were mostly similar to that of human enamel, around 4.0 wt %, except for *Moeritherium* enamel, which had the lowest content (Fig. 5).

In the TGA-DTA (Fig. 6a) and IR (Fig. 6b) patterns for *Stegodon* enamel, the presence of calcium carbonate, probably the calcite form, was distinguishable by its endotherm at 850 °C accompanied by a weight loss and a weak absorption peak at 720 cm^{-1}, respectively. A minor amount of calcium carbonate was also detected in *Mammuthus, Dinotherium* and *Palaeomastodon* enamels.

To determine any diagenetic effects on the enamel samples, the outer and inner layer enamels were removed from the same teeth of *Palaoloxodon, Stegodon, Dinotherium* and *Palaeomastodon*. The relationship between a-axis length and crystallinity, measured by full width at the half maximum of peak, FWHM, between the outer and inner layer

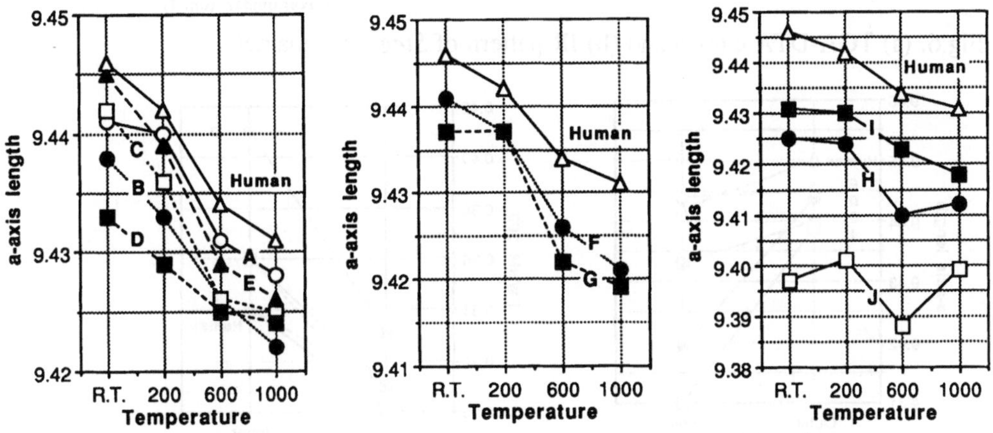

Fig.3. Changes in a-axis length upon heat treatment for (a) Group 1, (b) Group 2 and (c) Group 3 Proboscidea enamel.

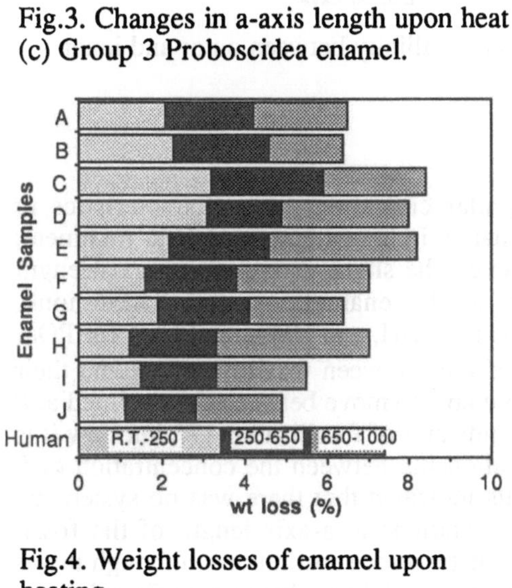

Fig.4. Weight losses of enamel upon heating.

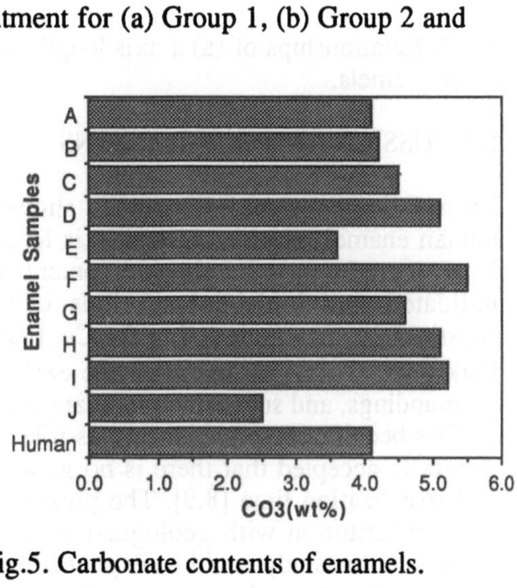

Fig.5. Carbonate contents of enamels.

enamels was quite similar to that of human enamel (Fig. 7a, b) and in harmony with the reference data [2,6]. These results indicate that fossil enamel has not suffered severe diagenetic action.

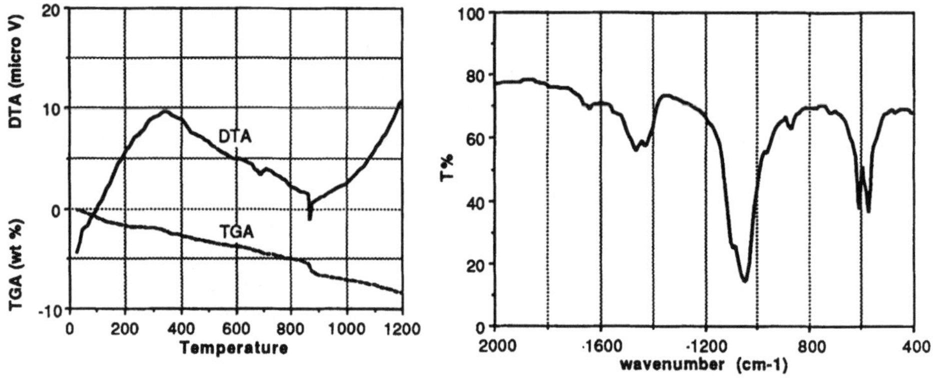

Fig.6. (a) TGA-DTA curves and (b) IR pattern of *Stegodon* enamel.

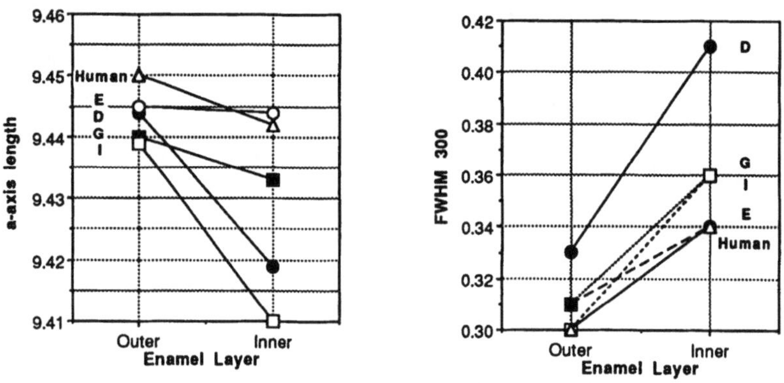

Fig.7. Relationships of (a) a-axis length and (b) crystallinity between outer and inner layer enamels.

DISCUSSION AND CONCLUSIONS

Enamel from recent Proboscidea showed similar crystallographic characteristics to human enamel with regard to a-axis length, change in a-axis length by heat treatment, heating products and carbonate content. However, the slight variation in a-axis length indicated that ionic substitutions occurred in the enamel apatites. These ionic substitutions may include Mg for Ca, F and CO_3 for OH, and HPO_4 and CO_3 for PO_4. Parker and Toots [1] discussed the exchange of ions between fossil materials and their surroundings, and suggested that many ions were able to move between the two media. F ion has been discussed regarding its role as an indicator of fossilization [7], but now it is generally accepted that there is no general relationship between the concentration of F and fossilization time [8,9]. The present results indicated that there was no systematic ionic substitution with geological time. The deviations in a-axis length of the fossil enamel in this study could be explained by diagenetic ionic substitutions, although it was observed that the carbonate content of the fossil enamel did not change significantly.

Table 1. Geological times and localities of Proboscidea enamels used in this study, together with a-axis lengths and crystallite sizes.

Sample	Animal species	(Time, Locality)	a-axis length	crystallite size (A)
A	*Loxodonta africana*	(Recent)	9.441	250
B	*Elephas maximus*	(Recent)	9.438	230
C	*Mammuthus primigenius*	(Pleistocene, Siberia)	9.442	215
D	*Palaeoloxodon naumanni*	(Pleistocene, Japan)	9.434	195
E	*Stegodon orientalis*	(Pleistocene, Japan)	9.449	230
F	*Stegolophodon shinshuensis*	(Pleistocene, Japan)	9.441	225
G	*Dinotherium lyonsi*	(Pliocene, Europe)	9.438	225
H	*Gomphotherium sp.*	(Miocene, Sywarik)	9.401	175
I	*Palaeomastodon wintoni*	(Oligocene, Egypt)	9.428	230
J	*Moeritherium lyonsi*	(Eocene, Egypt)	9.431	235
	Human[2]		9.446	300
	Macaca[3]		9.446	275
	Equus[4]		9.444	285
	Kangaroo[4]		9.462	185
	Hippopotamus[5]		9.450	410
	Sea cow[5]		9.442	-

The fact that the IR study indicated that the carbonate contents of the fossil enamels were mostly within the same range suggests that they had not suffered severe diagenetic action. This was also supported by the XRD studies of crystallite size and of the relationship between the outer and inner layer enamels. This suggests that carbonate in enamel apatite remains in its original state, although an explanation is needed for the presence of calcium carbonate in some fossil enamels.

Among the studied characteristics the thermal decomposition patterns were most closely related to the geological age of the samples. The thermal stability shown by thermal analysis could be related to the chemical constituents of enamel. These thermal decomposition patterns could be explained by assuming that the enamel samples retained their original nature and that they had been fossilized. Although there is no direct evidence, the former cannot be denied at the present time.

Finally, Table 1 also shows that the a-axis lengths of tooth enamels differed among the examined animal species. It has been shown repeatedly that tooth enamel structure has differentiated in the course of evolution [10]. Thus it is possible that enamel crystals might also have differentiated in a similar manner, developing different characteristics among animal species. Further analysis of tooth enamel crystals is underway in our laboratory.

REFERENCES

[1] Parker RB, Toots H (1980) In: Behrensmeyer AK, Hill AP (eds) Fossils in the Making. Univ Chicago Press. Chicago, pp.197-207.
[2] Sakae T (1988) Arch oral Biol 33:707-13.
[3] Sakae T, Sekikawa M, Hirai G, Ozaki T (1988) Nihon Univ J oral Sci 14:441-445.
[4] Sakae T, Sekikawa M (1990) J Nihon Univ Sch Dent 32:99-103.
[5] Trautz O (1967) In: Miles AEW (ed) Structural and Chemical Organization of Teeth. Academic Press, New York, vol.2, pp.165-200.
[6] Sakae T, LeGeros RZ (1990) IADR annual meeting Abstr. #311.
[7] Middleton J (1844) Geol Soc London 431-433.
[8] Brophy P, Hatch M (1962) Amer Mineral 47:1174-1180.
[9] McConnell D (1973) Apatite. Springer-Verlag. Wein, p.111.
[10] Kozawa Y (1988) In: Omori M, Suga S, Goto M (eds) Biomineralization and Phylogeny of Marine Organisms. Tokai Univ Press. Tokyo, pp.247-260.

CHAPTER 4.24

Development of Enamel Layar observed in Some Fossil Proboscideans

H. Kamiya

Department of Geology and Mineralogy, Kyoto University, Kyoto, 606 Japan

Key Words: Proboscidea, Stegodont, Enamel, Bilayering.

INTRODUCTION

Proboscidea (group of elephant) is well traced in its morphological change through the process of evolution. The primitive proboscideans in the stage of Mastodon have bunodont molars, and the advanced one like recent *Elephas* has high crowned elasmodont molars. It is very important subject to make clear the relationships between such a typical morphological change and the internal structure or the histological characteristics of the molars. Studies on the histology of enamel, especially on the Hunter-Schreger's bands and enamel prisms have been made by some workers [1],[2],[3]. Observation on *Stegodon* and *Eostegodon* which have an intermediate form, lophodont, has been also made[4],[5]. In this paper, histological characteristics of enamel of the intermediate form molars were investigated. **I am very thankful to Drs. Y. Kozawa, H. Saegusa, Mr. H. Taruno and the members of Research Group for Proboscidea for their advices and discussions.**

MATERIALS AND METHODS

Materials used in this study are shown in the following.

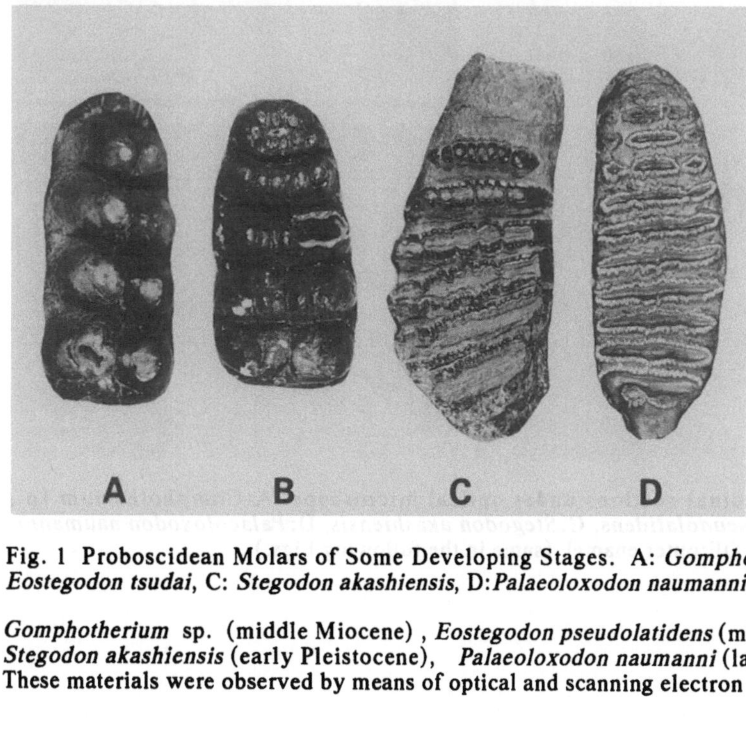

Fig. 1 Proboscidean Molars of Some Developing Stages. A: *Gomphotherium* sp. B: *Eostegodon tsudai*, C: *Stegodon akashiensis*, D:*Palaeoloxodon naumanni*

***Gomphotherium* sp. (middle Miocene) , *Eostegodon pseudolatidens* (middle Miocene), *Stegodon akashiensis* (early Pleistocene), *Palaeoloxodon naumanni* (late Pleistocene). These materials were observed by means of optical and scanning electron microscopy.**

OBSERVATION AND CONSIDERATION

Single Enamel Layer and Bilayered Enamel : Under the optical microscope, *Stegodon* enamel apparantly consists of two layers, inner and outer. The thickness of each layer is nearly half of whole enamel in the case of *S. akashiensis.* In the inner layer, the Hunter-Schreger's bands can be recognized, and it is obvious near the dentinoenamel junction. In the outer layer, the Hunter-Schreger's bands are not visible, and often it looks chalky whitish in its color. In the molar of the advanced type, enamel consists of two or three layers apparantly. In the molars of the primitive form, *Gomphotherium*, enamel has a large thickness, approximately 10 millimeters. The Hunter-Schreger's bands well develop throughout the whole enamel layer, and they slightly curve in their running direction at the part near the outer surface of enamel. However, bilayered structures can not be observed. These figures are shown in Fig. 2 A-D. Enamel of *Eostegodon pseudolatidens* of which molar has an interemediate form between bunodont and lophodont, and which is considered an ancester of *Stegodon*, has two layers. Hunter-Schreger's bands are observable in the inner layer, but they are not recognized in the outer layer. Such a pattern is similar to that of *Stegodon* , but it is not so obvious in comparison with *Stegodon.* On the fractured surface, it also can be seen the arranging of the enamel prisms.

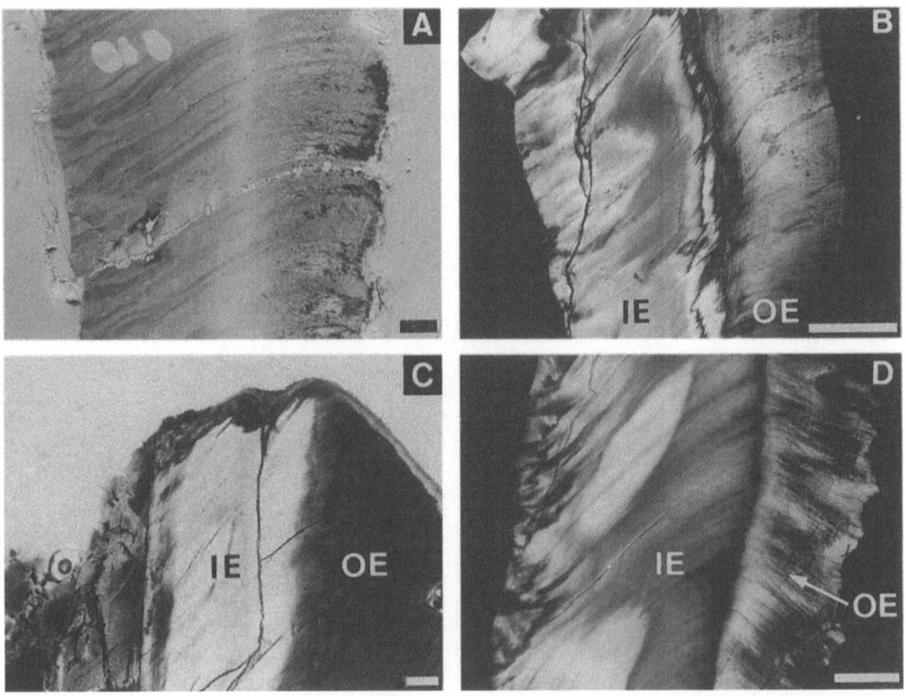

Fig. 2 Longitudinal sections under optical microscope. A:*Gomphotherium* sp., B:*Eostegodon pseudolatidens*, C:*Stegodon akashiensis*, D:*Palaeoloxodon naumanni.* IE:inner enamel, OE:outer enamel (same in the following Figs.)

The boundary between the inner enamel layer and the outer one is not a distinct line, but a zone where the arranging direction of enamel prisms suddenly changes. Under scanning electron microscopy, it is very clearly recognized. In *Stegodon*, enamel prisms or Hunter-Schreger's bands are running along the direction by 30 or 50 degrees to the dentinoenamel junction in the longitudal section, then they turn to the dirction nearly parallel to the occulusal surface.

Topography of Enamel on the Occulusal Surface : On the worn surface of enamel of some proboscideans, the step like structure is observed. It has been well known , but recently this character is reestimeated[6]. According this, it is one of the important characters in stegodont molar. As shown in Fig.3, running direction of enamel prisms is quite different in each layer(Figs. 4 and 5). In the inner layer, the direction is nearly perpendicular or by high angle to the occulusal surface, while it is nearly parallel to the surface in the outer layer. It is easily supposed that the inner layer is more resistant against wearing stress than the outer layer(Figs. 4 and 5).

Shape of Enamel Prisms : Enamel prisms of *Stegodon* and *Eostegodon* were observed by scanning electron microscopy. In the molar enamel of *Stegodon akashiensis*, the enamel prisms often show the laterally elongated shape or a little similar shape to gingko-leaf like shape can be observed, but it is not common. Many of them are so called key-hole shaped one(Fig. 6 A, B).

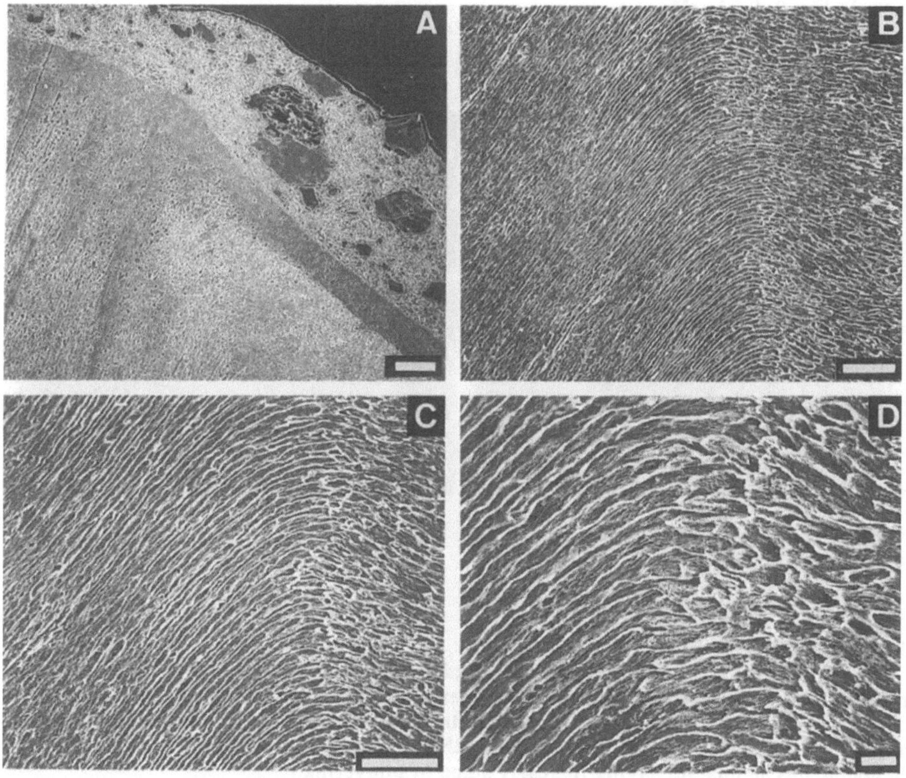

Fig. 3 Enamel prism arrangement in boundary between the inner layer and the outer layer. SEM images. Scale : 100 micronmeters(A), 50 micronmeters(B,C), 10 micronmeters (D)

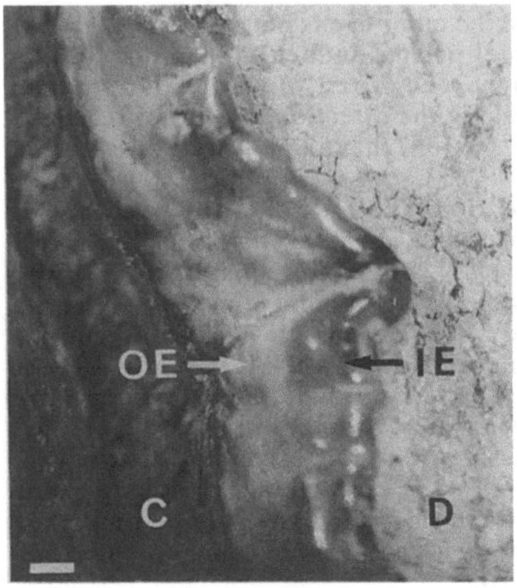

Fig. 4 Occulusal view of the enamel of . *Stegodon.* D:dentine C:cementum. Same in Fig.5. Scale shows 1mm.

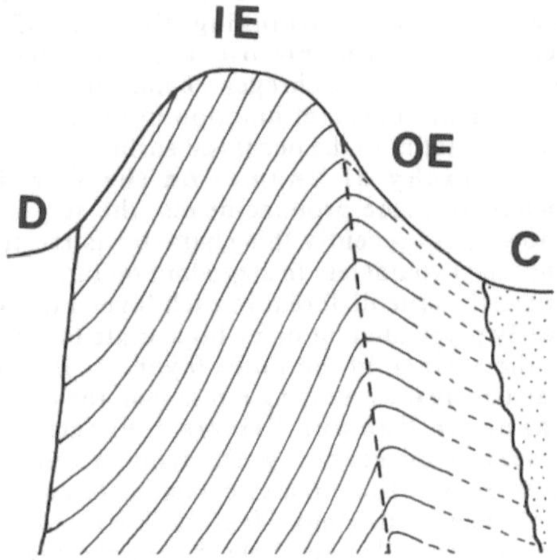

Fig. 5 Schematic diagram of two layered enamel and the arrangement of the enamel prisms.

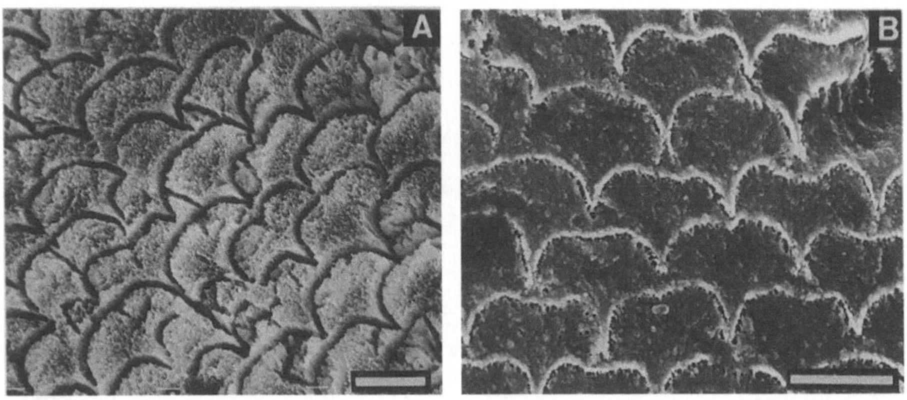

Fig. 6 Enamel prisms in transverse section. SEM image. A:*Stegodon*, B:*Eostegodon.* Scale shows 5 micronmeters.

CONCLUSION

1.The bilayered enamel is clearly observed in the stegodont molars which have an intermediate form between bunodont and elasmodont . It can be supposed that this differentiation of the enamel layer occurred through the process of evolutional change of molar form from the primitive elephants to the advanced one. *Eostegodon,* ancester of *Stegodon,* has an enamel having the early stage of bilayering.

2. The bilayering is basically due to the different arrangement of the enamel prisms and it is associated with presence of Hunter-Schreger's bands in the inner layer.

3. The step like structure on the worn enamel surface corresponds to the bilayering of enamel. The outer layer is more easily worn away than the inner layer due to parallel arrangiment of enamel prisms to the worn surface and no presence of the Hunter-Schreger's bands.

REFERENCE

1.Kozawa Y(1978) J Stmat Soc Japan 45;585-606
2.Kozawa Y(1985) Tooth enamel VI. Elsevier Science Pub., pp.437-441
3.Bertrand P(1988) Mem Mus natn Hist nat, Paris (serie C) 53:85-99
4.Kamiya H,Taruno H(1988) ditto 53:233-240
5.Kamiya H,Saegusa H(1988) Abst 95 Annual Meeting Geol Soc Japan:289
6.Taruno H(1985) Bull Osaka Mus Nat His 38:23-36

CHAPTER 4.25

Well-Preserved Collagen from a Late Pleistocene Elephant Ivory

H. OCHIAI[1] and M. AKIYAMA[2]

[1]Department of Biology, Hokkaido University, Sapporo, 060 Japan and [2]Department of Geology, Shinshu University, Matsumoto, 390 Japan

Key words: Collagen, Fossil ivories, Late Pleistocene, SDS-electrophoresis.

INTRODUCTION

Many reports have been published on collagens in fossils [1]. α1 and α2 peptide chains of fossil collagen molecules were identified from a Magadan frozen mommoth [2]. Tuross et al. [3, 4] reported a trace amount of α chains of collagen molecules from a fossil whale bone of Late Pleistocene age, and concluded that the fossil collagens were well preserved under favorable conditions, and were not affected principally by the presearved years.

No biochemical approach has been conducted on collagen molecules from fossil dentines, though dentine forms a compact structure and seems to be the most appropriate materials for biochemical studies of fossil collagens. For this reason we examined dentine collagens of fossil elephants found abundantly in and around Japan, using SDS-electrophoresis.

MATERIALS AND METHODS

1 Samples

Five fossil and two modern elephant tooth specimens used in this study are as follows.

(1) Mammuthus primigenius (Leningrad specimen): Leningrad Zoological Institute; Ivory; 24,260 ± 270 yr. BP (radiocarbon age of dentine collagen by accelerator mass spectrometry of Nagoya University).

(2) Palaeoloxodon naumanni (Hirosawa specimen): Sea bottom of the Sea of Japan, 31 miles off Hinomisaki, Shimane Pref. [5]; Ivory; 38,500 ± 600 yr. BP [6].

(3) Palaeoloxodon naumanni (Nojiri-ko specimen): Lower Nojiriko Formation III B2; Lake Nojiri, Tategahana, Shinano-machi, Nagano Pref. ; Ivory; ca. 3.5 x 10^4 yr. BP [7].

(4) Palaeoloxodon naumanni (Holotype specimen): Sahama mudstone; Sahama-cho, Hamamatsu, Shizuoka Pref.; Ivory; Middle Pleistocene.

(5) Stegodon orientalis (Fukakusa specimen): Upper part of the Osaka Group (Ma6); Fukakusa-Yaguchi-cho, Fushimi-ku, Kyoto, Molar (^{2}M); Middle Pleistocene.

(6) Elephas maximus: Ivory; Modern.

(7) Loxodonta africana: Ivory; Modern.

2 Analytical Methods

Dentine samples powdered less than 20 mesh size in a stainless mortar were washed with distilled water and then acetone using a magnetic stirrer. The washed samples were decalcified with 0.5M EDTA-Tris buffer (pH = 7.5) at 4°C for two days. This decalcification was repeated five or six times until no detection of Ca^{2+} by a calcium C-test method. The residues were washed with distilled water and were freeze-dried. These freeze-dried samples were dissolved in a SDS-sample buffer at 100°C for 5 min for SDS-polyacrylamide gel electrophoresis (SDS-PAGE). SDS-PAGE (slab-gel, 8.5 x 5.4 cm in size) was conducted in a concentration gradient gel of 4.5 - 6.5 % with the Laemmli system[8]. Peptides were stained with Coomassie Brilliant Blue.

Cyanogen bromide (CNBr) digestion of the freeze-dried samples was conducted to compare primary sequences of peptides of collagens as follows. To 1-7mg of freeze-dried samples was added 1.0 ml of 0.5M acetic acid with 2-mercaptoethanol. This solution was kept in an oven at 37°C after flashing with N_2 gas. After freeze-drying with 5ml of degassed distilled water, the sample was put in 0.5-1.0 ml of 70% (v/v) formic acid, followed by flashing with N_2 gas to remove dissolved air. After a few mg of CNBr was added, the sample was flashed with N_2 gas and kept at 30°C for 4 hr. To the peptides cleaved with CNBr were added 5ml of distilled water and then, the solution was lyophilized. This step was repeated several times to remove the residual formic acid. An SDS-PAGE was conducted in the same procedure as described above in a concentration gradient gel of 7.5 - 17%.

For amino acid analysis collagen samples were hydrolysed at 110°C for 24 hr under N_2 atmosphere. Amino acid analyses were conducted by a Hitachi 835 Amino Acid Analyser at the Center for Instrumental Analysis, Hokkaido University.

RESULTS

1 Soluble Fraction of Collagens

Figure 1 shows $\alpha 1$, $\alpha 2$ and β chains with faint smear staining in the Hirosawa specimen, and also those of collagens in two modern elephant dentine, bovine achilles tendon (type I, Catalogue no. C9879, Sigma Co.), rat tendon (type I, 234132, CALBIOCHEM Co.) and mouse skin (type I, 234148, CALBIOCHEM Co.) .

Smear staining without any clear band is found in the Nojiri-ko and the Leningrad specimen, while no staining is found in the Middle Pleistocene Holotype and Fukakusa specimens (Fig. 2).

2 CNBr-Cleaved Peptides

Cyanogen bromide (CNBr) selectively cleaves peptide bonds on the carboxyl side of methionine residues. SDS-electrophoresis patterns of CNBr-cleaved peptides from two modern elephant dentine collagens and the bovine achillles tendon collagen are almost identical each other. CNBr-cleaved peptides of mouse skin collagen show a similar pattern with the three collagens mentioned above with an additional clear band around 30K daltons. $\alpha 2$ chains of mouse are cleaved into six peptide [9]; 1, 0, 4, 2, 3 and 5, while those of bovine achilles collagen into five peptide 1, 0, 4, 2, and 3-5 since one methionine residue does not exist at the position between 3 and 5 [10]. A clear band around 30K on SDS-PAGE (mark a) of mouse skin collagen could be identified to 3 and/or 5 peptides (Fig. 3). Therefore, dentine collagens of modern elephants are identical to bovine achilles collagen as for the position of methionine residues. The fossil dentine collagen of the Hirosawa specimen, however, has much less numbers of SDS-PAGE bands than those in the other four modern collagens. This evidence suggests that strong closs-linkages between α chains have already disappeared during fossilization.

3 Amino Acid Analysis

Amino acid compositions of fossil and modern elephant dentine collagens are shown

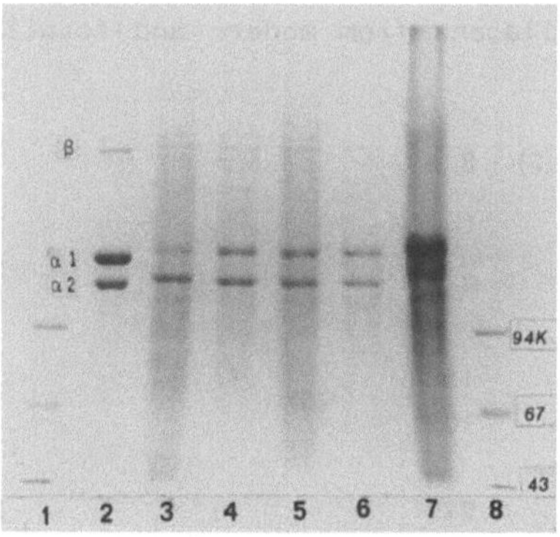

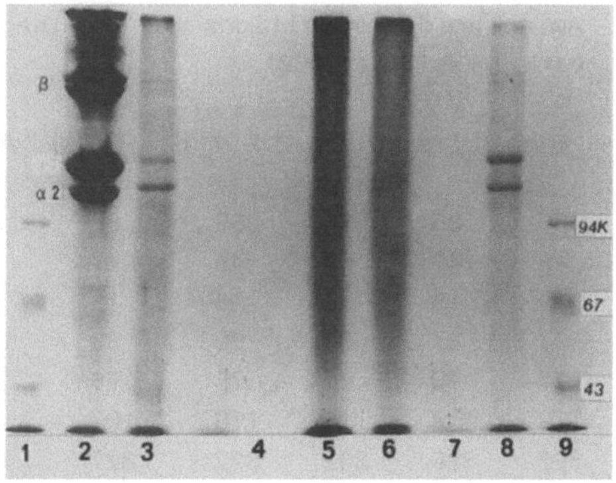

Fig. 1 SDS-polyacrylamide gel electrophoresis of SDS-soluble fraction of collagens.
1 and 8 : Protein size markers
2 : Mouse skin collagen
3 : Bovine achilles tendon collagen
4 and 6 : Dentine collagen of Loxodonta africana (Modern)
5 : Dentine collagen of Elephas maximus (Modern)
7 : Dentine collagen of Palaeoloxodon naumanni (Hirosawa specimen)
α1, α2 and β : Refer the text.

Fig. 2 SDS-polyacrylamide gel electrophoresis of SDS-soluble fraction of collagens.
1 and 9 : Protein size markers
2 : Rat achilles tendon collagen
3 : Bovine achilles tendon collagen
4 : Dentine collagen of Stegodon orientalis (Fukakusa specimen)
5 : Dentine collagen of Palaeoloxodon naumanni (Nojiki-ko specimen)
6 : Dentine collagen of Mammuthus primigenius (Leningrad specimen)
7 : Dentine collagen of Palaeoloxodon naumanni (Holotype specimen)
8 : Dentine collagen of Loxodonta africana (Modern)

in Table 1. The Hirosawa specimen shows the same amino acid composition as in the two modern elephants, though the total amino acid content of the fossil sample is only one-fifth of that of the modern. The Nojiri-ko and the Leningrad specimens, which contain less than one-hundredth of total amino acid contents found in the modern specimens, show the composition characteristic of collagen. However, the other fossil specimens, Holotype and Fukakusa specimens of Middle Pleistocene age, contain less than one-thousandth of total amino acid contents of the modern, and do not show any indication of collagen nature in their composition.

CONCLUSION

1) Three preservation states were recognized in five fossil specimens used in this study.

(a) Well preserved and α chains of collagen are found from a fossil ivory of Palaeoloxodon naumanni collected from the sea bottom of the Sea of Japan (Hirosawa specimen).

Table 1 Amino acid compositions of dentine collagens from modern and fossil elephant teeth (residues/1000).

	E. m	L. a	M. p	P. n (H)	P. n (N)	P. n (S)	S. o
Hypro	85	85	86	89	88	-	-
Asp	49	47	46	46	47	80	92
Thr	19	18	18	18	19	51	-
Ser	37	37	32	35	33	102	106
Glu	77	75	77	75	76	86	103
Pro	119	118	115	119	107	62	-
Gly	322	326	332	322	327	163	279
Ala	115	115	130	118	126	110	88
Cys/2*	-	-	5. 4	-	6. 5	38	152
Val	25	25	25	26	28	73	114
Met	5	1	2. 6	3. 2	3. 6	-	-
Ileu	11	11	9. 3	11	13	40	-
Leu	28	28	23	25	28	65	68
Tyr	5. 2	5. 1	-	1. 9	2. 7	19	-
Phe	15	15	12	14	14	36	-
Hylys	2. 9	2. 5	2. 3	3. 7	2. 5	-	-
Lys	30	31	39	31	31	41	-
His	4. 7	4. 8	1. 0	3. 6	1. 9	-	-
Arg	50	50	45	48	47	36	-
Total (m mol/g)	2. 96	2. 70	0. 02	0. 52	0. 01	0. 0001	0. 001

Cys/2* : Ninhydrin-positive matter at the Cys/2 position on a chromatogram
E. m : Elephas maximus (Modern)
L. a : Loxodonta africana (Modern)
M. p : Mammuthus primigenius (Leningrad specimen)
P. n (H) : Palaeoloxodon naumanni (Hinomisaki specimen)
P. n (N) : Palaeoloxodon naumanni (Nojiri-ko specimen)
P. n (S) : Palaeoloxodon naumanni (Holotype specimen)
S. o : Stegodon orientalis (Fukakusa specimen)

(b) Amino acid composition shows a characteristic nature of collagen in the Nojiri-ko specimen of P. naumanni and Leningrad specimen of Mammuthus primigenius. However, no clear bands but only smear patterns are observed on an SDS-PAGE.
(c) A faint amount of amino acids shows a different composition from that of collagen in Middle Pleistocene specimens.
2) As for methionine residues in α2 chain, the dentine collagen of the modern elephant ivories are identical to bovine achilles collagen, but not to mouse skin collagen.
3) Numbers of peptides cleaved with CNBr of the Hirosawa specimen are much smaller than those of modern collagens. This fossil specimen shows an extraordinarily excellent preservation, though the cross linkages between α chains have already been cleaved during diagenesis.

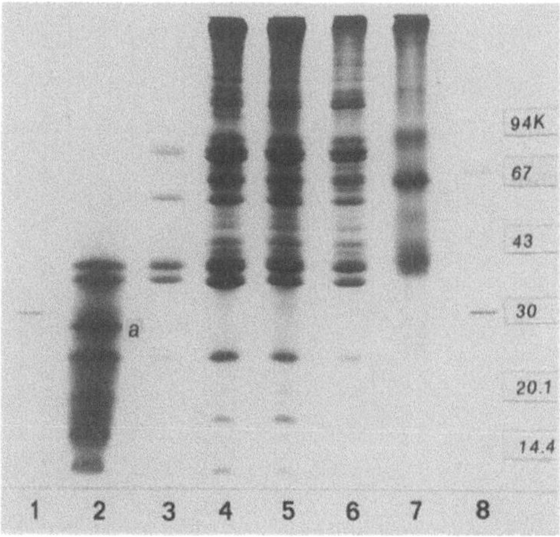

Fig. 3 SDS-polyacrylamide gel electrophoresis of CNBr-cleaved peptides of collagen. The samples and their numbers are same as those shown in Fig. 1.

Acknowledgements - We are grateful to Prof. S. Akagi of Tottori University , Dr. H. Kamiya of Kyoto University and Dr. Y. Kozawa of Nihon University for their kind offering of the specimens used in this study, and also to Prof. N. Nakai and Dr. T. Nakamura of Nagoya University for conducting a radiocarbon isotope age determination. This study was supported by a Grant-in-Aid for Scientific Research from the Ministry of Education, Science and Culture of Japan (Project No. 60304014).

REFERENCES

[1] Wyckoff RWG (1972) The biochemistry of animal fossils. Scientechnica Ltd., 152p.

[2] Goodman DE, Birk AE, Romero-Herrera MA, Lande HD, Barnhart MI (1980) FEBS Lett 114 : 30-34.

[3] Tuross N, Hare PE (1978) Carnegie Inst. Year Book 77 : 891-895.

[4] Tuross N, Eyre DR, Holtrop ME, Glimcher MJ, Hare PE (1980) In :Hare PE, Hoering TC, King KJr (eds) Biochemistry of Amino Acids. Wiley, pp. 53-63.

[5] Akagi S (1981) Sci Rep Fac Sci Tottori Univ 30 : 57-64.

[6] Akiyama M, Kamei T, Nakai N (1988) Earth Science (Chikyu Kagaku) 42 :29-31.

[7] Nojiri-ko Excavation Research Group (1990) The Nojiri-ko Excavation 5 (1987-1989), Monograph Assoc Geol Collab Japan 37, 181p.

[8] Laemmli UK (1970) Nature 227 : 680-685.

[9] Ramachandran GN, Reddi AH (eds) (1976) Biochemistry of Collagen. Plenum Press.

[10] Japanese Biochemical Society (ed.) (1979) Biochemistry Data Book I, Connective Tissues, 1657-1664.

CHAPTER 4.26

Comparative Biochemistry of Bone Matrix Proteins in Bovine and Fish

Y. KUBOKI, T. WATANABE, M. TAZAKI, and H. TAKITA

Department of Biochemistry, School of Dentistry, Hokkaido University, Sapporo, 060 Japan

Key words: Bone collagen, Noncollagenous proteins, Bovine, Fish, Comparative biochemistry

INTRODUCTION

Bone and dentin collagen fibrils are the matrix proteins within which the hydroxyapatite crystals deposit during the formation of these tissues. It is believed that characteristic cross-linking patterns and subunit composition of these collagens must be related to the calcifiability of these proteins [1-4]. On the other hand, some of the noncollagenous proteins of the hard tissues are believed to regulate the calcification process [5,6]. If these hypotheses are to be verified, evidence must be presented from a wide variety of vertebrate species from mammalia to fish. However, previous studies on chemical structure of bone matrix have been mostly done about the materials from mammalian [1-5,7-9] and avian [10] species, but very limited from fishes. In this paper we attempted to elucidate the universality of the characteristics in bone collagen and noncollagenous proteins by comparing the matrices of pollack and of bovine bone.

Dehyro-dihydroxylysinonorleucine and pyridinoline are both lysine-derived collagen cross-links and comprise the principal cross-links of bone collagen [3]. While, histidinoalanine has recently been established to be the cross-link amino acid in the certain phosphoproteins in bovine bone and dentin as well by Kuboki et al.[11, 12]. In this study these three cross-links were adopted as the markers of comparative studies of fish and bovine. Also osteonectin which is one of the major noncollagenous proteins in bovine bone with a molecular weight of about 40 kDa was studied whether there is an homologous protein in fish bone.

MATERIALS AND METHODS

<u>Extraction of Bone Proteins</u> Metacarpal and metatarsal bones of adult bovine (average three-year-old) and vertebral bones of pollack were pulverized in a stainless steel mortar and pestle under liquid nitrogen after removing soft tissues as previously reported [12]. The bone powder (20-100 mesh) was extracted first with 1 M NaCl/0.05 M Tris-HCl, pH 7.4, and then with 0.5 M EDTA/1 M NaCl/0.05 M Tris-HCl pH 7.4 at 4 °C. Both extraction solvents contained protease inhibitors (50 mM 6-aminohexanoic acid, 5 mM benzamidine-HCl and 1 mM phenylmethanesulfonylfluoride). The extract with 0.5 M EDTA/1 M NaCl was extensively dialyzed against distilled water and lyophilized. This material was designated as the EDTA/NaCl extract, while the insoluble residues after EDTA/NaCl extraction was designated as

bone collagen. Bovine osteonectin in the EDTA/NaCl extract was purified by the calcium-induced precipitation method reported previously [13].

Sepharose CL-6B Chromatography Gel filtration chromatography was performed by a method essentially the same as that described by Termine et al. [5], except for the column size (2.2 x 94 cm).

Cation-Exchange Chromatography with carboxy-methyl Cellulose The CNBr peptides were prepared as reported [3] and applied on a Whatman CM-52 column (16 x 150 mm), which was run with a 0.02-0.14 M NaCl linear gradient in 300 ml of 0.02 M sodium citrate buffer, pH 3.6.

Anion-Exchange Chromatography with DEAE Cellulose A Whatman DE-52 column (16 x 160 mm) was equilibrated with 6 M urea/0.05 M Tris-HCl, pH 8.3. Pollack EDTA/NaCl extract was applied on a DE-52 column, which was then run with a 0-0.5 M NaCl linear gradient in 400 ml of the same buffer. The flow rate was 60 ml/hour.

High-Performance Gel-exclusion Chromatography A YMC-Pack column Diol-300 (8 x 500 mm) obtained from Yamamura Chemical Laboratory Co,. Kyoto, Japan. The EDTA/NaCl extract in 4 M guanidine HCl/50 mM Tris-HCl, pH 7.4, were loaded on the column and eluted with the same buffer as previously reported [12]..

Reverse-Phase High-Performance Liquid Chromatography An M & S-Pack C-18 column (4.6 x 150 mm) was obtained from M & S Corp., (Tokyo, Japan) and was equilibrated with 0.1% (v/v) trifluoroacetic acid. Elution was performed for the first 5 ml with 0.1% trifluoroacetic acid, then with a linear gradient between 0.1% trifluoroacetic acid and 60% acetonitrile in 0.1% trifluoroacetic acid, totaling 30 ml. The flow rate was 1 ml/minute.

Cross-Link Analysis Bone collagens were reduced with [^{3}H]-$NaBH_4$ as previously reported by Kuboki et al.[1-3]. Hydrolyzed samples (5 X 10^5 dpm) were analyzed chromatographically on the column (0.9 x 50 cm) of a Hitachi 034 amino-acid analyzer, eluted for first one hour with a citrate buffer, 0.2 M in Na^+ ions, pH 3.25, and for following three hours with 0.35 M, pH 5.28 at 60°C. The flow rate was 60 ml/hour. By using a split-stream device, 60% of the column effluent was monitored for tritium radioactivity by mixing it with a scintillation cocktail (PCS solution, Amersham Japan, Tokyo) and pumping through a Teflon coil in the well of a scintillation counter [1].

Trypsin Digestion of Bone Collagens Bovine and pollack bone collagens were digested with trypsin accompanied by heating at 60°C before and after the digestion. This method gave complete trypsin peptides of hard tissue collagens as reported by Kuboki et al. [4]. Fluorescent pyridinoline peptides were separated on reverse-phase HPLC.

Amino Acid Composition and Sequence Analysis The hyrdolyzed samples were analyzed on a Hitachi L8500 high-speed amino acid analyzer by a standard lithium buffer system for usual analysis and by a special buffer system for histidinoalanine [12]. Automatic Edman analysis by using a Applied Biosystems Model 477A Gas Phase Sequencer was done for the purified 40 K protein and their trypsin peptides.

RESULTS and DISCUSSION

Quantitation of Cross-Links It was clearly demonstrated that histidinoalanine (HAL) is present in 0.5 M EDTA/1 M NaCl extractable (noncollagenous) proteins in bovine bone, but is not present significantly in pollack and bovine collagens, and pollack noncollagenous proteins. While, DHLNL and pyridinoline are definite components in bone collagens from both species. The contents of DHLNL and pyridinoline in bovine bone collagen (0.71 and 0.45 moles/collagen mole respectively) were much higher than those in pollack bone collagen (0.10 and 0.33 moles/ collagen moles). It is of interest that the ratio of DHLNL to HLNL was extremely higher in pollack bone than in bovine bone.

In general, cross-link patterns of both collagens share the common features being rich in pyridinoline and DHLNL compared to that of the mammálian soft tissues collagen.

Peptide Mapping and Distribution of Cross-Links in the Fibrils

Limited pepsin digestion could solubilize only a few per cent of the insoluble collagen of pollack and bovine bone as well. Thus, we had to use other two methods of specific solubilization of collagen, cyanogen bromide (CNBr) cleavage [3] and trypsin digestion. The latter method has been developed in this laboratory and proved to be highly effective to solubilize bone and dentin collagen [4]. Pollack bone collagen was also completely solubilized by this trypsin digestion. Carboxy-methyl cellulose chromatography and reverse-phase HPLC of CNBr peptides were considerably different between pollack and bovine collagens. This is coincided with the different methionine content between pollack (15 residues/1000 amino acids) and bovine (6 residues/1000 amino acids) collagen. Again, the reverse-phase HPLC of trypsin digest of both collagen were diferent, indicating a high degree of variation in amino acid sequence between the two collagens from different species.

Subunit Composition of the Bone Collagens

Although the amount of pepsin-solubilized fraction of pollack bone collagen was small, SDS gel electrophoresis of this fraction suggested that α-chain composition of pollack bone collagen is different from that of the typical type I collagen. Instead of showing $\alpha 1(I)$ band which is usually twice as much as $\alpha 2(II)$ band, pollack bone collagen showed an additional band above $\alpha 1(I)$ band.

Although the data are still preliminary, the pollack bone collagen may contain $\alpha 1 \cdot \alpha 2 \cdot \alpha 3$ heterotrimer in addition to $[\alpha 1]_2 \cdot \alpha 2$ as Kimura and Ohno have shown in skin collagen of Alaska pollack [14].

Purification of 40 K Protein from Pollack Bone Matrix Protein

In the comparative gel-exclusion HPLC patterns of pollack and bovine noncollagenous proteins, some equivalent peaks in terms of molecular weights were found between both proteins. They were 62 K, 40 K, 24 K and 10 K proteins. One of the noncollagenous protein in pollack bone which possessed an equivalent molecular weight (about 40 K) to that of osteonectin was purified and sequenced. The 40 K protein was first separated on DEAE-cellulose column, and then on Sepharose CL-6B column. A single peak was obtained on the final step of purification by reverse-phase HPLC. First seven amino acid sequence of the purified 40 K protein from pollack was as follows: Ala-Leu-Pro-Pro-Ala-Val-Pro. Also, the internal amino acid sequencing of this protein was attempted by digesting with lysine-specific Acromobacter protease I. Sequence of one of the isolated peptide was Phe-Leu-Leu-Ser-Ala-Ser-Lys. No homologous sequence to both portion was found by a search of the NBRP (National Biomedical Research Foundation) protein database.

It was concluded that collagen of pollack bone has considerably different primary sequence from that of bovine bone as judged by amino acid analysis and by fingerprinting of both CNBr and trypsin peptide. But essential features of cross-links pattern were similar being rich in DHLNL and pyridinoline. These cross-links are very scarce in mammaliam soft tissue collagens. Also, subunit composition of molecule may be different from typical mammalian type I collagen. Noncollagenous proteins comprise at least four groups in terms of molecular weight. Each component seemed considerably different from that of bovine bone, although there are some equivalent proteins in regard to molecular weight.

ACKNOWLEDGMENTS

This work was supported by Grants-in-Aid from the Ministry of Science, Edication and Culture of Japan (61480383, 62870073).

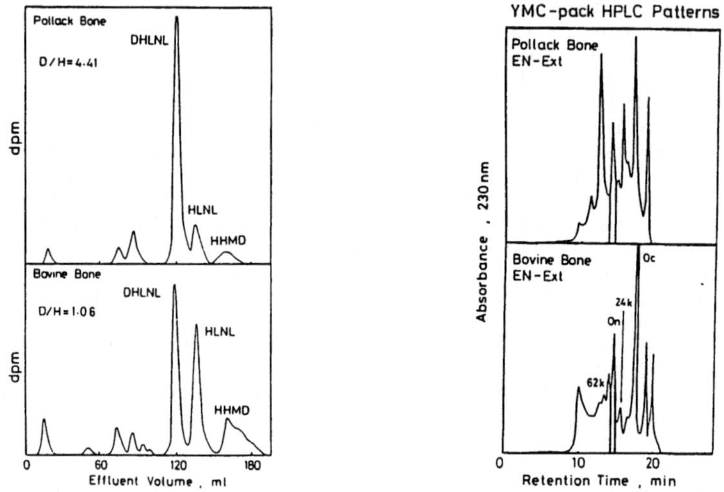

Fig. 1A Chromatograms of reducible collagen cross-links of pollack (upper) and bovine (lower) bone. Ratio of dihydroxylysinonorleucine (DHLNL) to hydroxylysinonoleucine (HLNL) of pollack bone is much higher than that of bovine. HHMD, histidinohydroxymerodesmosine.
1B Comparison of noncollagenous protein in pollack (upper) and bovine (lower) bone in gel-exclusion HPLC. See text for conditions.

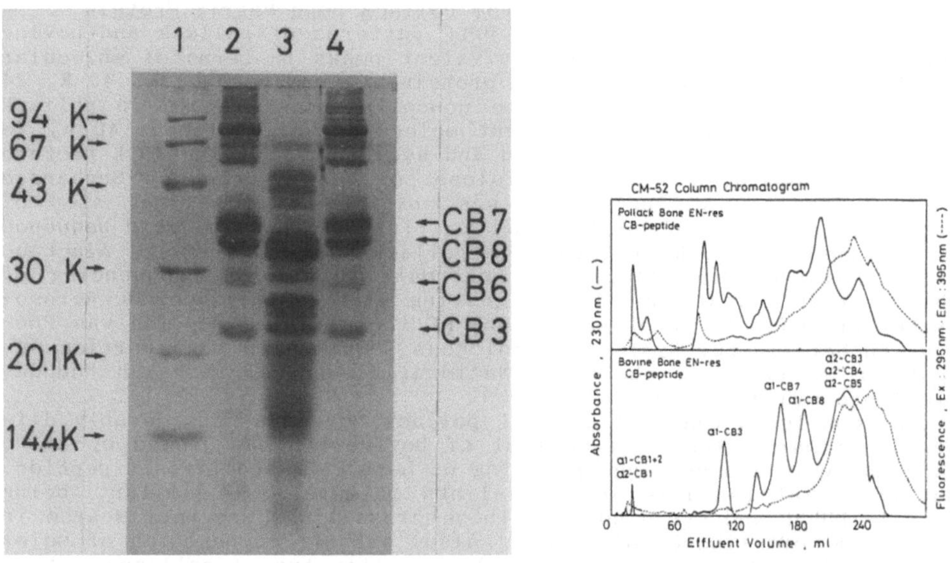

Fig. 2A SDS gel-electrophoresis (15%) of cyanogen bromide peptide of bone collagens from bovine and pollack. Lane 1, authentic proteins; lane 2, bovine bone; lane 3, pollack bone; lane 4, bovine dentin. 2B Carboxy-methyl cellulose chromatograms of the cyanogen bromide peptides of collagens from pollack (upper) and bovine (lower) bone.

REFERENCES

1. Kuboki Y, Takagi T, Sasaki S, Saito S, Mechanic GL (1981) J Dent Res 60: 159-163
2. Kuboki Y, Mechanic GL (1990) Calcif Tissue Int (1982) 34:304-308
3. Kuboki Y, Takagi T, Shimokawa H, Oguchi H, Sasaki S Mechanic GL (1981) Connect Tissue Res 9: 107-114
4. Kuboki Y, Tsuzaki M, Sasaki S, Liu CF, Mechanic GL (1981) Biochem Biophys Res Commun 102: 119-126
5. Termine JD, Belcourt AB, Conn KM, Kleinman HK (1981) J Biol Chem 256:10403-10408
6. Kuboki Y, Fujisawa R, Aoyama K, Sasaki S (1979) J Dent Res 58:1926-1932
7. Bolander, ME, Young MF, Fisher LW, Yamada Y, Termine JD (1988) Proc Natl Acad Sci USA 85:2919-2923
8. Fisher LW, Hawkins GR, Tuross N, Termine JD (1987) J Biol Chem 262:9702-9708
9. Franzen A, Heinegard D (1985) Biochem J 232:715-724
10. Gotoh Y, Gerstenfeld LC, Glimcher J (1990) Eur J Biochem 187:49-58
11. Kuboki Y, Fujisawa R, Tsuzaki M, Liu CF, Sasaki S (1984) Calcif Tissue Int 36:126-128
12. Kuboki Y, Takita H, Komori T, Mizuno M, Furu-uchi E Taniguchi (1989) Calcif Tissue Int 44: 269-277
13. Kuboki Y, Takita H, Fujisawa r, Yamaguchi H, Yamada H, Tazaki M, Ohnuma Y, Mizuno M, Tsuzaki M (1990) Calcif Tissue Int 46: in press
14. Kimura S, Ohno Y (1987) Comp Biochem Physiol 88B: 409-413

CHAPTER 4.27

Effect of Phosphorous-Containing Compounds on Polymorphic Formation of Calcium Carbonate

T. OOMORI[1] and Y. KITANO[2]

[1]Department of Chemistry, College of Science, University of the Ryukyus, Okinawa, 903-01 Japan and
[2]Sugiyama-Jogakuen University, Nagoya, 464 Japan

Key Words: Calcium Carbonate, Polymorphism, Mineralization, Magnesian calcite, Phosphate, Organic Phosphate.

INTRODUCTION

Occurrence and persistance of Mg-bearing carbonates such as magnesian calcite formed in biological fluids or in sea water are one of the most interesting problems in the field of biomineralization. To solve this problem, it is required to elucidate the controlling factors of the polymorphic crystallization from aqueous solution and the Mg content of calcium carbonate.

It has been known that the formation of magnesian calcite is favored by the presence of dissolved organic materials such as citrate, malate and pyruvate[1], Ba^{2+} ions and high carbonate alkalinity of a parent solution[2]. It was also found that low dielectric constant waters favor the formation of a solid solution between calcitic $CaCO_3$ and $MgCO_3$[3]. Still, the occurrence of magnesian calcite in biological fluids and especially in interstitial waters of anerobic marine sediments can not be understood enough for the formation process. Regarding this problem, the behaviors of phosphate ions, during the course of carbonate crystallization from aqueous solution, are interesting; because it was known that phosphate ions favor the calcitic carbonate formation from aqueous solution, and also the ions act as a strong inhibitor for crystal growth or crystal poison of calcium carbonate [4][5]. It is expected that a strong inhibitor for crystallization may play an important role in the polymorphic crystallization from an aqueous solution kinetically, even in the presence of a large amount of Mg^{2+} ions in a parent solution. Phosphorous is one of the most important bio-elements, and a significant amount of phosphorous is contained in marine organisms. In blood or extrapallial fluids of marine organisms, 50 to 500 μM of phosphorous are dissolved as both inorganic and organic forms[4]. There are little studies on the polymorphic formation of $CaCO_3$ in the presence of phosphate ions under marine water conditions.

The purpose of this paper is to present the evidence that the various dissolved phosphate ions in a parent solution strongly affect the polymorphic crystallization of $CaCO_3$ and the incorporation of Mg^{2+} ions from a solution.

EXPERIMENTAL PROCEDURE

Calcium carbonate was synthesized according to the following procedure: $Ca(HCO_3)_2$ solution was prepared by bubbling CO_2 gas(P_{co2}=1atm.) into a suspension of reagent grade $CaCO_3$(Kanto Chem. Co.) and then filtering through Toyo filter paper(No. 5c). This $Ca(HCO_3)_2$ solution contains ca.10 mM of Ca^{2+} ions. Parent solutions were prepared by adding $MgCl_2$ and KH_2PO_4 solutions or organic phosphate solution such as adenosine tri-phosphate disodium salt trihydrate(ATP), adenosine di-phosphate disodium salt trihydrate (ADP), adenosine mono-phosphate disodium salt trihydrate(AMP), α-glycerophosphoric acid disodium salt(α-GADS) or phytic acid sodium salt (PASS) into $Ca(HCO_3)_2$ solution before precipitation started. These organic phosphates are common either in biological fluids or in some marine environment.

The 1000 ml of the parent solution containing 35 or 53.5 mM Mg^{2+}, were taken in a glass vessel in the presence of inorganic phosphate or above shown organic phosphate ions from 0 to 1500 μM, and stand still for about 5 weeks at 28 $\pm$3 ^{o}C. The solution was stirred occationally. With the lapse of time, carbon dioxide gas was escaped to the air and calcium carbonate was formed from the solution. The pH value in the solution ranges from 6.5 to 7.8. It took a couple of days for calcium carbonate to begin to precipitate. Phosphate ions are coprecipitated with calcium carbonate during the course of precipitation. After several weeks, more than 70% of dissolved calcium are precipitated. Obtained precipitates were filtered off through Millipore filter (0.45 μm) and washed with distilled water.

Initial and final concentrations of Mg^{2+} and Ca^{2+} ions in a parent solution were determined by EDTA titration with the lapse of time. Phosphate-P in parent solution and formed carbonate precipitate was determined by colorimetry[6] after decomposition of organic materials if it is necessary. Calcium and Mg contents and mineral composition of the precipitate were measured by atomic absorption(HITACHI 170-50A type) and x-ray diffractmetry (RIGAKUDENKI; 35 KV, 15 mA), respectively.

RESULTS AND DISCUSSION

Effect of inorganic and organic phosphates dissolved in parent solution on crystal form and Mg content of carbonate precipitate

(A) Inorganic phosphate.

The experimental results on the effect of KH_2PO_4 dissolved in a parent solution on the mineral composition, the diffraction angle of (104) spacing in a calcite type carbonate and the Mg/Ca ratio of carbonate are shown in Fig. 1. It is apparent that phosphate ions in a parent solution favor magnesian calcite precipitation even when Mg^{2+} ion concentration is the same as that of sea water. Magnesian calcite is precipitated at the addition of phosphate ions from 20 to 500 μM to a parent solution, with a small amount of aragonite or monohydro-calcite. Aragonite and monohydro-calcite are formed at

low or high phosphate ion concentration in a parent solution. The diffraction angle(2Θ) of (104) spacing of a calcite type carbonate is around 29.7 °(CuKα).
The Mg/Ca molar ratio of the formed precipitate increases with increasing phosphate ion concentration in a parent solution up to 100 μM. This fact is consistent with the proportion of magnesian calcite in the precipitate. Above 500 μM/l of phosphate ions, a small amount of amorphous-like precipitate appeared. X-ray diffraction analysis of the surface floating minerals showed no characteristic diffraction peaks.
The Mg/Ca molar ratio of the precipitates is around unity and the precipitates contain phosphate.
The experimental results at 35 mM of Mg^{2+} ions show the formation of amorphous carbonate and dicalcium phosphate dihydrate(DCPD; brushite) at higher phosphate concentration(PO4-P=1300 μM).

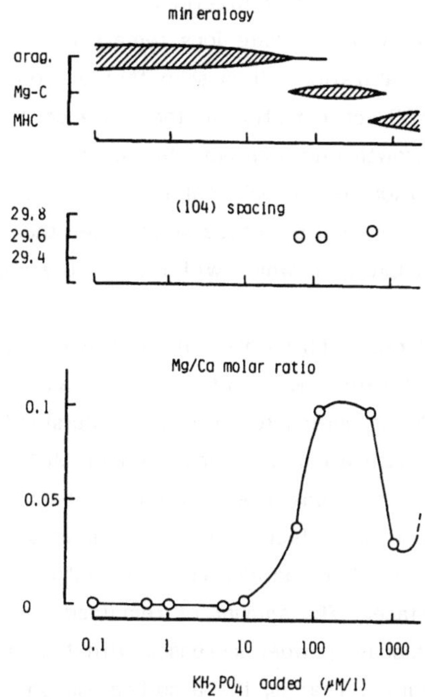

Fig. 1. Effect of KH_2PO_4 dissolved in the parent solution on mineral composition, diffraction angle (2θ) of (104) spacing of calcitic carbonates and Mg/Ca molar ratio of precipitates. Carbonates are formed from the 10 mM $Ca(HCO_3)_2$ solution (Ca^{2+}, 400 mg/l) containing 35 mM $MgCl_2$(850 mg/l) and KH_2PO_4(0.1 to 1500 μM).

(B) Organic phosphate.
The results for ATP, ADP and AMP are analogous to that for inorganic phosphate. ATP and ADP are poly-phosphates which ave three or two $-PO_4$ functional groups in their molecular structures. It is found that organic phosphate ions in a parent solution also have a very significant influence on the magnesian calcite formation and the Mg content of the precipitates as inorganic phosphate ions. Magnesian calcite containing around 6 mole percent of magnesium in the lattice is formed in the presence of 5 to 105 μM of ATP, 5 to 100 μM of ADP and 10 to 50 μM of AMP in a parent solution. Above these concentrations, monohydrocalcite and/or amorphous-like carbonate were formed.
In the α-GADS or PASS containing systems, almost similar results to that of ATP group were obtained. Magnesian calcite was formed in the presence of 10 to >50 μM of GADS and 2 to >150 μM of PASS from the solution containing 55 mM of Mg^{2+} ions.
It is expected that phosphate ions in a parent solution favor the Mg-bearing carbonate formation possibly in biological fluids and probably in interstitial waters of anerobic marine sediments, as dissolved organic materials such as citrate, malate or pyruvate, where the former is more than ten times stronger than the latter in activity of influence.

(c) Rate of crystallization.

Dissolved phosphate ions remarkably decrease the rate of calcium carbonate crystallization. It is seen that a growth rate was decreased with the increase in α-GADS concentration in the presence of Mg^{2+} ions in a parent solution. On the other hand, a growth rate was not changed by the increase in α-GADS in the absence of Mg^{2+} in solution, where only calcite was formed. It is suggested that an interaction of α-GADS with aragonite surface is stronger than that with calcite and inhibit aragonite precipitation, which will lead to the magnesian calcite formation from a solution.

Effects of phosphate ions on the interaction of Mg^{2+} ions with calcite surface

Phosphate ions are clear to play an important role in the polymorphic formation of calcium carbonate. For the understanding of their role in the polymorphic crystallization, the adsorption of Mg^{2+} ions on calcite surface in the presence of phosphate ions in a solution was examined, together with the the presence of other ions which are effective on polymorphic crystallization of $CaCO_3$.

When calcite powder is poured into a solution, Mg^{2+} ions are adsorbed onto calcite surface. The amount of remained magnesium ions in a solution after the addition of carbonate powder increases with the increase in the added amount of phosphate ions in a solution. That is, the adsorbed amount of magnesium ions on calcite surface decreases with increasing phosphate ion concentration in a solution, and finally it becomes zero. It is clear that the presence of phosphate ions in a solution inhibits the adsorption of Mg^{2+} ions on calcite surface.

The required concentration for phosphate ions(1 mM) to inhibit completely the Mg^{2+} ion adsorption is one-tenth of that of fluoride ions(10 mM). Barium ions(10 mM) and Ca^{2+} ions(100 mM or more) in a parent solution favor calcitic carbonate formation even in the presence of Mg^{2+} ions, and inhibit Mg^{2+} ion adsorption on carbonate surface. The sequence of the inhibition effect of ions on Mg^{2+} ion adsorption is as follows: phosphate > Ba^{2+}, F^- > Ca^{2+}. This sequence is consistent with that of the activity of ions forming calcitic carbonate formation from the solution containing Mg^{2+} ions(53.5 mM): phosphate(0.05-0.5 mM) > citrate(0.2 mM[1], Ba^{2+}(0.25 mM[2], F^-(0.5-2.5 mM) > alkalinity(10-50 mM)[9], > Ca^{2+}(50 mM)[7].

Magnesium ions form ion pairs such as $MgH_2PO_4^+$ and $MgHPO_4^0$ in a solution containing phosphate ions. A dearease in the activity of Mg^{2+} ions owing to the ion pair formation is around 50 % at most, even the sufficient amount of phosphate ions inquired to inhibit the Mg^{2+} ion adsorption is added.

It is noted that the presence of Mg^{2+} ions in a solution favors aragonite formation, whereas that of phosphate, fluoride, barium, calcium or citrate ions favors calcite formation. The effects of phosphate ions on magnesian calcite formation and also on the inhibiting role in Mg^{2+} ion adsortion are especially large. Phosphate ions in a solution may change the surface condition of carbonate precipitate and may inhibit Mg^{2+} ion adsorption on carbonate surface.

SUMMARY

The results of the present paper are summarized as follows;

(1) Inorganic ions or organic phosphate such as ATP, ADP, AMP, α-GADS and PASS dissolved in a parent solution containing Mg^{2+} ions affect remarkably the polymorphic crystallization of calcium carbonate. Magnesian calcite was favorably precipitated at a moderate phosphate concentration ranging from 10 to 500 μM. This concentration range is commonly observed in biological fluids of marine organisms and in interstitial waters of anerobic marine sediments. Magnesian calcite is possibly formed under such conditions of interstitial waters.

(2) Phosphate ions in a parent solution decrease remakably the rate of calcium carbonate (aragonite) precipitation in the presence of Mg^{2+} ions, whereas do not affect so much on calcite precipitation in the absence of Mg^{2+} ions.

(3) It is suggested that particular ions, which inhibit strongly the adsorption of Mg^{2+} ions onto $CaCO_3$ surface, favor either the calcite formation even in the presence Mg^{2+} ions.

(4) The present results will be useful for the understanding of carbonate formation in biological fluids and interstitial waters in marine anerobic sediments.

REFERENCES

[1] KITANO Y. and KANAMORI N.(1966) Synthesis of magnesian calcite at low temperature and pressure. Geochem. J. **1**, 1-10.

[2] KITANO Y., TOKUYAMA A. and ARAKAKI T.(1979) Magnesian calcite synthesis from calcium bicarbonate solution containing magnesium and barium ions. Geochem. J. **13**, 181-185.

[3] OOMORI T. and KITANO Y.(1987) Synthesis of protodolomite from sea water in the presence of dioxane. Geochem. J. **21**, 59-65.

[4] SIMKISS, K. (1964) Phosphate as crystal poisons of calcification. Biol. Rev. **39**, 487-505.

[5] REDDY M. M. and NANCOLLAS G. H.(1973) Calcite crystal growth inhibition by phosphonates. Desalination **12**, 61-73.

[6] MURPHY J. and RILEY J. P. (1962) A simplified single solution method for the determination of phosphate in natural waters. Anal. Chim. Acta **27**, 31-36.

[7] OOMORI, T., KYAN, A. and KITANO, Y. (1988) Magnesian calcite synthesis from calcium bicarbonate solution containing magnesium ions in the presence of fluoride and phosphate ions. Geochem. J. **22**, 275-283.

[8] TOMIYAMA C. and KITANO Y.(1984) Calcite formation in calcium chloride rich water. Japan. J. Limnol. **45**, 1-5.

[9] OOMORI T., KANESHIMA K., TAIRA T. and KITANO Y.(1983) Synthetic studies of protodolomite from brine waters. Geochem. J. **17**, 147-152.

List of Participants

Akiyama, Masahiko
Department of Geology,
Shinshu University
3-1-1, Asahi, Matsumoto-shi,
Nagano 390, Japan
Phone 0263-35-4600 EXT. 4170
Fax 0263-33-5323

Allemand, Denis*
Centre Scientifique de Monaco,
c/o Musee Oceanographique-
Avenue Saint-Martin
MC 98000, Monaco
Phone 93-15-36-00
Fax 93-50-52-97

Aoba, Takaaki
Forsyth Dental Center
140, Fenway, Boston,
MA 02115, USA
Phone 617-262-5200
Fax 617-262-4021

Azumi, Reiko*
National Chemical Laboratory
for Industry
1-1, Higashi, Tsukuba,
Ibaraki 305, Japan
Phone 0298-54-4582
Fax 0298-54-4487

Campana, Steven
Marine Fish Division, P.O. Box 1006,
Bedford Institute of Oceanography
Dartmouth, Nova Scotia
B2Y 4A2, Canada
Phone 902-426-3233
Fax 902-426-2256

Carr, Stuart*
Unilever Research,
Port Sunlight Laboratory
Quarry Road East,
Bebington, Wirral,
Menseyside L63 3JW, UK
Phone 051-645-2000
Fax 051-645-3249

Cheng, Pei-Tak
Mount Sinai Hospital,
University of Toronto
600, University Avenue
Toronto M5G 1X5, Canada
Phone 416-586-4468
Fax 416-586-8589

Chetail, Monique
Universite Paris VII
2, place Jussieu 75251
Paris Cedex 05, France
Phone 43-36-25-25
EXT.35-43 and 31-34

Crenshaw, Miles
Dental Research Center,
University of North Carolina
Chapel Hill, NC 27599-7455, USA
Phone 919-966-4581
Fax 919-966-3683

Cuif, Jean-Pierre
Bat. 504 Geologie, Faculte
des Sciences
91405 Orsay, France
Phone 1-69-41-67-41
Fax 1-64-46-59-38

Cusack, Maggie
Department of Geology
& Applied Geology.
University of Glasgow.
Lilybank Gardens. Glasgow
G12 8QQ Scotland, UK
Phone 041-339-8855 EXT. 5476
Fax 041-330-4817

Dauphine, Yannicke
URA 723-Laboratoire de petrologie
sedimentaire et Paleontologie
bat 504-Universite Paris II
Orsay, F 91405 Orsay, France
Phone 1-69-41-61-21
Fax 1-64-46-59-38

Dean, John M.*
Belle Baruch Institute,
University of South Carolina
Columbia, South Carolina
29208, USA
Phone 803-777-3917
Fax 803-777-3935

Deutsch, Dan
Dental Research Unit,
Hebrew University of Jerusalem
-Hadassah Faculty of Dental Medicine
P.O. Box 1172, Jerusalem, Israel
Phone 02-428565
Fax 972-2-784010

Didymus, Jonathan M.
School of Chemistry,
University of Bath
Claverton Down,
Bath BA2 7AY, UK
Phone 0225-826826
Fax 0225-826231

Donachy, Julie E.
Department of Biological Sciences,
University of South Alabama
LSB 124
Mobile, Alabama 36688, USA
Phone 205-460-7309
Fax 205-460-7357

Dubois, Philippe
Laboratoire de Biologie marine
C.P. 160,
Universite Libre de Bruxelles,
av. F.D. Roosevelt, 50, B-1050
Bruxelles, Belgium.
Phone 32-2-642-29-70
Fax 32-2-642-28-72

Fearnhead, Ronald W.
Department of Anatomy,
School of Dental Medicine,
Tsurumi University
2-1-3, Tsurumi, Tsurumi-ku,
Yokohama 203, Japan
Phone 045-581-1001
Fax 045-573-9599

Fujisawa, Ryuichi
Deptartment of Biochemistry,
School of Dentistry,
Hokkaido University
N13 W7, Sapporo 060, Japan
Phone 011-716-2111 EXT. 4234

Fukae, Makoto
Department of Biochemistry,
School of Dental Medicine,
Tsurumi University
2-1-3, Tsurumi, Tsurumiku,
Yokohama 230, Japan
Phone 045-581-1001
Fax 045-573-9599

Gaspard, Daniele
Department des Sciences de la Terre,
Universite de Paris-Sud
Bat. 504, F-91405 Orsay Cedex,France
Phone 33-1-69-41-67-69
Fax 33-1-64-46-59-38

Goto, Masatoshi
Department of Anatomy,
School of Dental Medicine,
Tsurumi University
2-1-3, Tsurumi, Tsurumi-ku,
Yokohama 230, Japan
Phone 045-581-1001 EXT. 349
Fax 045-573-9599

Hashimoto, Iwao*
Tsurumi University
2-1-3, Tsurumi, Tsurumi-ku,
Yokohama 230, Japan
Phone 045-581-1001
Fax 045-573-9599

Hidaka, Michio
Department of Biology,
Faculty of Science,
University of the Ryukyus
1, Senbaru, Nishihara-cho,
Okinawa 903-01, Japan
Phone 09889-5-2221 EXT. 2671
Fax 09889-5-2247

Honda, Eisuke*
Third Department of Internal
Medicine, Faculty of Medicine,
Tokyo University
7-3-1, Hongo, Bunkyo-ku,
Tokyo 113, Japan
Phone 03-815-5411 EXT. 8274
Fax 03-815-2087

Humbert, Willy
Universite Louis Pasteur, URA CNRS
1332,"Neurobiologie des fonctions
rythmiques et saisonnieres".
Laboratoire de Zoologie12, rue de
1'Universite, 67000
Strasbourg, France
Phone 88-35-85-07
Fax 88-24-04-61

Iijima, Mayumi
School of Dentistry,
Asahi University
1851-1, Hozumi, Hozumi-cho,
Motosu-gun, Gifu 501-02, Japan
Phone 05832-6-6131 EXT. 1437
Fax 05832-7-4364

Ino, Michiyo*
Department of Pharmacology,
Faculty of Dentistry,
Tokyo Medical and Dental University
1-5-45, Yushima, Bunkyo-ku,
Tokyo, Japan
Phone 03-813-6111 EXT. 5143
Fax 03-812-4576

Inoue, Takao*
Department of Anatomy,
School of Medicine,
Tottori University
Yonago, 683 Japan
Phone 0859-34-8004
Fax 0859-34-8080

Inoue, Takeshi*
Kao Coporation Research Laboratory
2606, Akabane, Ichigai-cho,
Haga-gun, Tochigi 321-34, Japan
Phone 0285-83-7570
Fax 0285-83-7418

Isa, Eishin
Department of Biology,
Faculty of Science,
University of the Ryukyus
1, Senbaru, Nishihara-cho,
Okinawa 903-01, Japan
Phone 09889-5-2221 EXT. 2666
Fax 09889-5-2247

Ishiki, Testuo*
Department of Oral Pathology,
School of Dentistry at Niigata,
The Nippon Dental University
1-8, Hamaura-cho, Niigata
951, Japan
Phone 025-267-1500
Fax 025-267-1134

Ishiyama, Mikio
Department of Histology,
School of Dentistry at Niigata,
The Nippon Dental University
1-8, Hamaura-cho, Niigata
951, Japan
Phone 025-267-1500
Fax 025-267-1134

Ito, Kiyoshi*
Lion Corporation
100, Tajima, Odawara-shi,
Kanagawa 255, Japan
Phone 0465-48-3211
Fax 0465-48-4079

Kakei, Mitsuo
Department of Anatomy,
School of Dentistry,
Meikai University
1-1, Keyakidai, Sakado,
Saitama-ken, Japan
Phone 0492-85-5511
Fax 0492-87-6657

Kalish, John M.
Fisheries Research Centre,
Ministry of Aguriculture and
Fisheries, MAF Fisheries Greta Point
P.O. Box 297, Wellington,
New Zealand
Phone 4-861-029
Fax 4-863-179

Kamegai, Akihide
Department of Oral and Maxillofacial
Surgery, School of Dentistry,
Asahi University
1851-1, Hozumi, Hozumi-cho,
Motosu-gun, Gifu 501-02, Japan
Phone 05832-6-6131 EXT.1472
Fax 05832-7-4364

Kamiya, Hidetoshi
Department of Geology and Mineralogy,
Faculty of Science, Kyoto University
Kitashirakawa-oiwake-cho,
Sakyou-ku, Kyoto 606, Japan
Phone 075-753-4166 or 4150
Fax 075-753-4189

Katsura, Nobuhiko
Depatment of Oral Biochemistry,
School of Dentistry,
Nagasaki University
7-1, Sakamoto-machi,
Nagasaki 852, Japan
Phone 0958-47-2111
Fax 0958-43-1060

Kawamata, Seiichi*
Department of Anatomy,
Toyama Medical and
Pharmaceutical University
2630, Sugitani, Toyama-shi,
Toyama 930-01, Japan
Phone 0764-34-2281
Fax 0764-34-4656

Kawamoto, Tadafumi*
School of Dentistry,
Tsurumi University
2-1-3, Tsurumi, Tsurumi-ku,
Yokohama 230, Japan
Phone 045-581-1001

Kawasaki, Kenzo*
Tsurumi University
2-1-3, Tsurumi, Tsurumi-ku,
Yokohama 230, Japan
Phone 045-581-1001
Fax 045-573-9599

Kishino, Masaru
Department of Oral and
Maxillofacial Surgery,
School of Medicine,
Tottori University
86, Nishi-machi, Yonago 683, Japan
Phone 0859-34-8131
Fax 0859-34-8071

Kitano, Yasushi*
Sugiyama Jogakuen University
17-3, Hoshigaoka Moto-machi,
Chikusa-ku, Nagoya 464, Japan
Phone 052-781-1186
Fax 052-781-4466

Kobayashi, Iwao
Department of Geology and Mineralogy,
Faculty of Science, Niigata University
8050, Ninomachi, Ikarashi,
Niigata 950-21, Japan
Phone 025-262-6192
Fax 025-262-6194

Kozawa, Yukishige
Department of Anatomy,
School of Dentistry at Matsudo,
Nihon University
2-870-1, Sakae-cho-nishi,
Matsudo-shi, Chiba 271, Japan
Phone 0473-68-6111
Fax 0473-64-6295

Kuboki, Yoshinori
Department of Biochemistry,
School of of Dentistry,
Hokkaido University
N-13, W-7, Kita-ku,
Sapporo 060, Japan
Phone 011-716-2111 EXT. 4231
Fax 011-758-4074

LeGeros, Racquel Z.
New York University College
of Dentistry,
345, East 24th Street, New York,
NY 10010, USA
Phone 212-998-9580
Fax 212-995-4080

Lowenstam, Heinz
California Institute of Technology
MS 170-25
Pasadena, CA 91125, USA
Phone 818-356-6136
Fax 818-568-0935

Mann, Stephen
School of Chemistry,
University of Bath
Bath BA2 7AY, UK
Phone 0225-826122
Fax 0225-826231

Mataki, Shiro
Department of Pharmacology,
School of Dentistry,
Nagasaki University
7-1, Sakamoto-cho,
Nagasaki 852, Japan
Phone 0958-47-2111
Fax 0958-43-1060

Matsumoto, Mutsuyoshi*
National Chemical Laboratory
for Industry,
1-1, Higashi, Tsukuba,
Ibaraki 305, Japan
Phone 0298-54-4582
Fax 0298-54-4487

Matsunaga, Tadashi
Department of Biotechnology,
Tokyo University of Agriculture
and Technology
24-16, Nakamachi-2, Koganei,
Tokyo 184, Japan
Phone 0423-81-4221
Fax 0423-85-7713

Mishima, Hiroyuki
Department of Anatomy,
School of Dentistry at Matsudo,
Nihon University
2-870-1, Sakaecho-Nishi Matsudo,
Chiba 271, Japan
Phone 0473-68-6111 EXT. 386
Fax 0473-64-6295

Moreno, Edgard C.
Forsyth Dental Center
140, Fenway, Boston, MA 02115, USA
Phone 617-262-5200
Fax 617-262-4021

Mori, Kimie
First Department of Oral and
Maxillofacial Surgery,
School of Dentistry, Showa
University
2-1-1, Kitasenzoku, Ohta-ku,
Tokyo, Japan
Phone 03-787-1151
Fax 03-787-1229

Moriwaki, Yutaka*
School of Dentistry,
Asahi University
1851, Hozumi, Hozumi-cho,
Motosu-gun, Gifu 501-02, Japan
Phone 05832-6-6131 EXT. 1436
Fax 05832-7-4364

Mugiya, Yasuo*
School of Fisheries,
Hokkaido University
3-1-1, Minato-cho, Hakodate,
Hokkaido 041, Japan
Phone 0138-41-0131
Fax 0138-43-5015

Mutvei, Harry
Swedish Museum of Natural History
S-10405, Stockholm, Sweden
Phone 08-666-41-87
Fax 46-8-666-41-84

Nakahara, Hiroshi
Oral Anatomy Department,
School of Dentistry,
Meikai University
1-1, Keyakidai, Sakado-shi,
Saitama 350-02, Japan
Phone 0492-85-5511
Fax 0492-87-6657

Nakashima, Syozi*
Lion Corporation
100, Tajima, Odawara-shi,
Kanagawa 255, Japan
Phone 0465-48-3211
Fax 0465-48-4079

Nancollas, George H.
Chemistry Department,
State University of New York
at Baffalo
Acheson Hall, South Campus, Baffalo,
New York 14214, USA
Phone 716-831-3264
Fax 716-831-2960

Ochiai, Yoshihito*
Lion Corporation
100, Tajima, Odawara-shi,
Kanagawa 255, Japan
Phone 0465-48-7317
Fax 0465-48-4079

Ohno, Kohsuke*
First Department of Oral and
Maxillofacial Surgery,
School of Dentistry, Showa
University
2-1-1, Kitasenzoku, Ohta-ku,
Tokyo, Japan
Phone 03-787-1151
Fax 03-787-1229

Ohya, Keiichi
Department of Pharmacology,
Faculty of Dentistry,
Tokyo Medical and Dental University
1-5-45, Yushima, Bunkyou-ku,
Tokyo 113, Japan
Phone 03-813-6111
Fax 03-812-4576

Oishi, Masamichi*
Electron Microscopy Center,
University of South Carolina
Coker Life Sciences Building,
Columbia, SC 29208 USA
Phone 803-777-7058

Okazaki, Masayuki
Faculty of Dentistry,
Osaka University
1-8, Yamadaoka, Suita,
Osaka 565, Japan
Phone 06-876-5711 EXT. 2228
Fax 06-876-7931

Okazaki, Megumi
Department of Biology,
Tokyo Gakugei University
4-1-1, Nukuikita-Machi, Koganei-shi,
Tokyo 184, Japan
Phone 0423-25-2111 EXT. 2667
Fax 0423-24-9832

Okoshi, Kenji
Department of Science and Technology,
Faculty of Biotechnology,
Senshu University of Ishinomaki
1, Shinmito, Minamisakai, Ishinomaki,
Miyagi 986, Japan
Phone 0225-22-7716 EXT. 3163
Fax 0225-22-7746

Omori, Masae
Faculty of General Education,
Azabu University
1-17-71, Fuchinobe, Sagamihara,
Kanagawa 229, Japan
Phone 0427-54-7111
Fax 0427-53-3395

Oomori, Tamotsu
Department of Chemistry,
University of the Ryukyus
1, Senbaru, Nishihara-cho,
Nakagami-gun,Okinawa
903-01, Japan
Phone 09889-5-2221 EXT. 2655
Fax 09889-5-2247

Pentecost, Allan
Division of Biosphere Sciences,
King's College London,
Campden Hill Road,
London W8 7AH, UK
Phone 071-836-5454
Fax 071-937-5396

Prostak, Kenneth S.
Forsyth Dental Center
140 Fenway, Boston,

MA 02115, USA
Phone 617-262-5200
Fax 617-262-4021

Saito, Kouichi*
Lion Corporation
100, Tajima, Odawara-shi,
Kanagawa 255, Japan
Phone 0465-48-3211
Fax 0465-48-4079

Saitoh, Setsuo*
Hokkaido Institute of Mariculture
539-112, Honbetu, Shikabe,
Hokkaido 041-14, Japan
Phone 01372-7-2234
Fax 01372-7-2235

Sakae, Toshiro
School of Dentistry at Matsudo,
Nihon University
2-870-1, Sakae-cho-nishi,
Matsudo-shi,
Chiba 271, Japan
Phone 0473-68-6111 EXT. 386
Fax 0473-64-6295

Saleuddin, Abu
Department of Biology,
York University
4700 Keele Street, North York,
Ontario M3J 1P3, USA
Phone 416-736-2100 EXT.33832
Fax 416-736-5698

Samata, Tetsuro
Faculty of General Education,
Azabu University
1-17-71, Fuchinobe,
Sagamihara 229, Japan
Phone 0427-54-7111
Fax 0427-53-3395

Sasagawa, Ichiro
Department of Anatomy,
School of Dentistry at Niigata,
The Nippon Dental University
1-8, Hamaura-cho,
Niigata 951, Japan
Phone 025-267-1500 EXT. 223
Fax 025-267-1134

Sasaki, Satoshi
Depatment of Biochemistry,
School of Dentistry,
Tokyo Medicine and Dental University
1-5-45, Yushima, Bunkyo-ku,
Tokyo 113, Japan
Phone 03-813-6111
Fax 03-812-4576

Sasano, Yasuyuki*
2nd Department of Oral Anatomy,
School of Dentistry,
Tohoku University
4-1, Seiryo-machi, Aobaku,
Sendai 980, Japan
Phone 022-273-9330
Fax 022-263-9867

Sawamura, Hiroshi
Department of Anatomy,
School of Dental Medicine,
Tsurumi University
2-1-3, Tsurumi, Tsurumi-ku,
Yokohama 230, Japan
Phone 045-581-1001
Fax 045-573-9599

Secor, David*
Belle W. Brunch Institute for Marine
Biology and Coastal Research,
University of South Carolina,
Columbia, SC, 29208, USA
Phone 803-777-3930

Shimamoto, Masanori
Institute of Geology & Paleontology,
Tohoku University
Aoba, Aramaki, Aoba-ku,
Sendai 980, Japan
Phone 022-222-1800 EXT. 3417
Fax 022-262-6609

Shimizu, Masaharu
Department of Biochemistry,
School of Dental Medicine,
Tsurumi University
2-1-3, Tsurumi, Tsrumi-ku,
Yokohama 203, Japan
Phone 045-581-1001
Fax 045-573-9599

Shimizu, Motohiro
Faculty of Fisheries,
Hokkaido University
3-1-1, Minato-cho, Hakodate,
Hokkaido 041, Japan
Phone 0138-41-0131
Fax 0138-43-5015

Simkiss, Kenneth
School of Animal & Microbial Sciences,
University of Reading
Reading, RG6 2AJ, UK
Phone 0734-318463
Fax 0734-310180

Slavkin, Harold C.
Center for Craniofacial Molecular
Biology, School of Dentistry,
University of Southern California
Health Sciences Campus, CSA, 103,
2250, Alcazar Street, Los Angeles,
CA, 90033, USA
Phone 213-743-6095
Fax 213-749-5837

Somiya, Hiroaki
Department of Aquatic Biology,
Research Institute of Biosciences,
Azabu University
1-17-71, Fuchinobe,
Sagamihara 229, Japan
Phone 0427-54-7111
Fax 0427-53-3395

St. Pierre, Tim G.
School of Mathematical and
Physical Science,
Murdoch University,
Perth W.A. 6150, Australia
Phone 61-9-332-2831
Fax 61-9-310-5005

Suga, Shoichi
Department of Pathology,
School of Dentistry,
The Nippon Dental University
1-9-20, Fujimi, Chiyoda-ku,
Tokyo 102, Japan
Phone 03-261-8311
Fax 03-238-1289 or 03-264-8399

Suzuki, Seiichi
Department of Earth Sciences
and Astronomy,

Fukuoka University of Education
729-Akama, Munakata,
Fukuoka 811-41, Japan
Phone 0940-32-2381

Takagi, Yasuaki
Faculty of Fisheries,
Hokkaido University
3-1-1, Minato-cho, Hakodate,
Hokkaido 041, Japan
Phone 0138-41-0131
Fax 0138-43-5015

Takahashi, Masashi
School of Dentistry at Niigata,
The Nippon Dental University
1-8, Hamaura-cho, Niigata 951, Japan
Phone 025-267-1500 EXT. 223

Takeyama, Haruo*
Department of Orthodontology,
School of Dentistry,
Meikai University
1-1, Keyakidai, Sakado-shi,
Saitama 350-02, Japan
Phone 0492-85-5511
Fax 0492-87-6657

Takita, Hiroko*
Department of Biochemistry,
School of Dentistry,
Hokkaido University
W7 N13 Kita-ku, Sapporo, Japan
Phone 011-716-2111 EXT. 4234
Fax 011-758-4074

Tanabe, Takako*
School of Dentistry,
Tsurumi University
2-1-3, Tsurumi, Tsurumi-ku,
Yokohama 230, Japan
Phone 045-581-1001 EXT. 480
Fax 045-573-9599

Taylor, Marina G.
School of Animal & Microbial Sciences,
University of Reading
Reading, RG6 2AJ, UK
Phone 0734-875123 EXT. 7061
Fax 0734-310180

Tazaki, Kazue
Department of Geology,
Shimane University
1060, Nishikawatsu,
Matsue 690, Japan
Phone 0852-21-7100
Fax 0852-31-0812

ten Cate, J. M.
Department of Cariology
and Endodontology,
Academic Centre for Dentistry
Louwesweg 1, 1066EA Amsterdam,
The Netherlands
Phone 31-20-5188440
Fax 31-20-5188333

Termine, John D.
Bone Research Branch,
National Institute of Dental Research
National Institutes of Health,
Bethesda, Maryland 20892, USA
Phone 301-496-4563
Fax 301-402-0824

Togo, Yoshihiro
Department of Earth Science,
Hokkaido University of Education
2-34, Midorigaoka, Iwamizawa,
Hokkaido 068, Japan
Phone 0126-22-1470
Fax 0126-24-3484

Tuross, Noreen
Conservation Analytical Laboratory,
Smithsonian Institution,
Museum Support Center
Washington, D.C. 20560, USA
Phone 301-238-3700
Fax 301-238-3709

Wada, Koji*
National Research Institute
of Aguriculture
422-1, Nakatsuhamaura, Mansei-cho,
Doai-gun, Mie 516-01, Japan
Phone 05996-6-1830
Fax 05996-6-1962

Wakamatu, Hironari*
Department of Pharmacology,
Faculty of Dentistry,
Tokyo Medical and Dental University
1-5-45, Yushima, Bunkyo-ku,
Tokyo, Japan
Phone 03-813-6111 EXT. 5141
Fax 03-812-4576

Waller Andrew*
Unilever Research,
Port Sunlight Laboratory
Quarry Road East, Bebington, Wirral,
Menseyside L63 3JW, UK
Phone 051-645-2000
Fax 051-645-3249

Watabe, Norimitsu
Electron Microscopy Center,
CLS 001,
University of South Carolina
Columbia, SC 29208, USA
Phone 803-777-4834
Fax 803-777-3935

Webb, John
School of Mathematical
& Physical Sciences,
Murdoch University
Perth, W.A. 6150 Australia
Phone 0011-61-9-332-2148
Fax 0011-61-9-310-5005

Webb, Mary Alice*
Department of Botany and
Plant Pathology,
Purdue University
West Latayette, IN 47907,
USA
Phone 317-494-0598
Fax 317-494-5896

Weiner, Stephen
Isotope Department,
Weizmann Institute of Science
Rehovot 76100 Israel
Phone 972-8342552
Fax 972-8-466966

Whyte, Martin A.
Earth Sciences Unit,
Sheffield University

Beaumont Building, Brookhill,
Sheffield S3 7HF, UK
Phone 0742-768555 EXT. 4784
Fax 0742-739826

Yamada, Juro*
Faculty of Fisheries,
Hokkaido University
3-1-1, Minato-Machi, Hakodate,
Hokkaido 041, Japan
Phone 0138-41-0131
Fax 0138-43-5015

Yamaguchi, Hisao
Department of Pathology,
School of Medicine,
Keio University
35 Shinano-machi, Shinjuku-ku,
Tokyo, Japan
Phone 03-353-1211 ext. 2677
Fax 03-353-3290

Yamaguchi, Keiko*
Department of Geology and
Mineralogy, Faculty of Science,
Kyoto University
Kitashirakawa-Oiwake-cho,
Sakyo-ku, Kyoto 606,
Japan
Phone 075-753-7531

Yamashiro, Hideyuki
Radioisotope Labolatory,
University of the Ryukyus
1, Senbaru, Nishihara-cho,
Okinawa 903-01, Japan
Phone 09889-5-2221 EXT. 3166
Fax 09889-5-2247

Yoshioka, Sayoko
Aichi University Of Education
Horosawa, Igaya-cho,
Kariya 448, Japan
Phone 0566-36-3111
Fax 0566-36-4337

Index

MIX
Papier aus verantwortungsvollen Quellen
Paper from responsible sources
FSC® C105338

If you have any concerns about our products,
you can contact us on
ProductSafety@springernature.com

In case Publisher is established outside the EU,
the EU authorized representative is:
Springer Nature Customer Service Center GmbH
Europaplatz 3, 69115 Heidelberg, Germany

Printed by Libri Plureos GmbH
in Hamburg, Germany